MULTIPLEXING AND NETWORKING VOLUME 2

PT-128

Edited by

Ronald K. Jurgen

Published by
SAE International
400 Commonwealth Drive
Warrendale, PA 15096-0001
U.S.A.
Phone (724)776-4841
Fax (724)776-0790

All SAE papers, standards, and selected books are abstracted and indexed in the Global Mobility Database.

For multiple print copies contact:

SAE Customer Service
Tel: 877-606-7323 (inside USA and Canada)
Tel: 724-776-4970 (outside USA)
Fax: 724-776-0790
Email: CustomerService@sae.org

ISBN 0-7680-1715-7
Library of Congress Catalog Number: 99065331
SAE/PT-128
Copyright © 2006 SAE International

Printed in USA

MULTIPLEXING AND NETWORKING VOLUME 2

Other SAE books in this series:

Automotive Software
by Ronald K. Jurgen
(Order No. PT-127)

Electronic Engine Control Technologies, Second Edition
by Ronald K. Jurgen
(Order No. PT-110)

Sensors and Transducers, Second Edition
by Ronald K. Jurgen
(Order No. PT-105)

Electric and Hybrid Vehicles
by Ronald K. Jurgen
(Order No. PT-85)

On- and Off-Board Diagnostics
by Ronald K. Jurgen
(Order No. PT-81)

Electronic Transmission Controls
by Ronald K. Jurgen
(Order No. PT-79)

Multiplexing and Networking
by Ronald K. Jurgen
(Order No. PT-78)

Electronic Steering and Suspension Systems
by Ronald K. Jurgen
(Order No. PT-77)

Electronic Braking, Traction, and Stability Controls
by Ronald K. Jurgen
(Order No. PT-76)

Navigation and Intelligent Transportation Systems
by Ronald K. Jurgen
(Order No. PT-72)

INTRODUCTION

Multiplexing and Networking on the Rise

The ever increasing electronics content of modern vehicles, and the resulting complexity of controls and systems, has been accompanied by a comparable increasing need for multiplexing and networking. Not too many years ago, three classes of networks were thought to be sufficient for any future needs, but that no longer seems to be the case. With additional networks has come a need for greater conformance testing of those networks and a greater number of gateways among them.

This book, the second in the Automotive Electronics Series dealing with multiplexing and networking, contains 63 papers, none of which appeared in the first book (PT-78). The papers are presented in five categories: Multiplexing and Networking Systems, Protocols, Gateways and Middleware, Network/Protocol Testing, and Viewpoints on Future Trends.

The following thought-provoking ideas are but a few of the many such ideas expressed herein by the expert authors of those papers:

- "Although there are many multiplex wiring systems being used to simplify the car wiring harness, very few are low cost, fault tolerant, and redundant at the same time. Most of the systems address mainly the protocol and software issues and neglect the reliability of the multiplex wiring system." – (2004-01-1572)

- "Vehicle functions requiring high-speed, low-latency communication are driving the recent proliferation of optical data network systems in the vehicle." – (2004-01-0197)

- ". . . at least seven in-vehicle networks may be necessary—at least on high-end vehicles in the next ten years. These categories include, besides the existing SAE classes, diagnostics, airbags, mobile media, and X-by-Wire." – (2001-01-0060)

- "Of the currently available open standard network protocols, CAN has become the most prominent across the world's automotive industry. The probable reasons for this have been its huge support from major semiconductor manufacturers, tool suppliers, and automotive OEMs." – (2005-01-1539)

- "LIN provides a cost-efficient bus communication where the bandwidth and versatility of CAN are not required. It will be the enabling factor for the implementation of a hierarchical network in order to gain further quality enhancements and cost reduction in vehicles, appliances, and equipment." – (2004-01-1742)

- "FlexRay is a new communication standard for future in-vehicle controls. This protocol provides flexibility and determinism by providing scalable static and dynamic message transmission." – (2005-01-1279)

- "Because of the rapidly increasing amount of electronic components and buses in a vehicle, the use of gateways in electronic control units (ECUs) becomes more important. The upcoming question is how to design an optimal gateway." – (2005-01-1696)

- "Unfortunately, ten years of practical experience show that implementations of communication protocol specifications done by different manufacturers in most cases are not compliant *a priori*. Luckily, the experiences of automotive OEMs, together with their suppliers and silicon manufacturers, have proved that conformance testing is a very effective means to avoid interoperability problems in mixed suppliers' automotive systems." – (2005-01-1535)

- "Visionaries present during the infancy of multiplex wiring often predicted that smart sensors and actuators and fully distributed systems would be the norm rather than the exception by the year 2000. In spite of the promising possibilities observed back then, the fully distributed system has not yet reached a level of ubiquity in 2005. Many factors have contributed to a slower pace of adoption than originally forecast, not the least of which is the automotive industry's insistence on controlled and deliberate innovation." – (2005-01-1283)

- "This paper has presented an approach toward meeting future requirements in vehicle networks using redundant high-speed communications buses with software voting and time synchronization. The example implementation demonstrates true distributed control, a target that many OEMs are seeking, to provide the levels of advanced functionality with a minimum number of high-performance ECUs." – (2004-01-0206)

- "This paper proposes the idea that at least eight in-vehicle networks may be necessary—mainly on high-end vehicles in the next ten years. These categories include (besides the existing SAE classes) diagnostics, airbags, mobile media, X-by-Wire, and wireless. Each area needs its own protocol and one or more networks running that protocol." – (2003-01-0111)

*   *   *   *   *   *   *   *   *   *   *

This book and the entire Automotive Electronics Series are dedicated to my friend Larry Givens, a former editor of SAE's monthly publication, *Automotive Engineering International.*

Ronald K. Jurgen, Editor

TABLE OF CONTENTS

INTRODUCTION

MULTIPLEXING AND NETWORKING SYSTEMS

PROTOCOLS

Controller Area Network (CAN) Protocol

Local Interconnect Network (LIN)

Other Protocols

GATEWAYS AND MIDDLEWARE

NETWORK/PROTOCOL TESTING

VIEWPOINTS ON FUTURE TRENDS

MULTIPLEXING AND NETWORKING SYSTEMS

Low Cost Fault Tolerant and Redundant Multiplex Wiring System for Automotive Applications

Kelvin Shih
Lawrence Technological University

ABSTRACT

A low cost fault tolerant and redundant multiplex wiring system specifically designed for automotive applications is described in this paper. Although there are many multiplex wiring systems are being used to simplify the car wiring harness, but very few are low cost, fault tolerant and redundant at the same time. Most of the system address mainly the protocol and software issues and neglected the reliability of the multiplex wiring system. This paper addresses the fault tolerant and redundancy of the system and use hardware based integrated circuit to convert from parallel to serial at the transmitter side and serial to parallel at the receiver side.

INTRODUCTION

Modern airplanes have used redundant and fault tolerant multiplex system to transmit large amount of data from one location to another with very high reliability. This is because modern airplane is designed to fly by wire. Between the control panels to the actuators there are only wires between them. Military aircraft uses quadruple redundancy to make sure when one harness is lost the other multiplexed wiring harness can automatically taking over. To design a fault tolerant and redundant multiplex wiring system for automotive application however poses a completely different challenge. Because the large amount of cars and trucks made in the world each year, the cost of the system has to be extremely competitive against the wiring harness it replaces. This paper describes my attempt to solve this problem.

MAIN SECTION

The automobiles are gradually moving toward the drive by wire idea just like their counter part in the airplane industry. At the present time as many as four thousand lead ends can be found in a luxury vehicle and two thousand lead ends in a compact or sub-compact car. This makes the wiring harness very complicated and very heavy. Automotive companies have used some form of multiplex systems to simply the wiring harness. But most of the multiplex system is neither fault tolerant nor redundant in its implementations. Some of the multiplex system uses a special fault tolerant circuit to protect each node in the multiplex system. This could increase the cost by $5.00 or more per node. With a multiplex system that have ten nodes the cost of the fault tolerant circuit alone will drive the multiplex system out of competition. The multiplex system described in this paper propose a simple, fault tolerant and redundant multiplex system that will satisfy most of the needs of low speed class A multiplex applications. This system can be applied to the lighting of the car or the power windows, power mirrors, power seats and power door lock functions of the car. The best location for the power windows, power seats and power door locks is still at the driver side arm rest. With a luxury car that controls all those functions a bundle of sixty-five wires may be needed. And a connector that connects all these wires from the car door to the car body may cost over $30. It also makes the door closing and opening very stiff. By using this fault tolerant and redundant multiplex system the reliability is greatly increased and yet the cost can still be kept to a level that it can compete with the wires it replaces.

Figure 1 below is block diagram of the redundant and fault tolerant multiplex system.

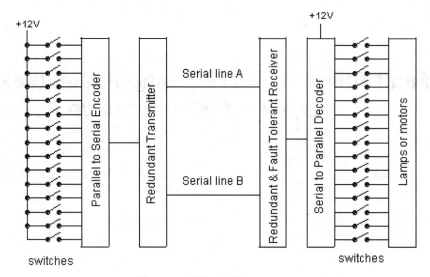

Figure 1 Block Diagram

In this implementation fifteen switch positions is fed into a parallel to serial encoder. These switch positions are converted into serial bit stream with its self-synchronizing clock signal at the front of each frame. The Manchester code encoder used in the VCR remote control or alarm system can be used here satisfactorily and at extremely low cost. This approach also completely eliminated the programming time and cost of the microprocessor. This serial bit stream is fed into a transmitter that consist of two small MOSFETs and two 0.25 W resistors. The outputs from the transmitter go through two wires connected to the front or rear receiver. One of wires is placed at left side of the car and the other wire is placed at right side of the car. This will prevent both wires being broken during a side impact accident. After the receiver received the redundant signals a simple circuit is used to combine the two redundant signals into a single bit stream. The serial to parallel decoder recovers the fifteen switch positions. These switches position through either a MOSFET or relay driver will drive the load. The load can either be a lamp or a motor.

Figure 2 shows the fault tolerant and redundant schematics.

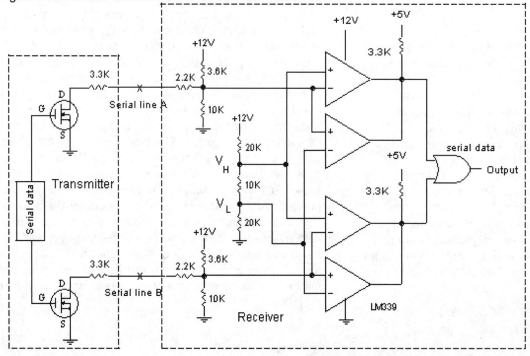

Figure 2 Fault Tolerant & Redundant System

4

In one of the example the transmitter consist of two small MOSFETs such as 2N7000 and two 3.3K resistors. The resistors are used to protect the MOSFETs when the serial data lines are shorted to the +12V supply. The serial data is fed to both gates of the MOSFETs. At the receiver a high reference voltage V_H and a low reference voltage V_L are established.

$$V_H = \frac{12 * (10K + 20K)}{20K + 10K + 20K} = 7.2V$$

$$V_L = \frac{12 * (20K)}{20K + 10K + 20K} = 4.8V$$

The V_H is fed into non-inverting inputs of both receiver channels. The V_L is fed into the inverting inputs of both channels. The input is fed into the non-inverting inputs of the top comparator in each channel. The same input is also fed into the inverting input of the bottom comparator of each channel. The quad comparators are connected as two window comparators. The outputs of each window comparator are wired "OR" through a 3.3K pull up resistor to 5V. At no fault condition, both serial lines are transmitting data from the transmitter to the receiver.

When a "0" is present on the gates of the MOSFETs, the MOSFETs are off.
The inputs to both channels V_{in} can be found.

$$V_{in} = \frac{12 * (10K)}{10K + 3.6K} = 8.82V$$

$V_{in} = 8.82V > 7.2V$ Thus the outputs for both channels are "0"

Two "0" fed into an "OR" gate will generate an output of "0".

When a "1" is present on the gates of the MOSFETs, the MOSFETs are on.
The inputs to both channels V_{in} can be found.

$$R_{eq} = \frac{(2.2K + 3.3K) * 10K}{(2.2K + 3.3K) + 10K} = 3.55K$$

$$V_{in} = \frac{12 * (3.55K)}{3.6K + 3.55K} = 5.96V$$

Since $7.2V > V_{in} = 5.96V > 4.8V$ Thus the outputs for both channels are "1".

Two "1" fed into an "OR" gate will generate an output of "1".

Receiver receives the correct serial data from the transmitter.

When one of the two serial is broken the other serial line will continue to transmit data. The V_{in} of the failed channel is calculated below.

$$V_{in} = \frac{12 * (10K)}{10K + 3.6K} = 8.82V$$

$V_{in} = 8.82V > 7.2V$ Thus the outputs for this channel is "0"

After combine the failed channel, which is "0" all the time with the other functioning channel the receiver, still receives the correct data.

When one of the two serial is shorted to the ground the other functioning serial line will continue to transmit data. The V_{in} of the failed channel is calculated below.

$$R_{eq} = \frac{2.2K * 10K}{2.2K + 10K} = 1.8K$$

$$V_{in} = \frac{12 * (1.8K)}{3.6K + 1.8K} = 4.0V$$

$V_{in} = 4.0V < 4.8V$ Thus the outputs for this channel is "0"

After combine the failed channel, which is "0" all the time with the other functioning channel the receiver still receives the correct data.

When one of the two serial is shorted to the +12V the functioning serial line will continue to transmit data. The V_{in} of the failed channel is calculated below.

$$R_{eq} = \frac{2.2K * 3.6K}{2.2K + 3.6K} = 1.37K$$

$$V_{in} = \frac{12 * (10K)}{10K + 1.37K} = 10.55V$$

$V_{in} = 10.55V > 7.2V$ Thus the outputs for this channel is "0"

After combine the faulted channel, which is "0" all the time with the other functioning channel the receiver still receives the correct data.

During any of the fault condition the MOSFETs will not see more than a current more than 3.64mA. The receiver will not see a voltage outside 0 to 12 Volt boundaries. Which make this system fault tolerant. This multiplex system can keep on transmit serial data when one of the two redundant wires is broken, shorted to ground or shorted to the 12 V supply. The diagram below shows a proposed implementation of the lighting system with this fault tolerant and redundant multiplex scheme.

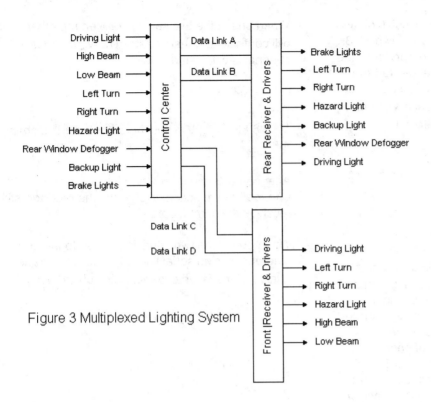

Figure 3 Multiplexed Lighting System

The control center sends out two data links to the rear of the car and two data links to the front of the car. All four data links carry the same serial data. The front and rear receiver only chooses the channels it needs to control the lights. For example, the rear receiver receives the high beam and low beam signals also but do not use them to drive anything. After the receivers received the signal it turns on or off the relays to control the lights.

The following diagram shows proposed power seats, power windows, power door lock and power mirror implementation.

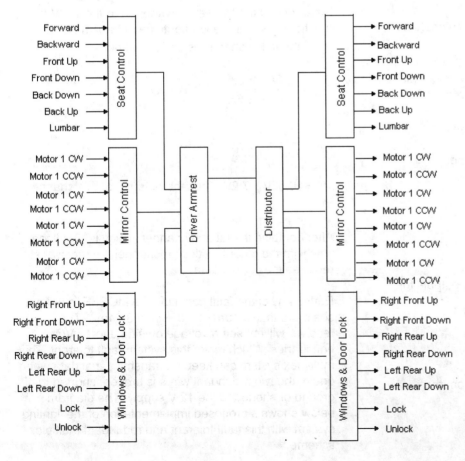

Again the transmitter will transmit all the signals needed to control all the functions. After break them up to different groups, the signal will be latched and feed into drivers. This fault tolerant and redundant multiplex system is hardware-based system thus it can work with any PCM (pulse coded modulated) signal. The limitation is only the speed of the voltage comparators used.

The possible saving could come from the following areas.

- Saving on the number of wires used in the car.

- Do not need large, high current connectors.

- Switches used to control the load will be low current signal switches.

- Some of the functions such as flasher can be implemented in electronics rather than electro-mechanical devices.

- Simplify the harness thus reduce the assembly cost during manufacturing.

- Simplify the harness thus reduce the warranty cost after sale.

- Because it is a redundant and fault tolerant system the reliability of the system is greatly enhanced. This can save a lot of money and grieve by reduce the possible lawsuits.

In the future a high-frequency low cost fault tolerant and redundant is under development for the other types of signal used in an automobile.

CONCLUSION

This paper is written to raise the attention for the need of a more reliable and still low cost multiplex system. Automotive industry is very cost sensitive. This is my attempt to design an affordable, redundant and fault tolerant multiplex system for automotive industry. If I have raised the interests of the engineers working for the automotive companies slightly, my goal is reached.

ACKNOWLEDGMENTS

I like to thank my wife Lily to put up with my insane working habits and hours.

I also want to thank my daughter Anna M. Shih of Carlson, Gaskey & Olds for writing all my patents for me for free. Without her help I cannot afford to continue inventing.

REFERENCES

USP 6,043,688 Ratio Metric Fault Tolerant and Redundant Serial Communication Link.

CONTACT

Kelvin   Shih   kelvinshih@comcast.net

A Sampling Period Decision for Robust Control of Distributed Control System using In-Vehicle Network

Kyunghan Chun and Myoungho Sunwoo
Hanyang University

ABSTRACT

This paper presents a preliminary study of a sampling period decision for robust control of a distributed control system based on an in-vehicle network with three types of data (real-time synchronous data, real-time asynchronous data, and nonreal-time asynchronous data). The architecture of automotive systems is currently changing from a number of standalone electronic control units (ECUs) to a functionally integrated distributed system which is linked by a network. The control performance of the integrated networked control system can be changed by the characteristics of time delays among the application ECUs. A basic parameter for a scheduling method of the networked control systems, a maximum allowable delay bound is used, which guarantees stability of the networked control system, and it is derived from the characteristics of the given plant using presented theorems. The presented theorems are derived from sliding mode technique for input-delay systems and a sampling period can be selected from the computed maximum allowable delay bound. The proposed method is shown to be useful through simulation examples.

INTRODUCTION

There are two alternative approaches for entering new control functions into an automotive electronic system: distributed or locally integrated (Figure 1). The distributed approach emerged with the first in-vehicle networks in the '80s. By this approach, different electronic control units (ECU) are connected by a communication link through equivalent network interfaces [1]. In addition, autonomous intelligent sensors [2] and actuators can be connected to other units through the network. Networking is the basis for new top-down control approaches, independent of local ECU platforms. Examples are traction control, vehicle dynamic control and electronic torque control.

The introduction of control network "bus" architectures can improve the efficiency, flexibility, and reliability of these integrated applications, reducing installation, reconfiguration, and maintenance time and costs. The change of communication architecture from point-to-point to common-bus, however, introduces different forms of time delay uncertainty between sensors, actuators, and controllers. These time delays come from the time sharing of communication medium as well as additional functionality required for physical signal coding and communication processing. The characteristics of time delays could be constant, bounded, or even random, depending on the network protocols adopted and the chosen hardware. This type of time delay could potentially degrade a system's performance and possibly cause system instability.

This problem is related to input-delay systems. To tackle this problem, we suggest a sliding mode control. A sliding mode control has attractive features such as fast response and good transient response [3][4][5]. It is also insensitive to variations in system parameters, and external disturbances. Other sliding mode control schemes are proposed for uncertain linear systems with time delay [6][7][8][9]. Especially, sliding mode control with delay compensation are proposed for robust stabilization of uncertain input-delay systems [8] and the control law is derived to ensure the existence of a sliding mode and to minimize the effects of the delay and uncertainty in the sliding mode

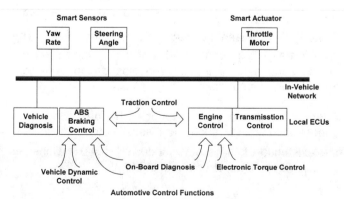

Figure 1 Distributed architecture of automotive electronics.

This paper presents a sampling period decision method for the distributed control system in a vehicle network. Derived from sliding mode control for input-delay systems,

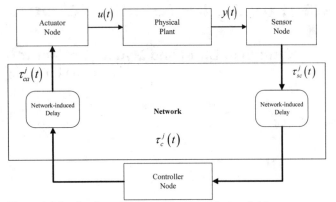

Figure 2 A feedback control loop with network-induced delays.

we suggest a guideline for the sampling period decision. A maximum allowable delay bound (MADB) is considered as a reference value of the sampling period, in which stability of the overall system could be guaranteed. A numerical simulation is given to illustrate the results.

A MAXIMUM ALLOWABLE DELAY BOUND

In Figure 2, many control loops can be connected using a single network medium. The MADB is defined as the maximum allowable interval from the instant when sensor nodes sample sensor data from a plant to the instant when actuators output the transferred data to the plant. If the sampling period in the j th loop exceeds the given MADB, then stability of the overall system could not be guaranteed. In this case, the output of the plant could deviate from the desired trajectory, or the controlled system could behave in an unpredictable manner. Hence it is necessary to derive the MADB from the parameters and configurations of the given plant and the controller.

Let us consider a linear input-delay system with uncertainties described by

$$\dot{x}(t) = Ax(t) + Bu(t-\tau) + f_0(x(t),t) + f_1(x(t-\tau),t) \tag{1}$$

where $x \in \Re^n$, $u(t) \in \Re^m$ and $\tau \in \big([0,\infty),\Re\big)$ are the state vector, the input vector and the known delay time, respectively, and A and B are constant matrices with appropriate dimensions. The uncertainties, $f_0(x(t),t)$ and $f_1(x(t-\tau),t)$, represent the nonlinear parameter perturbations with respect to the current state and the delayed state of the system, respectively. In addition to equation (1), the initial conditions are given by

$$x(0) = x^0, \; x_0(\theta) = \phi(\theta), \; u_0(\theta) = \upsilon(\theta), -\tau \leq \theta \leq 0 \tag{2}$$

where $x_t(\theta) = x(t+\theta)$ and $u_t(\theta) = u(t+\theta)$. It is assumed that the pair (A,B) is stabilizable and the states are available for feedback. It is also assumed that the unknown functions $f_0, f_1 : \Re^n \times \Re_+ \to \Re^n$ satisfy the matching conditions, i.e.,

$$\begin{aligned} f_0(x(t),t) &= Be_0(x(t),t) \\ f_1(x(t-\tau),t) &= Be_1(x(t-\tau),t) \end{aligned} \tag{3}$$

where

$$\begin{aligned} \|e_0(x(t),t)\| &\leq \rho_0 \|x(t)\| + k \\ \|e_0(x(t-\tau),t)\| &\leq \rho_1 \|x(t-\tau)\| \end{aligned} \tag{4}$$

for $\rho_0, \rho_1, k > 0$. In sliding mode control, the matching condition is a suitable condition for which the system in the sliding mode is insensitive to uncertainty. Input-delay systems are not controllable by the control input for initial time, $t + \tau \leq 0$. During the initial time, non-zero initial conditions and uncertainties can affect the stability of the system. In these types of systems, it is usually assumed that the initial condition $\upsilon(\theta) \in L^1\big((-\tau,0),\Re^m\big)$ exists.

DELAY BOUND BY USING A PREDICTOR

Consider a predictor (predictive sate), $\overline{x} \in \Re^n$, as

$$\overline{x}(t) = e^{A\tau}x(t) + \int_{-\tau}^{0} e^{-A\theta}Bu(t+\theta)d\theta \tag{5}$$

where $u_0(\theta) = \upsilon(\theta) \in L^1\big((-\tau,0),\Re^m\big)$. The sliding surface is defined as

$$\sigma(\overline{x}) = S\overline{x} = 0 \tag{6}$$

for $\sigma = [\sigma_1,\cdots,\sigma_m]^T \in \Re^m$ and the sliding matrix, $S = \big[S_1^T,\cdots,S_m^T\big]^T \in \Re^{m \times n}$. It is assumed that the matrix S is of full rank and the matrix SB is non-singular. The sliding matrix S is chosen so that the dynamics on the sliding surface have the desired closed-loop behaviors irrespective of input delay.

After selecting the sliding surface, the next step is to choose a control law such that it satisfies the condition for the existence of the sliding mode; $\sigma^T \dot{\sigma} < 0$. Consider the following control structure of the form:

$$u(t) = u_{eq} + u_N \tag{7}$$

where u_{eq} is an equivalent control for the nominal system of equation (1) without the uncertainty and u_N is a switching control to overcome the uncertainties of the system.

$$u_{eq} = -\left[SB\right]^{-1} SA\left[e^{A\tau} x(t) + \int_{-\tau}^{0} e^{-A\theta} Bu(t+\theta) d\theta\right]$$
$$= -\left[SB\right]^{-1} SA\overline{x} \tag{8}$$

$$u_N = \begin{cases} -\dfrac{(SB)^{-1} \sigma S e^{A\tau} B}{\|\sigma\|} \hat{\delta}(x,t) & if \; \|\sigma\| \neq 0 \\ 0 & \text{otherwise} \end{cases} \tag{9}$$

$\hat{\delta}(x,t) = \rho\|x\| + k + \beta$, for $\rho = \rho_0 + \rho_1 q$, $q > 1$, $\beta > 0$, is the upper bound on the norm of the total uncertainty of the system. Theorem 1 ensures the existence of the sliding mode of system (1) for the controller (7) by using the Razumikhin's theorem [8].

Theorem 1 [8]: If the control law (7) is used for system (1), then the sliding mode always exists.

To estimate a maximum delay bound, theorem 2 is developed to guarantee the asymptotic stability of the system in the sliding mode.

Theorem 2 [9]: If the Lyapunov equation

$$P(A_{11} - A_{12}K) + (A_{11} - A_{12}K)^T P = -Q \tag{10}$$

and the following condition

$$\lambda_{\min}(Q) - 2\tau\rho\|PB_2\|\left\|T^{-1}\begin{bmatrix} I \\ -K \end{bmatrix}\right\| > 0 \tag{11}$$

are satisfied for the positive constant and positive definite matrix P and positive definite symmetric matrix Q, then system (1) under control law (7) is asymptotically stable.

From theorem 2 it is possible to estimate a maximum delay bound that guarantees the asymptotic stability of the system in the sliding mode as follows:

$$0 < \tau \leq \tau_1 = \frac{\lambda_{\min}(Q)}{2\rho\|PB_2\|\left\|T^{-1}\begin{bmatrix} I \\ -K \end{bmatrix}\right\|} \tag{12}$$

Remark 1: the reduced order dynamics in the sliding mode are written as

$$\dot{z}_1(t) = (A_{11} - A_{12}K)z_1(t) + \hat{f}_0(t, T^{-1}z_1(t+s)) + \hat{f}_1(t, T^{-1}z_1(t+s-\tau)) \tag{12}$$

by transformation

$$TB = \begin{bmatrix} 0 \\ B_2 \end{bmatrix}, \quad TAT^{-1} = \begin{bmatrix} A_{11} & A_{12} \\ A_{21} & A_{22} \end{bmatrix} \tag{13}$$

Remark 2: find K so that K assigns eigenvalues of the reduced order system to the left-half plane. Choosing $\overline{S}_2 = I_m$, leaves $\overline{S} = \begin{bmatrix} K & I \end{bmatrix}$. Then, the sliding matrix is obtained as follows:

$$S = \overline{S}T \tag{14}$$

DELAY BOUND BY SMALL VALUE APPROACH

The aim of this subsection is to provide a time delay upper bound such that the asymptotic stability of system

$$\frac{dz(t)}{dt} = Ez(t), \quad E = A + B \tag{15}$$

ensures the asymptotic stability of the corresponding delay system

$$\frac{dz(t)}{dt} = Az(t) + Bz(t-\tau) \tag{16}$$

for any $\tau > 0$ less than this upper bound. An upper bound [10]

$$\tau_2 = \frac{\sqrt{\dfrac{\lambda_{\min}(P)}{\lambda_{\max}(P)}}}{2\|PB\|(\|A\| + \|B\|)} \tag{17}$$

is expressed in terms of the smallest and the largest eigenvalues $\lambda_{\min}(P)$, $\lambda_{\max}(P)$ of the real symmetric positive definite matrix solution of the Lyapunov equation

$$(A+B)^T P + P(A+B) = -I \tag{18}$$

with the identity matrix $I \in \mathfrak{R}^{n \times n}$.

DELAY BOUND BY LYAPUNOV-KRASOVSKII APPROACH

Suppose that

$$(A+B)^T P + P(A+B) = -Q \qquad (19)$$

with $P, Q \in \Re^{n \times n}$ which are symmetric and positive definite.

Let Q_1 be the square root of the matrix Q

$$Q = Q_1^T Q_1 \qquad (20)$$

Theorem 3 [11]: Let system (15) be asymptotically stable, then (16) is asymptotically stable for all $\tau \leq \tau_3$.

$$\tau_3 = \frac{1}{2\sqrt{\lambda_{max}(G)}} \qquad (21)$$

where $G = Q_1^{-T} E^T P B Q^{-1} B^T P E Q_1^{-1}$.

To apply τ_2, τ_3, adding m integrators to the system (1), the system without uncertainties is written as

$$\begin{aligned} \dot{z}_1(t) &= Az_1(t) + Bz_2(t-\tau) \\ \dot{z}_2(t) &= \tilde{\upsilon}, \quad \tilde{\upsilon} \in \Re^m \end{aligned} \qquad (22)$$

Consider the sliding surface

$$\sigma = z_2 + K_1 z_1 \qquad (23)$$

where $K_1 \in \Re^{m \times n}$. And assume that

1. $rank(B) = r \leq m$
2. (A, B) is controllable

Theorem 4 [11]: If the assumptions hold, then the control

$$\tilde{\upsilon}(t) = -K_1(Az_1(t) + Bz_2(t-\tau)) - g\,sgn(\sigma) \qquad (24)$$

with $g \in \Re_+$ makes the surface $s = 0$ attractive and invariant in finite time so that, for $\tau \leq max(\tau_2, \tau_3)$ where is given by (17),(21) applying to the system $\dot{z}_1(t) = Az_1(t) - BK_1 z_1(t-\tau)$, the system (1) with the control (24) is asymptotically stable.

Now we propose the maximum allowable delay bound, τ_{max} for robust control as follows.

Theorem 5: If τ_{max} is chosen as

$$\tau_{max} = \max_{i=1}^{3}(\tau_i) \qquad (25)$$

then the system (1) is asymptotically stable.

EXAMPLES

In order to illustrate the procedure, we consider an unstable plant

$$\begin{aligned} \dot{x}(t) = &\begin{bmatrix} 0 & 1 \\ 3 & -2 \end{bmatrix} x(t) \\ &+ \begin{bmatrix} 0 \\ 1 \end{bmatrix} \{u(t-\tau) + e_0(x(t),t) + e_1(x(t-\tau),t)\} \end{aligned}$$

where the initial condition, $x(0) = \begin{bmatrix} -1.6 & 1 \end{bmatrix}^T$ and the nonlinear parameter perturbations with model uncertainties are given by

$$e_0(x(t),t) = 0.3x_2(t)\sin(x_2(t))$$
$$+ 0.2\sin(2\pi\omega t), \quad \omega = 30$$
$$e_1(x(t-\tau),t) = 0.3x_2(t-\tau)\sin(x_2(t-\tau))$$

It can be easily seen that $\rho_0 = 0.3$, $\rho_1 = 0.3$, $k = 0.2$. Assigning -5 as the eigenvalue of the reduced order dynamics results in $K = 5$. Choosing $q = 1.1$, $T = I_n$ and $\bar{S}_2 = I_m$, yields $\bar{S}_1 = K$, $\|PB_2\| = 0.1$ and $\left\| T^{-1} \begin{bmatrix} I \\ -K \end{bmatrix} \right\| = 5.099$ for $P = 0.1$, $Q = 1$. Then, the sliding surface $S = \begin{bmatrix} 5 & 1 \end{bmatrix}$ and the delay bound $\tau_1 = 1.1814$.

To compute τ_2 and τ_3, $K_1 = \begin{bmatrix} 8 & 4 \end{bmatrix}$ so that $\dot{x}(t) = Ax(t) - BK_1 x(t-\tau)$ is asymptotically stable. $\tau_2 = 0.0089$ and $\tau_3 = 0.0791$ by equation (17), (21) for $P = \begin{bmatrix} 1.1 & 0.1 \\ 0.1 & 0.1 \end{bmatrix}$, $Q = I$. Therefore $\tau_{max} = \tau_1 = 1.1814$ and it can be known that τ_1 is less conservative than τ_2, τ_3.

SIMULATIONS

For the verification of the proposed theorem, the plant used in the example is considered. The simulation results are shown in Figures 3 and 5. We show the output of the control system in which a controller, sensors, and an actuator are connected directly or connected by a network. The proposed controller and the conventional sliding mode controller [3] are compared by using parameters of the example.

Figure 3 shows the result of the proposed controller by direct connection (no input delay) which is same with result of conventional sliding mode control system (designed without consideration of input delay). In Figure 4, input-delay induced by network connection affects the stability of the conventional sliding mode control system. Figure 5 shows that the system is asymptotically stable against uncertainties in the sliding mode although there is input-delay because of network-induced delay.

Here, we assumed that the bandwidth being able to transmit all data packets in all MADBs of loops is allocated in the vehicle network with three types of data (real-time synchronous data, real-time asynchronous data, and nonreal-time asynchronous data) based on a scheduling method such as the earliest deadline first algorithm [12],[13] and the plant in the simulation is a distributed control system. This assumption ensures that all periodic data are transmitted within the respective sampling period to guarantee stability of control loops, while guaranteeing real-time transmission of real-time asynchronous data and minimum transmission of nonreal-time asynchronous data. That is, transmissions of three types of data should be allocated in the sampling period.

CONCLUSION

In this paper, the MADBs are obtained for the robust stability of a distributed control system in a vehicle network. Using the results of a sliding mode control with an input-delay system, we find the MADB which can be used for sampling period decision. The presented method is a guideline, as it provides a solution for determining the sampling period of each control loop. An example is presented to show the usefulness of the proposed method for the distributed control system.

The sampling period is obtained from the plant model independent of network protocols, while the network-induced delays depend on network configurations. In addition, a faster sampling is said to be desirable in sampled-data systems because the performance of the discrete-time system controller can approximate that of the continuous-time system. But in distributed control systems using in-vehicle network, the high sampling rate can increase network load, which in turn results in longer delays of the signals. Therefore finding a sampling rate that can both tolerate the network-induced delay and achieve desired system performance is important. This will be investigated in future studies.

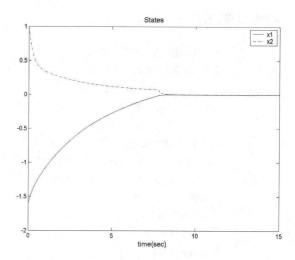

Figure 3 States (standalone controller with proposed SMC).

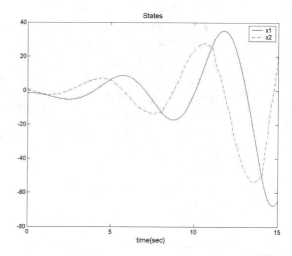

Figure 4 States (network based controller with conventional SMC).

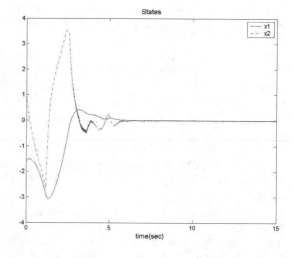

Figure 5 States (network based controller with proposed SMC).

13

NOMENCLUTURE

ECU	Electronic control unit
τ_{ca}^j	Communication delay from a controller to actuators of the j th loop
τ_{sc}^j	Communication delay from sensors to a controller of the j th loop
τ_c^j	computation time in the controller j
τ_{max}	Maximum allowable delay bound (MADB)
f_0, f_1	Nonlinear parameter perturbations
$\overline{x}$	Predictive state
σ	Sliding function
S	Sliding matrix
u_{eq}	Equivalent control input
u_N	Switching control input
$\hat{\delta}$	Upper bound on the norm of the total uncertainty of the system
P	Positive definite matrix
Q	Positive definite symmetric matrix
K	State feedback matrix
T	Transformation matrix
λ_{min}	Minimum eigenvalue
λ_{max}	Maximum eigenvalue
$\underline{S}$	Transformed sliding matrix
Q_1	Square root of the matrix Q
K_1	State feedback matrix
g	Switching control gain

REFERENCES

1. Mathony, H.-J., Kaiser, K.-H. and Unruh, J., (1994) "Serielle Kommunikation zwischen Steuergeräten", *Proceedings of Elektronik im Kraftfahrzeugwesen*, Vol. 437

2. Heintz, F. and Zabler, E., (1989) "Application Possibilities and Future Chance of Smart Sensors in the Motor Vehicle", *SAE Technical Paper No. 890304*

3. Hung, J.Y., Gao, W. and Hung, J. C., (1993) "Variable Structure Control: A Survey", *IEEE Transactions on Industrial Electronics*, Vol. 40, pp. 1-22

4. Ryan, E.P., (1983) "A Variable Structure Approach to Feedback Regulation of Uncertain Dynamical System", *International Journal of Control*, Vol. 38, pp. 1121-1134

5. Slotine, J.-J.E. and Li, W., *Applied Nonlinear Control*, Prentice-Hall

6. Koshkouei, A.J. and Zinober, A.S.I., (1996) "Sliding Mode Time-delay Systems", *Proceedings of the International Workshop on VSS*, pp. 97-101

7. Shyu, K.K. and Yan, J.J., (1993) "Robust Stability of Uncertain Time-delay Systems and Its Stabilization by Variable Structure Control", *International Journal of Control*, Vol. 57, pp. 237-246

8. Roh, Y.-H. and Oh, J.-H., (1999) "Robust Stabilization of Uncertain Input-delay Systems by Sliding Mode Control with Delay Compensation", *Automatica*, Vol. 35, pp. 1861-1865

9. Roh, Y.-H. and Oh, J.-H., (1999) "Sliding Mode Control with Delay Compensation for Uncertain Input-delay Systems", *Proceedings of American Control Conference*, pp. 309-313

10. Gopalsamy, K., (1992) "Stability and Oscillations in Delay Differential Equations of Population Dynamics", *Mathematics and Applications*, Vol. 74

11. Gouaisbaut, F., Perruquetti, W. and Richard, J.P., (1999) "A Sliding Mode Control for Linear Systems with Input and State Delays", *Proceedings of Conference on Decision and Control*, pp. 4234-4239

12. Levi, S.-T. and Agrawala, A.K., (1990) *Real Time System Design*, McGraw-Hill

13. Audsley, N.C., Burns, A. and Wellings, A.J., "Deadline Monotone Scheduling Theory and Application", *Proc. IFAC J. Contr. Eng. Practice*, Vol. 1, No. 1, pp. 71-78

Polymer Clad Silica Optical Data Communication System

Brian E. Johnson and Eric J. Olsen
Yazaki North America

ABSTRACT

High technology vehicle applications are expanding functionality and connectivity, enabling increased information access, and improving occupant safety. In-vehicle data transmission systems are also evolving to meet the requirements of these applications, which include diverse node locations, an increased number of interacting nodes, information transfer capacity, and data transmission rates. The Yazaki Polymer Clad Silica (PCS) Optical Data Communication System is leading this evolution. This next generation optical communication system integrates gigabit communication technology, including glass fiber and VCSEL light sources. It exceeds current vehicle network system demands, and anticipates future requirements to enable the automotive optical communication networks of tomorrow.

INTRODUCTION

Vehicle functions requiring high speed, low latency communication are driving the recent proliferation of optical data network systems in the vehicle. The convergence of consumer-driven applications within the vehicle, such as high quality digital audio and video, combined with the need for remote device location and control, provide compelling reasons to move from point-to-point architectures to optimized network structures[1]. Networks enabling this functionality must feature high bandwidth capability and high electro-magnetic immunity. In addition, such systems must be compatible with automotive system design and assembly requirements, such as design and application flexibility, compatibility with common assembly procedures, and environmental robustness.

AUTOMOTIVE OPTICAL NETWORKS

Introduced in 1998, the Domestic Digital Data Bus (D2B) was the first high-volume automotive optical network. The system provides network communication at a data rate of 5.6 megabits per second (Mbps). Recently, many automotive original equipment manufacturers,

including DaimlerChrysler, BMW, and Audi, have begun implementing the Media Oriented Systems Transport (MOST®) network. The network was introduced in July of 2001, and is being used on 11 production vehicles, with more than 60 approved devices currently in production[2].

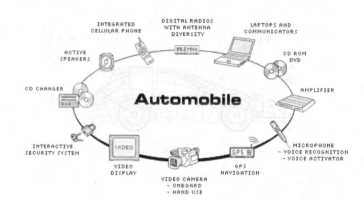

Figure 1 – MOST® Optical Fiber Network [3]

D2B and MOST® utilize an optical physical layer in a ring topology for single-fiber node-to-node communications, requiring only a single fiber optic transceiver (FOT) pair at each node. Optical signals are created by a transmitter which receives electrical data from a network interface controller (NIC) residing on the device printed circuit board (PCB), and converts these signals into 650nm digital light pulses using a light emitting diode (LED). An optical connector enables the light to be coupled into a 1.0 mm core diameter plastic optical fiber (POF), typically made of polymethyl methacrylate (PMMA), for transmission to a downstream node receiver. The receiver contains a positive-intrinsic-negative (PIN) photodiode, which converts the digital optical signal back into an electrical signal that can be understood by the network interface chip. The information is then translated by the NIC for use within the device. The MOST® system currently operates at data transmission speeds up to 22.5 Mbps in infotainment applications.

The Byteflight® system was introduced by BMW on its 2001 7-Series to enable high-speed data communication between the 13 distributed passive safety system electronic control units. Information from accelerometers, pressure sensors, seat occupation sensors, and belt buckle switches is transmitted, as well as body and convenience equipment data, and part of the powertrain controls[4]. This system also operates on a 1.0 mm POF-based physical layer, using LED light sources and PIN photodiode-based FOTs. One primary implementation difference is the active star topology, which relies upon a powered central node to actively direct and distribute information. This topology is used to minimize the impact of a single node failure on remaining node interoperation.

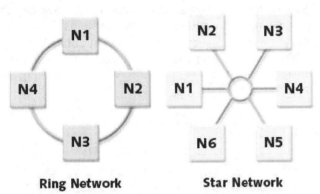

Ring Network   **Star Network**

Figure 2 – Illustrated Network Topology Examples

EMERGING NETWORK REQUIREMENTS

Even as POF-based systems become more prevalent in automotive applications, there is a growing recognition of their performance limitations by OEMs and suppliers. Current systems are restricted to specific passenger compartment applications in which maximum cable operating temperature does not exceed 85°C. Maximum data transmission speed is limited by the performance of the LED, as well as the data transmission capacity of the step-index POF fiber. Length-based fiber attenuation introduces limitations on node-to-node separation limits and fiber routing paths, restricting system architecture and partitioning design flexibility. In addition, POF mechanical performance limitations, such as the minimum long-term bending radius of 25mm, drive special handling and assembly requirements, adding to implementation complexity.

Next generation optical data communication systems must overcome the limitations of POF-based systems while improving environmental robustness and remaining cost competitive. These systems should also support emerging optical network applications, including IDB-1394 and MOST 2, which are targeting network data rates greater than 150 Mbps, and FlexRay, which may be applied in all areas of the vehicle[5]. When possible, pieces of the same technology should be used across applications, to leverage volume-based cost advantages.

SYSTEM DESCRIPTION

The Yazaki PCS Optical Data Communication System is a protocol-independent solution that integrates several key technologies to meet and exceed current and future optical physical layer and communication requirements.

PCS CABLE

Yazaki has worked with key partners to develop a custom PCS cable that is applicable for installation in all areas of the automobile. Standard PCS fibers are available 'off the shelf' from many sources, and have been used in several applications including industrial, medical, sensor, and data communications for many years.[6] Inherent fiber properties include very low signal attenuation (<0.01 dB/m compared with <0.3 dB/m in POF) in the relevant spectral range. This reduced attenuation nearly eliminates the need to consider node-to-node separation length.

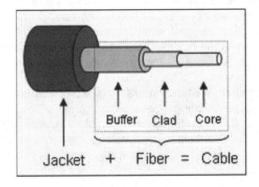

Figure 3 – PCS Cable Construction

The fiber also provides high flexibility and mechanical performance (15mm long-term bending radius). This increased flexibility translates into improved handling and routing capability, significantly reducing special assembly requirements. When the proper buffer and jacket combination is used, the cable assembly can withstand extreme temperature cycling exposure (-40°C to 125°C) for extended periods of time without incurring irreversible changes in optical or mechanical performance[7], allowing the fiber to be installed in all relevant application areas of the vehicle.

VERTICAL CAVITY SURFACE EMITTING LASER

One design challenge involves coupling light into the PCS fiber. Because the PCS core diameter is 5 times smaller than standard POF (200μm vs. 1.0mm), light sources with wide beam divergence, such as the LED, make efficient power coupling difficult. For this reason, the Yazaki PCS Optical Communication System uses a laser light source known as a VCSEL (Vertical Cavity Surface Emitting Laser) in the transmitter FOT.

The VCSEL was introduced commercially by Honeywell in 1996. Performance and reliability have been improving continuously since then, encouraging widespread use in broad applications. Industry volumes for 2003 are estimated above 8 million units for high-speed optical data communication applications[8].

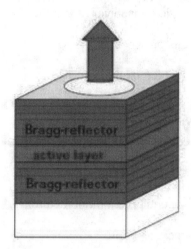

Figure 4 – VCSEL Construction

The VCSEL offers many advantages when compared with LED light sources. Commonly available chips are specified for 1.25 gigabits per second (1250 Mbps) or 2.5 Gpbs transmission speed, but can operate effectively over a range of 10 Mbps to 10 Gbps. In addition, VCSELs offer higher efficiency, meaning that more electrical energy is converted into light and less into heat, resulting in lower operating current consumption. This contributes to better performance over a range of higher ambient temperatures.

Figure 3 – Prototype PCS Transceivers (FOTs)

The small beam divergence of the VCSEL (typically 15°) enables efficient coupling of optical power into the 200μm PCS cable over all operating conditions. In addition, the VCSEL chip size is approximately the same as currently used LEDs. In addition, the VCSEL is more efficient that the LED, reducing junction temperatures and thermal concerns at higher operating temperatures. Taken together, these factors allow for the same or smaller transmitter package sizes.

PRECISION CONNECTION SYSTEMS

Yazaki has leveraged technological advances driven by the high-bandwidth, low-attenuation demands of the telecommunication industry to create robust, high performance connection systems that enable the PCS architecture. Precision design and molding techniques are utilized to increase molded component accuracy and reduce assembly tolerances. High-performance engineered polymers, such as liquid crystal polymer (LCP), are applied in critical alignment areas to minimize losses due to misalignment. Together, these factors enable connector performance that meets or exceeds existing POF system connection performance. As a result of continuous efforts throughout the design process, this is accomplished without sacrificing robust performance or cost competitiveness.

Figure 5 – Prototype PCS Header Connector

PCS SYSTEM FEATURES

The Yazaki PCS Optical Communication System technology exceeds all existing optical network requirements, enabling flexibility in application and offering many system level advantages.

ROBUST ENVIRONMENT APPLICATION

Yazaki PCS Optical Communication System components are designed to withstand operating temperatures up to 125°C, while offering high electro-magnetic immunity. This allows system application to be extended into all vehicle locations, enabling Yazaki to provide robust data communication to components distributed throughout the vehicle.

HIGH BANDWIDTH COMMUNICATION

The Yazaki PCS Optical Data Communication System utilizes next generation optical network technology such as 200μm PCS cable and VCSELs, which make it capable of satisfying current and future in-vehicle bandwidth requirements from the Mbps to the Gbps range. This system technology exceeds existing network bandwidth requirements, and delivers the information carrying capacity to enable emerging high performance systems and applications.

OPTIMIZED DATA TRANSMISSION ARCHITECTURES

By providing decreased attenuation, increased mechanical flexibility, and a significant expansion of application areas, the Yazaki PCS Optical Communication System enables unparalleled flexibility in optical data transmission architecture design and application. This new flexibility increases opportunities to optimize network configuration, enables additional system expandability, and provides added flexibility in determining network device location.

CONCLUSION

The Yazaki PCS Optical Communication System is a protocol independent solution built on technology that offers clear benefits to existing optical network applications. In addition, the basic building blocks of the system – PCS cables, VCSEL transceivers, and high precision connections – can be brought together by Yazaki to provide optimized data communication solutions for networks of the future.

REFERENCES

1. http://www.mostnet.de/news/Conferences+&+Prese ntations/2002/1/26/files/2002+Telematics+Update+C onference+2002-05-15.pdf
2. Ibid;
3. http://www.mostcooperation.com/technology/index.p hp; this figure is the property of the MOST® Cooperation, and is used with permission.
4. www.byteflight.com
5. www.wmrc.com/businessbriefing/pdf/data_2002/refe rence/ref21.pdf
6. J.P. Clarkin, "Hard Clad Silica Fibers for Automotive Applications," ITG-FG-541 Conference Proceedings, Munich, June 25th 2003, p. 10-11.
7. Ibid;
8. http://content.honeywell.com/vcsel/pdf/vcsels_are_c oming.pdf

CONTACT

Brian Johnson has worked in Research and Development at Yazaki North America in Canton, Michigan since 1997. He received a Bachelor of Science degree in Physics from Michigan State University in 1992, and a Bachelor of Science degree in Mechanical Engineering from Lawrence Technological University in 2001. His is currently pursuing a Masters Degree in Business Administration at Walsh College in Troy, Michigan. Contact information follows:

bjohnson@yazaki-na.com; phone: (734)983-2368

DEFINITIONS, ACRONYMS, ABBREVIATIONS

dB – Decibels; used to express optical attenuation

D2B – Domestic digital data bus

FOT – Fiber optic transceiver (transmitter or receiver)

Gbps – Gigabits per second

LED – Light emitting diode

Mbps – Megabits per second

MOST® – Media Oriented Systems Transport

NIC – Network interface controller

OEM – Original equipment manufacturer

PCB – Printed circuit board

PCS – Polymer clad silica

PIN – Positive-intrinsic-negative

POF – Plastic optical fiber

PMMA – Polymethyl methacrylate

VCSEL – Vertical cavity surface emitting laser

2003-01-1096

Issues on Load Availability and Reliability in Vehicular Multiplexed and Non-Multiplexed Wiring Harness Systems

M. Abul Masrur
US Army TACOM

John She and Paul Richardson
University of Michigan-Dearborn

ABSTRACT

In military vehicles reliability can sometimes be a more important issue than cost. With that in perspective, this paper discusses the load availability and reliability issues in automotive multiplexed wiring harness systems, which are potentially useful in the military, and compares the same with a regular non-multiplexed system. For that purpose, a figure of merit or metric is introduced, and the load availability is described in terms of this metric, which depends on the architecture chosen.

INTRODUCTION

Automotive multiplex system is an elegant concept in terms of simplicity that it introduces, accompanied with reduced packaging and cost [1-5]. Furthermore, introduction of the multiplex system data bus opens the possibility of immediate access to various sensor information at each of the nodes by simply tapping those at any point on the signal bus, which could be difficult in a non-multiplexed system. This information could be advantageously used for various functionalities, thus leading to a better system performance. These systems are also important for any future combat systems (FCS) by achieving a lighter vehicle. Intuitively speaking, multiplex system is considered to be very reliable (compared to regular hard connected wiring harness) as well, since there are likely to be less connectors and wires running through the vehicle. For the purpose of this paper a non-redundant ring architecture will

be chosen, which will consist of a power bus carrying the high power battery lead around the vehicle, and also a communication bus which is the signal carrying medium. The integrated load modules will have both a communication module and also a smart power module, which can consist of solid state power switches or other kind of relays, interfacing the loads. These two modules (communication network and power) need not be physically separate hardwares, rather these can form a single entity. It will of course, have receptacles to connect to loads, and power source, and signal connectors from the signal bus. To drive a load, a signal can trigger a solid-state power semiconductor. Alternatively, it can drive an electromechanical relay through an appropriate driver circuit.

In this automotive multiplex system, although the wiring harness complexity is obviously less, some additional components like the intelligent modules at various load and source locations are introduced. Each intelligent load module does not necessarily control a single load, rather, depending on the physical location, it can control multiple loads as well. Hence the overall reliability and thus the availability of the loads are dependent on the reliability of each of these modules, and all the intermediate components in between, namely, the power bus, the communications bus, connectors (at both power and signal levels), and the reliability of the message transmission, particularly during heavy message load conditions on the communication bus. In particular, a failure of a single module can disrupt several loads, which may be unacceptable in certain safety critical military situations, and hence can be of important consequence to any future combat system (FCS) using those. It is the purpose of this paper to discuss the reliability and thus the availability of the loads in a system which uses multiplexing and the one which does not, and the effect of the architectural organization on these quantities. The discussion will be based on certain assumed probability of various components being good, or operating in a manner as they are supposed to. To facilitate the comparison, a figure of merit based on several important attributes of the system, will be introduced and the comparison will be made on the basis of this figure of merit, which the authors believe is one possible way to compare such system quantities.

MULTIPLEX SYSTEM ARCHITECTURE

Although different architectures for the multiplex wiring system exist for possible implementation,

e.g. ring, linear, star etc., for our discussion we will assume a ring architecture, with all the nodes having equal status in terms of operational capabilities.

Figure 1. shows a ring architecture, which will be the architecture chosen for illustration in this work. The diamond shaped joints indicate connection points between various cables (and signal wires) to the other components. In order not to clutter the diagram, the diamond shaped connectors are not shown at every load connection point. For communication purposes it is assumed that there will be some form of arbitration based on certain protocol, and eventually only one node will win the arbitration. These are well discussed in the references [1-5]. The square boxes indicate the intelligent nodes for processing the communication protocols and also managing the loads. These can contain power electronics based switching modules to connect to the loads, or these can also use electromechanical relays as well. In

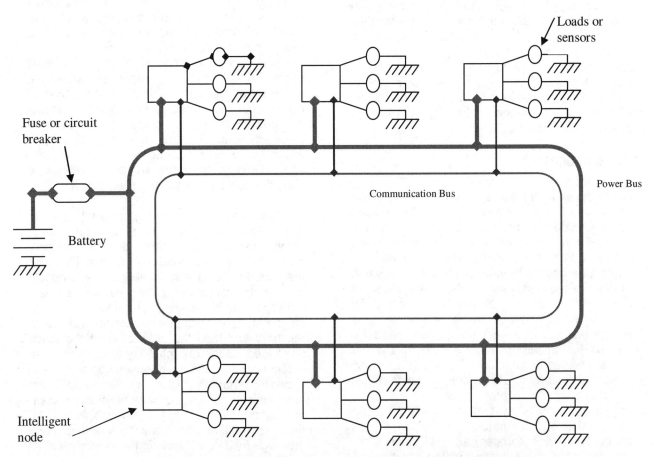

Figure 1. An automotive multiplex system architecture configuration with power and communication buses, loads, and intelligent nodes for processing communication protocol and load management.

20

addition, there can be one main circuit breaker or fuse for protecting the whole system. Additionally there can be fuses or mini-circuit breakers for protecting the individual loads segments. In the following, the hard and soft linkages from source to loads will be traced. It should be noted that loads can consist of lights, motors and the like, and similarly it could imply sensors at various points of the vehicle, which can communicate with the intelligent modules as needed.

A METRIC TO MEASURE SYSTEM LOAD AVAILABILITY

The ultimate purpose of the wiring harness (consisting of both the power and the communication buses) is to actuate certain loads, based on the information received from the driver of the vehicle and/or the various sensors, and to coordinate these in a certain desired manner. If this desire fails, due to whatever reason, we can say that the purpose has failed and the particular load will be unavailable. With this objective set, we can define the following terms.

λ_i = probability that the i-th load/sensor is up (or available) at a particular moment, -- which can range from 0 to 1.

C_i = criticality of the i-th load/sensor, -- a number which indicates how critical it is for this particular load to be up. A range of 1 to 10 is chosen for convenience, where 1 means not at all critical, to 10 meaning absolutely critical.

H_i = number of times on an average the i-th load is invoked or its status updated, during a given span of time (hr, min, sec etc.), -- a number which will depend on the nature of the load. The duration can be chosen to be in units of hour, if it is deemed that H_i may be a small fraction and inconvenient to use.

We now define a metric, which is an important figure of merit (or rather demerit) indicating how detrimental it is for a particular load/sensor (the i-th load/sensor) to be down, by introducing,

$$F_i = C_i H_i (1-\lambda_i) \left\{ \prod_{j=1,\dots n,\ j \neq i} (\lambda_j) \right\} \qquad \dots\dots (1)$$

where n is the total number of loads in the system. In equation (1), the term $(1-\lambda_i)$

indicates the probability of the i-th load being down. In this equation we make the assumption that the probability of more than one load being simultaneously down is of second order, and hence much smaller compared to the terms used in equation (1). This is accounted through the product terms in equation (1). This metric can be reciprocally defined as well, in terms of the probability of a load being up instead of down, in which case equation (1) has to be slightly modified by exchanging the roles of $(1-\lambda_i)$ and (λ_j). With the above background, the cumulative system figure of demerit can now be defined as:

$$F_s = \sum_{i=1}^{n} F_i \qquad \dots\dots (2)$$

assuming that the total number of loads = n. It should be noted here that loads do not imply merely various actuators, lights etc., they also imply various sensors which are running on their own and communicating with various systems in the vehicle continuously or intermittently.

APPLICATION EXAMPLE

The most important item involved in evaluating the cumulative system figure of merit (or demerit) requires finding the value of λ_i. If we want to evaluate this for the example architecture shown in Figure 1, the following items are to be accounted for, each of which can potentially lead to a failure. This list is just a possible example of items that can lead to failure, and depending on the specific architecture it will vary. Some additional comments on this are provided in the section prior to the conclusion. But the methodology described here will be valid regardless of the particular case being studied.

Hard items:
1. Battery to battery-cable connector
2. Battery-cable to main-fuse connector
3. Main-fuse to power-bus connector
4. Power-bus cable
5. Power-bus cable to intelligent-node connector (each node consisting of both the power module and also the communication module).
6. Signal-bus (twisted pair etc.)
7. Signal-bus to node connector
8. Node-module
9. Node to load-fuse connector
10. Load-fuse to load connector
11. Load to ground connector

12. Electromechanical relay or solid-state switches connecting the load (can replace the fuse depending on the system).

Soft items:
1. Network message overload and/or error at source end (at message initiating node) causing priority based queuing and leading to delay and/or error in transmission [6].
2. Failure to win contention with other nodes leading to delay for the message to reach destination, and/or error in message transmission.

It is to be understood that the probability of failure of hard items changes with time, starting from infant mortality [7-9] to deterioration with usage and age. For the soft items, the probability of failure will depend on the message loading and interval used in the system, and is directly related to the quantity $\sum H_i$, for i = 1 to k (number of loads), which was defined earlier. Since it is not the purpose of this paper to try to evaluate the various reliability numbers for various items listed above, rather to study the effect of system architecture on the reliability, we will assume arbitrarily chosen but reasonable quantities for the same.

Let us assume for now that there are six nodes in the system with three loads connected to each node. This is the configuration shown in Figure 1. The reliability of each item is indicated by the symbol ξ.

For hard items 1 to 7 let us choose:
$\xi_1 = 0.99999$ $\xi_2 = 0.99997$ $\xi_3 = .99997$
$\xi_4 = 0.99999$ $\xi_5 = 0.99998$ $\xi_6 = .99999$
$\xi_7 = 0.99999$
For node module itself let us choose
$\xi_8 = 0.99995$
For node to fuse connectors let us choose
$\xi_9 = 0.99996$ (same for other node to load connectors)
For fuse to load connectors let us choose
$\xi_{10} = 0.99996$ (same for other nodes to their respective load connectors)
For load to ground connectors let us choose
$\xi_{11} = 0.99996$ (same for other nodes to their respective ground connectors)
For the electromechanical (or solid-state) relays (can replace the fuse where applicable), let us choose $\xi_{12} = 0.99995$
For soft items 1 and 2, let us choose as follows:
$\xi_{13} = \xi_{14} = 0.99998$

In the above we just traced the items for one single node. The same will apply to the other loads. For hard items, only the items from 1 to 4 will be common to all nodes, and the rest of the items will be separate for each nodes.

For the 18 loads let us assume the following quantities for C_i and H_i :

$C_1 = 10$	$H_1 = 20$	$C_2 = 10$	$H_2 = 3$
$C_3 = 7$	$H_3 = 10$	$C_4 = 8$	$H_4 = 12$
$C_5 = 1$	$H_5 = 20$	$C_6 = 4$	$H_6 = 20$
$C_7 = 10$	$H_7 = 100$	$C_8 = 3$	$H_8 = 8$
$C_9 = 6$	$H_9 = 14$	$C_{10} = 2$	$H_{10} = 10$
$C_{11} = 1$	$H_{11} = 1$	$C_{12} = 2$	$H_{12} = 20$
$C_{13} = 3$	$H_{13} = 150$	$C_{14} = 9$	$H_{14} = 60$
$C_{15} = 10$	$H_{15} = 2$	$C_{16} = 10$	$H_{16} = 8$
$C_{17} = 5$	$H_{17} = 5$	$C_{18} = 6$	$H_{18} = 40$

For the example case here, the probability that a particular (n-th) load is available is given by the product of all the reliability terms ξ_i for i = 1 to 14.

$$\lambda_n = \prod_{i=1}^{14} \xi_i \qquad \dots\dots(3)$$

In the particular example, if the values of ξ_i are inserted, we get $\lambda_n = 0.99962$, for all n=1 to 18, assuming same component reliabilities in each node. Hence we can easily see that whereas the individual component failure probabilities (ξ_i) were chosen to be between 1 to 5 per 100000, after combining all the components together, we get a failure rate of about 38 per 100000. Although this number is quite small, it might still contribute significantly toward the overall system availability or lack thereof.

In addition to the above, since in general multiple loads are connected to a particular node, there is a reliability issue which leads to the failure of a cluster of loads all together, should any one or more of the linkages leading to the node fail. This can pose a potentially dangerous situation during failure. Thus we also define a group load reliability (or availability) in the following manner. For the n-th node, the group reliability is given by:

$$\lambda_n = \prod_{i=1}^{8} \xi_i \qquad \dots\dots(4)$$

In the example above, this becomes = 0.99983, or 17 failures per 100000, which is about half the individual load failure rate.

It is therefore noticed that the group reliability of the loads connected to node 1 and indicated by equation (4), is somewhat higher than the individual load reliabilities given by equation (3). This is naturally expected, since there are more linkages preceding an individual load than for the group of loads connected to a particular node.

Using these quantities (based on equations (3) and (4)) for all the nodes, we can easily compute that for this 6-node, 18-load system, the figure of merit (actually demerit to be more exact) according to equation (1) will be as indicated in the following.

Figure of demerit for the system considering individual load failure probabilities is computed by excluding multiple failure probabilities, which is normally of second order, compared to single component failures. So, we compute the probability of failure of load # 1 in node # 1, and assume that all the other loads and nodes are up (available). This leads to, with all the loads taken into account, a cumulative figure of demerit due only to individual load failures to be:

$$F_s = \sum_{k=1}^{18} C_k H_k (1-\lambda_k) \left\{ \prod_{m=2}^{17} (\lambda_m) \right\} \prod_{j=1}^{6} (\lambda_j) \qquad ....(5)$$

where λ_k for k-th load is evaluated using equation (3), and λ_j is evaluated by equation (4). λ_m is evaluated equation (3), but slightly modified, so that the reliability of elements already included in λ_j are excluded here (in other words, only components 9 to 14 are now included during this evaluation, rather than 1 to 14, as in equation (3)). Inserting all the necessary values, we get from equation (5),

$$F_s = 0.7677 \qquad ....(6)$$

In equation (5), the value of j runs from 1 to 6, for the six separate clusters of loads. Since this system also has the possibility of group failure as indicated earlier, we have to account for that as well. This is computed by using equation (1), except that we now use equation (4) instead of equation (3) to compute the λ_j and this λ_j (for the j-th cluster) will correspond to each cluster of loads corresponding to each node, there being a

total of six such clusters in the example case. During computation of the group failure, we assume that only one cluster can fail at a time. Thus the equation for computation of cumulative failure probability, assuming all the clusters can fail one at a time is given:

$$F_s = \sum_{k=1}^{18} C_k H_k [(1-\lambda_j)]_{j=1} \prod_{j=2}^{6} (\lambda_j) \qquad ....(7)$$

Here, first we compute the probability of cluster # 1 being down and others up, but since the result is symmetric, we can extend it to others to compute the cumulative figure of demerit. Thus, in our example case this group failure figure of demerit will become = 0.3448. Hence adding this with equation (6) we get the next level of figure of demerit for the system to be = 1.1125. Finally there is the probability of a total system failure due to the hard linkages 1 to 6. Following the same line of thought, it leads to an additional figure of demerit = 0.2233. Hence if we add this number to the previous ones, the final cumulative system figure of demerit will be:

$$F_s = 1.3358 \qquad ....(8)$$

which is a combination of items due to figure of demerit from individual load failures (0.7677), figure of demerit from group load failures (0.3448), and the figure of demerit from total system failure (0.2233). It can be immediately observed that the contribution to the total system figure of demerit comes mostly from the individual load failures, next from a group of loads failing together, and least from the total system failure. This is naturally expected. The authors believe that to best capture the phenomena of total system, group and individual failures, it is rational to add these numbers as has been done to get the equation (8), since all of these contribute towards the demerit of the overall system. This number can therefore be used as a metric for the purpose of evaluating the quality of a system, and the larger the number, the worse the system availability.

Let us now change the architecture to a non-multiplexed system, but retaining everything else the same. For an equitable and fair comparison, we will assume the same number of loads as in figure 1. However, since there is no soft method of communicating to the nodes anymore, we will use the hard architecture for the system as indicated in the figure 2.

In figure 2, the conventional scheme is used. In other words, we go from battery to a main fuse, and thereafter to a fuse box with all the fuses for different circuits. From there we go to the switches wherever applicable (sensors will normally not need a switch) and then to the load, and finally to the ground. In figure 2, only one

It can, therefore, be seen that in this system there are 12 linkages from source to load. Hence the total chance of failure of a particular load is somewhat less. Following the same line of approach used earlier for the multiplex system reliability analysis, we can get the figure of demerit here to be as follows: Figure of demerit

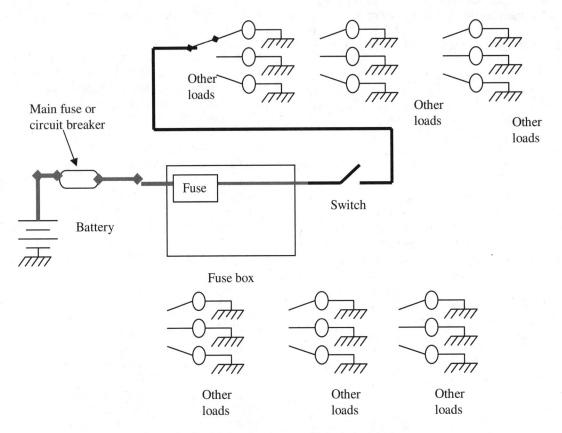

Figure 2. An automotive (non-multiplex) electrical
system architecture configuration.

particular load connection is indicated to avoid cluttering the diagram. Here, the hard items list will be as follows:

1. Battery to battery-cable connector
2. Battery-cable to main-fuse connector
3. Main-fuse to fuse-box cable
4. Fuse-box cable to fuse-box connector
5. Fuse-box to fuse connector
6. Fuse to switch-cable connector
7. Fuse-to-switch cable
8. Switch or relay
9. Switch to load cable
10. Switch to load-cable connector
11. Load-cable to load connector
12. Load to ground connector

due to individual load failure = 0.4247, due to total system failure = 0.1618. Adding the two numbers, the total figure of demerit for this non-multiplex system is 0.5865. The comparison between the two architectures is summarized in Table 1.

Table 1. Comparison between the architectures
of Figures 1 and 2.

TYPE → ARCH. ↓	INDIV. FAILURE	GROUP FAILURE	WHOLE SYSTEM FAILURE	TOTAL F_s
MPLX.	0. 7677	0. 3448	0. 2233	1. 3358
NON- MPLX.	0. 4247	N/A	0. 1618	0. 5865

24

Thus reliability-wise the architecture of figure 2 seems to be a better system with certain qualifying statements provided later. In this system there is no cluster of loads which are connected to nodes. However, there is a probability of total system failure due to the first 4 items in the hard component list above, leading to a figure of demerit due to this to be 0.1618, as noted earlier.

Sometimes, it is intuitively said that a multiplexed wiring system is more reliable than a regular completely hardware based system. But that statement needs some qualifications as detailed below. It is true, that a multiplexed system [3] helps reduce the number of conductors, connectors, crimps, and splices etc. in a wiring harness system from source to load. This definitely helps packaging and cost reduction. However, from source to load, as the number of conductors, connectors etc. are reduced by taking advantage of the commonality in usage of same conductor/s from source to multiple loads, the redundancy which existed prior to multiplexing is also thereby eliminated. The reliability of a system, in general terms, is enhanced by reducing serial components and increasing parallel components. The reason is obvious, since the failure of a serial component in a linkage from source to load leads to the failure of the load. However, addition of a parallel component reduces the chance of failure, by providing redundancy. It is this parallel redundancy in a non-multiplexed system that contributes to the enhancement of system reliability noted above.

SOME SPECIFIC ISSUES ON ARCHITECTURE

The following specific issues are to be noted in connection with a multiplexed system:

In the case of both the power bus and the signal bus, in a ring configuration as in figure 1, up to one cut (i.e. break, or open circuit fault) in the ring will still allow a satisfactory operation of the system. The analogous statement is obviously not valid for a short circuit type of fault, however. This situation does not apply to the non-multiplexed system since the source to load connections are topologically separate to a large extent. These issues are very much architecture dependent. If, instead of a ring architecture, one decides to select a linear architecture (bus architecture) where various node loads are connected to a linear bus as tree branches, then a break on the bus will, in general, disrupt the functionality. In our discussion above, we have tried to provide a fair comparison, in other words, we tried to compare the systems without introducing any additional features which could be naturally obtained as a beneficial feature in the multiplexed system. For example, as indicated in the introduction section, by introducing the multiplex system data bus, one opens the possibility of immediate access to various sensor information at each of the nodes by simply tapping those at any point on the signal bus, which could be difficult in a non-multiplexed system. This information could be advantageously used for various functionalities, thus leading to a better system performance. However, the scope of this paper is limited only to the reliability of the two systems, without regard to any by-product type of functionalities which might be achieved in one or the other.

In addition to the issues discussed above, there are certain manufacturing issues which need to be taken into consideration. It has been indicated [3], that in a multiplexed wiring harness (as opposed to a production wiring harness which is not multiplexed), the connector count was reduced by 34, the double crimps at connector pins were reduced from 66 to 43, the internal splice points were reduced from 48 to 18 which led to elimination of 143 wire terminations. In addition, total wire count was reduced by 147. These matters come under the category of manufacturing advantages and packaging. However, this still most probably does not give a fair comparison between the multiplexed and non-multiplexed system, due to the fact that those reductions indicated [3] could very well have been due to the comparison drawn between a very specific production wiring harness (non-multiplexed) and a multiplexed wiring harness. It is not necessarily fair to say that this particular production harness was designed in the best possible manner and was most optimized by all means. It is very likely that a better and most optimum design could yet be obtained for the existing non-multiplexed production wiring harness, leading to some improvement in the count of its connectors, crimps, splices, etc. even without a multiplex system. The authors could not trace any comparison of this nature in the literature. In general, in a vehicular wiring harness, there are lots of obstructions through which the harness has to be physically placed. This calls for practical reasonings, requiring one to take sub-optimal strategies for construction, thus leading to additional wires, connectors and so on. In our discussion, we did not take those into account. We assumed that there are no obstructions or other manufacturing issues. Due to the introduction of multiplexing system, some of these manufacturing advantages immediately become realizable, leading to the component reduction indicated in the reference [3].

A few additional comments regarding the hard items indicated earlier can be made here. It was mentioned earlier that the list of hard items is dependent on the architecture and certain other specificities of the system. For example, it is possible to include the following items in the hard items list: (1) Ground cable to splice or eyelet to chassis (if a chassis return system is used). If there are any redundant grounding strategies, those can be included as well, (2) Chassis to chassis or frame connections (which consider either oxidation or rust inhibition issues), (3) Chassis to eyelet of negative battery cable. However, inclusion or omission of these items will not change the overall conclusions indicated in the paper, with very slight change in the numerical probability numbers indicated. It should also be noted that we used a main fuse indicating a possible scenario when the complete system will be out (rendered unavailable) if that main fuse blows. In the existing legacy vehicle systems this may not be the case; it was shown in this paper as a possibility, dependent on the protection strategy in a particular design. Normally single point of failure is not allowed in the existing legacy systems. But if the protection coordination is properly done, then the question of the largest sized fuse blowing out will not arise, until the lower level fuses have blown out first. But if there is a short circuit (due to whatever reason) right after the main fuse, then indeed this main fuse will blow. In such a situation the system will definitely be benefited to have a temporary outage, rather than a fire or destruction of the power source. Such a protection scheme with a utility style protection coordination methodology can be beneficial for the overall load management and protection, even though it can come at an additional cost. Again, for new future designs one might consider such a protection to prevent any potential damage to the source. Similarly, in the systems shown in this paper, separate fuses were indicated for the loads. In an automotive system this can be prohibitive due to cost and other practical reasons, when it may be advantageous to let more than one load share a single fuse. But this paper indicated separate fuses, with an intention to make a fair basis comparison between the multiplexed and the non-multiplexed architecture.

CONCLUSION

Automotive multiplexing system reliability is very much architecture dependent, and is not necessarily higher than a non-multiplexed system on a fair basis comparison, as has been shown in this paper. This arises due to the fact that in the automotive multiplexed system, signal level bus and connectors, and also intelligent nodes, are introduced in exchange for certain components eliminated from the non-multiplexed system during this transition. More importantly, a multiplexed system can lead to the reduction in redundancy of parallel elements that exists in a non-multiplexed system. Advantages of multiplexed systems are obviously in the packaging and ease of scalability (i.e. ability to easily add or subtract loads to or from the system) without hardware reconfiguration, and the possibility of tapping various sensor signals from the data bus, which can be advantageously utilized, leading to enhanced system performance. Of course, the practical manufacturing issues in an actual vehicular body environment leads to component reduction (connectors, crimps, splices etc.) in the multiplex system as opposed to a non-multiplexed system, leading to additional cost reduction. It is important to take into account both the reliability and the cost, before deciding which architecture is best for a system. The authors believe that a partially multiplexed system might provide the best answer in a safety critical system.

ACKNOWLEDGMENTS

The authors gratefully acknowledge the comments and constructive inputs from several reviewers, which helped the authors enhance the quality of the subject matter and its presentation in this paper. The authors are particularly grateful to Mr. Jack Stein at Visteon for the personal interest he took in the paper and for coordinating the review, since originally this paper was planned for presentation in the automotive electronics session organized by him.

REFERENCES

1. M.A. Masrur, "Digital simulation of an automotive multiplexing wiring system", IEEE Trans. on Veh. Tech., Aug. 89, Vol 38, No. 3, pg. 140-147.
2. M.A. Masrur, "Studies on some alternative architectures for fault-tolerant automotive multiplexing network systems", IEEE Trans. on Veh. Tech., Vol. 40, No. 2, pp. 501-510, May 1991.
3. G.C. Wheat and W. J. Evans, "Vehicle Multiplex Wiring -- An implementation", SAE Paper No. 880591.
4. SAE-SP-1224, "Multiplexing", 1997
5. SAE-SP-1070, "Automotive Multiplexing Technology", 1995
6. P. Richardson, L. Sieh, and A. Elkateeb, "Fault Tolerant, Real-Time In-Vehicle Networks – An Adaptive Protocol Approach", (completed, planned to be submitted for publication in the IEEE Transactions on Vehicular Technology).
7. D. J. Smith, "Reliability, Maintainability and Risk", (book), 6th edition, Butterworth Heinemann, 2001.
8. SAE, "Automotive Electronics Reliability Handbook", (book), 1987.
9. J.H. Derr, C.M. Straub, and S. Ahmed, "Prediction of Wiring Harness Reliability", SAE Paper No. 870055.

CONTACT

Dr. M. Abul Masrur
US Army TACOM, AMSTA-TR-R, MS-264
6501 E. 11 Mile Road
Warren, MI 48397-5000, USA
E-Mail: masrura@tacom.army.mil

Dr. John Shen
Dept. of Electrical & Computer Engineering
University of Michigan-Dearborn
Dearborn, MI 48128, USA
E-Mail: johnshen@umich.edu

Dr. Paul Richardson
Dept. of Electrical & Computer Engineering
University of Michigan-Dearborn
Dearborn, MI 48128, USA
E-Mail: richarpc@umich.edu

2003-01-0226

Power Line Communication Implementation in Electrical Architecture

Anne Laliron, Olivier Maurice and Alain Gascher
Valeo

ABSTRACT

The implementation of the new electronic equipment makes the electrical and electronic architecture more and more complex. Thus technologies such as Power Line Communication constitute alternative solutions to reduce the number of wires and optimize the harness. Taking into account the know how and applications for household use, the power line communication feasibility in automotive domain had to be performed. This paper deals with the essential parameters to be identified in order to provide the appropriate power line modulation. Harness EMC behavior and EMI disturbances inside vehicle are developed in this paper.

INTRODUCTION

The distribution of data through power lines is one of the ultimate ways to reduce the number of communication wires. Technologies such as multiplexing are more and more used in order to distribute data while limiting unnecessary wires. Today multiplexing has reached its limit either one wire is used or twisted wires.

Several years ago, household companies started to implement this technology. The main advantages for home application is the modularity. With PLC, new electronic equipment implementation does not require adding specific wires inside walls. In addition to this advantage, PLC presents other useful features for automotive applications. Indeed, safety requirements can be fitted out by way of data redundancy on power lines without adding extra wires. Electronics function such as brake by wire or steer by wire require high dependability. Thus, the data distribution on power lines in addition to the conventional data wires implies separated sources for safe communication.

MAIN SECTION

This presentation is divided into three main parts: power line modulation technologies, harness characterization,

inside vehicle EMI model. These explanations will give an overview of the vehicle environment and on the different strategy concerning the use of PLC.

POWER LINE MODULATIONS

Depending on the context and on the data rate, different types of modulations can be used on power lines. All these modulations will not be exposed here, the aim is to give a global comprehension of the main possibilities offered by this technology. There are three main types of modulations: Amplitude, Frequency and Phase shift keying.

ASK Amplitude Shift Keying / OOK On/Off Keying

The ASK modulation is the simplest way to modulate the signal. As shown hereafter, when the signal is 0, amplitude of the signal is 0. When the signal is 1, the modulated signal follows the carrier signal.

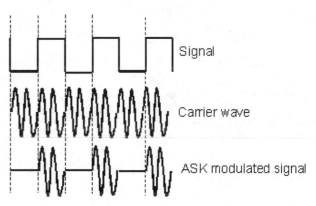

Signal

Carrier wave

ASK modulated signal

Figure 1

Model of the ASK modulation:

$$s(t) = \begin{cases} A\cos(2\pi fcT) \rightarrow 1 \\ 0 \rightarrow 0 \end{cases}$$

The modulation spectrum is translated to the carrier frequency.

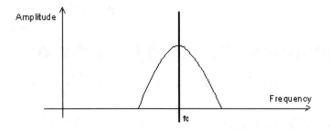

Figure 2

The bandwidth of the signal is unchanged.

FSK Frequency Shift Keying

FSK modulation uses two carrier signals instead of one. One frequency corresponds to 0 and the other is used for 1.

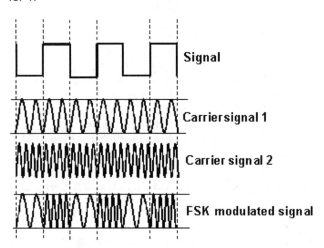

Figure 3

$$s(t) = \begin{cases} A\cos(2\pi f c_1 T) \to 0 \\ A\cos(2\pi f c_2 T) \to 1 \end{cases}$$

The frequency spectrum of the FSK modulation gives a larger bandwidth than ASK. As shown in figure 4 and 5, the size of the bandwidth depends, for one thing, on the difference between the two carrier frequencies.

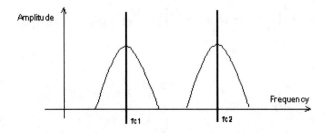

Figure 4

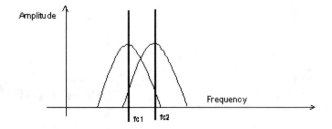

Figure 5

Even if the FSK modulation has a larger bandwidth, the noise immunity is better than ASK.

BPSK Binary Phase Shift Keying

The BPSK modulation uses different phases to separate the two logic levels. Thus, 0 and 1 logic levels are 180 degrees out of phase. This modulation is very robust concerning EMI disturbances.

$$s(t) = \begin{cases} A\cos(2\pi f c T) \to 0 \\ A\cos(2\pi f c T + \pi) \to 1 \end{cases}$$

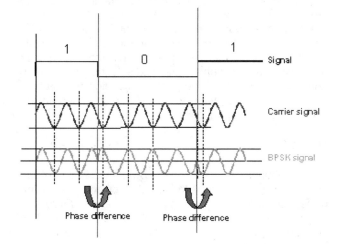

Figure 6

Spread spectrum / multi-carrier modulations

Spread spectrum is based on wide band modulation. This technology is used in military applications in order to secure data. Indeed, spread spectrum signals are comparable to noise signals.

Technologies such as OFDM Orthogonal Frequency Division Multiplexing are used when high data rate is needed. OFDM is a type of frequency hopping. This technology presents very good results concerning EMI immunity. OFDM is much more complex than single carrier modulation.

High speed household applications, due to wide band constraints, use, among others, OFDM technology. To fit out automotive economical constraints, technologies such as ASK, FSK and BSK are preferred to multi-carrier modulations for low speed functions. Nevertheless, high speed and safety applications, such as multimedia and X-by-wire, can require spread spectrum modulations.

HARNESSES CHARACTERIZATION

The EMI environment knowledge is essential for using the appropriate PLC modulation. As the harness is the physical support for the signal, its electrical characteristics are determinant. Thus impedance, splices and frequency response of the wires impact the quality of the transmission. The distance to the vehicle chassis is also one of the major elements.

In order to determine these parameters, several measurements have been done on different harnesses from different vehicles. TDR measurement, frequency measurement and coupling have been performed.

TDR Time Domain Reflection

TDR measurement allows determining the harness transfer function. The methodology used to determine the harness model consists in making a first model from the harness data sheet and then, measuring on real car in order to improve the model. To make it clear, we take the example of a door module. The three main functions concerned are external mirror, window lift and door lock. For each of them, a TDR measurement is done on the dedicated wires, from the door-through to the equipment connection.

- Passenger external mirror wire TDR measurement

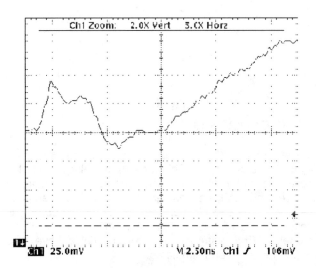

Figure 7

- Passenger external mirror wire model (from data sheet)

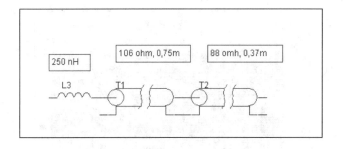

Figure 8

- Passenger external mirror simulation result

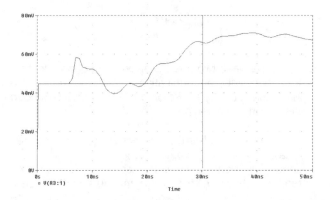

Figure 9

The model performed by way of wire data sheet corresponds to the result of measurement. The TDR method has been used to make the model of the several wires inside different cars. This measurement in passenger and engine compartment allows making wire models. These models are then used to compare different types of modulations.

Frequency transmission

Frequency measurement allows determining wires bandwidth. This parameter is one of the influent ones, it gives the ability of the media to distribute data. Typically, inside the vehicle, the bandwidth cut off frequency is from 3MHz to 30MHz, depending on the wire length and on the splices.

Example: left rear lights

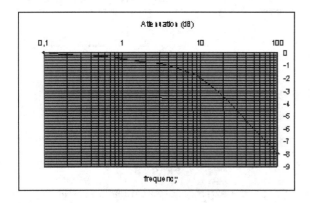

Figure 10

Coupling

Coupling between two wires is also an important element in the power line communication study. The impact of a signal transmitted by a wire on a neighbor wire has been measured inside a vehicle.

The example given here is the rear light. The two wires are taken from the same harness. The two main experiments done are explained on the scheme hereafter. A back wave coupling measurement consists in measuring the coupled signal at the wire source. The front wave coupling measurement consists in measuring the signal at the wire extremity.

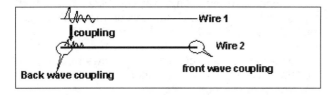

Figure 11

A 7V signal is generated in the wire 1 and this signal implies by coupling a signal in the wire 2. The back wave coupling measurement gives a 2V signal. The front wave coupling measurement gives very low signals, thus it is not one of the major disturbing elements.

Several experiments have been performed on different types of wires. The value of the measured coupling attenuation is around –10dB.

Harness model

Thanks to all these measurements and according to the wires data sheet, harnesses models can be realized and validated.

Figure 12

The elements of the driver door model given hereafter have been performed using this approach.

Door lock example:

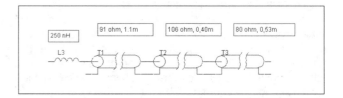

Figure 13

Equipment impedance

The impedance of the equipment is also an important parameter. In order to identify this element, several measurements have to be done on electronic functions. Combining the harness model and the measured impedance, the model of the door is given hereafter.

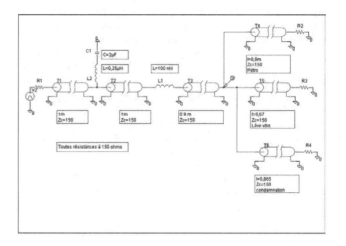

Figure 14

VEHICLE EMI MODEL

Inside vehicle, the electronic functions are faced to various disturbances. This equipment also generates noise. In order to make the appropriate modulation choice, the EMI environment needs to be identified. OEM's pulses specifications as well as measured pulses have been taken into account in order to make a global model of EMI disturbances. External fields have also been measured and simulated.

EMI pulse specification

In order to validate new equipment or new technology, pulse specifications are used. The equipment must overcome these disturbances. There are several types of pulses. The figures here below represent one of the pulses used for validation and its model.

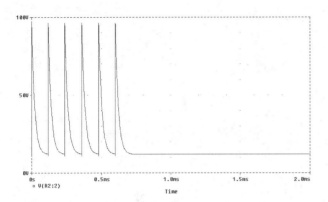

Figure 15

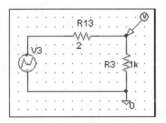

Figure 16

Measured pulses

The pulses used for validation should represent most of the disturbances of the power network. Nevertheless, actuators make a specific noise, which can be more severe than these standard pulses. Thus, in order to characterize these disturbances, direct measurement in a vehicle is necessary. The example of the mirror actuator is shown on the picture. The noise represents the starter phase of this actuator.

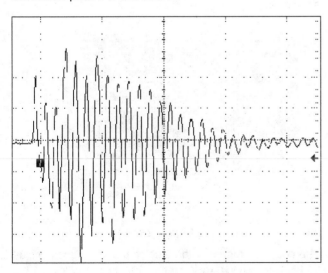

Figure 17

In order to integrate this noise source in the simulation, a model of this signal has been done.

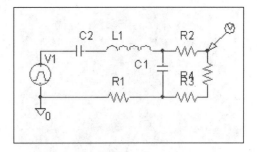

Figure 18

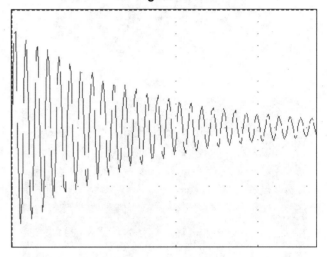

Figure 19

External fields: measurement and simulation

• Vehicle measurement

In order to measure the impact of external field inside vehicle, specific experiment has been done. Wires are placed over the car. A distance of about 30cm is maintained between the wires and the car by the way of cardboard boxes in order to make a strip line structure. The signal transmitted through the wires implies an incident field of 1V/m. This value is measured inside the car by the way of a 2m long wire maintained at 3cm of the vehicle body. Open and short circuits have been performed.

33

Figure 20

Figure 21

The car is considered as a metallic closed box with one split. The current in the split is calculated as described hereafter.

$$i_n := h \cdot \frac{Ei}{\dfrac{R_n}{2} + j \cdot L \cdot \dfrac{\omega_n}{4} - \dfrac{j}{C \cdot \omega_n}}$$

From this current, it is possible to calculate the electrical field radiated inside the car:

$$E = \frac{\mu.h.|i_n|.\omega_n}{4.\pi.r.10^{-1}}$$

This electrical field is represented in the figure given herein.

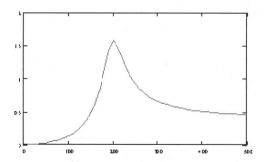

Figure 22

The coupling effect resulted from this electronic field on a 50 Ohms impedance is given on the figure 23. This figure gives the result of the field on an open circuit wire.

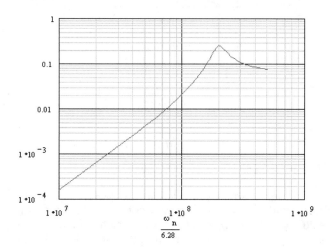

Figure 23

The vehicle measurement result is close to this simulation:

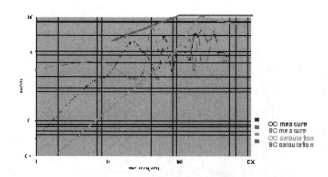

Figure 24

The coupling impact of the electrical and magnetic fields on the wire is given by the equations here below:

$$I = j.\omega.C.dx.V$$

$$e_E = \frac{1}{j.C.dx.\omega} I$$

$$e_H = j.\omega.\mu_0.\frac{V}{\eta_0}.\frac{S}{h}$$

dx : wire element

e_E : induced voltage from electric coupling

e_H : induced voltage from magnetic field

C: per unit length capacity

h: distance to the car chassis

S: surface

The electrical and magnetic fields are correlated to the distance to the car chassis.

SIMULATION MODEL FOR MODULATION COMPARISON

Thanks to the environment characterization described before, the choice of the appropriate modulation can be performed.

The harness model and the EMI environment are used to simulate the behavior of different types of modulations.

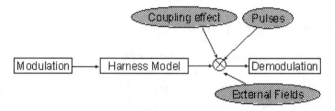

Figure 25

ASK, FSK and BPSK modulations can be tested using this simulation chain.

Several simulations have been performed to quantify either pulses disturbances or electromagnetic fields. Here above the example of ASK simulation disturbed by front wiper motor noise. The figure shows the disturbance generated by the wiper while starting.

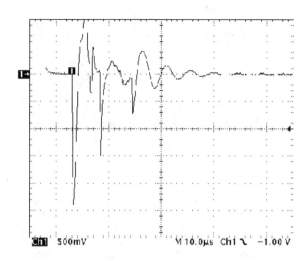

Figure 26

The model extracted from this measurement is given on figure 27.

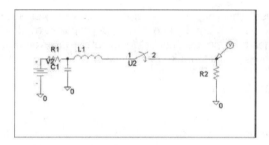

Figure 27

The simulation of the wiper starter phase model gives the result shown on figure 28. In order to take into account the repeatability of this phenomenon, the noise is generated twice (at to+20µs and to+70µs).

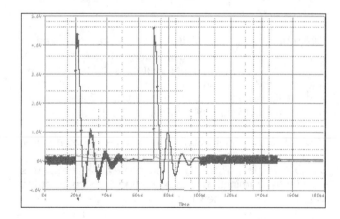

Figure 28

The effect on the demodulation result is given on figure 29.

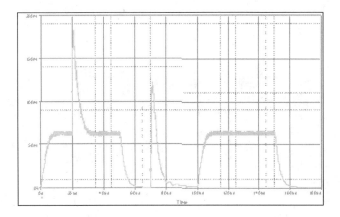

Figure 29

These results give the influence of such disturbances on ASK modulation.

For all types of electromagnetic fields or pulses, thanks to this simulation approach, it is possible to compare modulation technologies.

Noise immunity is one of the main elements to be studied in order to choose an appropriate modulation. Other parameters such as data rate and error rate are also essential to make good choice.

CONCLUSION

Power line communication is potentially an optimized way to transport data. Lots of parameters have to be identified in order to determine the feasibility of this technology on board vehicle. A system approach is necessary to ensure a complete comprehension of the difficulties to be overcome.

Simulation based on measurement is needed to identify the behavior of harnesses as well as the impact of disturbances.

Different types of modulations have to be compared to make the best choice. Depending on the results of simulation and on other parameters such as data rate and protocol constraints, the power line communication can be implemented inside the vehicle. There is not a unique modulation for all applications.

Power line communication can be used either as nominal communication layer or as redundant safety way to transmit data. This new technology, which offers substantial earnings, has to be taken into account at the very beginning of a new harness design.

System approach is necessary to evaluate technically and economically the power line communication technology.

REFERENCES

1. "Numerical and experimental study of the shielding effectiveness of a metallic enclosure" F.Olyslager & al. IEEE EMC Vol 41 N°3 August 1999

2. "La topologie électromagnétique" JP.Parmantier. REE n°4 Avril 1996

3. "Modélisation de la pénétration d'un champ électromagnétique à l'intérieur d'une automobile: simulation et validation expérimentale" A.Rubinstein & al. EMC Grenoble 2002.

4. "Electromagnetic compatibility handbook" Violette - White. VNR edition 1987

5. "Analysis and modeling of ipulsive noise in broad-bad powerline communication" M.Zimmermann & al. IEEE EMC Vol44. N°1 Feb.2002

CONTACT

Anne LALIRON: Electronic Engineer. Engineer degree obtained in 1999 from ESIEE Ecole Supérieure d'Electrotechnique et d'Electronique (Paris, France) She has been working at Valeo for 3 years. She started as development electronic engineer, and she is now Project Manager in the Innovation Department.

DEFINITIONS, ACRONYMS, ABBREVIATIONS

ASK: Amplitude Shift Keying

BPSK: Binary Phase Shift Keying

EMC: EletroMagnetic Compatibility

EMI: EletroMagnetic Interference

FSK: Frequency Shift Keying

OC: Open Circuit

OFDM: Orthogonal Frequency Division Multiplexing

OOK: On/Off Keying

PLC: Power Line Communication

TDR: Time Domain Reflexion

SC: Short Circuit

2002-01-0442

Vehicle Network Selection

Christopher B. Mushenski
OPM-Crusader, Mobility Systems PDT, U.S. Army TACOM

Amy Cartwright and Gregory Myrvold
General Dynamics Land Systems, Sterling Heights Complex

ABSTRACT

The subject of this paper is centered on presenting a Vehicle Network (Data Bus/Protocol) Selection process for a Common Software Architecture for Ground-based Vehicle Systems. Vehicle network data buses/protocols are analyzed/evaluated and selected to support the real-time requirements for the architecture. It is through data research, simulation and evaluation of architecture models that today's Crusader Combat Vehicle Program's RTCOE (Real-time Common Operating Environment) Software Architecture can be made THE BEST common software architecture for all Combat, Tactical and Utility Ground-based Military Vehicle Systems.

INTRODUCTION

The selected data bus/protocol will be installed into Crusader; the U.S. Army's next generation of indirect fire support cannon artillery and resupply vehicles that will be in service during the period from 2007 through 2030. The Crusader Self-Propelled Howitzer (SPH) vehicle features a 155 millimeter cannon with significantly increased capabilities over the current M109-series fleet. Two Crusader Re-Supply Vehicle (RSV) variants support the SPH by providing it with automated fuel, ammunition and propellant re-supply and manual re-supply of other consumables. The three vehicles are shown in Figure 1.

Figure 1. The Crusader Vehicles

OBJECTIVE

The goal of this Vehicle Network Selection is to <u>recommend</u> a data bus/protocol which best satisfies the documented and derived data distribution requirements for the Crusader vehicle architectures/applications. This selection process compiles and evaluates the technical and strategic aspects of various vehicle network protocols. These protocols were selected for detailed study based on down-select criteria which assessed the high-level suitability of a broad spectrum of network data buses in the Crusader's embedded, land-based military vehicle platform.

APPROACH

This data bus/protocol selection was conducted and documented in accordance with the Crusader trade study process.

Input to this trade study was provided by an integrated product development team consisting of representatives from multiple disciplines within General Dynamics Land Systems (GDLS/Crusader subcontractor) and data bus material suppliers.

REQUIREMENTS

At the request of GDLS, United Defense L. P. (UDLP/Crusader prime contractor) developed an initial cut of data bus interface board attributes which were used as a basis for this study. GDLS requested UDLP also identify which of these attributes were required, and which were high, medium and low priority. The list distinguishes any preferences or undesirable characteristics. For example, "predictable bus arbitration" is a requirement, "token passing" is a preference which exceeds the requirement, and "not collision detect" is an undesirable characteristic which does not meet the requirement. GDLS then separated the data bus/protocol from interface board attributes and created a list of key technical requirements and implementation constraints for the data bus/protocol only. Implementation constraints are considered board attributes which impact the bus selection and are therefore included. The key technical requirements and implementation constraints are:

1. Predictable bus arbitration protocol (Token passing rotating mastership preferred, Not collision detect)
2. Supports periodic and aperiodic messaging
3. Dual redundant
4. Token regeneration
5. Support a minimum of 24 nodes
6. Physical length to cover all nodes in vehicle
7. Sustained throughput on the order of 10 Mbps (includes 60% reserve)
8. Low latency for critical data on the order of 5 ms
9. Commercial off-the-shelf (COTS)

The initial cut of data bus interface board attributes was also used to create down-select criteria to access the high-level suitability of a broad spectrum of network data buses in Crusader's embedded, land-based military vehicle platform. Computer networking principles were applied to the attributes to derive "go / no-go" criteria used in down-selection. The down-select criteria are:

1. Predictable bus arbitration implies a broadcast channel; thereby eliminating unarbitrated point-to-point networks and hence layer 3 (network) reference models.
2. Physical length to cover all nodes in vehicle implies a network size of 10 to 100 m.
3. Token regeneration implies a ring bus allocation method which allows nodes from point-to-point networks containing an input and output copy buffer and eliminates contention oriented buses.
4. Open System Standard implies the bus must be in the public domain, widely used and easily expandable.
5. Sustained throughput on the order of 10 Mbps.

The broad spectrum of network data buses/protocols which were considered as potential candidates are shown below.

MIL-STD-1553B	MIL-STD-1773	ABUS
IEEE 802.3	IEEE 802.4	IEEE 802.5
USB	VNET	SAE J2058
Motorola MI Bus	SAE J1850	CAN
SAE J1939	VAN	SAE 4075A
SCSI	Firewire™	FDDI
IEEE 802.3u	B-ISDN ATM	HIPPI
Fibre Channel/Arbitrated Loop		

Buses/protocols which were eliminated from any further consideration did not pass down-select. Four (4) primary candidates remained after down-select. Through an iterative review cycle, the final data bus/protocol requirements for this data bus/protocol selection were developed and documented in Table 3. Beyond the technical requirements and implementation constraints already discussed, other programmatic issues and concerns were addressed as well.

RATING AND WEIGHTING

The data bus/protocol ranking results of this data bus/protocol selection culminated through a process of rating each network in the study against a specific parameter and requirement. The requirement and any preference or undesirable characteristic is grouped as a single parameter. Each parameter was assigned a weight by the design team to indicate the importance or priority of satisfying its requirement as deemed applicable to the Crusader vehicle's operation and environment.

RATING

A zero to four scale was used for rating each candidate data bus/protocol against a given parameter. The rating values and attributes are described below.

4 = Greatly exceeds requirement/preference
3 = Slightly exceeds requirement/preference
2 = Meets requirement/preference
1 = Slightly under requirement/preference
0 = Fails to meet requirement/preference

For each specific rating, a rationale was developed which equated candidate performance to an associated rating score. A summary of the established rating rationale is found in Table 2.

WEIGHTING

A weighting factor was assigned to each parameter to differentiate varying levels of importance or program impact amongst the requirements. This factor indicates the relative priority of a particular parameter. The weighting scheme used in this data bus/protocol selection is as follows.

8 - 15 = Highest priority
4 - 7 = Medium priority
1 - 3 = Lowest priority

The weighting of each parameter is shown in Table 2.

RANKING THE CANDIDATES

Table 4 contains the ranking worksheet. This worksheet contains the list of parameters and requirements for the data bus/protocol, the associated weight for each parameter, and the rating of each data bus/protocol against each parameter. The scores at the bottom of the worksheet represent the final score for each data bus/protocol. The scores were calculated by multiplying each data bus's/protocol's rating with the weight value for each parameter and summing the total. A higher score represents a higher ranked device.

RESULTS

The following table represents the ranking results with score totals for each data bus/protocol in this data bus/protocol selection.

RANK	DATA BUS (DB)/ Protocol	SCORE
1	DB/Protocol A	306
2	DB/Protocol B	215
3	DB/Protocol C	191
4	DB/Protocol D	159

Table 1. Data Bus/Protocol Scores

ANALYSIS OF RESULTS

The four data buses/protocols are evaluated against 14 parameters. For certain parameters the four data buses/protocols possess characteristic advantages and disadvantages when compared against each other. The weighting structure distinguishes between the importances of parameters. As such, a data bus/protocol accumulated more points when it ranked better on more highly weighted parameters.

The ranking worksheet in Table 4 indicates that DB/Protocol A scored the best followed by, DB/Protocol B, DB/Protocol C and DB/Protocol D. Although the 1553B data bus/protocol did not pass the Go / No-Go criteria, it was included in the scoring as a baseline reference scoring metric.

When normalized (total score / 380 total possible points), DB/Protocol A's overall score of 306 is 80% or 3.2 on the 4 point rating scale. When relating this to the original rating scheme, DB/Protocol A is considered as beyond slightly exceeding the requirement/preference. This can be attributed to DB/Protocol A being rated as greatly exceeding the requirement/preference in 10 of the 14 parameters. The only parameter which it was considered slightly under the requirement was cost.

Similarly, DB/Protocol B's overall score of 215 is 56% or 2.3 on the 4-point rating scheme, would also be considered to meet the requirement. Although meeting the individual requirement for most parameters, its scores were not as strong in the technical area. In particular, the bus allocation, messaging, fault tolerance/reliability, availability, migration path and maturity characteristics are only considered acceptable, as opposed to slightly exceeding the requirement. Cost and commonality are considered as slightly under the requirement.

DB/Protocol C and DB/Protocol D had similar scores to DB/Protocol B however failed to meet the requirement/preference for bus allocation, availability, migration path and commonality with other platform parameters. DB/Protocol D also failed to meet the cost criteria.

Remarkably, the 1553B baseline reference metric scored better than all but DB/Protocol A. This indicates that most but not all of the parameters can be met by 1553B. In particular, 1553B does better than newer technologies where it is expected to; availability, migration path, cost, schedule and commonality with other platforms. This phenomenon helps to explain the reason for the Go / No-Go screening criteria.

CONCLUSION

Technically speaking, any of the four data buses/protocols in this trade study could, with enough development work, meet the data distribution requirements of the Crusader vehicles. However, the goal of this trade study was to select the data bus/protocol which best satisfies not only the technical, but strategic aspects. This approach thereby minimizes implementation work and arrives at a solution best suited for todays as well as future needs. Many of the Crusader vehicle's data distribution needs can only be met if the commercial market, and community, accept and support a data bus/protocol so that it has the longevity to survive the evolution and long lifespan of the military application. Therefore, in addition to performance, the choice of the best data bus/protocol is also driven by market acceptance and potential. The commonality with other platforms measures helps to account for this factor, and complements the primary thrust of selecting a data bus/protocol that provides the most benefits to Crusader vehicles independent of other platforms.

The parameters are high order measures of the data buses/protocols. It was not deemed critical to assess the data buses/protocols to a degree of granularity such as counting variable message length. Instead, this trade study bound the assessment to encompass the entire gamut of data bus/protocol selection albeit at a somewhat higher than micro-level.

Due to its overall favorable ratings for both technical and programmatic measures, the DB/Protocol A is the top candidate to serve as a Crusader vehicle data bus/protocol. DB/Protocol B has reasonably good technical performance with promising potential, but finishes second due to marginal ratings for programmatic parameters mainly related to maturity. These parameters act as an added sanity check, which holds the more aggressive DB/Protocol B technology step in balance. DB/Protocol C and DB/Protocol D miss in both technical and programmatic arenas.

The total scores for the data buses/protocols under study can be manipulated up or down with a minor change to a parameter's weighting or a modification to the rating rationale. Inaccurate data bus/protocol data may account for score variances. As such, the total scores are more or less a normalized collation, in numeric form, of the large amounts of available data (Table 3) for each data bus/protocol. Nonetheless, by relatively comparing the data buses/protocols over 14 parameters the tally represents an accurate assessment.

RECOMMENDATION

Based upon the results of the ranking worksheet, the market acceptance, and overall strategic aspects and potential of the data bus/protocol, it is recommended that DB/Protocol A be selected for implementation in the current Crusader vehicle applications. DB/Protocol A is an acceptable solution for the Crusader vehicles' data distribution in lieu of the others due to it's:

1. predictable bus allocation method (token passing)
2. high fault tolerance / reliability of redundant dual ring topology
3. maturity factors such as availability, migration path, ability to meet schedule, and commonality with other platforms

REFERENCES

[1] DY 4 Systems Inc.; Preliminary White Paper on Local Area Network Options for Today's Vetronics Programs

[2] Andrew S. Tanenbaum; Computer Networks; Englewood Cliffs, NJ: Prentice Hall, 1996.

[3] Crusader Data Bus Evaluation and Recommendation Trade Study; GDLS-SHC, Sterling Heights, MI; 1997

[4] DJ Ackerman; Preliminary Report: Data Acquisition and Control Bus (DACB) Selection for Crusader Self Propelled Howitzer Vehicle Electronics; 16 Jan 1995.

[5] U.S. Army Contract (Crusader Program) (DAAE30-95-C-0009), (1995-)

Table 2. Weight and Rating Rationale

PARAMETER	WEIGHT	RATING	RATIONALE
Bus Allocation Method	5	4	Specified As Token Passing (Rotating Mastership Around Physical Ring)
		3	Token Passing (Rotating Mastership Around Logical Ring)
		2	Arbitrated Loop (Token Acquisition)
		1	Can Be Made Token Passing (Rotating Mastership Around Physical Ring)
		0	Master / Slave Linear Bus
Messaging	8	4	Specified As Supporting Periodic & Aperiodic
		3	None / Not used
		2	Can Be Made To Support Periodic & Aperiodic
		1	None / Not used
		0	Can Not Be Made To Support Periodic & Aperiodic
Sustained	8	4	> 100 Mbps
		3	20 to 100 Mbps
		2	10 - 20 Mbps
		1	1 to 10 Mbps
		0	Less than 1 Mbps
Data Latency	4	4	< .1 msec
		3	.1 to 1 msec
		2	1 - 5 msec
		1	5 to 10 msec
		0	> 10 msec
Fault Tolerance / Reliability	9	4	Specified As Dual Redundant
		3	None / Not used
		2	Can Be Made Dual Redundant
		1	None / Not used
		0	Can Not Be Made Dual Redundant
Fault Tolerance / Reliability	9	4	Specified To Support Token Regeneration
		3	None / Not used
		2	Can Be Made To Support Token Regeneration
		1	None / Not used
		0	Can Not Be Made To Support Token Regeneration

Table 2. Weight and Rating Rationale (cont.)

PARAMETER	WEIGHT	RATING	RATIONALE
Number of Nodes	7	4	More than 35
		3	28 - 34
		2	21 - 27
		1	14 - 20
		0	Less than 13
Bus Length	5	4	More Than 100 m
		3	65 - 90 m
		2	35 - 65 m
		1	10 - 35 m
		0	Less Than 10 m
Availability	5	4	COTS & MOTS Plug-In Products
		3	COTS Plug-In Products
		2	COTS Chipsets
		1	Under Development
		0	Not Being Developed
Migration Path	5	4	Product Exists To Upgrade Benchtop to Extended Temperature
		3	None / Not used
		2	Product Is Under Development For Extended Temperature
		1	None / Not used
		0	No Plans Exists For Extended Temperature
Cost	15	4	< 0.5 Times Lower
		3	0.5 to 1 Times Lower
		2	On Par With 1553B Per Dual Redundant Node
		1	1 to 5 Times Higher
		0	> 5 Times Higher
Schedule	5	4	Benchtop 1Q97
		3	Benchtop 2Q97
		2	Benchtop 3Q97
		1	Benchtop 4Q97
		0	Benchtop 1Q98
Maturity	5	4	Set Standards, Existing H/W and S/W
		3	Set Standards, Introducing H/W and S/W
		2	Set Standards, Introducing H/W, S/W Coming
		1	No Set Standards
		0	Aging Standards and Products
Commonality w/ Other Platforms	5	4	Wide Military (all branches) use
		3	2 or More Army Applications
		2	Planned Army Application
		1	Navy / Air Force Applications
		0	No Military Applications

Table 3. Data Bus/Protocol: Data Sheet

PARAMETER	REQUIREMENT	1553B [1]	Data Bus/ Protocol A	Data Bus/ Protocol B	Data Bus/ Protocol C	Data Bus/ Protocol D
BUS ALLOCATION METHOD	PREDICTABLE ARBITRATION	TDM	Token Ring	Token Acquisition	Hierarchical Point-to-Point Arbitration	NA (Point-to-Point Cell Switching)
MESSAGE TRANSFERS	PERIODIC & APERIODIC	Soft, Yes	Soft, Yes	Soft, Yes	Soft, Yes	Soft, Yes
SUSTAINED THROUGHPUT	10 Mbps Minimum	1	100	100	98.3	155
DATA LATENCY	5ms Maximum	.06	.036	.03	.354	.3
FAULT TOLERANCE / RELIABILITY	DUAL REDUNDANT	Yes	Yes	Possible	Possible	Possible
FAULT TOLERANCE / RELIABILITY	TOKEN REGENERATION	Possible	Yes	Yes	Possible	Possible
NUMBER OF NODES	24 Minimum	32	500	127	63	500
COPPER BUS LENGTH (between nodes / total)	10 to 100 m total	473/473	10/5000	10/1270	4.5/72	10/5000
AVAILABILITY	COTS	MOTS, ROTS, COTS	MOTS, ROTS, COTS	COTS	No VME Development	No VME Development
MIGRATION	PATH TO VER1	Product Exists	Product Exists	Product Planned / Under-development	No Development Plans	No Development Plans
COST	ON PAR WITH 1553B	1	2.5	2.5	0.5	8
SCHEDULE	EBTDS-3Q97	1Q97	1Q97	2Q97	N/A	N/A
MATURITY		Set Std.s, Product Available	Set Std.s, Product Available	Set Std.s, Introducing H/W, S/W Coming	Set Std.s, Introducing H/W, S/W Coming	Set Std.s, Introducing H/W, S/W Coming
COMMONALITY W/ OTHER PLATFORMS		ASEP, M1A2, HAB...	Navy Ship Control Room, AAAV	Navy F-18 TAMMAC, Air Force B-1B	None	None

[1] For Baseline Reference Only

Table 4. Data Bus/Protocol Ranking Worksheet

	PARAMETER	REQUIREMENT	PREFERENCE	WEIGHT	1553B [1]	Data Bus/ Protocol A	Data Bus/ Protocol B	Data Bus/ Protocol C	Data Bus/ Protocol D
		EVALUATION CRITERIA				SCORE			
T	Bus Allocation Method	Predictable Bus Allocation	Token Passing	5	0	4	2	1	1
e	Messaging	Supports Periodic and Aperiodic		8	2	2	2	2	2
c	Sustained Throughput	10 Mbps Minimum		8	1	3	3	3	4
h	Data Latency	5 ms Minimum		4	4	4	4	3	3
.	Fault Tolerance / Reliability	Dual Redundant		9	4	4	2	2	2
	Fault Tolerance / Reliability	Token Regeneration		9	2	4	2	2	2
P	Number of Nodes	24 Minimum		7	3	4	4	4	4
e	Bus Length	Cover Vehicle Nodes		5	4	4	4	3	4
r	Availability	COTS		5	4	4	2	0	0
f									
.	Migration Path	Benchtop Upgradable to Extended Temp for Ver1		5	4	4	2	0	0
	Cost	On Par with 1553B		15	2	1	1	3	0
	Schedule	Benchtop hardware by 3Q97		5	4	4	3	0	0
M	Maturity	Set Std.s, Introduc'g H/W, S/W Coming		5	0	4	2	2	2
i	Commonality w/ other Platforms	Army Application		5	4	3	1	0	0
s									
c									
.	**Results**		**Total Score**		245	306	215	191	159

[1] Shown for baseline reference only

Versatile Human Interface Utilizing the "Windows CE for Automotive" OS for Vehicle Computing Applications

Kenneth A. Schmitt and Raymond P. Ernst, Jr.
Yazaki North America

ABSTRACT

Recent trends in automotive electronics point towards a growing demand for sophisticated electrical & electronic systems in new vehicles. These systems are being implemented into vehicles at an ever-increasing rate and demanding more of the driver & passenger's attention to control and operate them. The incorporation of digital network buses, such as MOST, Media Oriented Systems Transport, builds the digital data infrastructure in the vehicle for some of these systems, while other systems exist in "stand alone" non-network configurations. Consequently, considerable attention must be given to the constraints around the control of each system in such a way that it is natural, intuitive and non-distracting for the driver / passengers, while at the same time is flexible enough to handle the application of additional network devices in a "Plug & Play" fashion. This paper will explore these constraints and discuss some of the approaches that are possible when utilizing the "Windows CE for Automotive" operating system for in-vehicle computing applications.

INTRODUCTION

Vehicle networks, such as MOST, are bringing a number of new devices into the car. These devices provide entertainment, information awareness and driving assistance to the driver and passengers. Along with these devices comes a very large problem, utilizing all these devices without creating a burden for the driver.

Yazaki North America, YNA, is researching and developing a vehicle infotainment network that integrates digital audio/video and can address the concerns of usability of these high tech devices and minimize the driver distraction issues. YNA is solving these issues in its advanced product development department with the use of Microsoft's Windows CE for Automotive.

VEHICLE COMPUTING APPLICATIONS

The car of tomorrow will have many devices in it. The driver and passengers will need to interact with them. These devices may include navigation, hands free cell phones, audio/video entertainment (video cassette players and DVD players), video game systems, and the list goes on. Now with all these devices comes a complex issue with using them while driving. This is where the need for a highly integrated device for controlling and interfacing to these information and entertainment devices.

There are a couple of ways the device control can be handled. The traditional method is switches and knobs. This is somewhat easy to design and relatively low cost. The usability is somewhat difficult for the driver, many buttons and knobs to distinguish from. YNA is solving this problem by developing a multimedia vehicle platform, an in-vehicle computer, to solve these issues. This in-vehicle computer, that has been dubbed MVP – Multimedia Vehicle Platform, can control any device on a MOST network. YNA has developed application software to handle the control of these devices in one central location and provide a user-friendly interface to the operator. This system has been built from the ground up with trying to leverage "off the shelf" software components as much as possible.

THE SOFTWARE SOLUTION

This solution utilizes a common computing platform ruggedized for automotive use along with software to perform the functions of a human machine interface, HMI. The software can be either embedded software or high-level software with the use of an operating system. Our goal here is to leverage "off the shelf" software components to minimize our development. So we started our development with operating system software and high-level development tools.

"What OS is best for this application"? To answer this we need to look at the requirements of this application.

- A graphical user interface
- Multimedia capability
- Open system for 3rd party application development
- Hardware supported
- Device driver availability

- Application Programming Interface, API
- OS Stability
- Are there real time requirements?

For the MVP project at YNA, we selected Microsoft's Windows CE for Automotive, WCEfA. For this project, WCEfA met all our requirements and allowed us to get a system up and running with relative ease. It provides many APIs, has device driver support for almost any type of third party hardware, its an open system based on the Windows programming model and also offers automotive centric APIs to boost the robustness of the OS.

The use of an operating system also brings in the issues of being an open system. By open system we are talking about the ability for the end user to install third party applications. With an open system, there is no guarantee that every application will be rock solid, i.e. it will not crash. WCEfA addresses this with a component called the Critical Process Monitor, CPM. This component acts as a watchdog over applications and monitors their resource usage. If it discovers that it's taking too much CPU time, the CPM has the authority to kill the application and restart it. If the situation is severe enough, it can re-boot the system automatically. It also monitors system resources and notifies all the applications when they are getting low. From this notification, applications can clean up their unused resources and stabilize the operating environment.

Another important factor that we looked at in WCEfA is power management. The typical PDA relies on a backup battery to keep memory alive. Backup batteries of this nature are not acceptable in automotive applications. WCEfA supports a zero current draw shutdown state by flushing memory pages to flash before current loss. Upon revival, the OS restores itself by retrieving the saved information from flash and returning it into the state it was in before the shutdown occurred.

NETWORK CONNECTIVITY

In order for an in-vehicle computer like MVP to control many different entertainment devices in a car, you need to have a common connection between them. YNA is a leader in MOST networking technology. YNA's MVP utilizes a MOST ring to control entertainment devices such as CD/DVD Players, AM/FM Radios, Entertainment Displays, Audio Amplifiers, and many others.

Why MOST? MOST stands for Media Oriented Systems Transport. It is an emerging standard for in-vehicle multimedia data for entertainment. MOST offers a purely digital solution for audio and video data. This information can be streamed to any point in the car with zero EMI interference. The audio/video streams are fully routable to any device or devices. MOST is a tight standard controlled by the MOST Cooperation. The MOST Cooperation is an organization of automotive manufacturers and suppliers working together to achieve a true multimedia network infrastructure standard.

YNA has a four-layer approach to vehicle networking, segregating vehicle network devices into 4 functional areas. The lowest layer is the power distribution network. This layer is the electrical distribution system for the vehicle, handling the main power, the junction block and fusing. All this is interconnected through wiring harnesses. The next layer is the chassis network, which handles the power train, chassis control, steering control, ABS brakes and other critical safety systems. The typical protocol on a chassis network is CAN. The third layer is the body network. This layer is less critical and handles devices like lighting, HVAC, some power features and basic displays. This is a relatively low to medium speed network for control data. Typical protocols used on this layer are J1850, CAN, LIN, etc. The final layer on our network architecture is the information/entertainment network. This is a digital high-speed network. Typical protocols that are used on this type of network are D2B, MOST, and IEEE-1394. Its primary use is for streaming audio/video data for entertainment and other information, such as traffic alerts, brought in by wireless devices. Figure 1 illustrates these four layers and the connectivity between them with gateways.

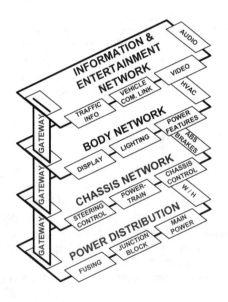

Figure 1

The gateway is a connection point between the network layers to allow communication and control between them. It can also be considered as an information sharing point. The gateway also acts as a firewall between the layers to safeguard the critical systems from any rouge actions. YNA also uses another type of gateway for multimedia applications. This multimedia gateway can be used to connect video game systems, camcorders, or any type of audio/video source into the

MOST network. This type of gateway helps us abstract the automotive constraints from the consumer electronic devices. It allows any consumer device to interact with the entertainment system in the vehicle.

THE INTERFACE

So far we have discussed the needs of an OS for application development and connectivity with devices that need to have an interface to control them. An In-Vehicle computer, such as MVP, can bring this all together into a human machine interface, HMI. In this discussion, we will refer to HMI as an interface that bridges the gap between a human and a machine. It is the operator interface to the devices on our MOST network.

One of the most challenging issues when developing an HMI is designing the user interface in a way that does not distract the driver. What type of things takes a drivers attention away from driving? Small text to read on a small display, cluttered displays with many objects on them, many knobs that are close together, ... In designing our HMI for MVP we have tried to keep these issues in mind and design a methodology that would keep us from rendering such screens. We've replaced many knobs and switches with the use of software type buttons on a touch screen. Now to keep these software type buttons easy and friendly to use, we've made sure that the graphic items representing the buttons are large and the spacing between them is adequate enough. We've also minimized the touch to several areas to reduce confusion and miss-presses. The following figure demonstrates our main interface.

Figure 2

In this particular screen shot, Figure 2, there are 4 touch zones to launch the Radio, CD Player, Video Player and Volume Control applications. Icons are arranged in a horizontal manner and only in one row. This reduces the amount of concentration of finding the area to touch. Figure 3 shows the screen of the radio application. The idea here is limited use of touch points to reduce the amount of time the driver has to focus on touching a screen element.

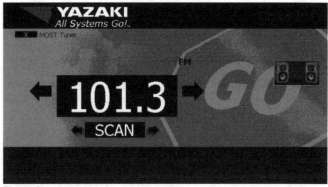
Figure 3

The screen shot in Figure 4 makes use of consumer electronic video devices. These devices are connected to a multimedia gateway and their audio/video is put on to the MOST network. This screen provides the connectivity of the consumer video device to a MOST network display. The user can route the video from any video source on the network to any entertainment display in the vehicle.

Figure 4

ALTERNATIVE INTERFACES

There are also other solutions to driver distraction. Speech recognition is becoming a popular method for command and control applications. The user speaks a command to the car rather than reaching forward to make an interaction with an HMI element. There are two main speech recognition types, discrete and continuous. In a discrete speech recognition environment, the driver would have to speak a command wait for acceptance notification then speak the next command. This can become tedious and causes the driver to pause between speech commands and listen for an acceptance response. This still requires some driver attention to think and process information to get the task done. Now the continuous speech engines provide more flexibility by allowing the user to speak the commands in a phrase and the speech engine does the work of parsing the commands and performing the operation. The user involvement is only to speak a phrase rather than several commands.

Speech engines are becoming more robust and are adapting to the noisy environment of a vehicle. Work still needs to be done with special microphones to pre-process an audio signal before speech processing can occur. The pre-processing is done by passing the audio signal through a digital signal processor, DSP, to remove noise, perform echo cancellation, beam forming and various other filters.

Another complement to speech recognition is text to speech. This allows an in-vehicle computer to speak information back to the driver rather than having the driver read it. This can be a very effective way of communicating information to the driver without the driver taking their eyes off the road. The text to speech feature can also enhance features of an in-vehicle computer. In today's Internet world, people can get access to information in the form of news, stocks, weather, and any information any time anywhere. The in-vehicle computer can be equipped with wireless communication services to provide this type of content to the driver.

CONCLUSION

The automobiles of today and tomorrow are becoming increasingly complex with the technology push of mobile computing. Yazaki North America is looking at ways to resolve this automotive dilemma by working with companies like Microsoft to develop and effective operating system for automotive use. Microsoft is making strong efforts in bringing a common OS to an automotive space and adapting it to fit its new use. Yazaki North America is designing with Windows CE for Automotive to launch its in-vehicle computing product development.

The use of consumer audio/video electronics and Internet devices in vehicles are driving the need for an infotainment network. The use of these devices and networks are making the long drive productive by entertaining the passengers and informing the driver. They are providing robust connectivity by abstracting the automotive constraints away from the consumer electronics. Microsoft's Windows CE for Automotive has helped us build our in-vehicle computer's human machine interface to help bridge this gap between automotive systems and consumer electronics.

ACKNOWLEDGMENTS

Windows CE for Automotive is a trademark of the Microsoft Corporation.

CONTACT

Kenneth A. Schmitt, Supervisor – Advanced Technologies, Yazaki North America, Canton, MI, kschmitt@yazaki-na.com

Raymond P. Ernst, Jr. – Chief Engineer – Advanced Technologies, Yazaki North America, Canton, MI, rernst@yazaki-na.com.

ADDITIONAL SOURCES

1. MOST – www.mostcooperation.com

2. WCEfA – www.microsoft.com/automotive

3. Yazaki North America – www.yazaki-na.com

DEFINITIONS, ACRONYMS, ABBREVIATIONS

API: Application Program Interface

CAN: Control Area Network

CPM: Critical Process Monitor

CPU: Central Processing Unit

HMI: Human Machine Interface

Infotainment: Information and Entertainment

MOST: Media Oriented Systems Transport

OS: Operating System

PDA: Personal Data Assistant

WCEfA: Microsoft's Windows CE for Automotive

YNA: Yazaki North America

CARTRONIC® Based Safety Analysis: Introducing Safety Aspects In Early Development Phases

W. Längst, A. Lapp, P. Torre Flores, J. Schirmer and D. Kraft
Robert Bosch GmbH

U. Kiencke
University of Karlsruhe

ABSTRACT

This article gives an overview of the CARTRONIC® based safety analysis (CSA) including an approach for the automatic determination of failure dependencies in automotive systems. CSA is a safety analysis in an early stage of product development. The goals are to identify safety critical components as soon as practicable in the product development process and to automate the analysis as far as possible. This implies that the system view is abstract, i.e. independent of a certain realization just regarding system functionality.

In the CSA so called global failure effects will be systematically identified and assessed regarding severity of potential injuries. Global failure effects are especially important because they reveal failures within the system to the outside world (see also definition 3.1). Additionally the CSA keeps track of failure dependencies and supports the integration of safety measures in the system structure. The analysis of a system structure for the determination of failure dependencies can be automated using MAX-PLUS algebra.

The result of the CARTRONIC® based safety analysis is also valuable information to a System-FMEA (Failure Mode and Effects Analysis, e.g. [Misra 1992], [Deckers et al. 1994]) later in a development process.

The approach summarized above is explained using a simplified example.

1. INTRODUCTION

The demand for innovative vehicle functions with respect to safety, fuel consumption, emissions reduction, comfort, driving fun, etc. results in more complex and networked system structures. Developing such systems is especially demanding considering confined cost limits and decreasing development periods. To meet these requirements a structuring concept regarding functionality termed CARTRONIC® was developed by Bosch. The CARTRONIC® methodology is an object-based approach to organize system structures and is based on formal rules for structuring and modeling. Systems will be decomposed into subsystems. Subsystems may contain other subsystems or functional components as a refinement of a more abstract or complex system part. These subsystems or functional components are not necessarily a physical part like an electronic component or software but rather an abstract functional unit.

The intention of CARTRONIC® is to manage system complexity, distributed development and to enforce reusability and thus exchangeability of subsystems and components without influencing other parts of a system compound. The methodology and results were presented at the SAE 1998 and SAE 2001 World Conference ([Bertram et al 1998], [Torre Flores et al. 2001]). For a formalized, tool supported specification of CARTRONIC® function structures the Unified Modeling Language (UML, [OMG UML 1999]) is used. [Torre Flores et al. 2001] describes the mapping of CARTRONIC® function structures into the UML which is an international standard for object oriented modeling.

The focus of this article is to describe a CARTRONIC® based safety analysis. For a simplified representation the CARTRONIC® function structure is used in this contribution. The CARTRONIC® function structure of our example system is depicted in Figure 3. However it shall be stated that already for a partly automated implementation of the methodology the use of a formalized representation, like the CARTRONIC®-UML model, is necessary. In this case an automated data exchange between different tools is possible, e.g. between a UML tool to represent the functional structure and a tool for safety analysis. An example for an open standard for defining, validating, and sharing documents is the XML (Extended Markup Language, [OMG XML 2000]).

For the development of all systems, but especially for the implementation of new technologies, e.g. X-by-Wire systems, it is crucial to develop them thoroughly, i.e. planning their structure thoroughly and considering safety aspects as early as possible in the development process.

2. DEVELOPMENT PROCESS

Modern automotive systems which are supplied from various sources have to be seamlessly integrated into an overall vehicle structure. A systematic development process is the key to find suitable solutions for this challenge. An integral part of a development process is the consideration of safety aspects in an early stage of a product development [Bertram et al 1998], [Main et al 1999]. Therefore the CSA is included into the CARTRONIC® based development process.

In the CARTRONIC® based development process the whole *system* automobile is decomposed into *subsystems*, which realize well defined functional tasks [Lapp et al 2001]. Engine control, brake system, transmission, adaptive cruise control (ACC), etc. are examples for such subsystems. Dependencies and communication interfaces between subsystems have to be considered and integrated into the networking concept of the overall system. For instance the ACC subsystem is interacting with engine control, transmission and brake system.

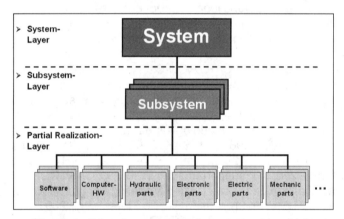

Figure 1: Generic modules of mechatronic vehicle systems

In the partial realization layer each subsystem is specified more precisely. Every subsystem is composed of different partial realizations (Figure 1). Partial realization are the implementation of networked mechatronic systems by different technical and physical principles i.e. different parts of the subsystem are realized as software, hardware, electric, electronic, mechanical and hydraulic devices, etc.

An adapted V-model is applied to get a systematic procedure based on the generic modules of Figure 1. The V-model gives obligatory and unified activities and results within the product development process. The V-model can be adapted to a specific project if necessary. This is called *tailoring* [Müller-Ettrich 1999]. The resulting structure for a CARTRONIC® based development process is shown in Figure 2. The V-model follows a clearly defined order. However iterations for system improvement are implemented as well.

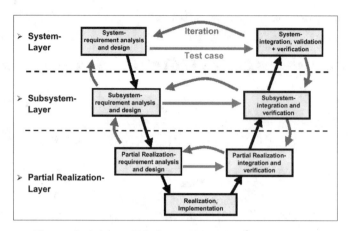

Figure 2: Adapted V-shaped development process

The left side of the V-model is the development area and the right side is the review area. On each layer the requirements are analyzed and realized as design. On the partial realization layer the realization and implementation is done. The results from the requirement analysis of the partial realization layer and subsystem layer are verified in the review area of the V-model. A verification examines and checks whether the designs fulfill the requirements obtained by a requirement analysis. On the system layer a verification and validation is done. The validation on the system layer checks the requirements against the customer demands. It is important to point out that this adapted V-model is not a rigid and straight forward approach but an iterative approach for finding optimized solutions. Therefore this adapted V-model is called IV-model and it is always possible to check an approach by performing one or more test cases i.e. leaving the development area and entering the review area, realize the results in the design part of the considered layer in the development area and continue with the IV-model. The way how to navigate in the IV-model depends on the problems identified in the review area.

A safety analysis is integrated into this CARTRONIC® based development process. The CARTRONIC® based safety analysis (CSA) is included in the IV-model of Figure 2 on the subsystem layer and is based on the CARTRONIC® function structure which is used to describe the results of the requirement analysis on the system and subsystem layer. The intention of the CSA is to identify safety critical areas within a given CARTRONIC® function structure. The results can be validated and verified without applying all steps of the V-model by utilizing one or more iterations on the subsystem layer indicated in the IV-model of Figure 2. The next section describes the CSA methodology.

3. CARTRONIC® BASED SAFETY ANALYSIS (CSA)

The CSA is an abstraction of the FMEA approach. In addition to a FMEA the CSA also considers structural failure dependencies. The intention of the CSA is to improve overall product safety and to avoid costly integration of safety measures in late development phases. Here lies the potential to reduce development time and thus expenditures. This section introduces the methodology on which the CSA is based and the display of the CSA results in a table. This CSA table is capable of assigning failures within the function structure to a malfunction of a single component. The methodology is explained by an example of a simplified braking system.

Before starting the explanation of the CSA-methodology and the structure of the CSA table the structuring concept of CARTRONIC® as well as the example system is introduced.

CARTRONIC® function structuring - The structuring concept according to CARTRONIC® is a hierarchical system representation with communication links between functional components. CARTRONIC® distinguishes between three types of functional components ([Bertram et al 1998], [Torre Flores et al. 2001]):

- components with mainly coordinating and distributing tasks, like the *torque distributor* or the *brake coordinator* in Figure 3,
- components with mainly operative and executing tasks like the *brake actuator* or the *brake lamp* in Figure 3 and
- components which exclusively generate and provide information. There is no such component in our example depicted in Figure 3, but the component *propulsion* may contain such a component in its refinement.

Interactions between components and subsystems are modeled using three different kinds of communication links (see Figure 3). These communication links are:

- *orders* (with response),
- *requests* and
- *inquiries* (with notification).

In order to make it possible to discern between the different communication links CARTRONIC® displays *orders* as solid line, *requests* and *inquiries* as dotted line (see Figure 3). The differentiation between *requests* and *inquiries* is realized by adding a prefix to the communication name. The prefix for *requests* is "!" (exclamation mark) whereas the prefix for *inquiries* is "?" (question mark). Some rules apply to the use of communication links [Bertram et al 1998]. *Orders* are obligations for the receiving component which have to be realized. If this is not possible for any reason then a response to the source component of an *order* has to be transmitted. The response must explain why the *order* could not be realized. A single component can only get one *order*. This rule prevents components from dead-locks which would arise if a component gets two or more contradictory *orders*. A *request* represents the wish of its source component that the destination component fulfills a certain operation. In contrast to an *order* a *request* is not an obligation which has to be fulfilled by its destination but the source component has a strong interest in fulfillment. An *inquiry* asks an information source to deliver certain information. Each subsystem has a single entry component, e.g. a coordinator, which redirects in-bound *orders* to embedded systems or components. *Requests* and *inquiries* can address subsystem components directly, i.e. they do not have to address the entry component of a subsystem before reaching their destination.

Example system - For the purpose of a descriptive example a simplified CARTRONIC® structure was chosen, showing a refinement of *propulsion and brake* and *light and signals* (Figure 3).

Propulsion and brake is a subsystem of *vehicle motion* which is on the highest level, the *vehicle layer*. Within the *vehicle layer* all functions of a vehicle are organized. The subsystem *propulsion and brake* comprises a *torque distributor*, as well as the components *propulsion* and *brake system*. The *brake system* is further refined into two components a *brake coordinator* and a *brake actuator*. The component *brake lamp* from a different subsystem is also addressed. The *brake lamp* is placed within the subsystem *light and signals*, which itself is placed in the subsystem *exterior illumination*. *Exterior illumination* is placed within *visibility and signaling* which is placed in the subsystem *body and interior*. *Body and interior* is on the *vehicle layer*.

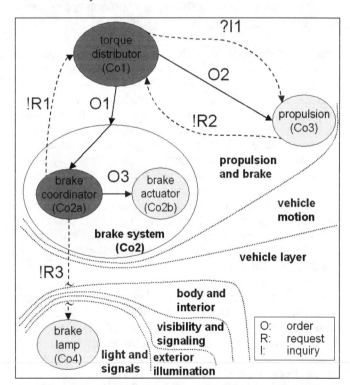

Figure 3: Simplified CARTRONIC® function structure with refinement of *propulsion and brake* and *light and signals*

In this example the *torque distributor* is responsible for coordinating the wheel torque. The components *brake coordinator* and *propulsion* are requesting torque from the *torque distributor* (!R1 and !R2). The *torque distributor* asks for information from the *propulsion* component (?I1) about minimum and maximum values for propulsion torque, checks the requests and all the available information and generates *orders* (O1 respectively O2) accordingly. The *order* O1 tells the *brake coordinator* the necessary brake torque, the *order* O2 tells *propulsion* the necessary acceleration torque. If the *brake coordinator* is receiving an *order* it will check the *order* and calculate a consecutive *order* (O3) and a consecutive *request* (R3) with the necessary parameters for the *brake actuator* and for the *brake lamp*.

The abbreviation for a functional component name or a subsystem name is given in brackets within Figure 3. These abbreviations will be used later for a compact description.

The following subsection explains the CSA basics, gives definitions of terms used for CSA, describes the CSA methodology and introduces a way for partial automation of the approach.

CSA BASICS - The CSA methodology is applied on the subsystem layer of Figure 2. However it would also be possible to apply CSA on the system layer but the results obtained there are less meaningful because at an early stage of product development only few details are known.

The CSA is based on the standard DIN V 19250 ([DIN V 19250], [Bertram et al. 1999]). The approach is also similar to the one described in IEC61508 [IEC61508].

The standard DIN V 19250 is applicable to general automation systems. Failure effects are evaluated in the standard by using "request classes". These "request classes" are defined for general automation systems and are not applicable to vehicles because it is not possible to determine a time duration a person spends within a danger zone and it is also not possible to distinguish whether one or more persons may be affected by a failure effect. Therefore it was necessary to adopt the general "request classes" to automobiles. If the above mentioned two points are disregarded for the evaluation of failure effects the resulting new "request classes" are obtained and are termed safety levels (SL) in the context of CSA. These safety levels are depicted in the risk graph of Figure 4.

The risk graph discerns between failure effects occurring regularly (normal case) and failure effects occurring rarely (special case). The assignment of safety levels to failure effects is basically a qualitative assessment. Nevertheless worst case failure rates are linked to the safety levels. In an early stage of product development it is not possible to deal with these failure rates however after a field test, i.e. a posteriori, it is possible to verify whether a component fulfills the necessary demands concerning its failure rate. This demand is fulfilled if the gained failure rate from field tests is lower than the assigned worst case failure rate for a certain safety level.

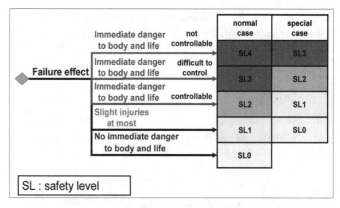

Figure 4: CSA risk graph

Definitions - At first some important terms will be defined before the methodology is explained using the afore introduced simplified braking system example.

Definition 3.1 (global failure effects)
Global failure effects are physical properties which affect the whole vehicle by malfunctions of actuators. Global failure effects can be recognized by the vehicle driver as a loss of elementary functions (e.g. loss of braking ability), comfort functions (e.g. ACC), etc.

Definition 3.2 (function-structure failure)
Function-structure failures (FS-failures) are failures which lead to a malfunction of a functional component or communication within the CARTRONIC® function structure.

Definition 3.3 (function-structure-failure causes)
Function-structure-failure causes (FSF-causes) give reasons for a malfunction of a component. The reasons for a malfunction are FS-failures. A FS-failure can be further refined into failure modes. These failure modes are the cause for the more general FS-failures. The failure modes on an abstract system view can be:

Component failures:

- Component dead
- Component is calculating wrong output
- Component is uncontrolled active
- Component produced output at an undesired time

Communication failures:

- Communication link broken
- Communication is supplying wrong information
- Communication is uncontrolled active
- Communication is supplying information at an undesired time
- Communication is misdirecting message

FS-failures can thus be refined to the failure modes described above. These failure modes can be interpreted as the causes of FS-failures on an abstract system level.

CSA-METHODOLOGY – This sub-section explains the CSA methodology stepwise. It can be partitioned into seven main steps:

Step 1: Identification of global failure effects.

The CARTRONIC® based safety analysis examines global failure effects. These global failure effects can be identified by examining the system interface to the surrounding environment, i.e. the whole vehicle. Actuators driven by the considered system are these interfaces of interest. The consequences of a failure are noticeable by the driver through these interfaces and therefore are global failure effects.

The actuators and thus the system interfaces used in the example of Figure 3 are:

- the _brake system_, respectively in the refinement the _brake actuator_,
- the _propulsion_ actuator and
- the _brake lamp_.

For these subsystems respectively functional components global failure effects have to be identified. The scope of the identification process is to detect only those global failure effects which are caused by the system of interest. For the example in Figure 3 some of the global failure effects are shown in Table 1.

Table 1: Examples of global failure effects

Actuator	global failure effect

propulsion actuator	• acceleration
	  ▪ uncontrollable acceleration
	    ○ acceleration too high
	    ○ acceleration too low
	  ▪ no acceleration

brake actuator	• deceleration
	  ▪ no braking action
	  ▪ braking action too low

brake lamp	• braking signaling
	  ▪ no signaling
	  ▪ continuous signaling

In order to get a clear presentation the refinement of uncontrollable acceleration is omitted in Table 2. The next step is the evaluation of the identified global failure effects.

Step 2: Evaluate global failure effects using safety levels.

For the evaluation of global failure effects the safety levels given by the risk graph of Figure 4 are applied. For the evaluation of the global failure effect "no braking action" of the _brake actuator_ displayed in Figure 3, one has to consider that the brake is a basic functionality of a vehicle and that the complete loss of retardation leads to an immediate danger to body and life. This effect can not be controlled by the driver in a normal case. Therefore the safety level SL4 has to be assigned to this global failure effect. For the evaluation of the failure effect "no acceleration" one has to consider that this effect could also lead to an immediate danger to body and life but it can be controlled by taking appropriate measures, e.g. warning following cars by switching on the hazard-warning system or to position a warning triangle. Therefore the worst case scenario happens only in special cases. The safety level SL1 can thus be assigned to this

global failure effect. The assigned safety levels for all global failure effects of the brake example can be seen in the upper left part of Table 2 which is called CSA table.

Step 3: Analysis of FSF-causes (see Definition 3.3) for potential failures in components.

The analysis of FSF-causes is looking for the cause of a malfunction in CARTRONIC® components. The analysis comprises components and their refinements. The intention of this step is to reveal failure dependencies within the function structure. The tracing of failure dependencies in the CARTRONIC® based safety analysis is a deductive [Misra 1992] or button-up approach because the reasoning is done starting with a failure consequence or a resulting event which is here a malfunction of a component or subsystem. Then the failures which are potentially able to cause the consequence are investigated.

The CSA table is divided into a left side and a right side by the column "SL" (Table 2). The right side of the CSA table is dealing with the analysis of FSF-causes. There are two columns for each component of the function structure. One representing the component itself abbreviated "c" and one representing in-bound communications abbreviated "m". A malfunction of a component itself causing a FS-failure is visualized in the CSA table by a "x" in the column "c" (see Table 2). In Table 2 the abbreviations for functional components and subsystems depicted in Figure 3 are used.

It is also possible that a malfunction of a component is caused by a communication link. The communication links _order_ and _request_ are considered at the component to which they point. _Inquiries_ are taken into account at the component from where they originate. This is reasonable for _inquiries_ because the information flow is directed from the target of the communication link to its source. The analysis of FSF-causes is done by tracking back the flow of communications for the search of potential failures. Subsequently the components which are sources of the communication links _order_ or _request_ respectively the destination components for _inquiries_ are searched. The search is continued until no more communication links have to be backtracked or an already visited component is reached. These communication dependencies are documented in the column "m". The prefix of the communication name in the CSA table indicates the kind of communication. The prefixes for communications are order "O_", request "R_" and inquiry "I_". They are followed by the name of the communication, e.g. order O1 is represented by "O_O1". In order to explain this procedure an example is given using the _brake actuator_ as initial component. In Table 2 the row "malfunction of component _brake actuator_ (Co2b)" has to be regarded. A malfunction of the _brake actuator_ can be caused by a FS-failure in the component itself. Therefore an "x" is written in the column "c" of the function structure for component _brake actuator_ (Co2b). The in-bound _orders_ or _requests_ respectively out-bound _inquiries_ to this component are considered in the column "m" of the same component. Here the only communication is the in-bound *order* O3. Therefore O_O3 is written in the column "m" of this component. The next step is the backtracking of the communication O3 which originates from the component *brake coordinator*. Hence an "x" is written in the column "c" of this component. The only in-bound communication is O1. The interpretation of this can be that a malfunction of component *brake actuator* can be caused by a FS-failure in the component *brake coordinator* or the communication O1. The outbound communications R3 and R1 are not considered here. The communication O1 is backtracked to its origin which is the component *torque distributor*. The communications which have to be considered here are the in-bound *requests* R1 and R2 and the out-bound *inquiry* I1. Subsequently these communications are backtracked. The backtracking of R1 leads to the *brake coordinator* which was already considered. The backtracking of this communication link is stopped there. The next communication link which has to be backtracked is R2. This leads to the *component propulsion* which will be marked in the table. The communication link O2 is the only communication to be further investigated here. This communication leads to the *torque distributor* which was already considered. We have to keep in mind that we still have to examine the communication link I1 originating from the *torque distributor*. We see that tracking back I1 leads to the *component propulsion* which already considered. Now there are no more communication links to be tracked back, thus we have finished the investigation for a malfunction of component *brake actuator*.

The part on the right side of column SL in Table 2 visualizes the analysis results of step 3. The analysis of FSF-causes comprises the scrutiny of:

- components itself,
- in-bound *orders*,
- in-bound *requests* and
- out-bound inquiries.

It can easily be seen that the search for failure dependencies is very time consuming. However this part of the CSA can be automated using MAX-PLUS-Algebra which is explained later in this section.

Step 4: Assignment of component malfunction to global effects.

A malfunction of a component is now assigned to global failure effects. We start here with the system interfaces which are the actuators of the system (see *Step 1*). A malfunction of the *brake actuator* (Co2b) can cause the global failure effects "no braking deceleration" and "braking deceleration too low", a malfunction of the *propulsion* (Co3) can lead to an "uncontrollable acceleration" or "no acceleration" and a malfunction of the component *brake lamp* (Co4) can lead to the global failure effects "no signaling" or "continuous signaling". These are the basic assignments visualized in Table 2 by "x". Now each column of the component structure is evaluated. An element of a column which is assigned to a malfunction of a system interface component can potentially cause the same global failure effects. For example the component *torque distributor* (Co1) is assigned to a malfunction of component *brake actuator* (Co2b), to a malfunction of component *propulsion* (Co3) and as well to a malfunction of component *stop light* (Co4). Thus a malfunction of the component *torque distributor* can potentially cause all of the basic global failure effects. This is also true for component *coordinator brake* and component *propulsion*. Whereas the component *brake actuator* is only assigned to itself, indicating that this component can not cause other global failure effects than the already identified basic global failure effects of itself. The same applies for the component *brake lamp*.

The assignment of a safety level to a malfunction of a component is a worst case estimation. The highest safety level assigned to a global failure effect is written in the column SL which evaluates a component malfunction altogether. This is illustrated by the gray shaded fields in Table 2. The interpretation of the safety levels is analogous to the risk graph of Figure 4. For instance a mal-

Table 2: CSA table for example in Figure 3

Global failure effects	propulsion acceleration		braking deceleration		deceleration signaling		SL	malfunction of components / function structure	Co1		Co2		Co2a		Co2b		Co3		Co4	
	uncontrollable acceleration	no acceleration	no braking action	braking action too low	no signaling at all	continuous signaling			c	m	c	m	c	m	c	m	c	m	c	m
SL	3	1	4	3	1	1															
	x	x	x	x	x	x	4	Co1	x	I_I1 R_R1 R_R2			x	O_O1			x	O_O2		
	x	x	x	x	x	x	4	Co2	x	I_I1 R_R1 R_R2	x	O_O1	x	O_O1			x	O_O2		
	x	x	x	x	x	x	4	Co2a	x	I_I1 R_R1 R_R2			x	O_O1			x	O_O2		
			x	x			4	Co2b	x	I_I1 R_R1 R_R2			x	O_O1	x	O_O3	x	O_O2		
	x	x	x	x	x	x	4	Co3	x	I_I1 R_R1 R_R2			x	O_O1			x	O_O2		
			x	x	x	x	1	Co4	x	I_I2 R_R1 R_R2			x	O_O1			x	O_O2	x	R_R3

function of the *brake actuator* is assigned to the two global failure effects "no braking action" and "braking action too low". The first global failure effect is evaluated with SL4, the second with SL3. Thus the maximum rating is chosen for the evaluation of the malfunction of the *brake actuator* which is SL4.

Step 5: Measures for failure detection and handling.

There are several generic measures which can be used for failure detection and failure handling on an abstract system level. Measures for failure detection can be:

- Acknowledgement
  The receiving component sends an acknowledgement to the sending component.
- Redundancy
  Two or more comparable results are available.
- Reference checks
  A question with a known answer is asked.
- Observation of transmission channels
  Typical examples for such measures are CRC-checks, parity checks or Grey-Codes.
- Observation of physical properties
  Observation of voltage, current, temperatures etc.
- Logical/temporal observation of program execution
  Watchdog-timers can be applied for this purpose.
- Dynamic communication
  Different coding algorithms can be applied to transmit measurement results whereas the sequence of coding algorithms is fixed.

Measures for failure handling can be:

- Redundancy (at least triple redundancy)
  Using several different results for elimination of wrong results
- Shut down of system part
  The system part that is influenced by a failure is turned off
- Complete shut down of electronics
  This measure can only be used for comfort systems, e.g. cruise control is such a system.
- System stays in faulty state
  This is possible for fault tolerant systems.
- Failure removal
  A repair mechanism has to be implemented, e.g. reboot of an electronic control unit (ECU).

It shall be mentioned that on a high abstraction level it is very difficult to find concrete measures. The reason therefore is the lack of detailed knowledge. However measures shall be implemented as early as possible but in many cases it is recommendable to postpone the search for concrete measures to a later stage of the development process, i.e. in the partial realization layer. Concrete measures can be identified in a development process as soon as the system topology is specified, considering costs and technical restraints.

Summarizing the steps of the CSA methodology described so far it can be stated that the CSA is capable of identifying critical system functionality. The systematic search for critical components creates an awareness of system parts where special attention is needed. This knowledge influences the system structure and thus contributes to the development of safe products. Measures shall be implemented in the development process as early as possible but in a stage where reasonable solutions can be found. Note that this depends on the kind of problem and the ideas for solution. Measures which are based on plausibility may be obvious in a very early stage of product development.

The CSA comprises two further steps which are not yet completely elaborated. Step 6 will be the implementation of failure mitigating measures in the CARTRONIC® function structure to become a function and safety structure. These measures extend the basic functionality of a system and are based on non-functional requirements. Step 7 will be the verification of the results obtained by the CSA.

The integration of the CSA in the CARTRONIC® based development process is discussed in the following subsection.

Integration of the CSA in the V-Model - Figure 5 shows the implementation of the CSA into the CARTRONIC® based development process. The CSA is not used at the system layer although it would be possible. The reason therefore is that little results can be gained in such an early development phase because few details are known. At the subsystem layer a CARTRONIC® function structure of the considered subsystem is used as basis for the CSA.

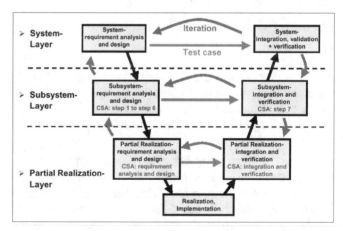

Figure 5: Integration of CSA in a CARTRONIC® based development process

The steps 1 to 6 of the CSA methodology are performed in the development area whereas step 7 is performed in the review area. The verification of the results include questions like "Is CSA done properly ?" or "Are failure mitigating measures chosen correctly ?". On the subsystem layer several iterations of the CSA can be completed. It is planed to extend the CSA methodology to the partial realization layer. In this case the CARTRONIC® function structure has to be complemented by descriptions of different realizations considering the complete system topology. The CSA will be used here to check a chosen design of a system topology. It shall be mentioned that within the partial realization layer also other risk analyses like FMEA or hazard and operability study (HAZOP) among others are used. The information collected with the CSA and its results can be used as input for these analyses.

Even in the short example shown in Figure 3 it is possible to see that the search for failure dependencies is very time consuming. Therefore a way to automate this part of the CSA was investigated. The methodology found is a formalized notation which uses a mathematical system representation. The mapping of CARTRONIC® function structures into the mathematical system representation is done by structure matrices. The approach outlined here is a kind of state space representation. The next subsection explains how this methodology the so called Max-Plus Algebra can be used for calculating failure dependencies.

INTRODUCTION TO MAX-PLUS ALGEBRA – The CARTRONIC® function structure can be interpreted as a graph in which the functional components are nodes and the communication links are edges. The search for failure dependencies in Table 2 is a similar problem to the search for paths within graphs. This is a very common problem in graph theory and can be resolved by using MAX-PLUS algebra. This algebra - as the name MAX-PLUS already implies – comprises only the operators "max" respectively "min" and "plus" ([Cunninghame-Green 1979], [Kiencke 1997]). Such an algebraic structure is termed semifield [Baccelli et al 1992]. Before continuing with the explanation how this methodology can be applied for automating the CSA methodology, a short introduction to MAX-PLUS algebra is given.

The MAX-PLUS algebra is defined on the set

$$\mathfrak{R}_{max} = \mathfrak{R} \cup \{-\infty\} \qquad (4.1)$$

and is equipped with the two operators $\oplus$, $\otimes$. Whereas the symbol $\oplus$ is used for the "max"-operator and the symbol $\otimes$ is used for the "plus"-operator. A brief introduction to the axioms of MAX-PLUS algebra can be found in the Appendix.

SYSTEM DECOMPOSITION IN STRUCTURE MATRICES – The system structure depicted in Figure 3 is mapped in the mathematical description of MAX-PLUS Algebra by using three different structure matrices. These matrices represent the system structure in a kind of state space representation and describe the following properties:

- A-matrix: Subsystem internal communication
- B-matrix: In-bound communications to a component or subsystem
- C-matrix: Out-bound communications from a component or subsystem.

The entries in the matrices are kept binary, i.e. for an existing communication link between two components the MAX-Plus one element e is used otherwise the MAX-PLUS zero element ε as defined by equation (A.6) and (A.7) in the Appendix.

For the functional components *torque distributor*, *coordinator brake*, *brake actuator*, *propulsion* and *stop light* from Figure 3 these matrices are:

$$A_{Co1} = Co1\begin{matrix}Co1\\[\varepsilon]\end{matrix}$$

$$B_{Co1} = Co1\begin{matrix}R_1\ R_2\ I_1\\[e\quad e\quad e]\end{matrix}$$

$$C_{Co1} = \begin{matrix}&Co1\\O_1&[e]\\O_2&[e]\end{matrix}$$

$$A_{Co2} = \begin{matrix}&Co2a\ Co2b\\Co2a&[\varepsilon\quad\varepsilon]\\Co2b&[e\quad\varepsilon]\end{matrix}$$

$$B_{Co2} = \begin{matrix}&O_1\\Co2a&[e]\\Co2b&[\varepsilon]\end{matrix}$$

$$C_{Co2} = \begin{matrix}&Co2a\ Co2b\\R_1&[e\quad\varepsilon]\\R_3&[e\quad\varepsilon]\end{matrix}$$

$$A_{Co3} = Co3\begin{matrix}Co3\\[\varepsilon]\end{matrix}$$

$$B_{Co3} = Co3\begin{matrix}O_2\\[e]\end{matrix}$$

$$C_{Co3} = \begin{matrix}&Co3\\R_2&[e]\\I_1&[e]\end{matrix}$$

$$A_{Co4} = Co4\begin{matrix}Co4\\[\varepsilon]\end{matrix}$$

$$B_{Co4} = Co4\begin{matrix}R_3\\[e]\end{matrix}$$

$$C_{Co4} = [\varepsilon]$$

The indices of the matrices are chosen according to the abbreviations introduced in Figure 3.

The interpretation of the matrices is depicted in Figure 6. The component or communication name written on top of the matrices is the source and the component or communication name written on the left side of the matrices is the destination. An e entry indicates that there exists a communication link whereas an ε indicates that there is no such link. In Figure 6 there is only one existing communication link between component Co2a and Co2b.

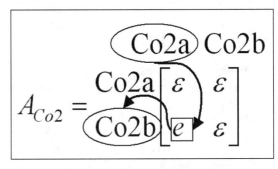

Figure 6: Matrix interpretation

It shall be mentioned that the direction of *inquiries* is reversed. As mentioned before the information flow of *inquiries* is counter-directed to the depicted communication link. The direction of the depicted communication link signals that the source component is interested in information however the information flow is from the target of the *inquiry* to its source.

This decomposition approach was chosen because it fits very well into the hierarchical, object based structure of CARTRONIC®. It is possible to replace a component respectively a subsystem or change the level of abstraction or refinement without affecting other parts of the system. This is possible as long as the interfaces, i.e. the communication links between the components do not change. This is in unison with the object based principles of information hiding and exchangeability. For instance the subsystem *brake system* in Figure 3 without refinement can be described as:

$$\tilde{A}_{Co2} = Co2[\varepsilon] \qquad \tilde{B}_{Co2} = O_1 \atop Co2[e] \qquad \tilde{C}_{Co2} = \begin{array}{c} Co2 \\[2pt] \begin{array}{c} R_1 \\ R_3 \end{array}\!\!\begin{bmatrix} e \\ e \end{bmatrix} \end{array}$$

This can be done if the refinement of subsystem *brake system* is not important for the actual analysis. Note that the structure matrices for all the other components remain unchanged.

For an optimized calculation of failure dependencies it is necessary to combine the structure matrices A_{Co1}, A_{Co2}, A_{Co3} and A_{Co4} to a global structure matrix A.

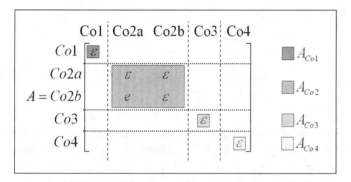

Figure 7: A-matrix composition

The same is necessary for B_i and C_i matrices with $i \in \{Co1,...,Co4\}$. The generation of these global matrices can be done automatically by combining the substructure matrices as illustrated in Figure 7 for the A-matrix. For the B- and C-matrix the same methodology is applied. All the empty elements of these matrices are filled in with ε.

The resulting matrices are:

$$A = \begin{array}{c} \\ Co1 \\ Co2a \\ Co2b \\ Co3 \\ Co4 \end{array}\!\!\begin{array}{c} \begin{array}{ccccc} Co1 & Co2a & Co2b & Co3 & Co4 \end{array} \\ \begin{bmatrix} \varepsilon & \varepsilon & \varepsilon & \varepsilon & \varepsilon \\ \varepsilon & \varepsilon & \varepsilon & \varepsilon & \varepsilon \\ \varepsilon & e & \varepsilon & \varepsilon & \varepsilon \\ \varepsilon & \varepsilon & \varepsilon & \varepsilon & \varepsilon \\ \varepsilon & \varepsilon & \varepsilon & \varepsilon & \varepsilon \end{bmatrix} \end{array} , \tag{4.2}$$

$$B = \begin{array}{c} \\ Co1 \\ Co2a \\ Co2b \\ Co3 \\ Co4 \end{array}\!\!\begin{array}{c} \begin{array}{cccccc} R_1 & R_2 & I_1 & O_1 & O_2 & R_3 \end{array} \\ \begin{bmatrix} e & e & e & \varepsilon & \varepsilon & \varepsilon \\ \varepsilon & \varepsilon & \varepsilon & e & \varepsilon & \varepsilon \\ \varepsilon & \varepsilon & \varepsilon & \varepsilon & \varepsilon & \varepsilon \\ \varepsilon & \varepsilon & \varepsilon & \varepsilon & e & \varepsilon \\ \varepsilon & \varepsilon & \varepsilon & \varepsilon & \varepsilon & e \end{bmatrix} \end{array} \text{ and} \tag{4.3}$$

$$C_{us} = \begin{array}{c} \\ O_1 \\ O_2 \\ R_1 \\ R_3 \\ R_2 \\ I_1 \end{array}\!\!\begin{array}{c} \begin{array}{ccccc} Co1 & Co2a & Co2b & Co3 & Co4 \end{array} \\ \begin{bmatrix} e & \varepsilon & \varepsilon & \varepsilon & \varepsilon \\ e & \varepsilon & \varepsilon & \varepsilon & \varepsilon \\ \varepsilon & e & \varepsilon & \varepsilon & \varepsilon \\ \varepsilon & e & \varepsilon & \varepsilon & \varepsilon \\ \varepsilon & \varepsilon & \varepsilon & e & \varepsilon \\ \varepsilon & \varepsilon & \varepsilon & e & \varepsilon \end{bmatrix} \end{array} . \tag{4.4}$$

These three matrices are a mathematical description of the CARTRONIC® function structure of Figure 3 considering its functional components and communication links. These matrices can be used for the calculation of failure dependencies which is explained in the next subsection.

CALCULATING FAILURE DEPENDENCIES – Before it is possible to calculate failure dependencies it is necessary to reorganize the sequence of rows in the C_{us}-matrix, where *us* indicates unsorted. The ordering of the rows of the C_{us}-matrix must match the ordering of columns in the B-matrix. The resulting sorted matrix C_s is

$$C_s = \begin{array}{c} \\ R_1 \\ R_2 \\ I_1 \\ O_1 \\ O_2 \\ R_3 \end{array}\!\!\begin{array}{c} \begin{array}{ccccc} Co1 & Co2a & Co2b & Co3 & Co4 \end{array} \\ \begin{bmatrix} \varepsilon & e & \varepsilon & \varepsilon & \varepsilon \\ \varepsilon & \varepsilon & \varepsilon & e & \varepsilon \\ \varepsilon & \varepsilon & \varepsilon & e & \varepsilon \\ e & \varepsilon & \varepsilon & \varepsilon & \varepsilon \\ e & \varepsilon & \varepsilon & \varepsilon & \varepsilon \\ \varepsilon & e & \varepsilon & \varepsilon & \varepsilon \end{bmatrix} \end{array} . \tag{4.5}$$

Using the matrix from equation (4.3) and equation (4.5) it is possible to calculate in-bound and out-bound communication links of all components in the following way

$$B \otimes C_s = \begin{array}{c} \\ Co1 \\ Co2a \\ Co2b \\ Co3 \\ Co4 \end{array}\!\!\begin{array}{c} \begin{array}{ccccc} Co1 & Co2a & Co2b & Co3 & Co4 \end{array} \\ \begin{bmatrix} \varepsilon & e & \varepsilon & e & \varepsilon \\ e & \varepsilon & \varepsilon & \varepsilon & \varepsilon \\ \varepsilon & \varepsilon & \varepsilon & \varepsilon & \varepsilon \\ e & \varepsilon & \varepsilon & \varepsilon & \varepsilon \\ \varepsilon & e & \varepsilon & \varepsilon & \varepsilon \end{bmatrix} \end{array} . \tag{4.6}$$

The matrix A from equation (4.2) is added to the matrix of equation (4.6). The result is given by equation (4.7). This result contains all paths of length 1 within the graph of Figure 3, i.e. component and subsystem internal and external links.

$$T = A \oplus B \otimes C_s = \begin{array}{c} \\ Co1 \\ Co2a \\ Co2b \\ Co3 \\ Co4 \end{array}\!\!\begin{array}{c} \begin{array}{ccccc} Co1 & Co2a & Co2b & Co3 & Co4 \end{array} \\ \begin{bmatrix} \varepsilon & e & \varepsilon & e & \varepsilon \\ e & \varepsilon & \varepsilon & \varepsilon & \varepsilon \\ \varepsilon & e & \varepsilon & \varepsilon & \varepsilon \\ e & \varepsilon & \varepsilon & \varepsilon & \varepsilon \\ \varepsilon & e & \varepsilon & \varepsilon & \varepsilon \end{bmatrix} \end{array} . \tag{4.7}$$

Finally the failure dependencies are calculated by respecting all possible paths within T of length zero up to $(n-1)$ where n is the number of components. This is calculated by

$$T^* = I \oplus T \oplus T^2 \oplus \cdots \oplus T^{(n-1)} \qquad (4.8)$$

where I is the identity matrix and the nth-power of a matrix T is the "multiplication" of n matrices T. It is not necessary to consider paths which are longer than $(n-1)$ because in a graph with n nodes there can only be paths without cycles of maximum length $(n-1)$. Since this approach is binary, i.e. we are just interested in whether a node was visited or not, cycles of arbitrary length are already taken into account by "adding" the identity matrix I. By applying equation (4.8) to our example system the result is

$$T^* = \begin{array}{c} \\ \text{Co1} \\ \text{Co2a} \\ \text{Co2b} \\ \text{Co3} \\ \text{Co4} \end{array} \begin{array}{ccccc} \text{Co1} & \text{Co2a} & \text{Co2b} & \text{Co3} & \text{Co4} \\ \left[\begin{array}{ccccc} e & e & \varepsilon & e & \varepsilon \\ e & e & \varepsilon & e & \varepsilon \\ e & e & e & e & \varepsilon \\ e & e & \varepsilon & e & \varepsilon \\ e & e & \varepsilon & e & e \end{array} \right] \end{array} \qquad (4.9)$$

Comparing the result given by equation (4.9) and the failure dependencies of Table 2 it can be seen that they are the same if the legend on the left side of T^* is interpreted as a malfunction of a component and the legend on top as the component structure. An e in the matrix of equation (4.9) equals an "x" in Table 2 and an ε equals a blank element of the table.

Equation (4.9) just shows the failure dependencies between components. But it is also possible to calculate which communication links have to be considered at which component. This information is necessary to fill in the "m"-columns of Table 2. In order to calculate these communication links a communication matrix K will be introduced. For the installation of such a global communication matrix K it is necessary to include the internal communication of subsystems, e.g. the internal communication of the subsystem *brake system*. Therefore a matrix L_i with $i \in \{Co1,...,Co4\}$ is used. This matrix describes the in-bound communication links to subsystem internal components. For the refined subsystem *brake system* (Co2) the matrix L_{Co2} is

$$L_{Co2} = \begin{array}{c} \\ \text{Co2a} \\ \text{Co2b} \end{array} \begin{array}{c} O_3 \\ \left[\begin{array}{c} \varepsilon \\ e \end{array} \right] \end{array}. \qquad (4.10)$$

Extended structure matrices $^{ext}B_i$ with $i \in \{Co1,...,Co4\}$ are composed by adding new columns to the B_i-matrices for respecting subsystem internal communication. These new columns are gained from the matrices L_i. For the example shown in Figure 3 the only refined

subsystem is the subsystem *brake system* (Co2), thus only matrix B_{Co2} has to be extended by L_{Co2}.

$$^{ext}B_{Co2} = \begin{array}{c} \\ \text{Co2a} \\ \text{Co2b} \end{array} \begin{array}{cc} O_1 & O_3 \\ \left[\begin{array}{cc} e & \varepsilon \\ \varepsilon & e \end{array} \right] \end{array}. \qquad (4.11)$$

$$\underbrace{}_{L_{Co2}}$$

The resulting communication matrix K is composed from the B - respectively $^{ext}B_i$-matrices in the same way as depicted in Figure 7. The so received matrix is

$$K = \begin{array}{c} \\ \text{Co1} \\ \text{Co2a} \\ \text{Co2b} \\ \text{Co3} \\ \text{Co4} \end{array} \begin{array}{ccccccc} R_1 & R_2 & I_1 & O_1 & O_3 & O_2 & R_3 \\ \left[\begin{array}{ccccccc} e & e & e & \varepsilon & \varepsilon & \varepsilon & \varepsilon \\ \varepsilon & \varepsilon & \varepsilon & e & \varepsilon & \varepsilon & \varepsilon \\ \varepsilon & \varepsilon & \varepsilon & \varepsilon & e & \varepsilon & \varepsilon \\ \varepsilon & \varepsilon & \varepsilon & \varepsilon & \varepsilon & e & \varepsilon \\ \varepsilon & \varepsilon & \varepsilon & \varepsilon & \varepsilon & \varepsilon & e \end{array} \right] \end{array}. \qquad (4.12)$$

By analyzing the communication matrix (4.12) it can be seen that the communication links R_1, R_2 and I_1 have to be considered at the *torque distributor* (Co1), the communication link O_1 at the component *brake coordinator* (Co2a), the communication link O_3 at the component *brake actuator* (Co2b) etc.

4. CONCLUSION

The CSA methodology described here is a systematic procedure for a safety analysis of a system on an abstract, function based system level.

The CSA methodology is based on CARTRONIC® function structures which are hierarchical, object based system representations, respectively the representation of these function structures in the UML. This system representation is a hierarchical decomposition of a system into subsystems, functional components and communication interactions. The CSA is applied at an early stage of a product development process to identify safety critical components and communications. Already at this stage failure dependencies of subsystems and components can be revealed. This can be automated by using MAX-PLUS algebra. So a time consuming and maybe incomplete manual search for failure dependencies can be avoided. Further a rating system of failure effects is supplied by the CSA. The knowledge of failure dependencies combined with the rating of failure effects is an essential information to supports decisions where measures to improve the safety of system functionality brings the highest level of improvement.

So the CSA produces an awareness where counter acting measures have to be implemented and which potential risks have to be mastered. However it can be quite challenging to implement appropriate safety measures at an early stage of product development. Therefore specific safety measures shall be implemented as early as possible but not before reasonable measures under technical and economic aspects can be found. Therefore an iterative development process is necessary to implement such specific safety measures into the system structure, including a possibility to reevaluate the supplemented structures until all necessary system parameters meet the specification. Therefore the CSA is integrated into a modified, incremental and iterative V-Model (IV-Model).

So the CSA can improve the overall product safety and can avoid costly integration of safety measures in late development phases.

REFERENCES

[Baccelli et al 1992] Baccelli, F.; Cohen, G.; Olsder, G.J.; Quadrat, J.-P. (1992): Synchronization and linearity, John Wiley&Sons (Chichester, England)

[Cunninghame-Green 1979] Cunninghame-Green, R. (1979): Minimax Algebra, Springer Verlag (Berlin, Germany)

[Bertram et al 1998] Bertram, T.; Bitzer, R.; Mayer, R.; Volkart, A. (1998): CARTRONIC – An open architecture for networking the control systems of an automobile (SAE98200), SAE International Congress and Exposition (Detroit, MI, USA)

[Bertram et al. 1999] Bertram, T.; Dominke, P.; Müller, B. (1999): The safety related aspect of CARTRONIC (SAE 1999-01-0488), SAE International Congress and Exposition (Detroit, MI, USA)

[Müller-Ettrich 1999] Müller-Ettrich, G. (1999): Objektorientierte Prozeßmodelle, Addison-Wesley-Longman (Reading, MA, USA)

[Deckers et al. 1994] Deckers, J.; Schäbe, H. (1994): FMEA und Fehlerbaumanalyse im Verbund nutzen, Qualität und Zuverlässigkeit (QZ), Vol. 39, No. 1, pp. 47-50; Hanser Verlag (Munich, Germany)

[Kiencke 1997] Kiencke, U. (1997): Ereignisdiskrete Systeme, Oldenbourg Verlag (Munich, Germany)

[Lapp et al 2001] Lapp, A.; Torre Flores, P.; Schirmer, J.; Kraft, D.; Hermsen, W.; Bertram, T.; Petersen, J. (2001): Softwareentwicklung für Steuergeräte im Systemverbund – Von der CARTRONIC-Domänenstruktur zum Steuergerätecode, VDI 10. Internationaler Kongress Elektronik im Kraftfahrzeug (Baden-Baden, Germany)

[Main et al 1999] Main, B. W.; McMurphy, K. J. (1999): Safety Through Design: The State of the Art in Safety Processes (SAE 1999-01-0421), SAE International Congress and Exposition (Detroit, MI, USA)

[Misra 1992] Misra, K. B. (1992): Reliability analysis and Prediction, Elsevier Science Publishers B.V. (Amsterdam, The Netherlands)

[OMG UML 1999] Object Management Group Inc. (1999): OMG Unified Modeling Language Specification V1.3, http://www.omg.org/UML/

[OMG XML 2000] Object Management Group Inc. (2000): OMG XML Metadata Interchange (XMI) Specification, Download Version 1.0 (June 2000); http://www.omg.org/technology/xml/index.htm

[Torre Flores et al. 2001] Torre Flores, P.; Lapp, A.; Hermsen, W.; Schirmer, J.; Walther, M.; Bertram, T.; Petersen, J. (2001): Integration of a Structuring Concept for Vehicle Control Systems into the Software Development Process using UML Modelling Methods (SAE 2001-01-0066), 2001 SAE World Congress (Detroit, MI, USA)

STANDARDS –

[DIN V 19250] (1995): Leittechnik. Grundlegende Sicherheitsbetrachtungen für MSR-Schutzeinrichtungen, Beuth-Verlag (Berlin, Germany)

[IEC61508] (2000): Functional safety of electrical/electronic/programmable electronic safety-related systems, IEC (Geneva, Switzerland)

CONTACT

Robert Bosch GmbH
Corporate Research and Development
New Systems
P.O. Box 30 02 40
D-70442 Stuttgart
Germany
Email: Laengst@iiit.etec.uni-karlsruhe.de

APPENDIX

The following axioms hold true for the two operators $\oplus$, $\otimes$ with $\forall a,b,c \in \Re_{max}$.

Axiom 1 (Associativity)

$$(a \oplus b) \oplus c = a \oplus (b \oplus c) \qquad (A.1)$$
and
$$(a \otimes b) \otimes c = a \otimes (b \otimes c) \qquad (A.2)$$

Axiom 2 (Commutativity of addition)

$$a \oplus b = b \oplus a \qquad (A.3)$$

Axiom 3 (Right and left distributivity)

$$(a \oplus b) \otimes c = (a \otimes c) \oplus (b \otimes c) \qquad (A.4)$$

This is the right distributivity of product over sum. Note that there also exists a left distributivity which in general is not equal to the right distributivity. The left distributivity of product over sum is given by

$$c \otimes (a \oplus b) = (c \otimes a) \oplus (c \otimes b) \qquad (A.5)$$

Axiom 4 (Zero and identity elements)

$$\exists \varepsilon \in \Re_{max} : a \oplus \varepsilon = a \qquad (A.6)$$
$$\exists e \in \Re_{max} : a \otimes e = a \qquad (A.7)$$

$\mathcal{E}$ is the zero element defined by equation (A.6) and e is the identity element defined by equation (A.7).

Axiom 5 (Idempotency of addition)

$$a \oplus a = a \qquad\qquad (A.8)$$

The aforementioned Axioms 1 to 5 also hold true for matrices. The matrix addition of two elements $A, B \in \Re_{max}^{nxm}$ is defined by

$$A \oplus B = [a_{ij} \oplus b_{ij}], \qquad\qquad (A.9)$$

which is the maximum of the two matrix elements a_{ij} and b_{ij}. The matrix multiplication of two elements $A \in \Re_{max}^{nxi}$ and $B \in \Re_{max}^{ixm}$ with $x \in \{1,2,...,m\}$ and $y \in \{1,2,...,n\}$ is defined by

$$(A \otimes B)_{xy} = \oplus\{(a_{x1} \otimes b_{1y}), \cdots, (a_{xi} \otimes b_{iy})\}. \qquad\qquad (A.10)$$

The result $(A \otimes B)$ is of the dimension $\Re_{max}^{nxm}$.

FANTASTIC : Fast Automotive Networking Through ARM-based System, Tools and Integrated Circuits

Sghaier Noury, Tristan Bonhomme, Jean-Pierre Demange and Nicolas Demoulin
Europe Technologies

ABSTRACT

Europe Technologies has developed a new concept, named *easy*SoC™ using the FCM® technology (Fast Configurable Microchip), allowing fast and foolproof development, emulation and validation of ARM-based System-On-Chip. This concept was applied to create an Automotive platform with special focus on in-car networking, demonstrated and validated with initial applications in the North American and European automotive markets.

Thanks to this platform, car and equipment manufacturers enjoy unprecedented benefits, particularly important for this industry :
- New systems development cycle time reduction of up to 50%.
- New systems total cost reduction of up to 30%.
- First-pass success (elimination of functionality, security and reliability risks in new systems)

INTRODUCTION

The Automotive segment, with focus on in-car Networking applications, is affected with the following constraint : the integration of new embedded services into the car is becoming the most important differentiation factor for car manufacturers. The vectors for such services are Networking and Multimedia applications. The complexity of automotive electronic applications is continuously increasing and can be achieved only through higher system integration chips i.e. "System-On-Chip". Time-to-market is a key competitive factor. It requires continuous shortening of new products development time. Cost reduction is another key competitive factor. It requires profound system optimization and perfect match of a dedicated chip with the application. Car manufacturers cannot afford any compromise or error in terms of functionality, security or reliability because of possible impact on "image" and cost in case of redesign or recall. It requires fast but intensive validation of new hardware and software as well as maximum reusability of already proven hardware and software components.

For all the above reasons there is a strong need to "do things differently" in automotive. It is an opportunity for significant changes in working methods, new product development strategies and project organization and management.

The *easy*SoC™ concept

an innovation

Facing the Automotive market constraints, Europe Technologies has developed a new concept, named *easy*SoC™ and integrating the FCM® technology (Fast Configurable Microchip), allowing fast and foolproof development, emulation and validation of ARM-based application specific microcontrollers.
The concept and technology has been demonstrated and validated with the first versions of the FCM® general-purpose platform and its main components.

Its basic elements are:

A emulation platform, the FCM®, allowing a full configuration of the required ARM-based ASSP (Application Specific Standard Product) and totally re-usable,

A dedicated development methodology adapted to the strict automotive constraints,

And some Value-added components, such as software tools, up-to-date Automotive oriented IP's and foundry support.

Europe Technologies, as an embedded system provider, has proven with practical experience, that its *easy*SoC™ concept provides the tools, technologies and components that meet all the requirements of in-car embedded services, cost reduction, high reliability and short cycle time thanks to its modular approach towards reusability and standardization.

What is the FCM® platform ?

Thanks to the FCM® FANTASTIC platform (Fast Configurable Microchip), which is the heart of the *easy*SoC™ concept, the user can configure an ARM based chip for its application, in order to validate this chip in his application, and to develop and debug intensively his application software and to validate his application board /system, before going to the final ASSP silicon.

The philosophy of the FCM® platform is to represent a monolithic ASSP in its exploded form. For this, three Integrated Circuits have been developed, linked on the same electronic board. The first chip, named the "Core Chip" is built around an ARM core. It allows external access to the high-speed internal bus (the ASB for ARM core) and the peripheral bus (APB). This core chip contains a fully configurable external bus memory (up to 64 Mbytes addressable memory), an interrupt controller (16 priority levels) and a DMA controller. It contains also all the necessary signals for standard emulation (bond-out chip) and debug (JTAG).

The second chip is a "Universal Peripheral Chip" including general purpose modules, both Digital and Analog, including Timers, USARTs, SPI, Watchdogs, PWM, I²C, Captures, ADC, DAC, Op Amps, Comparators, …

The third chip is an "Automotive Peripheral Chip", including dedicated modules such as CAN, LIN, J1850, Bluetooth, GPS, 802.11b, GPRS, MOST, …

All the modules of the Universal and Automotive peripheral chips are designed to be totally autonomous, having their own power management controller, as well as their own DMA and interrupt lines. They can be connected and disconnected to the core chip as needed, according the application needs. All these modules are duplicated several times, in each peripheral chip. Each module can be viewed as an independent block of the future ASSP. They are certified bug free, since they have already been well tried on several already existing different applications.

These three chips are manufactured in the same technology as the final ASSP, which guarantees the functionality as well as the timing.

For customs modules, non-available in the two peripheral chips, a four million gates FPGA is implemented on the platform with the same core interconnection as the peripheral chips.

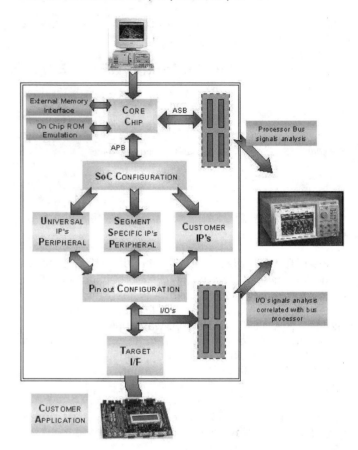

FCM® block Diagram

To interconnect all the blocks and provide to the user a real overview of the final ASSP, as well as providing powerful debug tools, a board has been developed. This board allows the configuration of all the connection between the core, the peripherals and the FPGA, and the connection of the different applicative modules (e.g. USART, Timer, CAN, …) on the user's application board. Indeed, the FCM® platform have an in-situ plug which allows the replacement of a non-existing chip directly on the user's development. The FCM® board interconnections are implemented in a way that allows the emulator to be configured or re-configured as needed in a few seconds, either to change the mix of peripherals (complete new ASSP configuration) in the chip being emulated and/or to change the pin-out of the device so as to optimize the PCB lay-out in terms of cost and EMI/RFI compatibility.

Mictor connectors are provided for a Logic Analyzer to be plugged-in, for the internal bus (ASB) and peripheral bus (APB) connection, in order to make a non-intrusive real-time trace of the code executed by the core. Many other connectors are connected to the internal signal (DMA signal and interrupt) and I/O pins. In total more that 300 signals can be spy and correlated with the source code.

In addition to all those emulation functionality's, the FCM® platform also offer a stand-alone mode. It has its own configurable clock (between 1 and 50 Mhz) and its own applicative memories. A first memory is used for the emulation of the on-chip ROM (1 Mbyte), with the same read access. The second is an external memory, for standard applicative software (1 Mbyte Flash and 2 Mbytes SRAM).

FCM® design flow and methodology

With the FCM® platform, the FCM® development flow changes the whole development process : the critical path is no longer the chip design (as in a normal ASSP development) but the application and software development.

The first step of the development flow is to define, according to the system specification, the preliminary ASSP specification, with a list of features, functions and modules, knowing that the FCM® platform helps validating some assumptions and tuning the specification in a real hardware environment. This work, the most important in the FCM® design methodology, has to be done by both parties with a System approach, in order to find the best compromise between functionality integration, cost, technology process and performance constraints. This is t_0.

The second step is the verification of the match between the ASSP Specification and the FCM® modules library. If required by the specification, new IP's are developed and implemented in the FCM® platform with a fast and secure validation process. Indeed, the new IP's are implemented in the FPGA and validated with the other elements of the platform (linked with the core chip, real-time trace through the Logic Analyzer, ...) In the majority of cases, the FCM® library matches 100% with the ASSP specification. If a new specific IP is required in the application, the implementation will need between 1 to 3 months, depending on the complexity of the new IP. This is t_0 +1 week.

In third step, the use of the FCM® resources (core, functions, modules, macrocells.....) available in the FCM® platform (Universal Peripheral Chip and Automotive Peripheral Chip), allows the immediate setting up of the chip configuration according to the application needs. The result is that, the exact "chip representation" tailored to the application comes out from the platform, as it will be configured to have the same performance and the same functionality as the final ASSP. This chip configuration set up takes less than 1 week to be complete. It's now t_0 + 2 weeks.

At this point in the flow, the user now has in his hands an immediate emulation tool of his future chip that meets his specification. Then, it is time to perform the configuration of the pin-out according to the application needs and constraints (easier production and test, better EMC / RFI performance...), in order to obtain a form factor pin to pin compatible with the application chip which then can be plugged via a target interface into the final application board/system. This takes 2 weeks including the shipment time. It's now t_0 + 4 weeks.

At the fifth step, the complete development, test and debug of the application can be made through the emulation platform. Indeed, with the FCM®, the development process is dramatically accelerated. Assuming the customer has already started his application software development he his now ready to download his software routines into the FCM® and start the software debug and validation in real time using the FCM® built-in Logic Analysis and trace capabilities, in conjunction with a Logic Analyzer. Therefore, after 4 weeks, the "future ASSP chip" is connected to the application hardware and the possibility is offered to run, debug and trace intensively the application software in the real hardware environment of the end application. All the elements are at that time of the process available to fully debug and validate the application software and hardware. It's now t_0 + 12 weeks.

Then, once (and only when) the FCM® end-user is confident enough that the ASSP specified for the application is well validated on the FCM®, in the application board, with the application software, there is then a ASSP design assembly of the different bricks (IP's) and a release of the chip to silicon fabrication. This flow allows a secure development process, and ensures that the ASSP will be right the first time. The development flow is now ended, at t_0 + 32 weeks.

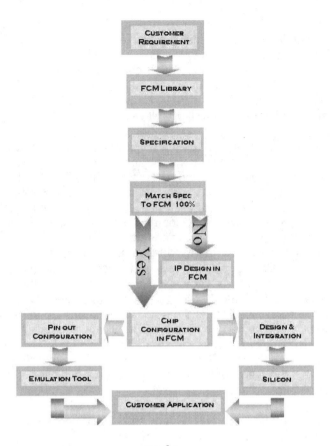

FCM® flow

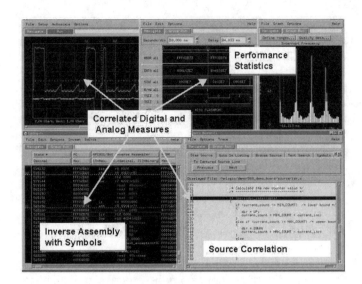

Powerful debug capability throughout the Logic Analyzer

With FCM®, in addition to the acceleration of the development flow, the user can introduce in his product development process several additional test and validation steps which were impossible to perform before. First of all with the possibility to correlate the test at the RTL and software level according to the HDL simulation models the same as the FCM platform modules. This allows the evaluation of the different module reactivity in a real application environment (a physical model in the same technology as the final ASSP is applied right from the start). Secondly, the non monolithic architecture of the platform (The core and peripherals chip are separated in different packages) allows to visualize internal signals such as Interrupt lines, DMA lines, some internal clocks, … Those signals can not be observed on a standard chip neither with a simulator. Thanks to the Logic Analyzer, the code executed by the CPU can be correlated in real-time with the C source code, in a non-intrusive way (i.e. without stopping the CPU, useful for the debug of the interrupt sub-routines and boot). In coupling the Logic Analyzer with a scope functionality, the FCM allows a unprecedented debug ; a simultaneous, correlated and interactive debug of the assembler trace is obtained, such as C source code, internal signals, digital and analog events on all the pins and serial protocol analysis.

This increases dramatically the security of the development, whilst reducing significantly the development cost and the time-to-market.

Value-added component

The automotive *easy*SoC™ concept provides either a hardware and a software environment : through the standardization of the low level drivers (CSP or BIOS) between the different ASSP developed, the integration of many software IP's (J1939, J1705 / J1587, LIN software layer, KWP2000, …), and also, from a higher level overview, by the RTOS adoption like OSEK / VDX, a reference in the automotive world.

The software tools are a complete set of tools integrated to the *easy*SoC™ concept to simplify the use of the platform. Indeed, the entire software environment can be configured simultaneously to the board. This offers to the user a physical overview of the chip as well as a software overview (debugger including all the registers, the memory configuration, …).

The *easy*SoC™ concept has a second source supply policy, which includes a privileged supply agreement with a few partners selected for their technologies, manufacturing process and reliability as worldwide semi-conductor supplier, as well as an actual possibility (built in the FCM®) to retarget a second source of supply when it is nece ssary for the customer support.

Implication of the *easy*SoC™ concept

Why Fast and Foolproof ?

The current state-of-the-art in terms of new microcontrollers development is still far from car and equipment manufacturers care-about. Pure simulation tools are now fully available but real emulation tools are only partially available or do not meet actual developers requirements, and the gap is even larger as regards emulation in the final application environment.

Standard cycle time
Using existing IP block

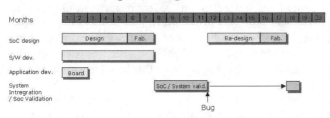

Minimum development time : 18 months

Standard cycle time
When integrating new IP block

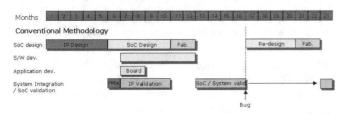

Minimum development time : 23 months

At the overall architecture level, the FANTASTIC platform made of a fully compatible set of boards, chips and software offers:

- Fast configuration of real-time emulation and fast prototyping of new ARM based ASIC chips dedicated to automotive networking applications: **development time reduction**

- Fast and intensive debugging of new IP or hardware becomes dedicated to automotive networking applications: **secure and predictable results**

- Fast development and intensive validation of ARM based ASIC software in hardware environment close to final application before availability of customer's application board: **reduced time-to-market and development cost**

*easy*SoC cycle time
Using existing IP block

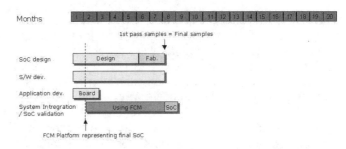

Development time : 8~1/2 months

*easy*SoC cycle time
When integrating new IP block

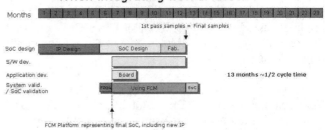

Development time : 13~1/2 months

The adjective that qualify the best the FCM® cycle time is certainly "simultaneous". The FCM® allows the possibility, as soon as the beginning of the new product development flow, to run simultaneously the three main steps or phases which are usually done in sequence in the development process : the EMULATION (of the new chip), the SIMULATION (or design and verification of the new chip) and the APPLICATION development (i.e. the development and the debug of the application software and hardware).

Continuous improvement strategy

The *easy*SoC™ concept has been developed in order to respond to the current but also future trends of the automotive market. Indeed, as this concept is open and can validate new technologies and new IP's, the users are able to test and validate new functionality in a short development time and in a secure way.

The *easy*SoC™ concept allows then to define and develop one FCM® platform for each targeted application, including the right ARM-core (ARM7TDMI, ARM946, ARM922, ARM926), the right peripheral (including software drivers), the right IP Modules (as CAN, J1939, Bluetooth, GRP, GPRS) and the OS support. This was done with the *easy*CAN® FCM® platform, dedicated for the body control and dashboard applications, and with the *easy*MEDIA FCM® platform, dedicated for the Multimedia and Telematics applications.

The targeted platform will then allows Europe Technologies partners (OEM and Car manufacturers) to derivate all needed ASSP with a full system integration (peripherals and IP integration), an emulation tool available at the project starts (for software and application validation), in a very short cycle time with very high chances of first pass success.

The continuous improvement strategy can be separated in two ways. The first one is technology driven ; new up-to-date technology processes are included in the improvement strategy of the platform, such as geometrical process (0.35µm, 0.25µm, 0.18µm, ...) as well as technology process (CMOS, Flash, E²PROM, Embedded, ...). The second improvement strategy is customer driven. In fact, new automotive IP's needed by the car manufacturers will be implemented, tested and validated in the FCM® platform through the FPGA. After validation, these new IP's will then be implemented in the market specific peripheral chip (*easy*CAN, *easy*MEDIA).

CONCLUSION

Some of the significant conclusions that may be drawn from this paper are the following :

- **Short development time** (Time-to-market)

- **Predictability** (security)

- **Reusability** (Control of own Destiny)

In the automotive market, these three key elements are essential to success in a long-term strategy. The configurable architecture and the flexibility of the FCM® allows the early availability of the three key development environments for an ASSP : the Simulation, the Emulation and the Application (the SEA) environment.

Simultaneous steps and concurrent development, thanks to the SEA, shorten strongly the development cycle time.

The FCM® built-in special validation features and the deep coupling with a logic analyzers allow unprecedented intensive, efficient and fast software debugging and hardware verification and makes the results of the new product development flow steps fully predictable, performance, time and cost wise.

Thanks to the intensive debugging and validation capabilities offered by FCM® and to the iterative process between the three environments of the SEA, FCM® secures the project planning : this is the predictability.

With FCM®, users shift from developing one application to developing a family of automotive derivative applications which allows a maximum reusability of the hardware and software IP's. When a new application becomes a derivative or a new configuration of the initial ASSP, its development is then an easier, smaller and cost-effective effort.

Hence, with the *easy*SoC® concept, the user has a high influence on the integration of his system on a chip ; he masters more his system architecture and future evolutions and therefore has bigger control of his own destiny, the aim in an automotive manufacturer strategy.

ACKNOWLEDGMENTS

The authors wish to express their gratitude to the following people for their continued encouragement and support : Rob Mathews of Europe Technologies, Rémond Sautreau of iNOVANDI.

CONTACT

Tristan Bonhomme
FCM Program Manager
Tel : +33 497 15 20 00
tristan.bonhomme@europe-technologies.com
www.europe-technologies.com

Nicolas Demoulin
Marketing Product Manager
Tel : +33 497 15 20 20
nicolas.demoulin@europe-technologies.com
www.europe-technologies.com

DEFINITIONS, ACRONYMS, ABBREVIATIONS

ADC : Analog to Digital Converter
APB : Advanced Peripheral Bus
ARM : Advanced Risk Machine
ASSP : Application Specific Standard Product
ASB : Advanced Serial Bus
CAN : Controller Area Network
CSP : Chip Support Package
DAC : Digital to Analog Converter
DMA : Direct Memory Access
FANTASTIC® : Fast Automotive Networking Through ARM-based System, Tools and Integrated Circuits.
FCM® : Fast Configurable Microchip
GPRS : General Packet Radio Service
GPS : Global Positioning System
HDL : Hardware Description Language
I²C : Inter Integrated Circuit bus
IP : Intellectual Property
KWP-2000 : KeyWord Protocol 2000 (Automotive Test protocol)
LIN : Local Interconnect Network
OEM : Original Equipment Manufacturer
Op Amp : Operational Amplifier
OSEK : Offene Systeme und deren Schnittstellen für die Elektronik im Kraftfahrzeug
PWM : Pulse Width Modulation
RTL : Register Transfer Level
SoC : System-on-Chip
SPI : Serial Peripheral Interface
USART : Universal Synchronous Asynchronous Receiver Transceiver
VDX : Vehicule Distributed eXecutive

Implementation of Fault Detection and Diagnostics in an Intelligent Vehicle Electrical System

Mark A. Thurber and **Marilyn R. Miars**
International Truck and Engine Corp.

ABSTRACT

With the added complexity of vehicle multiplexed systems, the task of accurately diagnosing vehicle electrical failures is daunting at best. Integration of smart controllers into the electrical system provide the added flexibility and capability of detecting and monitoring operating conditions. This paper discusses one approach to detecting and displaying vehicle electrical diagnostic information either directly to the driver/technician via on-board driver information systems or indirectly via off-board service tools.

INTRODUCTION

There are many advantages to vehicle multiplexed systems. One of these advantages is the ability to communicate system data efficiently. This data can include diagnostic information about various vehicle systems. The purpose of on-vehicle diagnostics is to capture, record and present the diagnostic data to the driver/technician.

Detecting a fault within a multiplexed environment requires advanced planning in developing an organized structure, which allows one to extract pertinent information. This leads to better and more accurate recording/reporting of faults, which in turn aids in finding and solving problems within this complex environment.

FAULT DETECTION AND RECORDING

FAULT TYPES

There are two types of faults, active faults and previously active faults. Newly recorded faults are considered active faults. Active faults exist as long as abnormal conditions have been detected within a system. An active fault will remain active until it becomes previously active. When a fault condition no longer exists for a given parameter, message, or event, the fault is automatically set to previously active. Previously active faults show that a problem had occurred but currently the conditions have been corrected.

RECOGNITION AND RECORDING OF A NEW FAULT

Fault recognition must be reliable to be of any use to the customer. If the fault detection system is too sensitive, e.g., reports a number of unstable faults that do not indicate any real failure in the vehicle, it will erode the customer's confidence in the system. Among the causes of these spurious faults are over sensitivity of the fault detection and monitoring of the system when conditions are such that it is electrically unstable.

Determining when we have a stable fault - For each of the systems that are monitored or controlled by the electrical system controller there is a portion of the software that is responsible for recognition of faults within that system. Recognized faults fall into three basic categories:

a. A measured parameter outside the established limits

b. A message or event that is not seen when expected

c. An unexpected message or event.

Faults that are measured parameters may not be stable enough to record on this instance. In this case the responsible software component may use one of the following methods to determine if the fault is stable prior to setting the status for that parameter to the fault value. A status is kept for each parameter, which will indicate whether a fault condition is present or not. The most common method of determining the stability of a fault is the debounce method. Similar to "debouncing" a switch, the software component will look for a number of measurements in succession that are out of limits prior to setting the status to the fault value. Another method that is used for slower, less critical parameters is the running average. Where the current parameter measurement is used in the calculation of a "running" average. When the "running" average then is out of the established limits, the status for that parameter is set to the fault value. The third method available, for determining the stability of a fault, is a percent of sample where a history of some defined number of samples is maintained and when a fixed percent of these samples is out of limits the status is set to the fault value.

When an expected message or event does not occur "on time" or when expected, there is no checking for the stability of the fault, but the latency time allowed before the event is considered "missed" can be adjusted. This gives the system a little bit of leniency for uncontrolled system characteristics.

Recording faults - There are two methods of fault recording. The first method is for the software component that detects the fault to call the fault software directly. At this point the fault software component will determine if this is a new fault and if so will record it in the fault database. If this fault has already been recorded the fault software will not record it again. The second method is for the fault software component to check for timed out message failures. If it is a new fault it will then record it, otherwise the fault will not be recorded. It is only after the fault is recorded that it is available for diagnosis and the operator is notified via an amber light on the instrument cluster. At this point the fault is considered an active fault.

Conversion of an active fault to inactive is done by one of two methods. The first method is for the fault software to be called by the software component that first detected the fault with the parameter status set to a fault free value. The second method is for the fault software component to scan the expected messages to determine if they have been received. In either case the fault software component will then set the fault to previously active. A record of the previously active faults is kept for future reference and is available to the operator or technician.

Recognition of conditions when faults should not be recorded - During the initial development and testing of the new fault detection system we experienced a number of faults that would occur, be recorded and immediately be set to inactive. This resulted in a disproportionate number of previously active faults being stored on the vehicle and an almost constant amber warning to the operator when starting the truck. Analysis of the conditions that lead to these faults occurring in the first place indicated that there were repeatable conditions where these faults would occur. These conditions are the initial power up of the system, during engine crank and when the battery voltage is low. There were several possible solutions. One solution would be

to prevent the faults from occurring during these conditions. This was not a viable option due to the extent of the hardware and software changes required. Another solution was to prevent the various software components from setting the status or calling the fault software when these conditions occurred. This solution would have meant extensive software changes and would risk reducing the reliability of the affected software components.

The solution that we chose was to have the fault software recognize the conditions that lead to the erroneous faults and prevent these faults from being recorded. To prevent the recording of faults during the initialization, a delay timer is triggered. The fault software checks that this timer is expired prior to recording a fault. This delay timer is also restarted when the key is turned from the OFF to the RUN position. This allows time for the electronic systems to complete their initialization before any faults are recorded. During engine crank the ignition voltage drops dramatically and will recover after several seconds. We saw erroneous faults during the time that the voltage was low. We are able to sense the ignition voltage and determined that the fault software component would not allow faults to be recorded if the ignition voltage fell below 9.5 VDC. This is high enough to assure that the Electrical system controller is well within its operating range, yet well below the normal ignition voltage. When the ignition voltage recovers (is above 9.5 VDC again) the fault software component will start recording faults again.

Any fault that occurs while the fault software component is not recording faults and continues to have a status set to the fault value, after the fault software component starts recording faults, will have its fault recorded at that point. This has eliminated the spurious faults from being recorded and reduced the incidence of alerting the operator needlessly.

FAULT REPORTING

J1939 DIAGNOSTIC MESSAGE

The ESC (Electrical System Controller) maintains a buffer of the active faults. The instrument cluster has memory set aside for longer-term storage of the active and previously active faults. The information necessary for the cluster to maintain its fault record is transmitted to it via an SAE J1939 bus. The DM1 message broadcast philosophy adopted for the ESC is to send only one fault code per broadcast message. The multi-packet transport protocol is not used. When a new fault is recorded a DM1 message is sent out immediately. As long as there are any active faults recorded in the buffer, a variable period timer will schedule successive fault messages for broadcast at such a rate that all faults will be broadcast every second.

We took a 3-step approach to identifying a system issue. Our diagnostic codes for this system consist of three numbers, one byte each. The first identifies the type of system or area that consists of a problem. The second diagnostic number points to a very specific wire, signal, or component where abnormal conditions have been detected. The third number in most cases describes the exact condition of that element which is diagnosed as "out of range." Unfortunately, some codes required two bytes to identify a particular point in the system, so the third number was set up as an extension of the second.

DISPLAYING OF FAULTS VIA CLUSTER DISPLAY

At this point we have captured and recorded the diagnostic data. Now for it to be of any use it must be presented to the driver/technician. The technician may have diagnostic tools that are capable of reading the fault codes, but it is unlikely that the driver will have this available. We chose to take advantage of an LCD display in our instrument cluster and the available multiplexed switches. The combination of switches had to be one that would be unlikely to occur in the normal operation of the truck.

Diagnostic codes can be very valuable to drivers as well as service technicians if the capability of an on-board display exists. The details provided depend on how well a vehicle is diagnosed, as described in the above section. Any additional information provided to a driver in the event of a problem can help to alert them of the condition of the vehicle. A service technician can use this information not only to know the condition of the vehicle, but also to locate and correct a problem with a particular vehicle system.

In the case of a problem, the driver/operator can better choose, for instance, to operate the vehicle, diagnose and fix a problem, disable one or more features, or shut down the vehicle system altogether. Of course, fixing a problem is limited to the necessary tools that a driver has available.

DIAGNOSTICS

SERVICE BAY DIAGNOSIS

With an intelligent vehicle system, increased diagnostic capability exists from a single point that used to take place by a service technician climbing all over and under a vehicle, sourcing pins with power and ground. Now diagnostic software can utilize the communication link to simulate physical actions on the vehicle. The diagnostic

tool has the ability to "get inside" the control module, view its signals and interact with it.

For example, consider a switch that controls a light, but the light does not come on. The problem could be at the switch or at the lamp or somewhere in between there that has to be tracked down. Now that there is a control module in the middle of the switch and the lamp, due to expanding the functionality of the lamp, there should be added diagnostics. On a simple example like this, the diagnostic code should tell you what the problem is. If it were a more complicated feature, the diagnostic software would show the status of the input side (i.e., the switch) and see if it changes when turning it on and off. On the output side, the software may allow you to activate/deactivate the output regardless of the position of the switch. Turn the output on and see if the lamp comes on. If it does, then the problem lies on the input side of the controller. Many features now have duplicate inputs and possess interlocks that are done in software instead of external components. Which means the above capability of a diagnostic tool can really help assist a technician when troubleshooting a difficult or complicated case.

CLEARING OF CODES

Removing codes always leads to questions of whom, what, and how. The answers to these questions lie in the judgment of the manufacturer and the limitations that they want to expend on the vehicle system. Inactive codes can be cleared totally. Active codes could be cleared, but depending on the diagnostics, it may already be inactive or if the fault condition is still active, then it will immediately reappear, thus not being able to clear it. How and who should have access to removing codes depends on the security that you want to put into place for this.

Clearing of the codes is normally done after the technician/operator has repaired the problem. This may be done via a service tool or using predefined controls on the vehicle. The ability to clear the codes without using a service tool is necessary for situations when a repair is completed and a service tool was not used.

CONCLUSION

With the addition of new advanced electronics, the system seems to be more complicated, but incorporating state-of-the-art diagnostics into the system takes the intricacy out. Even better is diagnosing problems that couldn't be detected before and providing advanced warning of faulty systems. This creates a safety standard above all others. The ability to provide diagnostic information at quick access on-board and in further details off-board will reduce a vehicle's downtime.

Keeping a truck out of the shop more and on the road longer is going to keep customers happy.

REFERENCES

1. SAE J1939
2. SAE J1939/73

CONTACT

DEFINITIONS, ACRONYMS, ABBREVIATIONS

The words "**vehicle**" and "**truck**" are used interchangeably and can apply to any multiplexed environment.

Previously active – inactive, occurred in the past.

ESC – Electronic System Controller
DM1 – Diagnostic Message 1
APPENDIX

2001-01-2740

A Vehicle Electrical System Architecture Based on a Multiplexed Design For Operator Controls and Indicators

Robert Dannenberg
International Truck & Engine Co.

ABSTRACT

Increased features content, with electrical effects, have produced tremendous complexity in the design and support of the electrical system for commercial vehicles. Using individual wires coupled with various electro-mechanical components, to implement electrical system features, is no longer a desirable solution. Adding more wires and connectors reduces reliability and makes diagnosing electrical failures rather cumbersome. Likewise, customizing a vehicle with a specific set of features, by using discrete wire designs, make the assembly process prone to error.

This paper explores a new approach to a vehicle electrical architecture that uses multiplexed wiring methods coupled with a programmable central control module. The most novel feature of the system is that the main electrical system control module may be reprogrammed with a unique set of vehicle features as required by each customer order.

Additional benefits of the system include an ability to add more input/output capability to the electrical system by adding various general purpose circuit modules, as nodes, on the back bone of various serial data communication links on the vehicle. Likewise, information gathered from these modules may be shared with any of the other intelligent electrical system components, which minimizes redundant sensors and lowers overall vehicle cost. Due to increased communication between intelligent circuit modules on the vehicle, more of the electrical system may now be diagnosed, either with on board displays or by off board diagnostic equipment.

This new approach allows the set of vehicle features offered and ultimately much of the electrical design, to be managed from a software database, instead of through thousands of combinations of wire harness designs using "hard wired relay logic". Building upon that capability, now the vehicle order to manufacturing process may be automated to further enhance the speed and agility of the OEM to provide exactly what the customer wants.

INTRODUCTION

Many approaches to multiplexed electrical wiring systems have now been developed by the automotive industry. Due to the benefits of using serial data links to communicate control signals and sharing data between circuit modules on the vehicle, automotive OEM's have achieved increased reliability, better diagnostics and reduced system costs. However, most of these systems are very application specific, meaning that the mission or function of each module, in a particular vehicle model, is well defined and variation in functional performance is limited to a few selections of programmable parameters. Furthermore, the components installed on any particular vehicle are completely under the design control of the automobile makers. The commercial vehicle market traditionally operates with very different set of rules. Each commercial vehicle can be a very customer customized platform that must integrate components from a variety of powertrain suppliers as well as other OEM designed components or modules. Accommodating customer feature variation is an everyday occurrence.

SYSTEM CONSIDERATIONS

The development of an electrical architecture based on a multiplexed design for a commercial vehicle must have many system considerations and tradeoffs evaluated. System design choices must not be made just for the benefit of one element of the vehicle OEM. The system considerations evaluated for the architecture discussed in this paper, affected manufacturing, reliability, diagnostic time, body builder integration, feature development and lifetime vehicle support.

REDUCE COMPLEXITY IN MANUFACTURING - A reduction in the complexity of wire harness manufacturing is an important goal in the design and development of a vehicle electrical system. Literally hundreds of wire harness designs are developed and maintained to implement the various features required by customers that spanned many diverse vocations. Interlocks and complicated switching functions were designed using "relay logic". Some features could only be implemented by adding special purpose modules. The cab and chassis harnesses were getting so large, due to increased feature content, that the harnesses are difficult to package and install in the vehicle. A means to reduce harness size became a requirement.

INCREASE RELIABILITY – Poor electrical reliability is often the number one complaint for a commercial vehicle. It is not the most expensive problem, but it is the most frequent. Complex wire harness designs employing literally hundreds of connectors, terminals and wire segments, provide a strong possibility for an electrical failure. A method to reduce the number of connectors and wire segments, in the vehicle wire harness, became another requirement.

SHORTEN TIME TO DIAGNOSE – Diagnosing problems in a vehicle electrical system has always been difficult. However, increased feature content, which leads to more wires and components in a "hard wired" system, makes the job even harder. When there is no module intelligence to help pin point the problem, the mechanic must rely on circuit diagram books and personal insight to isolate the failure symptoms to determine the root cause of the problem. Therefore, the electrical system needs a source of intelligence to help the mechanic diagnose electrical problems without requiring an engineering degree to accomplish the task.

MAKE CHASSIS TO BODY INTEGRATION EASIER – Many commercial vehicles are only partially complete when a truck OEM manufactures them. Vehicles must travel to another facility or company to have a specialized body installed to make it function in the vocation that it is intended. These "body builders" have many electrical requirements as they mate their bodies to the truck chassis. The body builder also encounters customer variation in the equipment that they supply. In this competitive market place, it is incumbent that the truck OEM needs to make integration with body builder equipment practical. Thus, making a particular truck, the chassis of choice. Therefore, a new electrical architecture needs to employ a means to provide easy interfacing to the vehicle chassis electrical system.

IMPLEMENT NEW FEATURES QUICKLY– The demands for more truck features are constantly requested from vehicle customers. Driven from a need to run their businesses more efficiently, customers need more features to make the commercial vehicle a better tool to make money for their business. Requests range from a desire to capture more information about the operating efficiency of the vehicle to better diagnostics for the electrical system to making the vehicle safer to operate as well as comfortable and convenient in the hopes of increasing driver retention.

PROVIDE A SYSTEM THAT IS UPGRADABLE IN THE FIELD – Though truck OEM's manufacture vehicles with a broad variety of options, many factors influence the need to be able to revise or upgrade the truck once it has left the factory. Customers change their mind about feature content after the vehicle is built. Dealers buy "stock trucks" to place on their lots so they will have a vehicle on hand that could be slightly customized if a customer has an immediate need for a vehicle. Body builders also customize stock trucks since they often will buy a common vehicle configuration, knowing that they can modify each truck to meet very different end customer requirements. So, it is a very important requirement that the vehicle electrical system architecture must be capable of field modifications or upgrades.

PROVIDE A SYSTEM THAT IS MAINTAINABLE – Any vehicle electrical system needs to be maintainable. In the past, this was accomplished by publishing electrical circuit diagram books and by supplying replacement parts and wire harnesses through Service Parts distribution outlets, such as is provided at dealerships. However, as electrical designs get more complex, the logistics of providing a customized wire harness is very difficult and expensive. Often, a cab harness may be constructed from a base harness and then, more than a dozen additional wire harness overlays are added to complete the required features. Supplying all of these piece parts in Service Parts is very complicated. Maintaining dozens of special purpose modules over the life of a vehicle is also cumbersome and expensive. If any element of a new electrical system design should employ modules that are programmable, then that introduces another level logistics that must be maintained and manageable.

SELECTED ARCHITECTURE

All of the above system considerations should be analyzed as an electrical system with a multiplexed design is developed. Many tradeoffs cause compromises in each of the areas considered. We will now explore the selected electrical architecture. Describing all the elements of the system is not a simple task. So, the following discussions shall be focused on the topics of Network Topology, describing the central control module, describing the purpose of the optional slave modules and finally providing a description of the control and communication strategies used on each data bus.

NETWORK TOPOLOGY - The chosen electrical architecture relies heavily upon a variety of serial data links. In the spirit of achieving system compatibility, now and in the future, SAE defined serial data links were employed throughout the design. See Figure 1 for an overall view of the complete electrical architecture that depicts a fully optioned vehicle. The first part of the design shows that the existing vehicle J1708/J1587 data link is present in this design. This data link has been used since the mid-1980's for diagnostics and programming of the major powertrain components. Later on, this data link has been used to supply various forms of gauge data and other information for use with gauge clusters and display modules. This link was only intended for sharing information and not intended for control purposes. The useable bandwidth of this data link is now fairly limited due to large volumes of information that are communicated on this link by various powertrain suppliers. Therefore, though this link is maintained in the over all design of the electrical system in support of legacy diagnostic and programming processes, it is not an integral part of operating strategies in the new electrical system architecture.

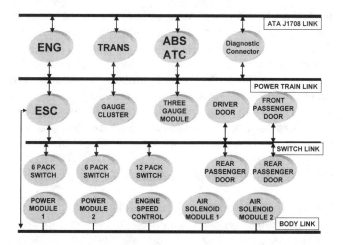

Figure 1: Fully Optioned Electrical System

The second major data link in the multiplexed electrical system is the powertrain SAE J1939 data link. This 250K baud data link has now been implemented by most all of the powertrain component suppliers. It is about twenty five times faster than the SAE J1708 data link and has the bandwidth to support the distribution of control data as well as informational data. This link connects the engine, transmission, anti-lock brake system, primary gauge cluster, secondary gauge cluster and a diagnostic connector, all on one media. This data link provides an interface connection for modules designed and/or managed by the truck OEM. It is intended that body builders and after market devices should not be connected to this data link. The successful operation of the vehicle depends upon the integrity of this data link. It

was desired that no other entity should be tapping into this link using unsupported wiring methods or bearing the effects of unproven functional operation from after market modules that might degrade the performance of this data link and likewise, the performance of the vehicle.

The third major data link is the switch data bus. This is a separate data link that is employed to communicate the status of various rocker switch assemblies and door module assemblies with a central control module. This data link is an SAE J1708 type data link that uses the proprietary PID 254 messages as well as standard J1587 data link messages to provide two-way communication between each module on the data link. A J1708 data link was chosen because of lower harness and connection costs for the vehicle. The powertrain J1939 data link was considered for this segment of the data bus, but was judged as being too expensive for the application. It was decided to stay with slower J1708 data link. In addition, if there are any faults on this data link, the powertrain J1939 data link will be unaffected and vice versa.

The fourth major data link in the system is another SAE J1939 data link. This link is referred to as the Body Data Link. This data link segment is primarily used to interface general purpose multiplexed modules with the central electrical system control module. This data link is to be the media that body builders would use to interface add on modules to the vehicle chassis with the express purpose of extending the electrical system input / output capability in a manner that is suited to their particular application. Therefore, the system may be customized specifically to a customer need, while providing the most flexibility at the lowest system cost.

CENTRAL CONTROL MODULE – As all of the system considerations and requirements were analyzed, it became apparent that one or more intelligent control modules were needed. Many variations were considered for the system. It was of course possible to place multiple control modules throughout the vehicle as standard equipment. However, there are cost trade-offs, which have to be addressed in a centralized – de-centralized architecture design. A compromise was chosen to design a central control module that had enough inputs and outputs to perform all standard vehicle electrical functions, plus about twenty percent of normal regular production options. Every input and output was rationalized for inclusion into this central control module. The main requirement for a feature to be designed into this the control module was that the feature must require some sort of logic processing. Therefore, features like the cigar lighter or back up lights was not included in the features performed by the central control module. A more detailed description of this module may be found in the MAJOR COMPONENTS section.

SLAVE MODULES TO EXTEND I/O – As the system requirements were documented, it became evident that a single central control module would not satisfy all customers' needs. A single central control module would be too large and too expensive. Furthermore, it was observed that adding additional central control modules, as the input/output requirements of a vehicle exceeded the capabilities of one module, was not viable either. Studying the feature content on present production, it was obvious that specific modules should be developed to best suit major groupings of features. A brief highlight of each module is provided in the MAJOR COMPONENT section.

COMMUNICATION & CONTROL METHODOLOGY – So far, we have seen that the proposed electrical system has four major communication data links, a central control module and a variety of other intelligent modules connected to these links. The powertrain modules have predefined control and communication methodologies. Each of the powertrain modules typically communicates in an independent peer to peer fashion. Though in some instances, there are closed loop control methods used between powertrain modules to perform special features like traction control or shifting of a transmission.

The remainder of the electrical system operates in more of a master / slave type configuration. This was done for a variety of reasons due to system requirements. The first major system requirement was that a broad variety of functions must be operable with the ignition key off. This by itself is not difficult. However, when there are multiple modules on the data link and any data link traffic is capable of waking up any other module from an inactive "sleep mode", then it is a challenge to perform an orderly shut down of the system when the ignition key is off. A central control module provides the ability to terminate data transmissions with slave modules when a system shut down sequence is desired. Slave modules have the ability to request a system start up at any time, but the central control module determines whether the request will be honored. Since the central control module logic is implemented with re-programmable feature software logic, the system start up and shut down scenarios can be modified to meet varying performance requirements.

Another requirement that drove a master / slave architecture was the desire to have the central control module detect when any of the other modules on the data link are either inoperative or have lost communication. Therefore, system operates with the central control module sending a command message to each respective remote module and then a response status message is expected within a specified time period. If communication is lost with a remote module, a diagnostic fault is logged in the central control module.

A third reason for a master /slave system was due to a requirement to minimize the quantity of programmable databases in the overall system. As stated before, the central control module is the heart of the electrical system. It was decided that there would be only one programmable database in the electrical system, excluding the powertrain components. The design controls of those modules are placed in many different companies. Therefore, the central control module was designated as the master controller for all modules except the powertrain.

In support of limiting the system to one programmable database, the central control module is tasked with providing configuration data to other semi-intelligent modules in the system. The central control module is designed to execute a "teach routine" for semi-intelligent modules such as the gauge cluster. In this case the gauge cluster is taught where each gauge is located in the cluster and which warning lights or information displays are to be enabled. Since all of the configuration data relevant to remote modules are handled by the central control module through a vehicle specific software database, the remote modules do not need DIP switches or some other Electro-mechanical switch contacts to establish the functionality of the module. This design approach reduces cost and increases reliability.

In addition to the teach data, the central control module has a vehicle specific, custom software program that identifies which modules should be present on each vehicle. It also has a list of features that are to be executed for that vehicle configuration. It performs the features pertinent to it's own inputs and outputs, but also issues commands to remote modules as well as processes status messages from all the remote modules. A detailed description of the control module software architecture is found in the CUSTOMIZING THE SYSTEM section of this document.

The proposed electrical architecture is extremely flexible. From previous discussions, it is obvious that in a base form, just a few modules will perform all of the standard electrical features of a commercial vehicle. Yet the input / output capability of the system can easily be expanded by attaching more specific modules to various data links and reprogramming the central control module to perform the necessary additional features.

Therefore, a master / slave control system is very effective when all of the above system requirements have been considered. Complete electrical system flexibility is achieved through a single software database. Changes in vehicle functionality can be implemented from one truck to the next by changing the software instead of changing multiple wire harnesses throughout the vehicle.

MAJOR COMPONENTS

ELECTRICAL SYSTEMCONTROLLER – The electrical system controller (ESC) is the central control module for vehicle electrical system.

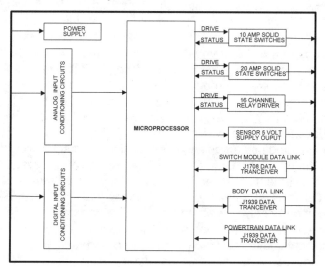

Figure 2: ESC BLOCK DIAGRAM

The processing capabilities of this single module include two J1939 data links, one J1708 data link, multiple analog & digital inputs, low current low side driver out puts and two current levels of high side solid state switch outputs. See Figure 2 for a block diagram of the ESC. See Figure 3 for a summary diagram of the functions directly interfaced to the ESC. The control program software for the module is stored in flash memory.

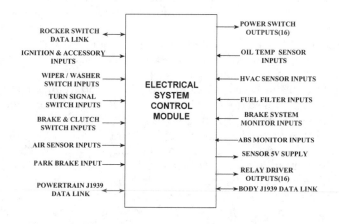

Figure 3: ESC Interface Functions

GAUGE CLUSTER – The gauge cluster is a node on the powertrain J1939 data link. It consists of a single module that can be populated with up to ten individual gauges, thirty-one warning lights and a digital information display. All gauges and all but four warning lights are driven via the J1939 data link. The gauge cluster communicates the status of three rocker switches as well as other system information to the ESC through the data link. The gauge cluster is the primary storage and display location for system fault codes.

AUXILIARY GAUGE & SWITCH PACK – An auxiliary gauge pack provides a means to display three additional gauge functions and houses up to six additional rocker switches. All gauges and switches are interfaced to the vehicle through the powertrain J1939 data link.

ROCKER SWITCH PACKS – The rocker switch packs are a multiplexed module design for what appear to be traditional rocker switch assemblies. However, the rocker buttons are actuating micro switches soldered on a printed circuit board and the status of the switch elements are reported on a J1708 data link to the ESC. Each switch location employs LED back lighting for nighttime viewing. In addition, the top section of each switch has the capability to display an indication of ON as dictated by the feature configuration software in the ESC. Rocker switch assemblies are individually replaceable and may be installed in any location. However, the ESC configuration software must match the location of installation of each respective rocker switch. The switch packs are designed in single DIN and double DIN housings.

REMOTE POWER MODULE – The remote power module is an optional device that contains six high side solid state switch outputs and six remote switch inputs that are active both high and low, since each input is biased to one half of the vehicle battery voltage. The unit is packaged in a potted enclosure with weather sealed connectors. It is suitable for installation anywhere on a vehicle. See Figure 4 for a block diagram of the remote power module.

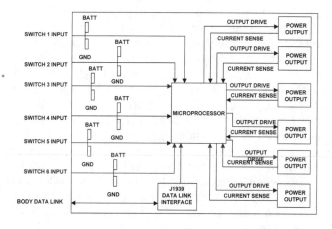

Figure 4: Remote Power Module Block Diagram

The remote power module receives commands from the ESC to control the state of each high side, solid state, switch output. Each output channel is capable of supplying twenty amps. A programmable fuse setting is implicit in the command message for each output.

The remote power module reports the status of each tri-state switch input for subsequent processing by the ESC. In addition, the current flowing in each power output is reported back to the ESC. The initial production release of the remote power module feature allows a bank of six rocker switches in the cab and six remotely mounted body switches to control each solid state switch output respectively, in a three way switching fashion. However, the functionality of controlling the outputs or processing the switch inputs may be re-defined through a different software feature installed in the ESC. The remote power module is designed with maximum flexibility for diverse usage in the future.

All remote power modules are identical. When installing multiple remote modules on a single vehicle, wire harness jumpers on a module connector determine the module address. No programming is required when a remote module is replaced in the field.

REMOTE AIR SOLENOID MODULE – The remote air solenoid module is another optional multiplexed module. It was developed to provide control of air accessories without routing individual air hoses into and out from the cab. Capitalizing on the multiplexed capabilities of the rocker switch packs and the ESC, the remote air solenoid was a natural addition to the family of multiplexed modules. The module has a J1939 data link interface for control of up to seven air solenoids. An air source is connected to the first solenoid in the pack. Additional solenoids are ganged in a row and mounted to a mounting base plate. Each solenoid has a connector-less air hose fitting for connection to the air accessory to be controlled. Messages from the ESC command each respective air solenoid to be opened or closed as dictated by the software feature logic running in the ESC configuration data. Rocker switch inputs, plus other interlock inputs, may be assembled in any combination in an ESC feature before an air solenoid is opened or closed. Up to two, air solenoid modules may be installed on a vehicle by resetting a single address jumper on the base plate of the air solenoid module. This provides the capability of controlling up to fourteen air solenoid accessories from the multiplexed architecture. The sum total of all the features of the air solenoid module, provide a very advanced and efficient solution for controlling vehicle air accessories.

REMOTE ENGINE SPEED CONTROL MODULE – The remote engine speed control module provides the ability to remotely control engine speed in multiple modes, plus monitor specific engine warning lights and engine data such as engine RPM or vehicle road speed. In the past, this interface was available from the engine ECM directly in a hard-wired fashion. In the future, such hard-wired interfaces may be removed from the engine ECM and this module shall perform this functionality.

DIAGNOSING THE SYSTEM

The multiplexed architecture described so far is very flexible and yet also powerful in providing customer features in a reliable and cost-effective manner. However, the system is complex such that features and functions are performed using software with microprocessor modules, instead of using hard-wired relay logic. Many signals are being communicated over a few wires by way of the various data links previously described. These signals cannot be measured or evaluated using traditional test equipment such as a test light or a volt Ohmmeter. Therefore, it was a system requirement that a diagnostic capability must be designed in from the start. The diagnostics must be comprehensive with minimal false indications. The diagnostics need to be simple to use and provide status indications from all module nodes on each data link. Two varieties of diagnostics were developed. One is "on-board" that provides encoded fault codes on a digital display. The second is off board diagnostics that provides a much more in depth analysis using a personal computer platform.

ON BOARD DIAGNOSTICS - On board diagnostics were developed as a quick go-no-go indication to display the status of the vehicle's electrical system. Diagnostic fault codes may be viewed on the multi-purpose digital odometer display in the gauge cluster. A single warning light in the gauge cluster alerts the driver or mechanic that system faults have been logged into an EEPROM location. Faults are stored as being either active or previously active. The diagnostic mode may only be initiated as long as the vehicle is parked with the park brake set. The input that initiates the diagnostic routine employs specific switch actuations of existing switch mechanisms. As an example, pressing both the Cruise On and the Cruise Resume switches at the same time will start the diagnostic mode. Encoded numerical fault codes may be interpreted from look up tables in a diagnostic service manual. Actuating the left turn signal switch and the Cruise On and Cruise Set switches pressed at the same time clears fault codes. The act of initiating the diagnostic mode for the multiplexed electrical system components also sends the appropriate command to the major powertrain components to initiate a self-test sequence. In response to this request each powertrain component provides an on board using flash codes on the respective warning lights of each system. The benefits of the on-board diagnostics are that they are quick and easy to use and that no special tools are required to perform a very detailed diagnostic check out of the vehicle.

OFF BOARD DIAGNOSTICS – Off board diagnostics were developed to help the mechanic diagnose the electrical system more easily. Using a personal computer platform, the mechanic can view text descriptions of faults instead of looking up numerically encoded fault codes. Furthermore, the status / values of inputs and outputs for all signals configured in the system may be viewed. If desired, the mechanic can over-ride an input or an output to test the integrity of the component or wire harness under test.

COMMON SENSE – Diagnosing an electrical system requires that the mechanic must exercise a certain amount of common sense. Whether the system is a simple hardwired design or a complex multiplexed module design, the mechanic needs to use good troubleshooting methods to help isolate the problem. The proposed architecture provides detailed diagnostic detection capabilities, but it requires that certain electrical requirements be met. For instance, we have to assume that all circuit modules have reliable power and ground connections. We have to assume that the integrity of the data links is solid and that all wiring rules for stub lengths and data bus terminations have been properly followed.

It should not be necessary to get out personal computers to run diagnostic programs because a light bulb is out. Common sense should be exercised by merely inspecting the light bulb first, instead of probing connectors and wiring harnesses for output voltages. A non-functioning gauge does not mean that you should replace the gauge cluster. The gauge data may have been supplied from some other module from some remotely located sensor, which has been formatted and transmitted across one or more data links. Also, it is not wise to blame every problem on the computer box or the software because they are easy targets to replace. Therefore, detailed training should be developed to provide the proper insight and processes for the mechanic to use in troubleshooting a multiplexed system. A thorough understanding of the system will help the mechanic to make the best decisions about diagnosing system problems.

CUSTOMIZING THE SYSTEM

The proposed electrical architecture has proven to be very flexible in implementing customer features. The architecture relies on ESC configuration data to define gauge function and location, switch function and location, functional assignment of discrete analog and digital inputs, remote module usage and definition of the type of powertrain components expected for a given vehicle. All of these configuration data selections are managed in a software database instead of using jumpers, DIP switches or some other combination of unique hardware to define the function of module. The ESC contains a microcontroller based design that has re-programmable memory organized into specific functional segments. The ESC memory is organized with a boot loader section, a core program section, a feature configuration section and a programmable parameter

section. See Figure 5 for a diagram of the ESC memory segmentation. All of the segments are non-volatile FLASH memory except the RAM section.

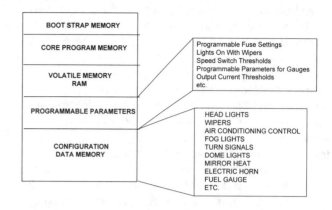

FIGURE 5: ESC SOFTWARE SEGMENTATION

The boot loader section provides enough intelligence for the remaining software sections to be programmed through the powertrain J1939 data link. The core program contains the main operational program allows the module run the basic I/O functions and access programmable parameters and perform customer features described in the configuration data. The programmable parameter section contains variables that may be modified by off board programming tools to change the performance of specific features. For example, the trip point of an electronic fuse setting may be adjusted with these parameters. The configuration memory section contains the software to perform all the standard and customer specific features for the vehicle. Only features pertinent to the vehicle are loaded into the ESC. This section is compiled automatically for each vehicle by the order to delivery information technology systems.

Therefore, generic multi-function modules may be reliably configured with maximum usage of the available module resources.

AT TIME OF VEHICLE MANUFACTURE – Knowing that the multiplexed architecture is open and flexible, it was necessary to develop a means to automate the order to delivery process, so that hundreds of vehicles could be programmed each day at a vehicle OEM. A software database and a programming infra-structure was developed to automatically convert individual customer sales features into a summary file. Next these features are linked to pre-engineered software feature files that are complied into a single configuration file for each truck on the fly as it comes down the assembly line. Since there is only one programmable module in the body and chassis electrical system, managing one file is very straightforward. The program memory of the ESC contains flash memory so revisions anywhere in the supplier chain or the OEM assembly plant can be accomplished in just a few minutes.

CHANGES TO THE SYSTEM IN THE FIELD – The system does not require that all features need to be present at the time of vehicle manufacture. Dealers order commonly equipped stock trucks that will be available for immediate sale as long as they can be customized for a particular application quickly. Body builders order trucks with common feature content, knowing that they will need to modify the trucks to meet their customer's specific requirements. The proposed electrical architecture provides the ability for upgrades in the field. Since the software is organized with a core software program plus pre-engineered packets of feature logic released as configuration data each feature, a service technician can add or delete software features from a list of pre-engineered features for a given vehicle and then re-compile the configuration database. The newly compiled file may then be down loaded into the ESC through the main diagnostic connector.

MAINTAINING THE SYSTEM

It was recognized very early that the configuration of each vehicle had to be maintained in a robust location. The configuration data could have been merely assumed to be stored in the ESC for the life of the vehicle, but all the data would be lost if the ESC ever had a catastrophic failure. Placing the data on a floppy disk or CD ROM and leaving it in the vehicle was considered, but it was certain that the disk would be lost or destroyed over the life of the vehicle. Furthermore, knowing that the vehicle might be modified in the assembly plant, a body builder location, a dealer location or used in a different vocation in a second life, it was decided that a much more disciplined approach was required. An archiving system was developed with dial up modem access. Trucks may not be modified unless the requested changes are stored in the archive system. Improvements to features or the core system are automatically available the next time a mechanic calls in. Configuration data is stored in the ESC and is unique to each vehicle identification number. Compatibility is maintained in the improvements to feature files must be backward compatible or else a new feature type must be established. The latest version of all files is invoked each time a re-compile process is initiated.

CONCLUSION

The multiplexed electrical architecture addresses all of the system requirements established at the beginning of the program. The system employs a distributed electrical architecture structure that was developed and implemented for high volume production of commercial vehicles. The selected architecture reduces wiring complexity, increases electrical system reliability, provides increased diagnostic capability, provides automated vehicle configuration files from the order entry system and is maintainable over the life of the vehicle. The architecture provides a flexible platform to provide customer features well into the future.

REFERENCES

1. SAE J1587 - "Joint SAE/TMC Electronic Data Interchange between Microcomputer Systems in Heavy-Duty Vehicle Applications", SAE J1587 Revision July '98.

2. SAE J1708 - "Serial Data Communications between Microcomputer Systems in Heavy-Duty Vehicle Applications", SAE J1708 Revision Oct. '93.

3. SAE J1939 - "Recommended Practice For Serial Control and Communication Vehicle Network"

CONTACT

Robert D. Dannenberg
Manager, Cab & Body Control Systems
Vehicle Electronics
International Truck & Engine Company
2911 Meyer Road
Ft. Wayne, Indiana 46801

DEFINITIONS, ACRONYMS, ABBREVIATIONS

OEM – Original Equipment Manufacturer

ESC – Electrical System Controller

RAM – Random Access Memory

CD ROM – Compact Disk Read Only Memory

PID - Parameter Identifier

LED – Light Emitting Diode

I/O - Input /Output

Multiplexed Operator Control Switches in a Modular Package

Gary Pomeroy
International Truck and Engine

ABSTRACT

The increase in the number of electronics in Trucks, and the resultant increasing in wiring, has led to an inevitable emphasis on reducing wiring harness size and complexity. The use of Multiplexing allows the application of additional electronics without proportional increases in wiring. Multiplexing also adds the benefits of increasing the flexibility in electronics placement (modularity) and thereby reducing incorrect wiring. By utilizing the J1587/J1708 SAE standards in modular switches, these capabilities can be achieved, but with many challenges not normally seen in these traditionally simple devices.

The keys to acceptance are to keep the capabilities of present control switches, reduce the wiring complexity in manufacturing, provide flexibility for bodybuilders, and to gain the acceptance of truck operators. This paper is an overview for the architecture of one method to implement multiplexed control switches, discuss the hardware design challenges, and an overview of the software implementation necessary to provide modularity and bodybuilder flexibility.

INTRODUCTION

It has become apparent that wiring harnesses have grown too large for ease of assembly, troubleshooting, or replacing. Normally, manufacturing creates these wire harnesses and tests them prior to installing them on a complete vehicle. Even with the capability of continuity testing, which checks for assembly troubles before vehicle installation, there are more problems, including: improper routing, improper installation, and difficulty in troubleshooting, these large harnesses. One solution to decrease wire harness size is to multiplex the vehicles' electrical and electronic systems utilizing a data link, such as SAE J1587/J1708.

Conventional switches in commercial vehicles, directly driving a load, are the bulk of the wiring on these vehicles. Typically, these wires go directly from the object to be controlled, to the switch that is designed to control that object, through a complex web of wire harnesses. Through the natural progression of electronic/electrical advances, more and more controls, and optional electronic controls, are being added to trucks from OEMs (Original Equipment Manufacturers), not to mention the after-market electronics installations. All of these items need to be controlled and the easiest way has generally been through the use of a switch, mounted somewhere on the instrument panel.

ARCHITECTURE OVERVIEW

There are many different ways to design the hardware architecture, with the software supporting the hardware setup. A good choice is to utilize the SAE J1708/1587 architecture, which is prevalent on most vehicles of this type; it is a relatively flexible and inexpensive system to implement for this function. In order to utilize this data link, some intelligence has to be introduced into the system, which is the first unique challenge to this type of design. Other unique challenges are the communication method (and protocol) to other devices; specifically diagnostics and switch identification, and the hardware setup, including the size of the switchpacks and new hardware for the switches.

SYSTEM DESIGN CONSIDERATIONS

There are several choices to solve the dilemma of where to put the intelligence on the vehicle in order to multiplex the switches. One choice is to build in enough intelligence into the switch or switchpack, to communicate to any device that is to be electronically controlled. Another choice is to put the intelligence in the instrument cluster and simply hardwire the switches to the cluster. A third choice is to utilize a separate intelligent device that can be used for more than this application, such as an ESC (Electronic System Controller), or another microcomputer solution. This third choice allows the switches to communicate with this separate intelligent device, thereby reducing the complexity of the switch/switchpack.

The option discussed, and used for this design, is to incorporate a small amount of intelligence in the switchpack(s), and task the ESC with the job of controlling the outputs via switch information. This quasi-intelligent switchpack option is the most viable for several reasons: it gives the switchpack enough intelligence to know when other switchpacks are present through wire harnesses. From data link communication the ESC knows which switches are present on those switchpacks, and in what state those switches are set.

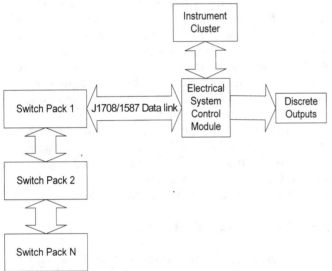

Figure 1: Basic Diagram of System Architecture

SWITCHPACK COMMUNICATIONS AND DIAGNOSTICS

The ESC monitors the switches and controls via the SAE J1587/1708 data link, and controls the status lights from software programming of the ESC that is done by a configuration system before the vehicle is built. Switchpacks are assigned specific addresses, so as not to confuse the ESC, and the software programming in the ESC corresponds to the switchpack address. However, if a switchpack is removed, it can alter the address of another switchpack, as they are linked in series and this will cause a fault to be logged by the ESC onto the display of the Cluster Diagnostic messaging center. There can only be two switchpacks on a vehicle with the current design. A single bit, of the J1587 configuration message, helps to determine the switchpack address.

Each switchpack is unique because of a pull-up resistor, which changes the state of one bit on that message. The first switchpack has the bit set to one,

unless a second switchpack is attached and pulls that pull-up resistor to ground. The larger switchpack, the twelve-pack described in the switchpack hardware section, uses the same setup, but has unique addresses with another of the proprietary bits set to zero to positively identify it. When the switchpack conveys its' information to the ESC regarding switch state, via SAE J1587, (See SAE J1587 section 3.3 and A.254) the switch module sends a 2-byte (16 bits) message to the ESC. Twelve of the sixteen bits are reserved for switch data, while the rest of the bits are reserved for showing the switch status and the switchpack identification information. This information is decoded from the Parameter Identification (PID) for switch module data sent by the switchpack to the ESC. The ESC, in turn, processes this information with software algorithms and controls the output to the particular device that the switch is intended to control. This also means that if the owner/operator decides to change the switch location(s) or switch operation, then the software for the ESC can be reconfigured through a software tool.

SWITCHPACK HARDWARE

In order to facilitate putting the switchpack in any given location on the instrument panel, it was decided that the switchpack, and all other openings in the dash panel would be a single-DIN, or double-DIN (twice the size of a single-DIN) opening size. This allows for ease of manufacturing, and also makes it simpler to add, move, or remove switchpacks, using the same tools that are used on radios and the other DIN sized objects in the instrument panel. Switchpacks with up to six switches and up to twelve switches are called six-packs and twelve-packs, accordingly. There is room for six switches in each six-pack, and twelve switches in a twelve-pack, but there isn't necessarily that many switches in each. The switchpacks communicate their switch states on the data link to the ESC.

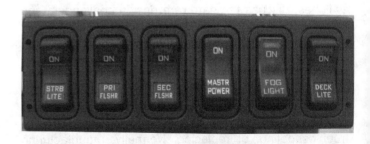

Figure 2: Six Pack, Single DIN-sized switchpack

The Double-DIN is basically a double set of switches in one large switchpack, but the ESC recognizes the twelve-pack as two six-packs with unique addresses and is used primarily for bodybuilders. This gives the vehicle purchaser the ability to order a vehicle with up to eighteen switches (one six-pack and one twelve-pack) that are configurable and modular.

Figure 3: Twelve Pack, Double DIN-sized switchpack

SWITCH HARDWARE

By multiplexing the switches/switchpacks, another design consideration (opportunity) presents itself, redesigning the hardware of the switches themselves. By putting all of the electronics onto the PCB, including dual micro switches, new wiring connection designs allow for a whole new set of design innovations.

Figure 4: Six-Pack PCB showing new switch design

The switch actuator is separate from the electrical connection for easier installation and removal. These new switch actuators move the switch via a cam lever approach, which then actuates the appropriate electrical switch on the PCB.

Figure 5: Back of switchpack showing actuators

The switch actuators come in several varieties to meet several functions including momentary and latched types, just like in today's vehicles. The different types of switches are color-coded on the rocker cantilever to aid in correct manufacturing placement into

the switchpack. The six-pack switches are designed to give the owner/operator the specific functionality and feedback necessary and are marked specifically for its intended purpose, however, the designated functions on the twelve-packs are often generically labeled so as to give the bodybuilder the flexibility to assign switch functionality as desired via a software tool to communicating with the ESC. Typically twelve-packs are only ordered for trucks that have specialized tasks and work in conjunction with the ESC and another device called a Remote Power Module (RPM).

Figure 6: Side view of new switch showing actuator

The switches are backlit, just like today's truck switches, but most also come with software driven status LEDs. These status LEDs can convey problems with the switch itself, or convey hardware problems, by blinking in a predetermined manner, which is described in the Communications and Diagnostics section. In the normal mode, however, the status LED doesn't convey the position of the switch, but rather the true status of the output, thereby enabling an operator to know exactly when a controlled device has been activated by the ESC.

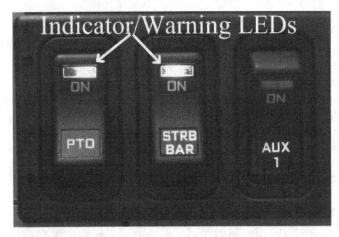

Figure 7: Output Indicators/Warning LEDs

The main design intent of the switches themselves was to have the look, feel, and capability of hardwired switches, while providing additional capabilities of giving diagnostic functions through the switch lights (on some switches), easy replacement of the switches themselves, and modularity (add, delete, or alter the switch actuators) on the switchpack. Changing the switch hardware in the field is easily accomplished because there is no electrical connection from the switch actuator to the PCB. However, changing the functionality of a switch is now a software issue as much as it once was a hardware issue. Since most of the intelligence for this system resides in the ESC, it requires a software change in the ESC to change the functionality of a switch in a particular switch location or switchpack.

CONCLUSION

The increase in the use of Electronics on all vehicles, including Heavy Duty Vehicles, has given rise to a need for more complicated and therefore larger wiring harnesses. By adding a small amount of intelligence and utilizing datalinks, the amount of wiring does not have to increase proportionally. The widespread use of J1708/J1587, and comparatively lower cost than J1939, on Heavy-Duty vehicles lends itself to be the data link of choice for today's trucks.

This simple hardware/software architecture, connected to an intelligent device for decoding that information, is all that is needed to help reduce wiring complexity and create modularity. This modularity comes with the advantages of adding switches without adding wiring, and being able to alter switch locations/functions through software programming. The only challenge is designing enough outputs from an ESC or equivalent device, to control all of the output devices.

REFERENCES

1. "Joint SAE/TMC Electronic Data Interchange between Microcomputer Systems in Heavy-Duty Vehicle Applications", SAE J1587 Revision July '98.

2. "Serial Data Communications between Microcomputer Systems in Heavy-Duty Vehicle Applications", SAE J1708 Revision Oct. '93.

3. "A Vehicle Electrical System Architecture Based on a Multiplexed Design for Operator Controls and Interfaces" Robert Dannenberg

CONTACT

Gary Pomeroy
International Truck and Engine Corporation
2911 Meyer Rd.
Fort Wayne, IN 46801
(219) 428-3623
Gary.Pomeroy@nav-international.com

ADDITIONAL SOURCES

"Application of a Digital Electronic Instrument Cluster as a Node in a Multiplexed Vehicle Electrical System" Deborah Payne 2001-01-2738

"One Example of a Software Architecture for a Configurable, Intelligent Vehicle Electrical Control System" Gregory Jean-Baptiste 2001-01-2741

"A Vehicle Electrical System Architecture based on a multiplexed design for operator controls and indicators" Robert Dannenberg 2001-01-2740

"Custom Configuration of a Vehicle Electrical Control System for Production and Service" Joe Kartje

"Implementation of Fault Detection and Diagnostics in an intelligent Vehicle ESC" Mark Thurber 01TB-54

DEFINITIONS, ACRONYMS, ABBREVIATIONS

OBC – On-Board Computer

ESC – Electronic System Controller

PID – Parameter Identification

RPM – Remote Power Module

DIN – Deutsches Institut für Normung

PCB – Printed Circiut Board

2001-01-2738

Application of an Electronic Instrument Cluster as a Node in a Multiplexed Vehicle Electrical System

Deborah M. Payne
International Truck and Engine Corp.

ABSTRACT

The primary function of a gauge and information display cluster is to report to the driver the current state of the different systems of the vehicle and particularly to report out of range performance conditions. This paper will examine one implementation of the instrument cluster as a node on the J1939 data link as it relates to that functionality. Traditionally clusters have received their information from a variety of sensors requiring a substantial number of connections. Once the cluster is installed in the vehicle, changing the configuration may only be accomplished by means of a hardware change. Today's multiplexed instrument cluster also receives and processes information from a variety of sources for display as gauge data, warning light status, and selected gear information, specifically from the engine controller, electrical system controller, and the transmission controller, but with a significant reduction in the number of electronic, hydraulic, and pneumatic connections to the module. This is ultimately due to the roll the instrument cluster plays as a node on the data link. Because the cluster is a node, sensors need not be directly connected to it. Instead the cluster receives data from other nodes in the system. Along with the reduction of sensors, a central benefit of the multiplexed system is an increase in the configuration flexibility. Additional gauges may be added to the cluster without adding any wiring. The cluster gauges and warning lights are configurable based on data from an electrical system control module. There is also an increased flexibility in messages configured based on engine and transmission. The gear selection graphic or PRNDL on the integral digital display can also be reconfigured for use with a variety of transmissions. The multiplexed cluster is also capable of displaying additional diagnostic information in its supplemental digital display along with storing and broadcasting diagnostic messages on the data link. Being a part of a multiplexed vehicle system greatly improves the instrument cluster's functionality as a fault-reporting device.

INTRODUCTION

A successful implementation of an information display or instrument cluster has several requirements. The instrument cluster must present information to the operator with gauge indications, warning lights, and text messages. For the owner or operator this means choosing which parameters are critical and need to be monitored. Because this is variable based on application, it is vital that the cluster be easily configurable both at the manufacturing facility and in the field. As a node on a multiplexed electrical system, a gauge cluster is easily configurable with little hardware change at the cluster. Information specific to the vehicle's engine and transmission is also available to the cluster for display. The cluster is also able to monitor the data link and report diagnostic information to the service technician.

MULTIPLEX APPLICATION OVERVIEW

DESIGN OBJECTIVES

The major objectives in the multiplexed cluster design were the development of a module that reduced the complexity and variations found at the vehicle assembly plant, increasing the reliability of the system by reducing the number of connections to the instrument cluster, and providing improved functionality in the area of reporting out of range conditions. Past cluster designs had suffered from having too many wiring connections, including individual wires for each warning light, along with the tubing required for hydraulic or pneumatic connections. The manufacturing facilities were also required to maintain more part numbers as each configuration was built individually based on which gauges and warning lights were required.

MULTIPLEXED GAUGE CLUSTER SYSTEM

The multiplexed system links the instrument cluster to other modules on the data link. The cluster receives data for display from an electrical system control module, engine control module, and possibly from a transmission controller and other external modules. Information sent in the standard SAE J1939 format are then decoded by the cluster and translated for display as gauge data, warning light state, audible alarm state, and digital display information. The cluster is also capable of communicating back to the data link with switch status, check sum value, and dimmer level. See Figure 1.

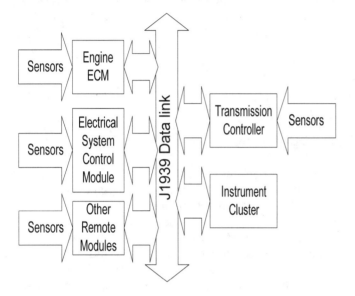

Figure 1 Overview of Data link Structure

DESIGN DESCRIPTION

The instrument cluster is an electronic module that consists of up to 10 gauges, over 25 data link driven warning lights, three user switches, a digital display, and an audible alarm. See Figure 2 below.

Figure 2 Data Link Driven Gauges and Warning Lights

As a node on the data link the instrument cluster takes data directly from other nodes for display in gauges and in the digital display (See Figure 3). This prevents problems that arise from having two different discrete sensors reading the same information and/or calculating different values due to rounding error. For example, historically, differences between mechanical odometer and engine hours displays compared to the value stored and broadcast by the engine control module have led to confusion in the field and warranty issues. Implementing

the odometer and hourmeter in the cluster's digital display with the data taken directly from the engine controller can eliminate this problem. Along with the odometer and hourmeter, the digital display also can display trip odometer, trip time, and fuel economy, among others. The digital display is very flexible and can also be used to report diagnostic messages to the driver, display PRNDL information, and indicate the software level of the cluster.

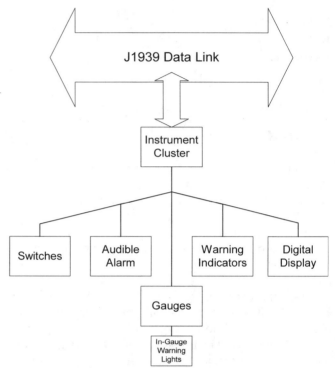

Figure 3 Diagram of Cluster Functionality

Many error conditions reported by the cluster also require an audible alarm. The alarm is integrated into the cluster but through proprietary messages on the data link can be activated by other nodes on the data link, such as the electrical system control module. The cluster prioritizes different alarm sequences so that safety related issues are alerted to the driver as soon as the fault is detected.

The cluster also integrates three user switches that are functional independent of the state of the ignition. The statuses of these switches (head/park light, dimmer level, and work light) are all communicated to the electrical system control module via a proprietary message. The cluster also has a wired output for the dimmer signal. The message, or the hard-wired signal, can be used by remote modules to track their back light levels synchronously.

BENEFITS

There are several benefits to implementing an instrument cluster as a node in a multiplex system. These include the reduction in the number of connections to the cluster, elimination of redundant sensors, improved serviceability, easy configuration of the system, and the increased diagnostic functionality.

WIRING REDUCTION – The total number of wires to a traditional instrument cluster varies with vehicle configuration. Individual wires for each warning light along with sensor outputs to gauges bring the wire count up to 50 or more per cluster. See Figure 4 below. With a cluster that acts as a node on the data link, this number can be reduced to nine or fewer depending on design requirements. Once this number is established, however, it is fixed for all applications. See Figure 5 below.

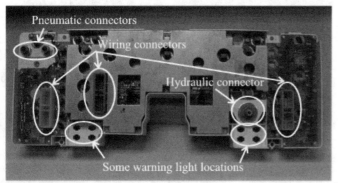

Figure 4 Traditional Instrument Cluster

Figure 5 Multiplexed Instrument Cluster

ELIMINATION OF REDUNDANT SENSORS – Since the multiplexed instrument cluster takes its data from other modules on the data link the cluster does not require its own uniquely assigned sensor inputs. This leads to a reduction in the overall cost of the vehicle. It also ensures that values reported to the driver in the instrument cluster are the same as those used by the power train modules to determine states and modes of the vehicle's operation.

SERVICEABILITY – The serviceability of the gauge cluster, and of the electrical system as a whole, is greatly improved in a multiplexed system. This is both due to limiting troubleshooting to fewer electrical connections and because diagnostics messages are more readily supplied to the driver or service technician (see DIAGNOSTICS below). This leads to shorter down time in the case of an electrical malfunction.

CONFIGURATION – The multiplexed instrument cluster may be configured easily during the manufacturing process and again as many times as necessary in the field as the vehicle's applications or work requirements change. This can be done with no wiring changes in the connections to the instrument cluster.

Methodology – The electrical system control module, as the primary node in the multiplexed system, acts as the source of the configuration data for the instrument cluster. This configuration includes the following information

- Gauge type (e.g. Fuel Level)
- Gauge range (min and max)
- In-gauge warning indicator trip points
- Alarm type (beep count and duration)
- Damping filter for gauge response
- Source address for each gauge
- Individual warning indicators available
- Transmission type
- Engine type

When a change is made to the configuration, such as a gauge being added to or removed from the cluster, it must first be registered in the electrical system controller. The system controller will then teach the cluster its new configuration. There is a checksum calculated in the system controller based on the configuration data. This checksum is compared to the one calculated and stored in the cluster. When these values do not match, the system control module configures the cluster with the new data.

Benefits – There are many benefits to the ability to reconfigure the instrument cluster easily. Many fewer part numbers are maintained since the potential for every configuration resides in each cluster's memory. That is, the data required by the instrument cluster to implement each of the over twenty gauge types is stored in the cluster along with multiple transmission and engine types. It is not necessary to track which instrument cluster is configured for use with specific engines, transmissions, or electrical system control modules. There are also several parameters that may be modified within each configuration. These programmable parameters allow the driver to precisely monitor whichever systems are most critical to a specific application by causing alarms and warning indications at different temperatures and pressures than are programmed at the factory.

Engine Specific – Because the cluster is taught what engine is on the truck, engine specific functionality is possible. This includes warning messages and service interval notices to the driver.

Transmission Specific – The cluster is able to display selected gear information (a PRNDL) based on the vehicle's configuration as well. The different display types are limited only by the number of segments in the digital display available for the purpose of gear display.

Programmable Parameters – Many of the parameters stored in the electrical system control module may be modified in the field. With the exception of safety related parameters that must be inflexible, parameters such as warning light trip points and alarm types may be changed to better support the vehicle's application.

DIAGNOSTICS – A primary benefit of the implementation of the instrument cluster as a node on a multiplexed system is the increase in diagnostic functionality. The new cluster reports internal problems as well as those experienced by other nodes on the data link, as is explained below.

Monitoring – The cluster monitors the data link for fault messages from other nodes. These nodes, such as the electrical system control module, the engine control module, and the ABS controller, for example, broadcast fault messages in standard formats that the gauge cluster can monitor for fault reporting.

Displaying – The instrument cluster is able to display fault messages on its integral digital display. Depending on requirements, these messages may be available for view to the driver at all times or they may be only available when the cluster is placed into a special diagnostic mode. Whatever the case, the decision on the format of the display is important. While fault reporting that incorporates English fault descriptions is ideal for easy interpretation it is difficult to implement due to the sheer scope of possible messages involved. A more realistic fault reporting mechanism is to assign numeric values to different fault types. These fault messages provide a great deal of information displayed in a very limited space. The diagnostic message indicates the SPN, FMI, and two values (bytes 7 and 8 of the diagnostic message) indicating specific information about the fault characteristics. See Figure 5 below.

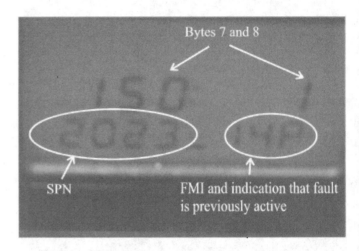

Storing – Local faults occurring in the cluster are stored in the cluster for retrieval when requested by another module or a service tool. Because all diagnostic messages are broadcast on the data link, the cluster may also store the fault messages experienced by other modules on the link. In this way the cluster may serve as a supplemental memory location.

CONCLUSION

The instrument cluster as a node in a multiplexed system provides for a more accurate display of gauge data, higher reliability through fewer connections, more features with built-in gauge warning lights and audible alarms, superior error reporting and quicker configuration changes. These improvements are not available to the traditional instrument clusters of the past. The reduction in wires, easy configuration updates and modifications with few hardware changes, and diagnostic functionality all are benefits of this system.

REFERENCES

1. SAE J1939/71: Recommended Practice for Serial Control and Communication Vehicle Network / Vehicle Application Layer
2. "A Vehicle Electrical System Architecture Based on a Multiplexed Design for Operator Controls and Interfaces" Robert Dannenberg
3. "Custom Configuration of a Vehicle Electrical Control System for Production and Service" Joe Kartje
4. "One Example of a Software Architecture for a Configurable, Intelligent Vehicle Electrical Control System" Michelle Cambron and Gregory Jean-Baptiste
5. "Implementation of Fault Detection and Diagnostics in an Intelligent Vehicle Electrical System Controller" Mark Thurber

CONTACT

Deborah M. Payne
Project Engineer, Vehicle Electronics
Electrical & Electronics Systems

International Truck and Engine Corporation
Truck Group

2911 Meyer Road
Fort Wayne, IN 46803

(219) 461-1795

Deborah.Payne@nav-international.com

DEFINITIONS, ACRONYMS, ABBREVIATIONS

FMI: Failure Mode Indicator
FMVSS: Federal Motor Vehicle Safety Standard
PRNDL: P (ark) R (everse) N(eutral) D(rive) L(ow)
SPN: Suspect Parameter Number

2001-01-2706

Can a Single Multiplex Data Bus Suffice to Serve all Functions on a Transit Bus ?

Mike Klaus
I/O Controls Corporation Project Engineer

ABSTRACT

There have been many developments during the implementation of multiplex data standards. Several proprietary data standards have been developed by various different vendors. A few industry standards exist such as the SAE J1939 and the SAE J1708. Although the proprietary data standards have been working well, there has been some attempt to use the SAE J1939 standard for all functions on all vehicle multiplex systems on the transit bus. This caused an attempt to use one single databus for all systems on the vehicle.

INTRODUCTION

Many complex systems the transit bus of today have been organized into distinct groups, each one having a separate databus or multiplex system. The design of the transit bus has evolved. Pneumatics, hydraulics, and electric relay logic have evolved separately. In more recent years, electronic 'ladder logic' controllers have been used in place of discrete relays. The electronic controller is able to perform the same functions as an array of electric relays and time delay relays. Design engineers had been connecting the discrete relays with wiring to make interlocking vehicle functions as needed. Those same engineers are now using symbols of relays to build a virtual ladder of logic on a desktop computer. Automation in the desktop computer reduces the ladder drawings to a PLC code program which is then loaded into a 'Brain Unit' in the vehicle, also known as the computer or PLC controller. All of the items which are being controlled by the PLC controller are connected together using a multiplex data bus of some kind.

MAIN SECTION

As the state of technology has progressed, the systems in a transit vehicle have become islands of automation. Electronic control units were developed for the engine, transmission, door controls, fire suppression controls, and housekeeping systems. Each of these systems was working well by itself, but there was no standard method of connecting these islands of automation together. Each vehicle manufacturer had a proprietary system which evolved and changed over the years. When two proprietary systems are connected together, the multiplex data is separated into a group of discrete signals which connects one system to another system through a couple of discrete-wired gateways. Modern vehicles have databus gateways which can translate the multiplex data directly from one proprietary data bus to another proprietary data bus, and a gateway to translate data from an industry standard databus (such as SAE J1939) to a proprietary data bus.

How should we connect the islands of automation on a vehicle? There has been some concern about the cost of each gateway compared with the amount of useable information that actually comes through the gateway, often expressed in terms of cost-per-bit. The discrete-wired gateway seems to be cheaper at first and easy to understand as it has one wire per signal, but it is often more expensive to install because of the many parallel discrete wires that are needed. Upgrading or changing a discrete wired system involves a lot of manual labor, adding wires to each vehicle in the fleet. Positive aspects of the discrete-wired gateway are: ease of troubleshooting, off-the-shelf availability, and no R&D cost.

The multiplex-to-multiplex gateway would seem to be a better choice. In theory, a vast amount of pertinent and useable information is available such that the actual cost per bit in a multiplex-to-multiplex gateway is expected to be less than the actual cost per bit in a discrete-wired gateway. This type of gateway is more expensive than a discrete gateway because of the burden of standardization. Extra costs for research, engineering, testing to standard, attending committee meetings, and dispute resolution are built in to the cost of any multiplex-to-multiplex gateway. Installation is generally cheaper because there are fewer connections, and those connections are generally made with industry standard plugs and sockets. Expansion or revision in the future involves a change to the programming only, as the multiplex wiring exists as a complete connection.

In practice, very little of the information available on an SAE J1708 data bus or on an SAE J1939 data bus is actually pertinent or useful to the body multiplex system. Up to the present day, the cost per bit of information on the SAE J1939 data bus has been high. The cost per bit of information on the SAE J708 data bus has become reasonable. The national sales forces are no longer selling the SAE J1708 data bus for any application. The national sales forces the have been pushing for global use of the SAE J1939 data bus for everything. A lot of pressure has been brought to bear on various manufacturers of in-vehicle systems to make a CAN-bus or SAE J1939 data bus interface for each system.

Although there are manufacturers whom have been pushing for exclusive use of the SAE J1939 data bus, those same manufacturers have actually been shipping a skeleton of a product. Very few of the SAE J1939 items have been implemented. The sales person shows a big list of SAE J1939 standard items which shows that a vast amount of information is supposed to be available on the SAE J1939 data bus for use by anyone that cares to listen to it. Then, that same company ships a product that puts out only so much SAE J1939 information as is needed for bare minimum functionality of the engine, transmission, and ABS/Traction control system. We are told that more information may become available at an extra added cost. This increases the cost per bit of information on the SAE J1939 data bus significantly as compared with the cost per bit of that same information on the SAE J1708 data bus.

There has been a blanket statement that the SAE J1939 data bus is the only data bus needed for any vehicle, no other data bus is acceptable, no other data bus should be allowed in any new design. Various customers and end users have been duped into specifying that the SAE J1939 data bus should be used exclusively of all others. Unfortunately, this data bus is inadequate for use on anything but the most simple truck. Adding items such as headlamps, marker lamps, interior lighting controls, cockpit switch inputs, and door position switches would seem to be too much for the J1939 data bus.

Any and all of those items that are not specified by a data bus standard (such as SAE J1939) are implemented in a 'Body Multiplex System' or else these things are implemented with discrete islands of automation using relays along with hydraulics and pneumatics. It is important to note that these items are required features of the vehicle, although these items are not specified in the J1939 standard. We are told that we could use some 'extensions' of the SAE J1939 data bus standard to encode the non-standard items. The data could be sent in a standard way, yet the data format itself would be proprietary and this goes against the whole idea is to stop using proprietary data formats. No procedure exists whereby two or more manufacturers could resolve a dispute arising from the use of the SAE J1939 data bus for proprietary data. The 'extensions' themselves appear to result in a stream of ambiguous data fields that cannot be resolved by a simple, one chip transceiver. All of these factors contribute to the increased cost per bit involved with using the SAE J1939 data bus standard exclusively.

Even if those difficulties could be resolved, there seems to be no pay-off for encoding proprietary data onto the SAE J1939 data bus. The proprietary data propagates along the (existing) proprietary multiplex system at a reasonable cost. The manufacturer of that multiplex system is totally responsible for it's design and functioning. A clear-cut procedure exists for resolving any dispute arising from the use of a proprietary multiplex system.

Many manufacturers have expressed dismay and objections to the proposal that another system could share the same SAE J1939 databus along with the engine, transmission, and ABS system. It simply is not possible to place all of the vehicle functions onto one J1939 data bus because of those objections alone. The consensus, if any, seems to be that two J1939 data busses would be needed. One J1939 databus would serve the engine, transmission, and ABS systems as things are built today. A second databus, whether it be proprietary, SAE J1939, or a new SAE standard, would serve all of the other functions in a vehicle such as cockpit switch inputs, cockpit displays, Door controls, sensitive edge/touch tape/treadle/pull cord; Destination

Sign/Stop Request Sign; ADDA items such as wheel chair lifts and Kneeling Bus Control; Air Conditioning Controls; Fire Suppression systems; Camera/surveillance/GPS systems; Built-in self test; Data Logger Storage and Retrieval; Tools and implements; and more to be added in the future. All of these things would end up on a list of proprietary functions rather than becoming standardized.

More and more pollution control standards are being brought to bear on the transit industry. In response, several manufacturers have been developing alternate means to propel the vehicle. Most of these designs incorporate electric traction motors to some degree. The Hybrid Drive vehicle does have a conventional engine which may utilize the SAE J1939 data bus. However, the electric drive portion has many different parameters that are not defined in the standard. It becomes necessary to monitor may different voltage or current values in a number of different batteries or fuel cells. Also, a Fuel Cell is a cryogenic system with many parameters that are not defined in the standard. For an electric trolley bus, there is almost no relevant item in the SAE J1939 standard hence it is unlikely that the J1939 databus would be used in this application at all.

CONCLUSION

Experience has shown that the SAE J1939 data bus lacks the definitions needed to implement a complete body multiplex system on a transit bus.

It remains an open question, whether a single data bus could serve all of the functions on a typical transit bus. It is clear that the J1939 data bus standard, while workable for a simple truck manufacturer, does not meet the needs of a transit bus manufacturer. It is also clear that certain manufacturers have been pushing the SAE J1939 standard as a cure-all item. Our own company has been put up against the wall and beaten for lack of SAE J1939 connectivity. We have spent a lot of time in research and development, and we have come to the unpopular conclusion that the SAE J1939 standard is inadequate for use in the body multiplex system of a transit bus as the standard is today.

Perhaps it might be possible to revise the SAE J1939 specification in the future. Perhaps a new SAE standard for 'Body Multiplex' systems is needed. Finally, as a group we must form a consensus as to whether one multiplex data bus can suffice to serve all functions on a transit vehicle or complex truck. In the real world, it would seem that two (or more) data busses per vehicle would be more appropriate.

2000-01-0812

A Distributed Java Architecture for Telematics Services

Parvathy Bhaskaran and Mark Clayton
Motorola Inc

ABSTRACT

This paper describes a robust, secure Java-based Telematics platform that enables the rapid development and deployment of distributed infotainment and safety service applications. Such a distributed architecture allows the Telematics client in the vehicle to offload data- and compute-intensive operations to the server, while executing location-sensitive and vehicle-centric services onboard. Further, by careful segmentation of the client and server modules, multiple tiers of the client product can be designed and rapidly assembled, ranging from inexpensive, thin clients to costlier, self-contained ones. By basing it on Java, we harness and empower a large development community, while reaping all the benefits of Java.

INTRODUCTION

The automobile space is one of the few remaining untapped frontiers of the information revolution. Currently, there is no standard architecture that allows safety, information and entertainment services to be delivered to the automobile. Most systems today limit themselves to simple emergency services that involve intervention by a human operator. More elaborate ones like navigation are autonomous; they are contained in the car and do not lend themselves to deliver timely information like traffic updates, road detours, etc.

In the following sections we describe and examine the proposed distributed Java Telematics platform in terms of its key enabling components, both on the client and the server. These components would be available as Java API classes and interfaces that applications would be built upon.

WHY JAVA

THE LANGUAGE – The Java language offers true portability across hardware platforms and different operating systems. Once an infotainment application or service is developed on this Java platform, it can be seamlessly made available on a multitude of client hardware with the availability of a standard Java VM and APIs as the only constraint. The ability to upgrade hardware without changing code results in huge cost savings in the development and time-to-market areas.

An extensive set of libraries currently available (security, UI, networking, database), and new ones being developed just for Telematics, provide a rich foundation for rapid application development. These libraries are common across vendors, ensuring that precious development costs are not locked into a proprietary platform. For example, the recently announced kJava, for small embedded devices, ran on several devices at its introduction including the Motorola Pagewriter and the 3Com PalmV.

The inherent strengths of the Java language are additional pluses. Its strong typing, enforcement of object oriented data protection, memory management are some examples of built-in protection against common programming errors. In addition, the security provided by the byte-code verifier, an intrinsic part of the VM, helps ensure that questionable applications are avoided.

THE TOOLS – Multi-vendor support and strong competition has resulted in a proliferation of tools, many of them free, and a rich programming and development environment. The breadth of the available tool-sets is impressive and includes pre-compilers as well as selective and JIT compilers. Apart from the standard Java Development Kit, complete development environments are available from many vendors. These include advanced scripting languages (which later generate standard Java), debuggers, applet libraries and front-ends to most commercial databases.

The Java VM area is also very vibrant. Currently, several JVMs are available from different vendors. While some of these are not officially certified as 100% Java by Sun, standard Java programs run interchangeably on most of them. More importantly, this competition has driven forward improvements in the JVM in terms of performance and features, and engendered more favorable licensing terms.

THE DIVERSITY – Java has gone beyond Web applications and garnered wide acceptance in diverse industries including telephony (JavaPhone [4], JTAPI [5]), smartcards (JavaCard [9]), speech and audio/video entertainment (JSAPI/JSML [7], Java Media Framework [6], JavaTV [8]). Java is used in a wide variety of devices (phones, pagers, set top boxes, printers, remote controls, hand-held computers) including the concept car at Java-One in 1999. While the latter may not be the how Java ultimately exists in the automobile, the diversity of Java makes a compelling case for its use.

Until recently, due to the processing requirements of multi-media programs and the performance constraints of Java, most multi-media APIs were available only in native form. But with availability of more powerful and specialized hardware coupled with improvements in Java performance, many voice recognition (VR) and text-to-speech (TTS) systems are now standardizing on Java as are audio and video software.

To ensure a rich and extensible framework for developing infotainment services in the automobile, many of the device classes (phones, pagers, streaming audio/video, smartcards) mentioned earlier need to be supported. Unless the current internal automobile architectures can standardize on a single internal operating system like Linux or WinCE, it is prudent to harness the necessary diversity through Java.

THE COMMUNITY – A Java based development framework, in addition to the other factors listed above, may be used to harness a large and enthusiastic development community that can rapidly offer a large and diverse portfolio of application services for the automotive space. Conventions like JavaOne (the largest programming convention in the world) and active internet newsgroups facilitate communication in the Java community. The Sun Developer Connection [1] program, a comprehensive set of resources and information tailored to meet the needs of Sun's Java developer community, boasts more than a million members, making it the fastest growing and second largest developer community in the industry (IBM [3] and HP [2] have their own Java developer groups). In addition, the Java Community Process [11] empowers the creation of open standards with input from diverse industry and academic sources, through draft reviews as well as expert, working and discussion groups.

While it may appear that multiple developers groups (Sun, HP, IBM, etc) fragment the community and pull it in different directions, the actual reality is that the community pressures from customers tasting the benefits of standardization and platform independence is forcing more convergence: IBM and HP are now working on aligning more closely with 100% Java efforts.

JAVA TELMATICS APIS

In addition to the general purpose Java libraries available today, the following telematics specific ones are needed to provide a complete and rich environment for developing distributed telmatics applications for the vehicle:

- Location Services API
- Abstract UI
- Profile Manager
- Network Interface Manager
- Service Discovery/Lookup
- Application Management
- Data Synchronization

These APIs are explained in more detail in the following sections.

CLIENT SERVER ARCHITECTURE

Figure 1 shows a distributed telematics architecture, where the mobile client (vehicle) uses a wireless link (typically cellular) to communicate with a remote server to access services. It also uses a wireless link (IR, Bluetooth [14]) to communicate with "accommodated" devices such as a PDA or laptop, as well as local proximity servers like a kiosk at a dealership.

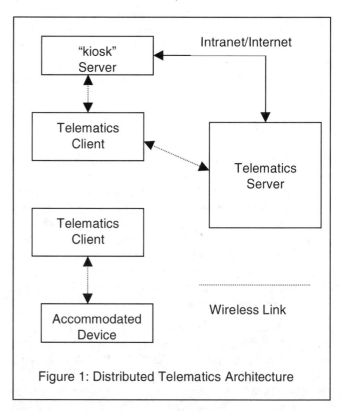

Figure 1: Distributed Telematics Architecture

Most automotive applications today are autonomous, i.e., the are fully contained in the vehicle and do not need a communication link to a remote server to provide services. While this is definitely an advantage in areas where there is no wireless coverage, a distributed architecture overcomes several limitations of an autonomous architecture:

- Provides support for time-sensitive services, e.g., traffic, stock, weather, sports updates.

- Provides up-to-date location services, e.g., navigation that takes into account current detours and road closures, point-of-interest (POI) information that includes news restaurants, shops, parking garages, theatres, etc.
- Offloads computing to a common server to reduce the cost of the client computing platform.
- Upgrades and configures both device and user profiles remotely.
- Allows user to set preferences from the comfort of a desktop.

TIERED CLIENTS – One of the main advantages of the proposed client-server architecture is that by the careful segmentation of the client and server modules of an application, multiple tiers of the client product can be designed and rapidly assembled. These can range from inexpensive, thin clients where the bulk of the work is done on the server to costlier, self-contained clients that are not dependent on a server and the associated communication channels.

In order to rapidly assemble tiered client products, an application should be designed in a modular form and segmented based on the planned tiers. For example, a navigation service that guides drivers from their current position to a desired destination is planned to be offered in three tiers: the low-end has minimal computing resources and no display, the mid-level has a two-line text display while the high-end has a nice multi-media display. To be used across all these products, one way to segment the navigation application is as follows:

- A module to extract the current location of the vehicle from an on-board position device such as GPS.
- A module to verbally interact with the user to determine the desired destination.
- A module to do the actual route calculation.
- Multiple modules to speak out the directions, display the directions as simple text or rich media (glyphs, sound cues overlaid on actual map data).

For the low-end product, just the first module that determined the current location would run on the client. The remaining modules would run on the server, which would interact with the user through server-based speech interaction. If the client supported VR and TTS on board, then the second and the final (spoken directions) modules could also be moved to the client.

In all cases, the compute intensive route calculation module would run on the server and only the final route would be sent back to the client in the desired form. For the mid-level product a simpler display module would be used than for the high-end product. If the high-end product had sufficient computing power and memory, even the route calculation module and associated map data could be resident on the client to eliminate any loss of service while roaming in an area with no wireless cover-age. The low-end and mid-level products would, unfortunately, have no such backup measures.

SECURE APPLICATION MANAGEMENT – An important component of the framework is the Application Manager that remotely controls the installation, removal and update of service applications on the client, as well as the management of critical resources during their execution. It is foreseen that in order to provide a wide range of services, applications from multiple vendors may be required to assemble a comprehensive service portfolio. Rigorous testing by trusted third-party developers and equally stiff subsequent acceptance testing will not be sufficient. More safeguards need to be in place to ensure that all these varied applications can play in the same sandbox together harmoniously

On the server, approved applications are preprocessed prior to loading. Bytecode verification, selective compilation for performance, attachment of digital signatures and encryption are some of the operations done before a profile is generated for the application. In addition, run-time characteristics of the application are also specified. These include required attributes such as anticipated resource (CPU, memory, threads) usage and dependencies on other services, as well as optional ones such as desired operating conditions for optimal execution. On the client, an application will adhere to this profile in order for it to be recognized by the client Application Manager and be allowed to load and run.

Under the aegis of such an Application Manager, the client device can be open enough to allow well-behaved third-party applications to execute in its controlled environment. This allows new and enhanced services to be made rapidly available to users from multiple sources in the industry. The Application Manager component is planned to be Open Services Gateway interface (OSGi) [10] compliant when the OSGi specifications are finalized and become available.

SECURITY – The mechanisms provided by Java 2 security [12] form the basis for this framework's security. These features include authentication, authorization, non-repudiation, containment, privacy and audit trails using a layer of appropriate policy modules on the client and the server. Functional areas where security is critical include:

- Interfaces to eCommerce modules
- Facilities for setting up Virtual Private Networks (VPNs). This is required in order to support productivity services such as email, applications that require access to enterprise data, etc.
- Gated access to sensitive vehicle information, location and diagnostic data through the Vehicle Information Gateway. For example, a trivia game application should be restricted from accessing any vehicle information by its profile specification.

- Resource monitoring to prevent unintended as well as malicious monopoly by run-away threads.

PROFILE/CONFIGURATION MANAGEMENT – An important attribute of the proposed framework is the ability for the extensive configuration of the final product:

- A service application on the client may have its own personality based on user profiles and preferences residing on the server. For example, one user of a navigation application may want elaborate turn-by-turn route guidance, while another user (maybe on the same car), would prefer all the directions to be displayed at once and minimal sound cues when a new maneuver is coming up.
- User profiles are also valuable in transporting the preferences of a user between vehicles.
- Client device profiles are maintained on the server so that content may be appropriately shaped for the device before transmittal. For example, if the client device just supports a simple two-line display, based on its profile the server can send a more compact route containing just text descriptions and not download associated map data.
- Client device profiles may also be used by the Application Manager to update the device with appropriate Java modules when a user adds or deletes a new service, upgrades the equipment in the car, etc.

To support these features, a secure Web interface would be available to provision these configuration profiles.

SERVICE DISCOVERY & LOOKUP – The mobile nature of the proposed framework requires a mechanism (possibly an existing one, like Sun's Jini or SLP) that detects the availability of new services and secures access to them. This would involve:

- Detection of devices (telephones, PDAs) brought into the automobile and access to associated services (address book, calendar, etc).
- Detection of "kiosk" servers (available at a dealership or home) that provide short-range, high bandwidth services. These could be Bluetooth (or similarly) enabled devices.

This facility would provide the client with a simple lookup service so that the user could select from the available list of services.

MOBILE NETWORK INTERFACE MANAGEMENT – To facilitate rapid development of service applications for the automobile, it is important to transparently handle the complexities of the underlying mobile communication. A simple API abstracts the underlying wireless interfaces and takes into account quality of service and security issues. The Mobile Network Interface Manager deals with:

- High bandwidth, short-range interfaces like Bluetooth, 802.11, IRDA.
- Low bandwidth, long-range cellular interfaces (circuit switched CDMA, AMPS, GSM).
- High bandwidth, long-range interfaces like satellite channels and, as they become available, 3G, GPRS.

Session Management – One challenge faced by this component is the management of a mobile session across a geographically distributed server network. For example, during a download/upload operation between the vehicle and the server, the wireless communication link may go down for several reasons:

- Urban canyons or cell boundary crossings may cause the call to be dropped.
- User drives outside the coverage perimeter.
- User or system (based on user preferences) deliberately shuts down the link if it becomes more expensive, e.g., when roaming.

When a session is suspended, the system keeps track of the session so that it can be resumed smoothly when the conditions are favorable again. In order to do this, after securing each session, its state is maintained with hand-offs between different types of sessions and between session suspensions and resumptions.

DATA SYNCHRONIZATION – The system will provide the framework for multi-point data synchronization. This would allow synchronization of PIM data between a user's phones, PDA or similar device.

Another important need that this framework would address is that of enterprise data synchronization. For example, fleet personnel or travelling salesmen would be able to synchronize their local data securely with the enterprise servers from the vehicle instead of waiting to reach a wired LAN access point.

TELEMATICS CLIENT

The telematics client architecture shown in Figure 2 shows the client Java framework on which the infotainment applications execute. The Java Native Interface (JNI) provides access to platform dependent native code from the pure Java applications. In the real world, there are always modules (communication stacks, device drivers and other compute intensive elements) that must be written in native platform code to meet performance requirements.

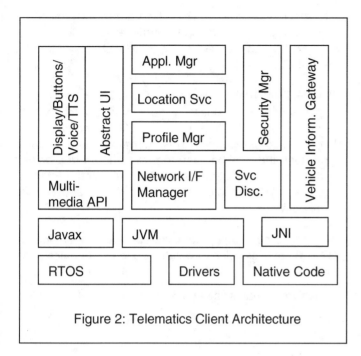

Figure 2: Telematics Client Architecture

EASE OF CUSTOMIZATION: ABSTRACT UI – While this framework enables rapid development of new telematics services, an important consideration is the ability for an OEM to easily customize the client product's look-and-feel to adhere to their branding requirements. Thus, the proposed framework provides the Abstract User Interface (AUI) on the client to allow application developers to design user interactions while sheltering them from the actual look-and-feel, which will be under the control of the OEM designers. The AUI consists of a set of widgets and a Dialog Controller to support multi-modal I/O.

The AUI also allows the same application to run on a broad range of client devices, since the mapping of the AUI widgets to the physical I/O elements is done on the client during execution. For example, a selection widget may be physically implemented using a touch-screen display, tactile buttons on the steering wheel or through spoken commands depending on the capability of the widgets on the client device. Thus using the AUI, the core application and its user interaction is shielded from rewrites required to achieve an OEM's look -and-feel.

MULTIMEDIA FRAMEWORK – A large number of the services available on this platform will require multimedia. The Multimedia Framework component is slated to deliver the required support. It will incorporate the Java Media Framework, which provides support for capturing, storing, and broadcasting time-sensitive/streaming media, using custom codecs, and manipulating media data before rendering it. In addition, it will provide the following extensions:

- A Tuner API to control a digital audio from multiple applications.
- A Content Management API to control broadcast filtering and trickle-in, background pre-loading of content.

VEHICLE INFORMATION GATEWAY – In the case of the automobile, for obvious safety and security reasons, a highly secure line needs to be drawn between facilities that support infotainment services and those that provide access to vehicle safety, information and control. The Vehicle Information Gateway (VIG) component provides this gatekeeper function and allows secure access to vehicle information, such as engine data, vehicle body and climate controls, through a firewall.

Such a Java abstraction also hides the variances in the internal buses and control units between different manufacturers (and even models) from the application service developers. For example, a safety application that runs on Motorola's MobileGT [15] platform using the VIG API could access similar information transparently from the underlying vehicle control systems on another manufacturer's platform if the latter also provided the standard VIG API.

Java 2 security is used to restrict access to this API from unauthorized applications.

NATIVE CODE SUPPORT – This framework will allow for simultaneous execution of stand-alone, non-interactive C or other native applications on the client. These would typically be daemons or monitors. Further, JNI support tools would be available to provide access to underlying native code libraries.

TELEMATICS SERVER

CONTENT MANAGEMENT – One of the more complex functions provided by the telematics server is that of content management. To take proper advantage of the distributed architecture, the system should provide timely and up-to-date content for the various applications to serve to the client:

- Map data with updates to reflect newest roads and latest road closures and detours
- Point-of-interest databases with updates to list the newest items (restaurants, shops, parking garages, etc) and delete obsolete ones.
- More frequent data feeds like stock/weather/news updates that constitute more transient data.

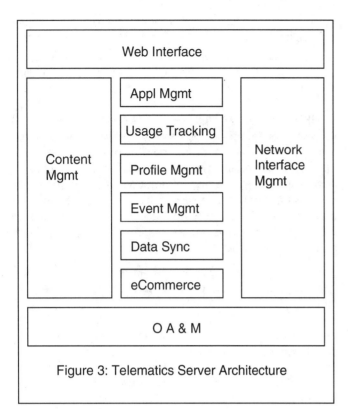

Figure 3: Telematics Server Architecture

Content Rendering – The framework provides facilities for Content Rendering based on any or a combination of the following:

- Content type
- User preferences
- Application configuration
- Client device characteristics

This is another enabling component that allows the same service application to be executed for multiple client devices. For example, weather reports generated by a service application as XML [13], can be shaped as a voice report for voice-enabled devices like phones. On the other hand, if the client device can support sophisticated graphics and an inexpensive, high-speed wireless channel is available, the information could be delivered superimposed on color satellite images.

Another example of controlled content rendering is email. The user may prefer to see messages displayed if the client has a display unit or the entire interaction could be in audio.

EVENT MANAGEMENT – The proposed framework will support an elegant Event Management system to allow users to receive notification when events subscribed to occur. This component will allow service applications to define event categories (traffic alerts, weather conditions, entertainment, sports or sales events). The list of event categories a user is subscribed to is maintained as part of the user profile.

E-COMMERCE SUPPORT – The framework provides a secure environment for conducting wireless eCommerce, which will be built on top of standard COTS systems for internet eCommerce.

USAGE TRACKING – An important enabler component of this framework resides on the server and tracks subscriber and application usage profiles under various demographic categories. This information is a valuable byproduct and may be used to provide input to a myriad of data-mining activities. A significant use of this information is to drive directed advertising, sales and promotional activities.

OA & M – Operational, Administrative and Maintenance (OA&M) facilities will be available on the Telematics servers for:

- Network availability at a 5Nines level.
- Billing
- Mail server
- Directory services
- General system administration

CONCLUSION

This flexible yet powerful distributed Java framework can play a central role in storming one of the last remaining bastions left untouched by the current information revolution. It harnesses the large and growing Java development community and enables them to rapidly design, build and securely deploy infotainment services and bring pervasive computing to the automotive space.

REFERENCES

1. Sun Java, http://java.sun.com
2. HP Java, http://www.chai.com
3. IBM Java, http://www.ibm.com/developer/java/
4. JavaPhone, http://java.sun.com/products/javaphone/
5. JTAPI, http://java.sun.com/products/jtapi/
6. Java Media Framework, http://java.sun.com/products/java-media/jmf/index.html
7. JSAPI, http://java.sun.com/products/java-media/speech/
8. JavaTV, http://java.sun.com/products/javatv/overview.html
9. Javacard, http://java.sun.com/products/javacard/htmldoc/overview-summary.html
10. OSGi: http://www.osgi.org
11. Sun Java Community Process, http://java.sun.com/aboutJava/communityprocess/
12. Java 2 Security, http://java.sun.com/products/jdk/1.2/docs/guide/security/

13. XML, http:// www.xml.com/
14. Bluetooth, http://www.bluetooth.org/
15. MobileGT Telematics Platform,
    http://www.mot-sps.com/automotive/mobilegt/
    index.html

DEFINITIONS, ACRONYMS, ABBREVIATIONS

5Nines:: A Motorola initiative to ensure 99.999% system availability.

Applets: : Java code that is downloaded and run locally inside a Web browser that may even interface with back-end servers.

COTS: : Commercial-Off-The-Shelf.

Infotainment: : A service that has both informational and entertainment value to a subscriber.

JVM: : Java Virtual Machine.

OEM: : Original Equipment Manufacturer.

OSGi: : Open Services Gateway interface.

PDA: : Personal Digital Assistant.

PIM: : Personal Information Management. Typically refers to address book, calendar and to-do data managed by PDA devices.

SLP: : Service Location Protocol, an IETF draft standard.

Telematics: : Delivery of mobile or location aware infotainment services.

VM: : Virtual Machine.

Basic Introduction to OSEK Communication - COM

Bruce D. Emaus
Vector CANtech

ABSTRACT

Because of the large software and networking architecture contained within the overall organization of OSEK, most beginners will find learning OSEK to be rather difficult. Additionally, if one starts with little or no experience in developing a distributed embedded system using small area network technology, the learning curve will be even longer. This technical paper aims at introducing the essential concepts behind one of OSEK's three major components - the OSEK Communication. The primary focus is to help the beginner understand the basic technical features of OSEK COM.

INTRODUCTION TO OSEK

Because the name of OSEK originated in Germany and the acronym does not translate into a traditionally recognizable series of four English words, it is much easier to describe what OSEK is rather than what the four letters mean.

OSEK is essentially the requirements for three re-usable software building blocks that may be used to construct distributed system applications.

The OSEK blocks are the Operating System (OS), the Network Management (NM), and the Communications (COM). By design, each block may be use independently of the others. Implementing one or all portions of OSEK are up to the user.

Note - Although OSEK now includes a suffix of VDX, this paper uses the name OSEK rather than OSEK/VDX to improve readability.

Although the five year old OSEK agenda still maintains strong ties to the European automotive electronics area, the OSEK architecture is not just for vehicles, its framework may be applied in virtually any industry using distributed product technology.

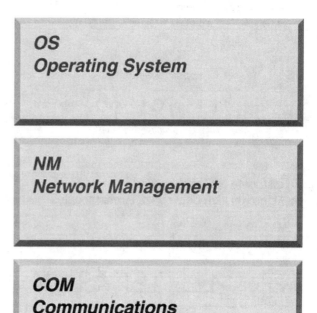

b99091718g

OSEK COM

As one of the three components of OSEK, OSEK Communications or OSEK COM primarily focuses on requirements related to information transfer across the distributed system. It is essentially a software and systems-level requirements document.

The current version 2.1 of OSEK COM, written by the OSEK VDX steering committee, is now available as SAE J2508-1.

Because vehicle network communications is an integral part of the distributed processing used in today's automotive electronic architectures and because it is one of the first documents to establish requirements for many of the necessary data transfer processes, the value of OSEK COM becomes sufficiently important to consider its combined technical and business value.

IN THE BEGINNING – In the beginning, all was centralized. And as vehicle system designers studied their architectures using large electronic modules with massive amounts of input and output wires, the idea of distributing became more and more attractive. Across the automotive industry, electronic and electrical architectures became distributed. This gave birth to the distributed function.

The Distributed Function – Beginning with a simple function that includes the traditional input, control, and output blocks, the distributed function results from the process of physically partitioning the general function between different modules.

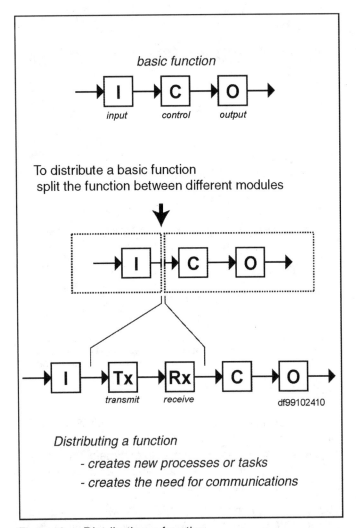

Figure 2.   Distributing a function

The process of distributing a function, as shown in Figure 2, creates several new requirements.

First, an information or signal transfer will be necessary to maintain the overall function's interconnectivity. This requires a suitable communications protocol. Additionally, two new blocks – a transmit and a receive block are added to the original function. Now, the overall distributed function includes five key elements. Each distributed function must transfer a signal and this signal must be

placed into a message. Notice that the arrow between the transfer and receive blocks represents the message.

THE OSEK COM BUILDING BLOCK – The OSEK COM portion is only one of two OSEK building blocks related to the use of network communications. The other related block, called the OSEK Network Management or OSEK NM is used to supervise network relationships. While OSEK COM, OSEK NM, and the user's APPLICATION are all capable of being interconnected to a possible OSEK OS implementation, it is the OSEK Communications that handles the primary activities of connecting the application to the network.

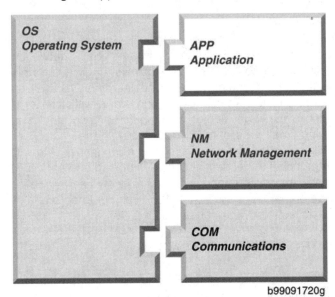

Figure 3.   OSEK Components with the Application

OSEK COM has intentionally been designed to be independent of communication protocol. Although it is possible to implement OSEK COM with any of the J1850 protocols, the specification gives the impression that OSEK COM is essentially constructed upon the CAN protocol, especially when one examines the eight byte transfer structure of the internal transport protocol. But this should not be any worry because there is no serious contender to the CAN protocol. With Detroit's abandonment of J1850 and Time Triggered Protocol (TTP) slowly merging as a possible drive-by Wire protocol, OSEK COM currently embraces the appropriate protocol.

If new emerging protocols warrant changes in OSEK COM, the governing committee will appropriately modify the specification.

APPLICATION NETWORK INTERFACE –   To better understand many of the various concepts within OSEK COM, it is best to start by examining the interface between the application software and the network software. Years ago when everything was centralized, no such interface existed. But, as the auto industry migration toward distributed-ness continued, it was realized that the creation of a common network software building block would be very beneficial to the system. Instead of each module developer creating their own vehicle network implementation based on their interpretation of the specification, it seemed possible to provide a proven network software implementation to all application developers in order to raise the system quality. This re-usable network software idea has become widely accepted across the automotive industry and has effectively created a new technical and business boundary between the application and the network software.

It is typical today for the automotive electronics module developer to integrate a third party vehicle network solution supplied by the OEM into the developer's module-specific application. For some vehicle network solutions, the use of all or portions of OSEK COM, OSEK NM, or the OSEK OS are possible choices.

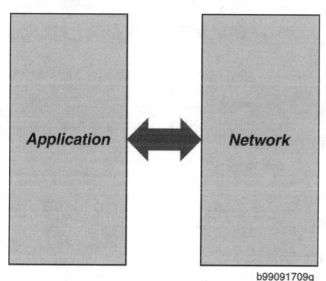

b99091709g

Figure 4.   The Application – Network Interface

While earlier application-to-network interfaces began as simple protocol driver software, more and more network related responsibilities are being moved into the network side of the interface to achieve higher levels of consistency.

BASIC NETWORK ACTIVITIES – The API between the application and the network supports three general activities - transmission, reception, and network supervision.

While supported transmission and reception activities may handle simple message-based transfers, some implementations may include advanced features like automatic periodic message handling or signal-based transfer handling. OSEK COM also includes many advanced features.

Network supervision commonly handles initialization of the protocol controller and usually includes capabilities to determine loss of network or loss of relationship (perhaps with a module or subsystem). Supervision activities might also include managing network sleep and wake-up operations. Relative to OSEK, many of the network supervision activities are not handled by OSEK COM. It is the OSEK Network Management (NM) that handles these tasks.

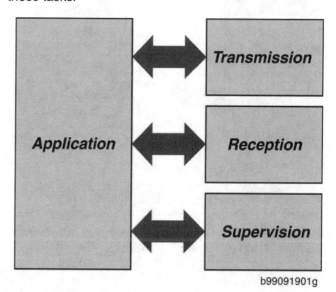

b99091901g

Figure 5.   Major Network Activities

OSEK COM HANDLES NETWORK TRANSFERS – OSEK COM specifies general and specific methods to handle network transfers for both message transmission and reception. It includes several software interfaces to provide the application a wide range of services from the sending or receiving of a single message to the complete transfer of a file.

Network transfer activities – The general network transfer activities handled by OSEK COM include the transmission and reception of information. These activities are asynchronous. The application may call the network software to initiate the transmission of a message without waiting for the entire process to be complete before returning back to the application. This means that the network transfer process is basically handled in two steps. First, "begin the transfer" and second "do something at the end of the transfer".

Advanced network transfer features may include options to handle file transfers (a transport protocol) or provide checking for a message transmission being successful.

The reception process may allow the receive data to be moved quickly into the application RAM without waiting for the application software to intervene. OSEK COM provides several of these optional features.

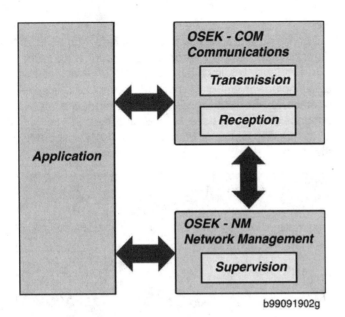

Figure 6. OSEK Network Activities

OSEK COM ESTABLISHES REQUIREMENTS – As a software interface document, OSEK COM establishes the requirements for managing the transfer of network information and allows choices between different transfer methods. Additionally, OSEK COM provides optional information transfer requirements which may or may not be appropriate for a given implementation. OSEK COM handles both necessary and optionally selected communication states and deals with the transitions between these states.

Developers using OSEK COM must make decisions. The process of scaling the requirements down is necessary in order to create an optimal-sized implementation. The resulting OSEK COM implementation becomes a discrete software component used in the module's application software development.

COMMUNICATION ACTIVITIES SUPPORT APPLICA-TIONS – Which OSEK COM activities or features are selected depends on the intended application. Developers of high-speed motion-control applications (like an ABS controller) may find the interface to the network software much simpler when compared to a body electronics module. Body electronic functions (like interior and exterior lighting, climate control, windows and seat controls) typically involve more complexity and require more software processing related to the network.

Communication Operational States – For example, the typical motion-control application uses network communications only while the engine is running. Compare this two state process to a body application which must deal with the additional states of sleep and wake-up and potential time delays between these multiple states, and it is clear that the software complexity is different.

MANAGING NETWORK TRANSFERS USING A TRANSFER DIALOG – When implementing a distributed function, the distributed system engineer must select an appropriate method of transferring the information. For automotive electronics the selected method depends on the application, the relative importance of this function when compared to all supported functions, and may depend upon which vehicle network is used. While each method must utilize at least one message, the method naming is more associated with the network conversation used to maintain the distributed function interconnectivity. From this point of view, the selected method is named as a transfer dialog. The three most common transfer dialogs use the event message, the periodic message, and conditionally periodic message to glue distributed functions together.

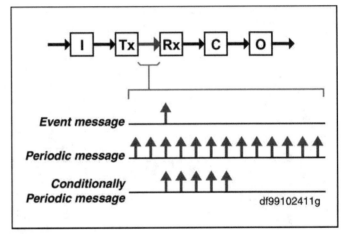

Figure 7. OSEK COM Transfer Dialogs

The Event Message

 When the single occurrence of a distributed function input change influences its related output, a single event message may be used to interconnect the function. Examples of distributed functions that might use the event message could include door unlocking, turning on/off the lights, and many of the other human-activated functions found within the driver's area. OSEK COM supports the use of event messages.

Handling of event messages may be simple or for some more critical functions may include additional processing. For instance, when sending an event message, it may be important to manage its transmission. Perhaps the application software needs to know if the message was actually sent. This requires a timer mechanism. If the message is sent successfully the application is notified and if the message is not sent, a timer will timeout and notify the application.

The Periodic Message

When a distributed function should maintain a continuous interchange with its related output, a periodic message may be more appropriate. Examples of automotive distributed functions that might use periodic messages could include continuous sending of vehicle RPM or speed.

One advantage of using the periodic transfer dialog is the high immunity to missing a message. If the information is continuously available, intermittent faults on the bus, in the hardware, or in the software should only cause limited short-term difficulty.

The use of periodic messages directly impacts the small area network bandwidth. As the number of periodic messages grows for a given distributed system, the network traffic increases. Heavy usage of periodic messages with short cycle times may impact the overall system performance.

Many early automotive electronics application-network interfaces placed the responsibility for managing periodic messages on the shoulders of the application software. But this is no longer the case. OSEK COM like many other new network software architectures provide mechanisms to manage the transmission of periodic messages. On the receiving side, OSEK COM also provides the option to manage the detection of a missing periodic message.

The Conditionally Periodic Message

When a distributed function should remain continuously interconnected during a specific set of conditions, the conditionally periodic message is an optional choice. Some examples of vehicle-based distributed functions that might use conditionally periodic messages include the driver's pressing of the speed control acceleration switch or pressing of the window up function. For each of these "more critical" distributed functions, operation is immediately stopped when the switch press is normally ended or if the interconnection is broken. OSEK COM allows the ability to implement conditionally periodic messages.

When it is used, timer-based supervision of the conditionally periodic message is a required reception process to determine the end of operation or the a potential disconnection of the distributed function.

MANAGING A FILE TRANSFER – File transfer operations similar to the download of a print file from the PC to the printer are also possible for vehicle-based networks.

While the movement of any sequentially organized information from one location to another is seldom used during normal operation for most vehicle network

architectures, it is common to see diagnostic and module flash programming activities use the file transfer mechanism.

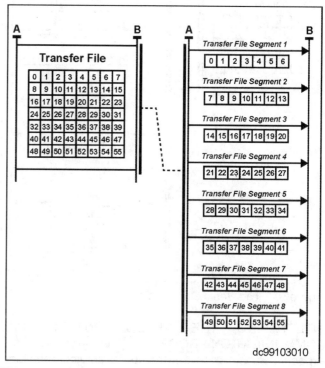

Figure 8.   Generalized File Transfer

Software capable of managing a file transfer is designated as a Transport Protocol (TP). OSEK – COM includes a transport protocol.

File Transfer Methods – Because most small area network protocols use messages with a limited amount of data, the method of transferring the file over a vehicle network requires that the entire data transfer sequence be segmented into as many messages as necessary. A general file transfer is shown in Figure 8.

If the file receiver is not capable of managing the entire transfer because of microcontroller resource limitations, some form of "flow control" mechanism may be appropriate to control the overall transfer process.

The file transfer method may include some form of acknowledgement process which could be potentially used for each message or after the complete transfer.

OSEK COM Includes a Transport Protocol – OSEK COM specifies a transport protocol suitable for large information transfers. This process is called the Unacknowledged Segmented Data Transfer (USDT).

The OSEK COM transport protocol includes a flow control mechanism that allows the receiver to manage both the transfer rate and the size of each segment. When used with the CAN protocol, the typical USDT transmit message transfer moves seven data bytes per segment.

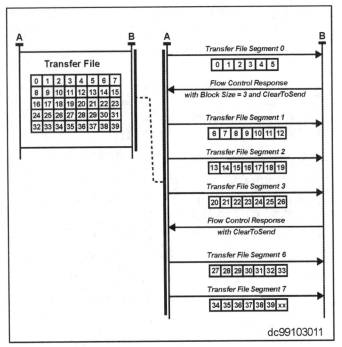

Figure 9.   The Basic OSEK COM File Transfer

INTERFACING THE APPLICATION TO THE COMMUNICATION PROCESS – In addition to the expected transmission and reception services provided to the application, a means to initialize the network software, the communication variables, the data transfer queues, and the communication protocol controller must be included in the application-to-network interface.

While many application interfaces to the network transmission process are message based, some alternate APIs allow the application to send data items directly to the network software for transmission in their corresponding message. In general, it is the application that initiates the transmit activity.

Most interfaces provide additional services for detecting and handling the loss of the network connection or the loss of a network relationship.

For network reception, several interface choices are available. The application could be made responsible for asking the network software if a message has been received. An alternate interface might allow a message reception to set a flag, which will be examined by application software during some upcoming scheduled time. Some interfaces also provide the ability for the network software to call an application function when a particular message is received.

THE OSEK COM API – Several application-to-network interfaces are defined for OSEK COM, but the number of interfaces used by OSEK COM is dependent upon which elements are selected for the implementation. For example, if the transport protocol is not used, then that particular interface is not implemented. If an OSEK Operating System (OS) building block is also used, the

OSEK COM is expanded to accommodate this additional interface.

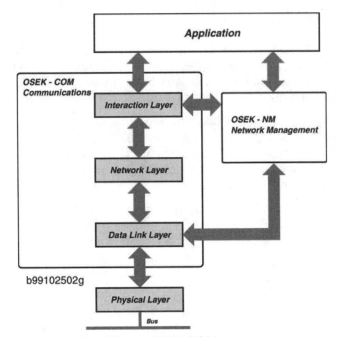

Figure 10. The APIs of OSEK COM

For the general-purpose event and periodic transfer dialogs of OSEK COM, a message-based interface rather than signal-based is used. This transmission and reception interface includes several C-language function prototypes. While these C-based implementation-specific interface concepts may be of value, it is unclear why the OSEK COM specification drifts into this software implementation area. Currently, it appears that an assembly language implementation would be a clear violation of OSEK COM's C-language interface. However, it does not seem likely that this was the OSEK committee's intention.

OSEK COM INTERDEPENDENCIES – Moving the various OSEK COM requirements into an implementation involves many interdependencies. OSEK COM requires that microcontroller resources be partitioned between the application and the various network software elements. To implement OSEK COM, the module developer must provide adequate program memory, RAM, timers, and perhaps some communication interrupt capability.

The network communication bit rate and network traffic rate influence the amount of time spent processing OSEK COM and the remaining time available to the application or other potential software components.

Whether the transport protocol within OSEK COM is used and how the transport protocol details are implemented will influence RAM requirements.

Whichever transfer dialogs are used - whether event, periodic, or conditionally periodic - the number of transfer dialogs, and the timing of transfer dialogs also impact microcontroller resource allocation.

OSEK COM may interconnect to an OSEK – Network Management (NM) or potentially connect to an OSEK – OS or functional equivalent. These additional interfaces also influence the implementation.

IMPLEMENTATION CONSIDERATIONS – The OSEK COM specification establishes requirements for a large amount of network software functionality. Should one implement all of these requirements? Of course this is one possibility. But for automotive electronic module development, it seems more appropriate to scale the software implementation down to handle only those requirements that are necessary.

For example, if the application does not require the use of periodic messages or does not need to support the file transfer capability provided by the transport protocol, there is no need to include this additional software.

It is important to understand that OSEK COM does not establish rules for implementations. Users are encouraged to scale and use only what is needed. This makes good business sense.

Faster network speeds, heavier use of periodic messages, and faster acting application software impacts a given OSEK COM implementation for motion control applications. One can expect that an OSEK-COM implementation to do a motion-control application will typically be a different scaled version when compared to a slower human-control (body bus) application.

OSEK COM USAGE – Because OSEK COM as a specification covers a wide territory of network software requirements, the actual usage of this OSEK component is difficult to measure. If one company implements a small portion of OSEK COM, does this constitute a user? If a company was using some of the general elements within OSEK COM without knowing the contents of the OSEK COM specification, does this mean they are or are not a user?

Some elements of OSEK COM will be used in upcoming DaimlerChrysler and GM vehicles that will be based on the CAN protocol. Both companies will use portions of the transport protocol as part of their respective diagnostic strategies.

WHAT OSEK - COM IS – OSEK COM is a large set of software definitions and concepts that deal primarily with small area networking information transfers. From single event and periodic message transfers through large file transfers, the OSEK COM specification attempts to be a reusable collection of comprehensive requirements.

OSEK COM is a set of communication transfer methods from which to choose from.

WHAT OSEK COM IS NOT – The OSEK COM requirements document is not an implementation. It is the whats and not the hows. Elements of OSEK COM are not necessarily required. Some items may be discarded and other cases, alternatives can be used. OSEK COM is not the complete answer. One should expect that new methods will probably be added in the future by the governing the OSEK committee. With so many choices and with all its possible options, OSEK COM is not easy to understand.

OSEK COM ADVANTAGES – There are several advantages to OSEK COM. The specification is a reasonably complete off-the-shelf set of requirements. The elements of OSEK COM cover the major software activities related to interfacing with a vehicle network solution, especially based on CAN.

From a business point of view, several scaled implementations are in or near production in Europe. OSEK continues as a growing standard and is making a transition into the ISO standards arena.

Off-the-shelf software that implements portions of OSEK COM is available. Module developers need not create specifications or implementations.

OSEK COM DISADVANTAGES – Like all existing good ideas placed onto paper, OSEK COM is not perfect. From the viewpoint of this author OSEK COM has a few significant disadvantages.

First, the document makes the assumption that the reader is already skilled in the use of small area networking technology. For the beginning distributed system developer, this produces the effect of making the document almost unreadable.

The scalability of OSEK COM is not emphasized enough. Many beginners and several of those skilled in this technology seem to think that all elements of OSEK COM need to be implemented. This unfortunate misperception impedes progress toward standardization.

Although the document appears to be too large for some readers, it is actually not big enough. Many concepts need to be presented first as generalizations (basic form) before hitting the reader between the eyes with the details.

From the American English point of view, the OSEK COM English translation and spelling is sometimes confusing and some of the European naming conventions are not consistent with the common usage in the U.S.

From the technical point of view, the specification does not encompass additional types of transfer dialogs, which are common to some companies. Also, the C-language function prototypes used to define portions of the application-to-network interface should be perhaps changed to a set of implementation recommendations. Since an alternate and equally adequate implementation is possible maybe this section could be moved to an appendix.

CONCLUSION

Distributed architectures need general and specific methods for handling network information transfers. OSEK COM is one of the first serious creations developed to handle this set of activities. An off-the-shelf prepackaged specification like OSEK COM that offers a wide range of small area networking services is certainly an advantage when one still needs to accomplish the implementation.

Across the automotive electronics industry other COM-type methods will be created, and perhaps some of these will be better. There is even the possibility that these new ideas may be incorporated into future OSEK COM revisions.

Overall, the business advantages of using OSEK COM may easily out-weigh the resource disadvantages of developing your own specifications.

REFERENCES

1. OSEK/VDX NM Specification V2.5, OSEK/VDX NM Working Group, May 1998
2. OSEK/VDX COM Specification V2.1r1, OSEK/VDX COM Working Group, June 1998
3. OSEK/VDX OS Specification V2.0r1, OSEK/VDX OS Working Group, October 1997

CONTACT

Bruce Emaus is Chairman of the SAE Embedded Software Task Force and may be contacted emaus@vector-cantech.com

2000-01-0152

Design of Intelligent Body Networks

Ralf Hadeler and Hans-Jörg Mathony
Robert Bosch Corp.

ABSTRACT

With additional new or more intelligent functions emerging in the vehicle, engineers are facing new challenges in the design of the Automotive Electric/Electronic System. New ECUs, actuators and sensors have to be provided with information, energy and control signals. In order to limit the weight of the wiring harness and to guarantee the electrical supply of the components an improved design and architecture of the body network is required.

Communication is being handled by several bus systems which are connected via gateways. Sensors and actuators are getting smarter. The existing 12V electrical supply system will be supplemented or replaced by a 42V network. Several new functions (e.g. extended systems diagnostics) will be added without additional actuators but just by additional software, requiring more computing power in the vehicle.

All this will require a systems approach for the body networks resulting in an improved functionality and reliability keeping the costs and complexity of the system limited.

INTRODUCTION

Body electronics offer a variety of well-known and new functions:

- basic functions: external and internal lighting, windshield washing and wiping, signalling
- comfort functions: seat, mirror and steering wheel adjustment, power window control, sunroof adjustment
- vehicle security functions: central door locking, keyless entry, immobilizer, anti-theft system
- heating and climate control systems
- driver information systems: user interface, instrument cluster, navigation, communication, multimedia.

The Intelligent Body Network – as we understand it – provides the sensors and actuators of the vehicle body electronics with the required **intelligence** to utilize them in an optimum way, plus the infrastructure: **communication** and **power supply**.

Today several trends are changing the world of body electronics:

- System and module approach: the vehicle is subdivided into systems and modules, reducing the complexity for development and manufacturing.
- Distributed Intelligence: additional functions will not necessarily result in additional ECUs, intelligence will be distributed.
- Networking: more and more systems and components are communicating via several networks.
- Innovative power distribution: the power distribution system is getting more sophisticated, a new voltage level (42V) is going to be introduced.

1 SYSTEM APPROACH

The challenge in vehicle E/E design has shifted from optimizing individual components like ECUs, sensors, actuators, mechanical or electrical elements to the optimization of complete systems within the vehicle. Examples are: signal and power distribution system, vehicle security system (immobilizer, keyless entry/go, anti-theft control) and cockpit electronic system (instrument cluster, driving access, driver information system, HVAC, etc.). The base technology of these systems is multiplexing: the electronic control units (ECUs) distributed across the car body are interconnected by means of a bus system. The objective is to minimize the number of wires, connectors and contacts, so even for the cars´ base equipment multiplexing can be introduced without additional costs. This is achieved by placing ECUs in those sections of a car, where electronics are required, e.g. dashboard, door, seat, roof and rear. Within a section sensors, actuators and switches are connected to the nearest electronic control unit, irrespective of the unit´s function. The systems are characterized by the following features:

- Multiple functions are integrated into a single ECU. In order to amortize the electronics´ costs multiple functions are integrated in a single ECU. Examples are functions like external and internal lighting, hazard warning control, front and rear wiper control, central door locking among others.

111

- Functions are distributed across multiple ECUs; for example the function "central door (un)locking via remote control" is typically realized by cooperation of the ECU in which the remote control receiver is integrated and the door electronic units integrating the lock function.

- Electrical distribution systems junction boxes are replaced by more expensive smart power distribution units (PDUs) including electronic functions and a bus interface. Additionally the distribution system is modularized by several PDUs into clusters. Each of these units add costs, but the cost reduction in the wiring harness and the easier handling of the system compensate these additional cost for the whole power distribution system.

- Smart sensors and actuators are naturally more expensive than simple ones, but they offer advantages in function (self-diagnostics, control), wiring and system availability, as several ECUs can utilize the same components. Actuators will include a power device, communication and maybe even intelligence. The smart actuator is reacting to the request of his „master ECU" given by e.g. serial communication and is controlling his inner control circuit independently. Functions like current limitation etc. can be handled by the actuator itself. The intelligence may be incorporated directly into the actuator or in the connector. With less control outputs on the ECUs the system becomes more flexible. An addition of a smart component is much easier than an addition of a simple component as less additional connectors, terminals etc are required. Just the software in the „master ECU" has to be adapted (control algorithm, CAN messages).

- Subsystems with sub-busses etc (e.g. doors, seats, cockpit) may include rather expensive ECUs and components, but reduce the complexity of the complete system and offer advantages in packaging.

- ECUs with rather high computing power will handle different tasks. The number of these multifunctional ECUs will be different in different cars. They will include software for „their" sensors and actuators as well as software with no dedicated component like gateway, diagnostics, personal profiling, power management etc. Thus additional functions can be added just by adding software.

There is a common denominator of all these features and examples: not an individual component is optimized (costs, packaging...), but a whole system. The optimized system may consist of non-optimum components.

Putting all these characteristics together, the most likely E/E architecture of the future is a hierarchical system, as shown in Figure 1.

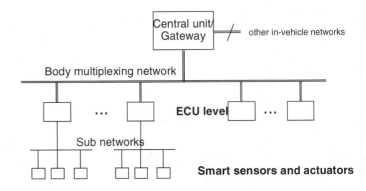

Figure 1.   Hierarchical electronics architecture

Dependent on the vehicle type, packaging, equipment and the OEMs philosophy the functions are implemented in different ECUs and integrated in a vehicle- or OEM-specific electronic architecture. This poses a major challenge to the suppliers requiring a lot of flexibility. They have to develop their software in a way that it can be transferred between different ECUs at a rather late state of a project. Even during the lifetime of a product major changes may occur caused by added features in the vehicle. And a full system supplier has to be able to include software developed and tested by another supplier or the OEM. This results in the following requirements:

- Clear decoupling of ECU hardware and software by means of a layered software architecture. The OSEK/VDX open architecture – comprising real-time operating system, communication protocols (intra-ECU and inter-ECU) and network management – is the fundamental basis for future ECUs software architecture and is the prerequisite for the introduction of software techniques and tools that support portability, reuse, software sharing and code generation. Figure 2 shows as an example a typical future ECU software architecture, including OSEK/VDX and additional hardware encapsulation software components like I/O, flash programming and diagnosis drivers.

- The functions and thus the software have to be structured in a modular way with very easy interfaces allowing easy reuse in different configurations. For this purpose Bosch has developed the CARTRONIC® approach. CARTRONIC® is an ordering concept for all vehicle control systems. The concept comprises modular expandable functions, safety and electronics architectures. Fundamental architectural structuring and modeling rules are established in order to organize networked systems independently of a special hardware topology from a functional point of view. The rules define components, their permitted interactions and patterns for recurring tasks independent from a special function. This means that the rules and patterns are a "common language" for engineers to analyze functions in the entire vehicle from the same view. CARTRONIC® is providing a consistent methodology to handle the challenges that arise by increasing system interconnections starting from the abstract level of function analysis [1, 3].

- Tools have to be available to enable software design independent of the hardware architecture. A step in this direction is the TITUS approach which was developed in a joint project of DaimlerChrysler and IBM [5]. TITUS uses the client/server model for the structuring of distributed applications. The TITUS architecture defines an application software framework with 6 different types of components which communicate with each other by use of well-defined interfaces and protocols. During the SW design phase ECU implementation aspects are not taken into account; the application software can be developed without knowledge of the target hardware configuration. The implementation phase is supported by code generators which generate the application (control) and the communication software based on the OSEK/VDX system software infrastructure. The TITUS approach makes a simple and tool-supported integration of software components possible. The software architecture is the basis to design software components which can be reused for different vehicle model lines.

2 NETWORKING

Most new vehicles nowadays offer at least one bus system for ECU communication and diagnostics.

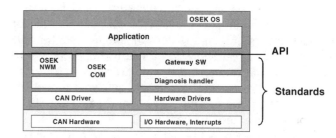

Figure 2.   Structure of future ECU architecture

2.1 IN VEHICLE NETWORKS – The requirements with respect to data transfer rate, protocol mechanism, reliability, fault tolerance and costs are dependent of their applications and have lead to the development and introduction of different network types (see Table 1). Network technology is further pushed by standards like CAN and OSEK/VDX [Kie98]. With OSEK/VDX, protocols and services in the areas of network communication, network management and ECU operating system have been specified and will be used from 2000 on in production cars by several European car manufacturers. The use of the OSEK/VDX standards simplifies significantly the development of network-based distributed applications and the integration of ECUs and software coming from different suppliers.

In future cars the communication across network limits will increase. Functions like HMI, user profile, power supply management, tele-diagnosis, etc. will require integrating all ECUs in a global vehicle network. This global network is the interconnection of the different vehicle networks by means of special gateways.

Table 1.   Characteristics of in-vehicle networks

Network	Interconnection of	Data transfer rate	Network features	Data transfer mechanism	Example
Powertrain network	Engine Man., ABS, TCS, VDC, Transm., Suspen., ACC, etc.	Typ. 500 Kbit/s	Multi-master bus, twisted pair line	Asynch., mess. priority based bus access, cyclic and event driven transfer	CAN High Speed
Body network	Door ECU, Seat ECU, Instrum. cluster, Climate control, Keyless go, etc.	< 125 Kbit/s	Multi-master bus, twisted pair line, fault-tolerant physical layer	Asynch., mess. priority based bus access, cyclic and event driven transfer	CAN Low Speed (Europe) J1850 (US)
Multi-media network	Navigation, Video, Audio receiver, Audio amplifierHMI panels, DisplaysCDC, Mobile phone	5 – 100 Mbit/s	Plastic optical fiber, Ring topology	Synchr. transfer of audio and video data, Asynch. transfer of control data	D2B optical MOST (Media Oriented Systems Transport) MML (Mobile Media Link)
X-by-wire network	x-by-wire ECUs, e.g. brakes, steering	1 Mbit/s, 2 Mbit/s	Fault tolerant physical layer, 2 wires	Time-triggered (TDMA), Static mess. scheduling, Global time	TTP/C

2.2 GATEWAYS – A gateway node is an ECU which is connected to at least two different networks. Its function is to transfer data from one network into another network and vice versa. Basically, there are two approaches for gateway architectures: a distributed system with several point-to-point gateways with each gateway connecting two networks, and a central multipoint gateway with all networks connected to a central ECU. A special case of a distributed gateway is the gateway bus, where each network is connected via a gateway node to the gateway bus, so that information can be exchanged between all networks via the gateway bus.

Distributed gateway – In a distributed gateway system several gateway nodes are implemented to achieve a global network. Typically the gateways are no standalone ECUs but integrated with other functions in a multifunctional ECU. To enable data exchange between each pair of N networks at least N-1 gateways are required. To have a minium distance of 1 between all N networks, $N*(N-1)/2$ gateways are necessary. In this case each network is directly connected with each other network by means of a gateway.

Central gateway – In a central gateway a single ECU integrates the gateway function. The microprocessor of the central gateway has access to several networks and is responsible for exchanging data across the different networks. Compared to the solution discussed above the central gateway realizes a minimum distance inter-network communication with the minium amount of hardware.

Gateway bus – The gateway bus solution is a special case of a distributed gateway system. Each network has access to the gateway bus via its own gateway node. In case of N networks, N nodes with a gateway function are required. The distance between two arbitrary networks is always 2.

Besides the general solutions described above, there are a lot of specific hybrid solutions which result from partitioning a set of networks into different clusters where each cluster is interconnected according to one of the described architectures.

Comparison – The central gateway has clear advantages in terms of system costs and latency time. However, in case of a gateway failure the corresponding networks are isolated. Fault-tolerance could be achieved by redundancy in form of additional gateway units or a central gateway with a multi-processor architecture.

The gateway bus is a very flexible solution, if new networks or messages are to be added; however it is a costly solution as an additional bus is required.

The distributed gateway system is feasible for a small number of networks. In case of a single gateway failure only a single network is isolated.

For more details see [6].

2.3 SUB-NETWORKS – Driven by semiconductor technology there is a continuous trend to integrate electronics into sensors, actuators and switch panels. The communication interface of smart sensors and actuators is rather simple compared to multiplexing networks like CAN. In most cases a low-cost SCI or UART-based master-slave communication scheme is used. The communication line is typically a single wire, the signal levels correspond to the 12V ISO 9141 physical interface (k-line), data transfer rate is 9.6 Kbit/s. Examples of such sub-busses are described in [7] and [8]. Application areas for sub-networks are for example:

Area	Sensors, actuators
Cockpit	Switch panels, steering wheel column stalk, radio satellite, air conditioning motors, displays, etc.
Door	Switch panel, lock, window lift motor, mirror (see Fig. 3)
Seat	Seat adjustment motors, switch panel
Roof	Sensors, roof adjustment motor, interior lighting control
Others	Alternator, headlamps, headlamps adjustment, smart wiper etc.

Currently there are activities driven by Motorola and european car manufacturers to establish a international standard for in-vehicle sub-networks. For more details, see [4].

Sub-networks will be connected to the entire vehicle network by means of gateways, which are integrated in the ECUs of the multiplexing bus, so that the resulting network is a hierarchical system (see Figures 1, 3). In a hierarchical electronics architecture, the smart sensor/actuator level components provide clearly defined functions and services which can be accessed by the higher levels in order to implement ECU- and cross-network functions.

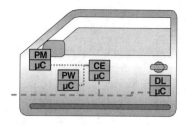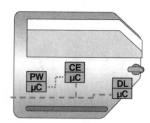

Advantages:

- Reduction of wiring harness and simple assembly
- Use of identical components
- Reduction of HW variants
- Higher reliability and lower costs because of mechatronics
- Easy integration of further functions

| – – – | CAN-Bus |
| .... | Sub-network |

PM: Power Mirror
PW: Power Window
CE: Control Elements
DL: Door Latch

Figure 3.   Sub-network within the door (example)

3. POWER DISTRIBUTION SYSTEM

With more and more functions being requested in upper class vehicles the power demand is increasing rapidly. Additionally efforts to reduce fuel consumption and emissions lead to the tendency of replacing direct drives by electronically controlled drives. This increases the demand of electrical power while reducing the demand of „mechanical power" – thus fuel. As the conventional 12V power supply system is hardly able to provide the required currents, the industry is discussing the introduction of a 42V PowerNet.

3.1 INTRODUCTION SCENARIO – Potential high power electric consumers include:

- Front window defroster
- Auxiliary electric heating
- Electromagnetic valve control
- Electric power steering
- Electromechanical brakes
- Electromechanical suspension control
- Heated catalyst
- Integrated starter/alternator
- Rear window defroster
- Electric pumps (ABS, cooling, water..)
- Electric motors (power seats...).

The introduction scenario of 42V will be different for each vehicle manufacturer. Main motivations are:

- Some loads require a higher voltage level (e.g. front window defroster).
- The current 12V (14V) system cannot supply the power requested by new loads (e.g. heated catalyst) anymore.

- Wiring harness or power dissipation are reduced.
- The redundant 12V/42V power supply is used for safety critical systems (e.g. electric steering/braking/ locking)
- The integrated starter/alternator is utilized as auxiliary drive to reduce emissions and fuel consumption.

With a great number of 12V loads already introduced in today's vehicle, a complete conversion to a pure 42V car is not likely. So a dual voltage power supply will be the common architecture for the next several years. The major additional costs (DC/DC converter) and the additional failure modes (shortcut between 14V and 42V) will in the long run drive automotive manufacturers and suppliers to a complete conversion to 42V. The new 42V PowerNet will then display 42V as the only common voltage level with some locally available lower voltages (12V, 5V..).

3.2 POWER DISTRIBUTION ECU – Back bone of the Power Distribution System are the wiring harness and the smart junction boxes or Power Distribution Units (PDU). The PDUs as the active part protect, switch and control the electric consumers and provide communication. Additionally they divide the E/E-system into smaller clusters reducing complexity of the system and the handling of the wiring harness (s. figure 4).

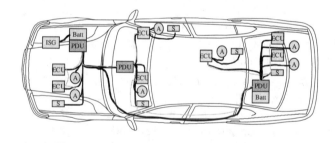

Figure 4: Clusters created by different PDUs

The transition from junction boxes to PDUs offers the chance to add intelligence to the Power Distribution System:

- improved diagnostics, including system diagnostics
- fail safe functions, keep alive functions
- personal profiling
- load management and management of the quiescent current.

Improved systems diagnostics will only partly be incorporated into the PDUs as they require a lot of computing power. But by diagnosing their subsystem and providing "meta-information" the PDUs ease the task of a central body computer responsible for system diagnostics and the proper communication with driver and garage. With improved on-board-diagnostics the PDUs identify defect actuators and sensors, and drive the system into a safe state or replace the defect function by a similar feature without adding cost to the original actuators. So a defect low beam lamp function may be replaced by pulse width modulating the high beam lamp; and with a detected failure of the lamp switch all lights may be turned on guaranteeing a safe vehicle.

Personal profiles allow to adapt comfort functions to the individual driver or passenger of the vehicle. The stored characteristics of the users are transferred to the respective actuators which are set to the predefined status.

Load management and management of the quiescient current shall prevent discharge of the battery and allow reduction of battery and alternator by taking better advantage of their abilities.

3.3 DUAL VOLTAGE SYSTEM ARCHITECTURE – Up
to now no standard architecture for a dual voltage network has been established. Some alternatives are:

- 42V alternator, DC/DC converter
- Alternator with dual voltage output
- 42V alternator, 12V alternator.

Most alternatives include a 36V as well as a 12V battery. Instead of a starter battery a "supercap" may be used as well. The approach with a DC/DC converter may incorporate several DC/DC converters to generate local distribution networks. The other approaches may include a (small) DC/DC converter as well for improved flexibility and jump starting.

Power generation by 42V alternator and DC/DC converter may be the best solution for a system with low power demand on 12V. If with the introduction of 42V only a small number of loads is on 42V and the majority stays on 12V, the DC/DC converter has to provide lots of power, creating a rather costly solution. In this case the introduction of two different alternators becomes an attractive solution.

The power distribution system is divided into several clusters by the addition of several PDUs. A very attractive solution appears to be the combination of PDU and DC/DC converter. The combination of a central PDU/Converter offers advantages in costs and failure handling, while the combination of local PDUs with DC/DC converters allows the reduction of the wiring harness and a very easy system approach (s. figure 5).

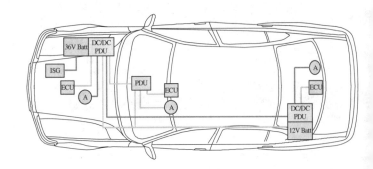

Figure 5: Dual voltage architecture with combined PDU/Converter

3.4 FAILURE HANDLING
– A new technical problem will be the potential shortcut between the 14V and the 42V voltage levels. Shortcuts are an issue in every electrical network, but the automotive dual voltage network displays a special problem. Whenever a shortcut between 42V and a 14V device switched by a MOS switch occurs, not only this 14V device, but the whole 14V network is connected to the 42V level. This is due to the unavoidable reverse diode of MOS switch S1 (s. figure 6).

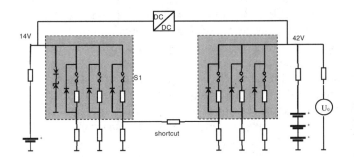

Figure 6: Shortcut between 14V and 42V

The most effective way to protect the 14V loads is detecting the "driving" 42V load and turning it off as soon as possible. In order to switch off the source of the short cut current unprotected 42V lines have to be avoided. In a very common architecture (see figure 7) the powerlines between starter/alternator, battery, DC/DC converters and PDUs can only be protected by switching off the smart battery post disabling the whole vehicle.

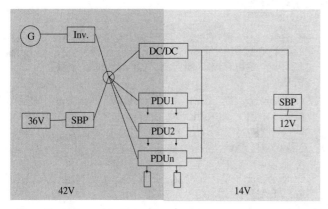

Figure 7: common architecture

To reduce the possibility of this major fault, some general rules should be applied:

- combine DC/DC converter and one PDU to a "Master-PDU/converter"
- mount Master-PDU close to battery
- mount battery close to starter/alternator.

Additionally

- protect all "satellite-PDUs" by the Master-PDU
- protect all 42V loads in PDUs
- minimize unprotected 42V lines
- avoid neighboring 14V and 42V lines.

To reduce the effect of the shortcut to the 14V loads, basically two different methods are possible. The first alternative avoids switches with a reverse diode, separating the shortcut from the 14V net. The 14V net is saved, but major problems may appear detecting the fault and the driving 42V switch, as the additional current through the 14V device may be small compared to the standard current through the 42V source.

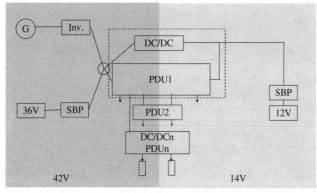

Figure 8: Proposed architecture

The second alternative is installing a "sacrificial device" on the 14V level. This device has to accept the whole current from the short cut reducing the overvoltage in the system and protecting the 14V loads until the short cut is detected. It has to stand currents close to 1000A and to limit the system voltage to a level of approximately 24V, still allowing the 14V and the 42V components to work. With this high current the source on the 42V level can easily be detected and turned off.

Different sacrificial devices are currently under investigation.

Both strategies turn off the minimum number of components or subsystems in case of a failure and protect all but one 14V device. Proper architectures, devices and strategies have to ensure the detection is fast and no device is destroyed by the current through short cut and other devices.

3.5 TECHNOLOGY – With the introduction of 42V using semiconductors is getting more (cost) attractive. Relays (with fuses) may be replaced by solid state relays, and fuses may be replaced by solid state relays or polyswitches. Tables 2 and 3 show some advantages, disadvantages and features of the different alternatives.

Table 2. Comparison of solid state and mechanical relays

Solid State Relays	Mechanical Relays
Current sensing, diagnostics	-
PWM control possible	-
-	Complete blocking, no reverse current
-	Different control and load circuits
Reliable	Robust (against overload)
Low control power	No quiescient current
Advantages in ECU construction	Easy servicing
Cheaper with small currents	Cheaper with high currents

Table 3.   Comparison of fuses, solid state relays and polyswitches

Fuses	Solid State Relays	Polyswitches
-	Additional functionality	-
-	Self healing	Self healing
Low power dissipation	Medium power dissipation	High power dissipation
-	Quiescient current required, Load management possible	-
High currents	Medium currents	Low currents
Low costs	High costs	Medium costs

It is expected that most relays (especially for low power devices) will be replaced by semiconductors, while fuses will still be prominently displayed in PDUs. Small loads may be protected by polyswitches while solid state relays will only be used for fusing in special cases.

CONCLUSION

In an effort to enhance the vehicles' comfort, safety and efficiency the electronics are getting more and more complex: huge software modules, a large number of ECUs, and different voltage levels are the flip side of the continuous evolution of the automobile.

A consequent systems approach is introduced in order to control the complexity of body electronics. Intelligent Body Networks organize the intelligence, communication and power supply in a novel way and open the door for more innovation and customer satisfaction.

REFERENCES

1. T. Bertram, R. Bitzer, R. Mayer, A. Volkart, " CAR-TRONIC – An Open Architecture for Networking the Control Systems of an Automobile", SAE Paper 980200

2. U. Kiencke, D. John, "On the way to an international standard for automotive applications – OSEK/VDX", Convergence 1998, Paper No. 98C012, see also: ftp://www.osek-vdx.org/

3. J. Weber, T. Bertram, T. Kytölä, F. Peruzzi, S. Thiel, "Information Technology Restructures Car Electronics", SAE Paper 99-01-0485

4. H.Chr. v. d. Wense, W. Specks, A. Krüger, "Anforderungen und Implementierungen von Sub-Bussen in hierarchischen Automobilnetzwerken", Conference "Embedded Intelligence 99", Nürnberg

5. P. Lanches, J. Eisenmann, et. al. "Client/Server Architecture – Managing New technologies for Automotive Embedded systems – A Joint Project of Daimler-Benz & IBM", Convergence 98, Paper No. 98C014

6. H.-J. Mathony, J. Maier, "Trends in Vehicle Body Electronics", International Symposium "ATA-EL 99", Como, Paper 99A1009

7. G. Bierbaum, "Multiplex bus system for sensors and actuators in automotive A/C systems", Conference "Elektronik im Kraftfahrzeug", Baden-Baden 1998, VDI Berichte 1415, 315-324

8. J.W. Specks, A. Rajnak, "Die skalierbare Netzwerkarchitektur des Volvo S80 mit Echtzeit-Betriebssystem und Mechatronikbausteinen", Conference "Elektronik im Kraftfahrzeug", Baden-Baden 1998, VDI Berichte 1415, 597-613

2000-01-0151

Intelligent Message Prioritization for System Integration Through Vehicle Networks

Y. Ni, C.O. Nwagboso and A. Zhang
University of Wolverhampton

J. Samuel
Dearborn Electronics (UK) Ltd.

ABSTRACT

With increasingly more in-vehicle systems and functions being implemented in a modern vehicle, wiring constraints demand that these features share resources. From a cost-effective systems integration viewpoint, it is required to connect these functions/devices on a single network such as CAN. As the number of systems and functions within the network grows, there is the need to intelligently priorities the messages taking into account the dynamic scenarios with the vehicle-operating environment.

This paper addresses the problem by presenting two message prioritisation methods, namely static grid association and a dynamic prioritization method with an incident detector. The static grid association method uses a message based systematic technique to assign message priority in the network design stage. The development of a fuzzy incident detector enables the dynamic assignment of message priority within the network. An application example is also presented.

INTRODUCTION

The ever-increasing number of electronic and electrical units and functions in a vehicle has greatly enhanced vehicle performance and value. The resulting proliferation of these electronic modules, however, has rendered the wiring harness, necessary to interconnect these systems increasingly large, heavy, complex and difficult to route. This further compounds the problem of limited package space and the demand for reliable and cost effective assembly. Moreover, as the result of this electronic proliferation, the amount of information shared among different modules has increased significantly. The introduction of on-line diagnostic techniques and remote monitoring and control for commercial vehicles has further increased the volume of data. These were the major reasons that prompted the industry to introduce the multiplexing and vehicle data network techniques to replace traditional wiring harnesses for information exchange on a vehicle.

Incorporation of vehicle data network systems into vehicle designs provides opportunities in reduced wiring complexity and packaging flexibility. Information sharing is made possible with the bi-directional communication capability of these networks. Potential benefits also include avoidance of duplicate components, provision for greater commonality of wiring harness for different vehicles, improvement of diagnostic capabilities, and easier addition of electronic features and options to existing product designs. Properly designed vehicle networks can result in reduced manufacturing and assembly costs and improvements in efficiency, safety, and comfort for vehicle users.

With drivers spending more time in their vehicles, and with the increasing potential of the vehicle being used as an extended part of the office and home, the demand for in-vehicle data communication will increase. Against this background, there will be increasing demand for high quality sound systems, Internet access, multiple playback devices, intelligent transport systems and other types of systems and devices. For example, in response to the increasing user requirements for these systems, many automotive systems manufacturers have started to introduce in-vehicle digital audio and video that requires high data transmission between 1Mbps - 10Mbps. From a systems integration viewpoint, it is the case that these devices and functions are connected on a single multimedia network This type of network will have to work with other types of in-vehicle network through an appropriate gateway. The issue of message priority within these networks will become more important as the number of functions, devices and systems increase in the network.

Vehicle network architecture strategy requires the integration of electronics into the sensors, actuators and telematic devices so that they can communicate over a single bus into the module that utilizes them. As a result, information latency could be introduced in this structure as all the information appears on a single bus and is exchanged on a media-sharing basis. Messages that require transmitting at the same time must compete for access to the network. In fact, when all the traditional signals of a system have been mapped into the single wire network, the assumption has been made that any of the previous individual signals require a rather low communication. This is not always the case, particularly for those time critical messages. They not only need a high communication rate for fast information throughput but also need immediate bus access granted by arbitration mechanism of a bus protocol.

CAN uses CSMA/CD (carrier sense multiple access with collision detect) with non-destructive bit wise arbitration for bus contention. The bit wise arbitration ensures that the highest priority frame is not delayed during collision. The arbitration process is normally implemented by hardware mechanism in a communication protocol. If bus competition occurs, it determines which message should have the priority for access to network bus at the given moment based on an individual message identifier. In the application layer of a vehicle network protocol, some bits of a message identifier are specified to indicate what the priority is for the message. The prioritization of a message is usually done by assigning default values through intuitive knowledge in design and will not be changed once the integration design is complete. This obviously does not reflect the importance of a message properly in terms of a time-critical requirement. Moreover, equal treatment for message priority may not be sufficient to meet the demand for rapid access of some time-critical information when some urgent actions need to be taken. In this paper, an intelligent message prioritization has been proposed.

A static method is first suggested which performs message prioritization by associating individual messages with the incident scenarios encountered in vehicle operations. By introducing a fuzzy incident detector, a dynamic message prioritization method has been proposed. The qualitative priority for the messages that are associated with a selection of monitored incidents can be dynamically determined in the suggested method when the monitored incidents occur and detected by the incident detector priority. An example has been used to show the effectiveness of the proposed message prioritization methods.

QUALITATIVE DESCRIPTION OF MESSAGE PRIORITY AND ARBITRATION MECHANISM

The message prioritization process examined in this paper can be viewed as a decision-making process based on a message's degree of urgency for obtaining bus access in given circumstances. Apparently, it is easier to describe the priority of a message in a qualitative way than in a quantitative way. Moreover, the qualitative description can be combined easily with the knowledge representation of human experience and

domain expertise. In view of this, a message's priority is interpreted in this paper as a linguistic variable P with a qualitative term set representing the different degree of message priority. Depending on how urgently a message needs to be transmitted on a data bus, the qualitative term set is designated as below:

$$P = \{EH, VH, HI, ME, LO, VL, EL\} \qquad (1)$$

These terms indicate that the priority of a message, when transmitted, is extremely high, very high, high, medium, low, very low, and extremely low, respectively.

The datalink and physical layer in most vehicle network protocols implement the bus access mechanism, which negotiates the bus access among the messages waiting to be transmitted according to their priority. These layers embody a message's priority usually by incorporating the information into a part of the message itself and then the arbitration process is physically performed based on the message contents in some non-destructive way. This ensures that only a message with the highest priority can get the bus access at the given moment when conflict occurs. Figure 1 shows this non-destructive arbitration mechanism used by CAN network, one protocol commonly used in automotive industry.

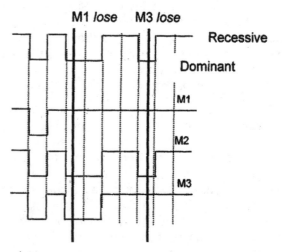

Figure 1.

In CAN protocol, message priority is embedded in the identifier part of a message. Figure 1 shows that three messages are requesting the bus access at the same time. Conflicts occur and are resolved by bit-wise arbitration on the identifiers in accordance with the "wired" mechanism. By this mechanism, dominant state (logical 0) overwrites the recessive state (logic 1); thus the competition will be lost by all these messages with recessive transmission and dominant observation. During the arbitration process, all messages that have been lost will automatically become receivers and do not re-attempt transmission until the bus is available again.

The message priority either predetermined or dynamically inferred from the methods described below, needs to be mapped as a part of message format for hardware arbitration. Some bits of a message are often reserved at the application layer of a protocol for this purpose. The mapping relationship between the qualitative descriptions of the message priority and these bits can be easily established and will be embedded as a part of the message. These, in conjunction with other parts of a message's identifier, will be used to resolve the conflicts that arise when multiple messages are requesting bus access at the same time. Table 1 shows a possible mapping between priority bits and the qualitative descriptions of the message priority, assuming that the message definition in the application layer has 3 priority bits in an identifier.

Table 1.

Priority	EH	VH	HI	ME	LO	VL	EL
Priority Bits	000	001	010	011 or 100	101	110	111

STATIC MESSAGE PRIORITIZATION

Priority for an individual message can be determined in the design stage of system integration once the devices and integration strategy have been specified. Depending on whether or not the assigned qualitative variable for the message priority is adjusted in response to the changes in vehicle conditions, driving environment, or incidents encountered during operation, prioritization process is referred as a static or dynamic process in this paper.

Static message prioritization process is a process in which the priority is determined during design stage and not changed after that. The method can be used where the bandwidth available in the prioritization field is sufficient for all known and anticipated messages and operating conditions. This may be true where not many devices are integrated in a vehicle, and thus the bus traffic is not so heavy as to have a significant influence on system performance, or in the case that the demand for latency optimization is not so critical.

The method for static message prioritization presented below is a message-based rather than device-based method. The reason for using a message-based method is that each device may contain a group of messages but these messages are not equally important from a system integration viewpoint. In many cases for a given incident, only parts of messages will be involved.

In the design stage, potential incidents to be encountered during vehicle operation are identified first. These incidents may include such events as reversing, skidding, light indicating, collisions as well as normal driving. According to the system integration strategy and desired functionality of an integrated system, each message will be associated with one or more incidents, which are frequently with different degree of urgency. For example,

messages for planar velocity signals from sensors such as a GPS receiver or radar might be used as information display in the normal driving condition, but these messages may also be required by a ABS device when a skidding incident occurs. The messages are obviously more urgently needed in the latter incident.

A grid association (GA) method is suggested to associate a message to the different incidents. The incidents identified in the design stage are represented in a incident set $I = \{I_i, i = 1,...,n\}$ with an incident priority term (w_i, i = 1,n) w_i takes value from the set $W = \{$ EH, VH, HI, ME, LO, VL, EL $\}$, which reflects the incident importance in terms of the degree of urgency for required information. Assume that a message set $M = \{M_i, i = 1,...,m\}$ has been used in the given vehicle network. For each incident, as shown in Figure 2, the messages it may acquire are associated with it by a solid circle. The associate mark also carries a weight factor $e_{i,j}$, $0 < \varepsilon_{i,j} < 1$, i = 1,...,n, j = 1,...m, to indicate how the information carried by the message j is urgently required when the incident i occurs.

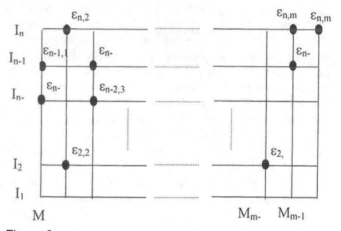

Figure 2.

Based on the information given by the GA table, a quantitative degree of urgency, u_k, for message k can be calculated as follows:

$$u_k = \frac{\sum_{i \in I_k} \varepsilon_{i,k} f'(w_i)}{\sum_{i \in I_k} \varepsilon_{i,k}} \quad (2)$$

where $I_k \in I$, and $I_k = \{I_i,$ Incident i associated with $M_k\}$, and $f'(w)$: $W \rightarrow R^1(0,1)$, is a mapping from W to a real value set (0,1) and can be defined as follows:

$$f'(w) = \begin{cases} 0.5 & \text{if } w = ME \\ 1 - \dfrac{i-1}{10} - 0.05 & \text{if } w = P_i, i = 1,2,3 \\ 0.7 - \dfrac{i}{10} + 0.05 & \text{if } w = P_i, i = 5,6,7 \end{cases} \quad (3)$$

where W_i is the ith element of **P** set. Other ways of assigning numeric values to members of the set P are clearly possible and equally valid.

The calculated u_k for message k in (2) needs to be mapped back to the linguistic variable **P**, in particular, to its qualitative terms so that the priority of the corresponding message can be embedded in the message when it is transmitted. This can be accomplished by using (4).

$$f(x) = \begin{cases} P_i & \text{if } (i-1)/10 \leq x < i/10 \text{ and } i = 1,2,3 \\ & \text{and if } (i+2)/10 \leq x < (i+3)/10 \text{ and } i = 5,6,7 \\ P_4 & \text{otherwise} \end{cases}$$

(4)

RULE-BASED DYNAMIC MESSAGE PRIORITIZATION

In the static message prioritization process described above, the priority of a message is determined in the design stage, and will not be changed during vehicle operation. Although the time-critical importance of a message has been considered in the prioritization method, such assessment is performed on the average sense in terms of all incidents in which the message will be involved. This may not sufficiently reflect the time-critical priority for some particular incidents and sometimes may result in situations where a message urgently required by some rarely occurring incidents such as a vehicle collision might be blocked by messages with higher priority on the average sense. For example, vehicle location and timing information from sensors may be viewed as not so time-critical in the normal driving situation as they may be used only for information display purpose or for route guidance. But when a collision incident occurs, these messages may be urgently required by other systems and be relayed to the collision avoidance devices and the accident recording device. These messages might be blocked by some other messages on the bus that have a higher priority in the average time-critical importance sense if the priority for these messages has been determined by the static method.

It is clear that the prioritization process can be improved if the message priority can be assigned according to different incident scenarios. A rule-based method has been presented here to dynamically prioritize a message in response to scenario changes in vehicle operation. The basic idea is that the message priority, initially determined by the static prioritization process using the grid association method described above, can be changed dynamically based on the knowledge of which incident the vehicle operation is involved in. A fuzzy incident detector has been introduced for selected incidents which require urgent bus accesses when they occurr. Figure 3 is a schematic diagram of the proposed rule-based dynamic message prioritization method.

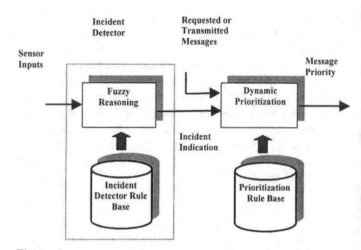

Figure 3.

The signals from specified sensors are fuzzificated as inputs of a fuzzy reasoning engine. Based on heuristic knowledge, an incident detection rule base can be established. This, in combination with the fuzzificated sensor inputs, allows the fuzzy reasoning engine to perform reasoning and to indicate what current incident the vehicle is involved in. A dynamic prioritization process is triggered and the status of the messages requesting bus access is fed to the prioritization reasoning engine. By means of the prioritization rule base, in combination of the indicated current incident, the reasoning engine dynamically determines the priority of the individual fed messages. This will then be dynamically mapped to the corresponding bits as the current priority of a message.

The establishment of the prioitisation rule base can be based on domain expert knowledge as well as statistical latance data. The elements of the rulebase may be in the form as follows:

If incident k is Π_k and message belongs to group M_j
Then priority is P_i

Where Π_k is the output of the incident detector, I_k belongs to the incedent set to be monitored, MG_i is the message group which has an association with incident I_k. P_i is priority qualitative variables and $P_l \in$ {EH, VH, HI, ME, LO, VL, EL}

Note that the content of the monitored incident set can be chosen in a flexible way. As the more incidents are included in the set, the more sensor signals will be required, and the heavier the computation cost for a reasoning engine will be, thus it is not necessary to take all incidents as monitored ones. A system designer can select incidents that need to be monitored based on the overall bus traffic load of a integrated system, hardware signals availability for an incedent detector, urgentness of action-taking, etc. It is also possible to combine the functions with that of a accident warning and on-line diagnosis devices which may be also available on a integrated system.

122

APPLICATION EXAMPLE

We take the priority assignment of the logitudinal velocity message, M_l, $l \in [1,m]$, as an example to demonstrate the effectiveness of the proposed methods. Assume that the message will be used in the following incidents in an integrated system: speed caculation for information display I_s, control signal caculation for automatic cruise controller I_u, and collision avoidance I_v, s,u,v, $\in [1,n]$. In the design stage, a default message priority is determined using the grid association method. The message association with the incidents mentioned above is shown in Figure 3, and for simplicity only the grid line relevant to the M_l is shown in the figure.

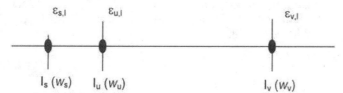

Figure 4.

Based on overall system performance consideration and heuristic knowledge, the priority w_s, w_u, w_v for the incidents of interest, I_s, I_u, and I_v are assigned as **{VL,ME,EH}** and weight factors for the message M_l for these incidents , which represent the message relative importance in a given incident, are 0.9, 1.0, 0.8, respectively. The resulting default message priority, determined by (2), (3), (4), is ME. Apparently this priority is weighted on all the incidents in which the message is involved and reflects a relative degree of urgency of the message in terms of overall performance requirement after the incorporation of human knowledge.

When a vehicle is in operation, the driving scenories are constantly changing. A dynamically assigned priority can ensure the most urgently demanded messages get the quickest access to the vehicle network bus. In the discussed problem, when the vehicle is dealing with the collision incident, it is desirable for the message's priority to be raised accordingly in consideration that the message may be urgently needed from a counter-measure algorithm. This can be achieved by using the dynamic prioritisation process described above.

We assume that the collision incident has been included in the monitored incident set of a vehicle. and that a radar sensor is available for measuring range R and range rate V from the host vehicle to a target vehicle. These measurements can be used to construct a fuzzy incident detector using the following membership functions and the rule table shown in Figure 4.

$R = \{R_1, R_2, R_3\}$ = {Small, Medium, Big}

$V = \{V_1, V_2\}$ = {Small, Big}

$\Pi = \{\Pi_1, \Pi_2, \Pi_3, \Pi_4\}$ = {UL (Unlikely), PO(Possible), LI(Likely), VL (Very Likely)}

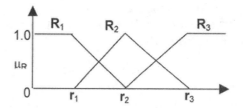

(1) Range Member Function

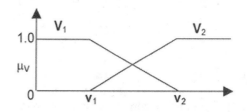

(2) Range Rate Function

	R_1	R_2	R_3
V_1	PO	UL	UL
V_2	VL	LI	PO

(3) Rule Table

Figure 5.

Based on the sensor measurements, conventional fuzzy reasoning mechanisms can be used to calculate the firing strengths, and the possibility measure π_i of each linguistic term of Π, which is defined as the maximum of the firing strength over the rules with the same linguistic term as their parts over the whole rule base.

The part of the prioritisation rule base corresponding to the collision incident and the logitudinal velocity message can be as follows:

If rear-end collision is UL then the p_{Ml} = Default

If rear-end collision is PO then the p_{Ml} = HI

If rear-end collision is LI then the p_{Ml} = VH

If rear-end collision is VL then the p_{Ml} = EH

The new quantitative degree of urgency of the message can be determined by:

$$u_{Ml} = \frac{\sum_i \pi_i f'(w_i)}{\sum_i \pi_i} \qquad (5)$$

where π_i is the possibility of a linguistic term Π_i which appears on the condition part of the rule base. The dynamic priority of the message can then be determined by mapping u_{Ml} using (4). The table shows results in light of different sensor measurements when a vehicle is operating in different scenarios. The default priority is chosen as ME. The membership function parameters used for this example are $r_1 = 150m$, $r_2 = 100m$, $r_3 = 25m$, $v_1 = 20$ mph, and $v_2 = 90$mph.

Table 2.

Sensor Outputs (R,V)	(50,40)	(200,65)	(100,85)	(30,85)
Priority p_{M1}	HI	ME	VH	EH

CONCLUSION

Integrating systems through a vehicle network has greatly alleviated the wiring harness complexity caused by recent electronic device proliferation in a vehicle. Determination of message priority is important, as all the signals will be transmitted on a time-sharing media, particularly for the integrated systems with a heavy busload on the vehicle network. Inappropriate assignment may result in the blockage of time-critical messages in some situations.

The paper has attempted to address the problem by presenting two message prioritization methods: Static grid association and dynamic prioritization with an incident detector.

The static grid association method presents a message-based systematic way to assign message priority in the design stage. The message priority in this method is basically weighed on the degree of importance of the incidents in which the message involved and the relative urgency of the message in the given incident. As message priority is not changed during the vehicle operation, this method may be limited to the situation where the busload is relatively light or the latency problem is not so critical.

Introduction of a fuzzy incident detector enables the dynamic assignment of message priority. The prioritization is improved in the sense that message priority can be changed in response to the changing scenarios. The example presented in the paper has demonstrated the method's effectiveness with a sensible priority assignment corresponding to the detected scenario changes. Note that this is achieved with additional computing cost. This burden could be relieved by limiting the number of monitored incidents and by combining with some other existing accident detecting mechanisms.

ACKNOWLEDGMENTS

The research in this paper is funded by the EPSRC and the authors gratefully acknowledge their support. We are also grateful for the support provided by the FRETSET research project consortium members.

REFERENCES

1. Y.Inoue, S.Uehara, K.Maguuro, Y.Hirabayashi, Multiplex System for Automotive Integrated Control, SAE Technical Paper Series 930002, 1993

2. Wolfhard Lawrenz, CAN System Engineering, Spinger-Verlag, 1997

3. ISO 11898, Road Vehicle-Inter change of Digital Information-Controller Area Network (CAN) for High Speed Communication, 1993

4. J.Unruh, H.J.Mathony and K.H.Kaiser, Error Detection Analysis of Automotive Communication Protocols, SAE Paper 900699, 1990

5. S.Dais and M.Chapman, The Impact of Bit Representation on Transport Capacity and Clock Accuracy in Serial Data Streams, SAE Paper 890532, 1989

2001-01-0072

LIN Bus and its Potential for Use in Distributed Multiplex Applications

John V. DeNuto, Stephen Ewbank, Francis Kleja, Christopher A. Lupini and Robert A. Perisho Jr
Delphi Automotive Systems

ABSTRACT

The increasing features and complexity of today's automotive architectures are becoming increasingly difficult to manage. Each new innovation typically requires additional mechanical actuators and associated electrical controllers. The sheer number of black boxes and wiring are being limited not by features or cost but by the inability to physically assemble them into a vehicle. A new architecture is required which will support the ability to add new features but also enable the Vehicle Assembly Plants to easily assemble and test each subsystem. One such architecture is a distributed multiplex arrangement that reduces the number of wires while enabling flexibility and expandability. Previous versions have had to deal with issues such as noise immunity at high switching currents. The LIN Bus with its low cost and rail-to-rail capability may be the key enabling technology to make the multiplexed architecture a reality.

INTRODUCTION

HISTORY OF MULTIPLEXING – The multiplexing of automotive electrical data onto communication buses dates back to the late 1970s (1). This technology is only beginning to reach its stride. It was originally hoped that a single bus protocol could handle the needs of any vehicle. Gradually that expanded to the SAE categorization of Class A, B, and C and the realization that up to three protocols and/or networks may be necessary.

Today, it is realized that at least seven in-vehicle protocols may be necessary (2). These include, besides the existing SAE classes - diagnostics, airbag, mobile media, and x-by-wire. Each area needs its own protocol and one or more networks running that protocol. Sometimes this is for safety reasons, such as with airbags or x-by-wire. Another reason for such a "niche"

protocol is specialty use. An example of a niche protocol is for the application of "smart sensor and actuator" technology, sometimes also known as "smart connector". This relatively new area of automotive electronics involves multiplexing low-level functions such as relays, high/low side drivers, etc. onto a data bus. For now smart connector functions are lumped into the SAE class A category. Perhaps in the future it will become its own class. Indeed, there are advanced vehicle electronic architectures being considered that utilize up to 200 multiplexed sensors, actuators, and lamps. Each one of these devices becomes a node on one or more smart connector buses. The challenge is to make it cost effective.

Multiplexing has always had an issue with cost. Replacing some discrete wires and connectors with a data bus interface tends to add cost. There are benefits associated with increased diagnostic capabilities and system expandability, but the bottom line is that the ICs necessary for the multiplex circuit add additional cost – making smart connector multiplexing appear unattractive. It is well known that electronic systems decrease their cost more quickly than traditional wiring. This makes the eventual cost effectiveness of multiplex inevitable, a question of when, not if.

Up to now the buses used in the first, basic implementations of smart connectors, have been custom or proprietary (e.g. CCD, BEAN, etc.), too complicated (e.g. TTP/A) or too slow and cumbersome (e.g. ISO 9141) (3).

The latest entry into the low-cost smart connector arena is Local Interconnect Network (LIN). With LIN lies the hope of a worldwide smart connector data bus protocol standard (4).

Specific advantages of LIN bus when applied to a distributed multiplex architecture include:

- Standardization
- Low cost silicon
- 12V single wire interface
- Self Synchronization without a crystal
- Guaranteed latency times
- Speeds up to 20Kbit/s

OBJECTIVES

A distributed multiplex architecture is currently being developed by Delphi Automotive Systems. This new architecture increases the electronic content while decreasing the conventional wiring content. The multiplex nodes are referred to as Smart Connectors. The goals of this approach are to offer the customer greater flexibility and better packaging at a reduced sub-system life-cycle cost.

BENEFITS

The potential benefits of a distributed multiplex architecture are listed below:

- Simplified vehicle assembly
- Common application on multiple vehicles resulting in higher volumes and lower costs
- Significantly increased flexibility – the ability to add features to a vehicle in a plug & play manner
- Improved quality and reliability due to a significant reduction in the number of wires
- Simplified wiring assembly and reduced mass

Many of these benefits are compounded when the same Smart Connectors are applied to multiple platforms.

CHALLENGES

The primary challenge to implementing a distributed multiplex architecture remains cost. The cost increase in electronics must be offset with a decrease in the cost of the wiring system and other structural cost improvements. A cost competitive solution for a distributed multiplex door sub-system has been proposed by Delphi.

In order to keep the electronic cost increase as small as possible, increased volumes, penetrations, and packaging improvements must be realized. Only then can cost parity be achieved.

The second challenge lies in the area of thermal management. Specific new packaging techniques are being applied to Smart Connector designs in order to minimize package size and thermal resistance. It is imperative to obtain the minimum thermal resistance with no increase in part cost. Advanced computer simulations are being employed to obtain the optimum packaging solution.

The third challenge is to demonstrate the performance of such technology with technical demonstrations and test data. With shortened product development cycles, OEM's are requiring this level of development prior to including a new technology into one of their programs. We are meeting these needs with functional demonstrations of distributed multiplex sub-systems in seats and doors and comprehensive testing.

Finally, we must overcome our natural resistance to change. The old adage "if it isn't broke, don't fix it" is often applied to technical innovations such as distributed multiplex. It takes a technology leader, with a real vision of the future, to see the many benefits that a distributed multiplex system has to offer but are difficult to quantify.

APPROACH

Figures 1 through 3 display three comparative architectures applied to a driver's door of a typical automotive application. In Figure 1, no multiplexing is used. Each device interfaces directly with a centralized body computer and / or bussed electrical center. This architecture possesses the least electronic content, the most wires, and the least flexibility to change.

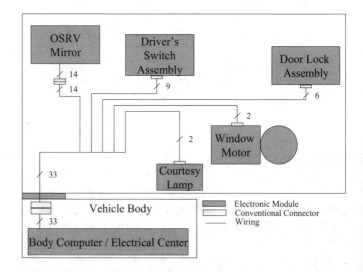

Figure 1: Driver's door with conventional (non-multiplex) architecture

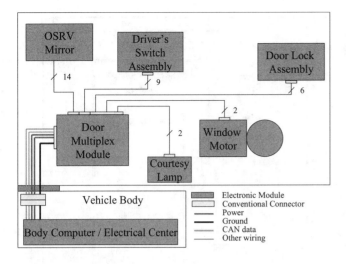

Figure 2: Driver's Door with centralized multiplex architecture

Figure 2 depicts the same door loads controlled using a centralized door multiplex module. This type of architecture normally utilizes the CAN data bus, allowing the door module to exist as a peer to the vehicle's body computer. Using this approach reduces the number of discrete wires but still limits flexibility. A change in any load can require a change to the entire door module. This technique usually results in different modules being applied to the driver's, passenger's, and rear doors as well as different modules on nearly every platform within an OEM.

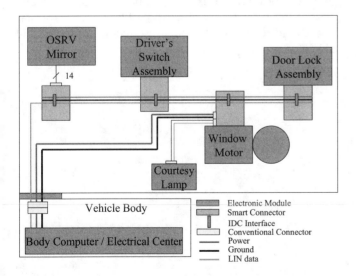

Figure 3: Driver's Door with distributed multiplex architecture using "Smart Connectors"

Figure 3 presents the same loads controlled using a distributed multiplex architecture employing Smart Connectors. In this approach, the functions of the central door module are distributed to four discrete Smart Connectors using the LIN protocol. This approach uses the most electronics and the least wiring of the three approaches presented. The four Smart Connectors defined in this system control the:

1. Window Motor
2. OutSide RearView (OSRV) Mirror Assembly
3. Driver's Door Switch
4. Door Lock Assembly

This approach offers the greatest flexibility to change and the greatest potential for inter-platform application. Four doors in a given vehicle can be executed using the four Smart Connectors detailed above. This four-door approach is shown in Figure 4.

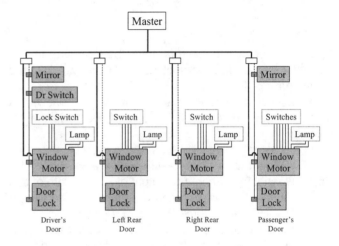

Figure 4: Four-door vehicle controlled using 4 types of Smart Connectors

Content changes in a vehicle can be executed more easily using this approach. A load can be added to a vehicle by plugging a Smart Connector to the bus and modifying the software in the master. This essentially makes the sub-system plug and play. Since the Smart connectors share a common three-wire bus structure, adding a Smart Connector need not change the wiring content. The distributed architecture also offers the possibility of automated wiring harness assembly through the use of Insulation Displacement Crimp (IDC) terminals

A comparison of the total wiring content can be seen in Figure 5. As seen in the figure, the distributed multiplex architecture uses the least number of discrete wires and enables automated assembly of the three-wire data bus. Of the remaining wires, 14 out of the 21 (66%) are associated with the mirror pigtail.

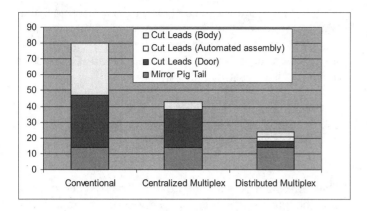

Figure 5: Comparison of wiring in conventional, multiplexed, and distributed Smart Connector architectures

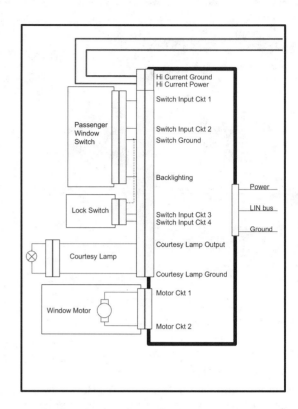

Figure 6: Window Motor Smart Connector I/O block diagram

The four Smart Connectors described above are all currently under development using Delphi's Advanced Development Process (ADP). The project recently completed a comprehensive requirements review. A more detailed description of each Smart Connector follows:

The first Smart Connector described is the window motor Smart Connector. An input / output block diagram of this Smart Connector can be seen in Figure 6. This Smart Connector is used in each door of a vehicle. It controls the window motor and several other miscellaneous door functions. These miscellaneous functions include the courtesy lamp and window and lock switch inputs. This functionality is included in the window motor Smart Connector because dedicated Smart Connectors controlling these functions were estimated to increase the sub-system's total cost.

The second Smart Connector defined is the OSRV Mirror Smart Connector. An input / output block diagram of this Smart Connector can be seen in Figure 7. This Smart Connector controls the mirror tilt motors, mirror fold motor, mirror heat, an exterior courtesy lamp, and reads sensors for mirror position and outside temperature. This Smart Connector would be used in each front vehicle door.

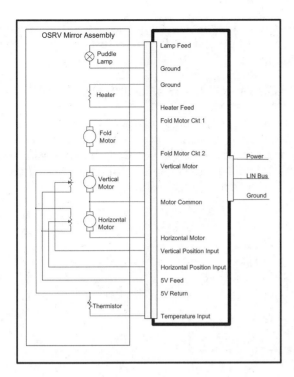

Figure 7: OSRV Mirror Smart Connector I/O block diagram

The third Smart Connector defined is the Driver's Switch Smart Connector. An input / output block diagram of this Smart Connector can be seen in Figure 8. This Smart Connector is designed to read window switch inputs for four windows, read the position of the mirror select / fold switch, read the position of the mirror joystick, read the position of the rear window lockout switch, and drive dimmable backlighting.

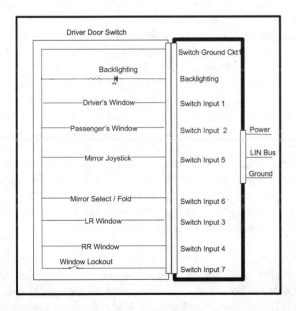

Figure 8: Driver's Switch Smart Connector I/O Block Diagram

The fourth Smart Connector defined is the door lock Smart Connector. An input / output block diagram of this Smart Connector can be seen in Figure 9. This Smart Connector drives the door lock and super-lock motors. It can also monitor the position of two digital input switches. It will be used in each vehicle door.

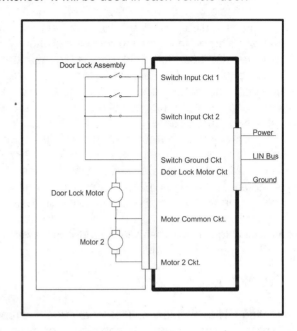

Figure 9: Door Lock Smart Connector I/O Block Diagram

PLANS AND TESTING

Extensive test plans have been developed to demonstrate the feasibility of the LIN bus for production applications. Since the bus protocol has just recently been defined (and is still subject to revision), there is little field experience to indicate how the bus will perform under real world conditions. Not only must the protocol be shown to be workable, but manufacturer's claims regarding integrated circuit functionality and development systems capabilities must be verified. Areas of focus include EMC, data bandwidth, latency, and bus operation under high current load switching. In addition to bench testing of individual components, an actual 11-node production intent system will be built and tested. All of this is being done under Delphi's ADP process, and will result in a system capable of being deployed on OEM vehicles throughout the world.

EMC

EMC will be an important issue for LIN, and requirements have been gathered from several different vehicle manufacturers. In keeping with its goal of being a low cost solution, the LIN bus uses a single unshielded wire as its transmission medium. Present day single wire systems run at speeds to about 10K bits/second, and use wave shaping on the signal to reduce radiation. But how will the LIN bus do? Bench testing has been performed to acquire LIN bus waveforms for use in mathematical modeling, and to measure radiated emission under worst case conditions -- highest bus traffic at the highest speeds (5). The LIN bus protocol specifies permissible data rates of up to 20K bits/second, faster than today's single wire systems. The 10-node test system will run as close as possible to this speed to provide another check on EMC performance. If 20K bits/second turns out to be unrealistic, it will be necessary to determine what the actual upper limit is. Speeds much lower will result in added cost, as more buses will be required to handle communications.

BUS LATENCY AND BANDWIDTH

In evaluating LIN performance, it is not enough to simply verify that the bandwidth is adequate for all the messages planned for the system. When a user generates a request through the system (by pushing a button, for example), he or she expects an immediate response to the command. A latency requirement has been established for Delphi's system to address this issue. Meeting this requirement has complicated the message strategy for the 11-node test system, since a single LIN bus is used for all communications. It would take over 100 ms for all messages to be transmitted if they were all processed sequentially.

A single master node initiates all communications in the LIN bus system. Slave nodes cannot initiate communications themselves, and must wait to be polled by the master before their data can be output. Typically, a schedule table is set up within the master, and this determines when each message will be output on the bus. A benefit of this method is that latency and bus utilization are very predictable. The problem is that using a static schedule table will result in unacceptable delays for the system. Using conditional branches in the schedule table can solve this (6).

Test and evaluation of these techniques is commencing using a bus development tool known as LINspector. This will be used to simulate a master node, and will be programmed to change the scheduled table based on data received from the slave nodes. Conditional branches result in some loss of predictability, but should reduce system delays to levels acceptable to vehicle users. Full development of the LIN protocol must also address variability of master nodes and slave nodes designed and built by different suppliers combined into a

sub-system. Worst-case analysis, as well as in-vehicle testing, will be used to demonstrate acceptable latency performance.

HIGH CURRENT LOADS AND BUS DISRUPTIONS

Earlier prototype systems suffered from communications disruptions when high current loads were switched off and on. Load currents (figure 10) in the three-wire smart connector configuration can result in a temporary (large) voltage shift in the ground wire.

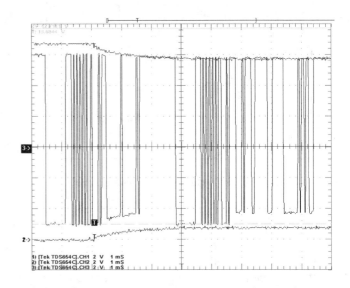

Figure 10: Ground shift at a window motor during inrush. LIN data communications are also plotted. Note the ground shift in the slave node's response.

This is due to long wire lengths and the sharing of a single ground wire by multiple smart connectors. These voltage drops can corrupt the data transmission in the commonly used single wire buses. The LIN physical layer uses a rail-to-rail (ground to battery) swing, which promises increased immunity to this problem. LIN requires a transmitting node to transmit within 20% of the local power and ground voltages. Receiving nodes are designed to interpret messages that are within 40% of its local battery and ground. This provides a significant margin for ground shift when a remote slave node responds to a poll by the master during a high current event. This is shown in Figure 10. As long as the node's transmission, which is required to be within 20% of its local ground reference, is received within the master's 40% tolerance, the message will be valid.

The 11-node test system will be used to verify the proper operation of the bus under realistic conditions. The system will be wired in a production intent configuration, using production intent loads. A tool such as LINspector will be used monitor the bus while various loads are energized. Electrical measurements will be taken to determine noise margins. Some of the smart connectors could have additional power and ground inputs, which may be used to supply current to the high power loads. This may be necessary in some vehicle applications.

SUMMARY: WHAT HAVE WE LEARNED?

The development of Smart Connectors requires that we:

- Define requirements to meet customer needs
- Benchmark current systems
- Assess the advantages and disadvantages
- Develop new concepts to capitalize on the strengths and resolve weaknesses

Historically multiplexed systems have suffered from various weaknesses. As these problems are resolved, the systems become viable. Some common drawbacks include:

- Custom non-standard protocols
- Complicated protocols
- Slow speed
- Lack of Development Tools
- Cost

The LIN Bus protocol overcomes these drawbacks by promoting a worldwide standard that is simple, fast and low cost. At present, the hardware and development tools are available but still working up the maturity curve.

At Delphi, we have completed a comprehensive review of global customer requirements and allocated those requirements into specific product attributes and performance specifications. We have an initial 4-node driver's door system functioning and undergoing test. We first showed this functional system to the public during Convergence 2000. The system is running at 19.2Kbit/s and meeting our latency requirements. Initial radiated emission testing run at 19.2Kbit/s has shown promising results. Our development team is extending from this 4-node driver's door system to a full 11-node vehicle architecture.

CONCLUSION

The viability of a distributed multiplex LIN bus appears very promising at this time. The development of such a system requires an organization with expertise in electronics, power and signal distribution, and connection systems. Without any one of these three competencies, successful deployment of such a system would be nearly impossible.

To comprehend the potential commercial viability of such a system, a thorough understanding of the underlying power and signal distribution system must be present. The majority of cost savings that can be realized in a distributed multiplex system are a result of a simplification of the wiring sub-system. Further cost savings and reliability improvements can be achieved through assembly automation. Process equipment that will automatically assemble Smart Connectors with IDC terminals onto a 3-wire bus is under development.

We are proceeding with packaging development of the Smart Connectors specified earlier in this paper. Figure 12 shows a concept for the OSRV Mirror Smart Connector. It mates to the three-wire power, ground, and data bus using IDC terminals. This allows the Smart Connector to be inserted onto a continuous piece of wire and eliminates the need to strip insulation.

Thermal evaluation of this Smart Connector is proceeding concurrently with the electronic and packaging development. Figure 13 shows a temperature contour plot of the OSRV Mirror Smart Connector with mirror fold and heat activated. The plot also shows the heat generated by the reverse battery diode, a requirement when using a solid state H-bridge. Concurrent electronic, packaging and thermal analysis offer the greatest opportunity for developing a commercially viable distributed LIN system with rapid speed to market.

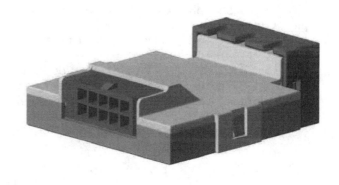

Figure 12: Smart Connector concept employing an IDC terminal interface to a three-wire bus.

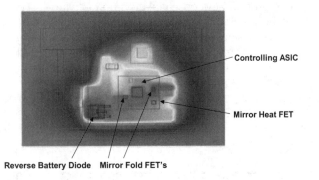

Controlling ASIC

Mirror Heat FET

Reverse Battery Diode   Mirror Fold FET's

Figure 13: Computer simulation of the heat transfer within a mirror Smart Connector with mirror fold and heat activated.

We are proceeding with the development of a distributed multiplex door subsystem using the LIN communications protocol. A detailed architecture study jointly conducted with a major OEM has shown the cost viability of such a system. Delphi is currently on track to develop this system for production application.

REFERENCES

1. Bell, "Multiplexing – Past, Present, and Future," SAE 760178, 1976.
2. Lupini, "Vehicle Multiplex Bus Progression," SAE 2001-01-0060, 2001
3. Koptez, Elmenreich, and Mack, "A Comparison of LIN to TTP/A," Research Report 4/2000 Institut fur Technische Informatik, TU Wien, Austria.
4. LIN Consortium, "LIN Protocol Specification 1.1, http://w.w.w.lin-subbus.org, April 2000.
5. Motorola, MC33399 LIN Interface, Product Proposal Rev 4.0.
6. Specks and Rajnák, "LIN – Protocol, Development Tools, and Software Interfaces for Local Interconnect Networks in Vehicles," 9th International Conference on Electronic Systems for Vehicles, Baden-Baden, Oct 2000.

CONTACT

John V. DeNuto is an Engineering Supervisor with Delphi Automotive Systems. He holds an MS degree in Mechanical Engineering from the University of Akron and a BS in Naval Architecture from the United States Naval Academy. He can be reached at (330) 306-1143 or at John.DeNuto@eng.ped.gmeds.com.

Steve Ewbank is a Senior Project Engineer with Delphi Delco Electronics Systems. He holds an MS degree in Electrical Engineering from the University of Kansas. He can be reached at (765) 451-0353 or at stephen.e.ewbank@delphiauto.com.

Fran Kleja is a Project Engineer with Delphi Packard Electric Systems. He holds a BSAS degree in Electrical Engineering Technology from Youngstown State University. He can be reached at (330) 306-1124 or at Fran.Kleja@eng.ped.gmeds.com.

Chris Lupini is a Senior Project Engineer with Delphi Delco Electronics Systems. He has an MSEE degree from Purdue University and a BSCompE degree from the University of Michigan and is a licensed Professional Engineer. He can be reached at (765) 451-0248 or at Christopher.A.Lupini@delphiauto.com.

Rob Perisho is a Senior Project Engineer with Delphi Delco Electronic Systems. He holds a BS in Electrical Engineering from Brigham Young University and did his graduate work at San Diego State University. He can be reached at (765) 451-0817 or at rob.a.perisho.jr@delphiauto.com.

2001-01-0060

Multiplex Bus Progression

Christopher A. Lupini
Delphi Delco Electronics Systems

ABSTRACT

The current SAE classification system for serial data communication protocols encompasses Class A, Class B, and Class C categories. Because of the proliferation of applications and new protocols these three groups are not enough. This paper will introduce and discuss several new categories which are Diagnostics, SafetyBus, Mobile Media, and X-by-Wire. The serial data protocols that fall under these categories are for the most part brand new and will serve distinct and unique tasks. All existing common vehicular multiplex protocols (approximately 40) will be categorized using the SAE convention plus the new groupings. Top contenders will be pointed out along with a discussion of the protocol in the best position to become the industry standard in each category. Future vehicle applications having up to seven different data networks will be presented.

INTRODUCTION

CURRENT MULTIPLEX CATEGORIES –The multiplexing of automotive electrical data onto communication buses dates back to the late 1970s. It was originally hoped that a single bus protocol could handle the needs of any vehicle. Gradually that expanded to the SAE categorization of Class A, B, and C and the realization that up to three protocols and/or networks may be necessary.

By 1995 the need for multiple buses per vehicle was becoming apparent [1]. The cost tradeoff, especially, was studied – do you put everything on one bus or split it up into several buses? Which is more economical? Which is more efficient?

This paper proposes that at least seven in-vehicle networks may be necessary – at least on high-end vehicles in the next ten years. These categories include, besides the existing SAE classes, diagnostics, airbag, mobile media, and X-by-Wire. Each area needs its own protocol and one or more networks running that protocol. Sometimes this is for safety reasons, such as with airbags or X-by-Wire. But regardless of vehicle function partitioning, we now have distinct classes of signals that will communicate over their own network, or networks (i.e. multiple sub-buses for smart connector) [2].

Although not discussed in this paper, there is a distinction between protocol and network. Conceivably one might have the same protocol running on several networks – say CAN for both a body bus and a powertrain bus. So even though there could be seven or more networks, there may actually be fewer protocols used. Also, not all protocols are complete – meaning they specify attributes of all seven layers of the OSI model[3]. Some are only physical layers (i.e. GM UART, J1708). Some are only higher layers (i.e. TTP).

CURRENT STATUS OF THE OLD CATEGORIES

CLASS A – Usage is for low-end, non-emission diagnostic, general purpose communication. Bit rate is generally less than 10 Kb/s and must support event-driven message transmission. Cost is generally about $0.50 to $1 per node. This cost includes any silicon involved (i.e. microprocessor module or transceiver, etc.), software, connector pin(s), service, etc. The "cost" data discussed in this paper is very crude and is only to be used as a "rule-of-thumb". It should be in no way construed to be an offer or quote.

Some examples of Class A protocols are listed in Table 1.

NAME:	USER:	USAGE:	MODEL YEARS:	COMMENTS:
UART	GM	Many	1985 - 2005+	Being phased out
Sinebus	GM	Audio	2000+	Radio steering wheel controls
E&C	GM	Audio/HVAC	1987 - 2002+	Being phased out
I²C	Renault	HVAC	2000+	Used little
J1708/J1587/J1922	T&B	General	1985 - 2002+	Being phased out
CCD	Chrysler	HVAC, audio, etc.	1985 - 2002+	Being phased out
ACP	Ford	Audio	1985 - 2002+	
BEAN	Toyota	Body	1995+	
UBP	Ford	Rear backup	2000+	
LIN[4]	many OEMs	Smart Connector	2003+	LIN Consortium developing

Table 1: Some Class A Protocols

Most of these Class A protocols are UARTs. UART is very simple and economical to implement. Most microcontrollers have the necessary SCI module built-in, or it can be implemented without a microprocessor. The transceiver is smaller and cheaper than those of other protocols. The transceiver IC may be a custom chip combining multi-protocol capability with regulators, drivers, etc. Right now the leading candidate for a Class A world standard is LIN.

Table 1a compares some of the major attributes of some of the Class A protocols from Table 1.

FEATURE	BUS NAME								
	UART(ALDL)	SINEBUS	E & C	I²C	SAE J1708	CCD	ACP	BEAN	LIN
AFFILIATION	GM	DELCO	GM	PHILIPS	TMC - ATA	CHRYSLER	FORD	TOYOTA	Motorola
APPLICATION	GENERAL & DIAGNOSTICS	AUDIO	GENERAL		CONTROL & DIAGNOSTICS	GENERAL & DIAGNOSTICS	AUDIO CONTROL	BODY CONTROL & DIAGNOSTICS	SMART SENSORS
MEDIA	SINGLE WIRE	SINGLE WIRE	SINGLE WIRE	TWISTED PAIR	TWISTED PAIR	SINGLE WIRE	TWISTED PAIR	SINGLE WIRE	SINGLE WIRE
BIT ENCODING	NRZ	SAM	PWM	AM	NRZ	NRZ	NRZ	NRZ	NRZ
MEDIA ACCESS	MASTER/ SLAVE	MASTER/ SLAVE	CONTENTION		MASTER/ SLAVE	MASTER/ SLAVE	MASTER/ SLAVE	CONTENTION	MASTER/ SLAVE
ERROR DETECTION	8-bit CS	NONE	PARITY	ACK bit	8-bit CS	8-bit CS	8-bit CS	8-bit CRC	8-bit CS
HEADER LENGTH	16 BITS	2 BITS	11 - 12 BITS		16 BITS	8 BITS	12 - 24 BITS	25 BITS	2 BITS/BYTE
DATA LENGTH	0 - 85 BYTES	10 - 18 bits	1 - 8 BITS			5 BYTES	6 - 12 BYTES	1 - 11 BYTES	8 BYTES
OVERHEAD	Variable	75 %	Variable	45 %	Variable	16.7 %	25 %	28 %	2 BYTES
IN-MESSAGE RESPONSE	NO	NO	NO		NO	NO	NO	NO	NO
BIT RATE	8192 b/s	66.6 KHz – 200 KHz	1000 b/s	1 - 100 Kb/s	9600	7812.5 b/s	9600 b/s	10 Kb/s	20 Kb/s
MAXIMUM BUS LENGTH	Not Specified	10 METERS	20 METERS	Not Specified	Not Specified	Not Specified	40 METERS	Not Specified	40 METERS
MAXIMUM NODES	10		10			6	20	20	16
μ NEEDED?	YES	NO	YES		YES	YES	YES	YES	NO
SLEEP/WAKEUP	NO	NO	NO		NO	NO	NO	NO	
H/W AVAIL?	YES	NO	YES		YES	YES	YES	YES (?)	NO
COST	LOW	LOW	LOW		MEDIUM	LOW	LOW	LOW	LOW

Table 1a: Comparison of Class A Protocols

CLASS B – Usage is for the vast majority of non-diagnostic, non-critical communication. Speed is between 10 Kb/s and approximately 125 Kb/s. Must support event-driven and some periodic message transmission plus sleep/wakeup. Cost is around $2 per node. Protocols used for Class B networks are listed in Table 2.

NAME:	USER:	USAGE:	MODEL YEARS:	COMMENTS:
GMLAN (SWC)	GM	Many	2002+	GM only user; J2411
GMLAN (mid)	GM	Infotainment	2002+	ISO 11898 CAN – might be IDB-C
ISO 11898	Europe	Many	1992+	Various speeds – 47.6 Kb/s to 500 Kb/s in use
J2284	GM,Ford, DC	Many	2001+	500 Kb/s; based on ISO 11898
Fault-tol CAN	Europe	Many	2000+	ISO 11519 CAN
Class 2	GM	Many	Until 2002+	J1850; being phased out
PCI	Chrysler	Many	Until 2002+	J1850
SCP	Ford	Many	Until 2002+	J1850
J1939	T&B	Many	1994+	Replacing J1708/1587/1922

Table 2: Some Class B Protocols

The world standard in this area is CAN. In particular, ISO 11898 at around 100 Kb/s for car applications and J1939 at 250 Kb/s for Truck & Bus applications. Both of these use the same digital circuitry. The transceivers may be the same but are usually different for truck applications because of their 24V electrical system. An OEM could use the 24 V transceiver in both cases to leverage economy of scale and reduce cost.

The ISO 11519-2 "fault-tolerant" low speed 2-wire CAN interface is becoming popular in some car applications. This CAN physical layer is slower and costs more than an ISO 11898 interface, but the bus fault detection capability is enticing.

Table 2a compares some of the major attributes of the Class B protocols from Table 2. GMLAN is not shown due to its exclusivity to GM.

FEATURE	BUS NAME					
	SINGLE-WIRE CAN (SWC)	CAN 2.0 ISO 11898 ISO 11519-2 ISO 11992 J2284	J1850 ISO 11519-4			SAE J 1939
AFFILIATION	SAE/ISO	BOSCH/SAE/ISO	GM	FORD	CHRYSLER	TMC - ATA
APPLICATION	DIAGNOSTICS	CONTROL & DIAGNOSTICS	GENERAL & DIAGNOSTICS	GENERAL & DIAGNOSTICS	GENERAL & DIAGNOSTICS	CONTROL & DIAGNOSTICS
TRANSMISSION MEDIA	SINGLE WIRE	TWISTED PAIR	SINGLE WIRE	TWISTED PAIR	SINGLE WIRE	TWISTED PAIR
BIT ENCODING	NRZ-5 MSb first	NRZ-5 MSb first	VPW MSb first	PWM MSb first	VPW MSb first	NRZ-5 MSb first
MEDIA ACCESS	CONTENTION	CONTENTION	CONTENTION	CONTENTION	CONTENTION	CONTENTION
ERROR DETECTION	CRC	CRC	CRC	CRC	CRC	CRC
HEADER LENGTH	11 BITS	11 or 29 BITS	32 BITS	32 BITS	8 BITS	29 BITS
DATA FIELD LENGTH	0-8 BYTES	0-8 BYTES	0-8 BYTES	0-8 BYTES	0-10 BYTE	8 BYTES
MESSAGE OVERHEAD	9.9 %	9.9 % - 22 %	33.3 %	33.3 %	8.3 %	9.9 % - 22 %
IN-MESSAGE RESPONSE	NO	NO	Optional Normally NO	Optional Normally YES	Optional Normally YES	NO
BIT RATE	33.33 Kb/s 83.33 Kb/s	10 Kb/s to 1 Mb/s	10.4 K b/s	41.6 K b/s	10.4 K b/s	250 Kb/s
MAXIMUM BUS LENGTH	30 METERS	Not Specified 40 (Typical)	35 METERS (5 Meters for scan tool)	35 METERS (5 Meters for scan tool)	35 METERS (5 Meters for scan tool)	40 METERS
MAXIMUM NODES	16	Not Specified 32 (Typical)	32	32	32	30 FOR STP 10 FOR UTP
µ NEEDED?	YES	YES	YES	YES	YES	YES
SLEEP/WAKEUP	YES	NO	YES	NO	NO	NO
H/W AVAIL?	NO	YES	YES	YES	YES	YES
COST	LOW	MEDIUM	LOW	LOW	LOW	MEDIUM

Table 2a: Comparison of Class B Protocols

CLASS C –Usage is for some safety-related, real-time systems such as engine timing, fuel delivery, etc. Bit rate is between 125 Kb/s and 1 Mb/s. Must support real-time periodic parameter transmission. Unshielded twisted pair is the medium of choice instead of shielded twisted pair or fiber optics. Cost is in the range of $3 to $4 per node, unless fiber optics is involved – which is typically necessary above 500 Kb/s.

NAME:	USER:	USAGE:	MODEL YEARS:	COMMENTS:
GMLAN (high)	GM	All	2002+	500 Kb/s CAN
ISO 11898	Europe	Most	1992+	Various speeds of CAN
J1939	T&B	Most	1994+	250 Kb/s CAN

Table 3: Some Class C Protocols

J1939 is commonly used for Class B and Class C applications for truck & bus, construction, agriculture, marine, and other industries. Most passenger car applications run ISO 11898 at 500 Kb/s for their Class C network. The big difference from CAN in Class B applications is the type of nodes that are connected.

Table 3a compares some of the major attributes of the Class C protocols from Table 3. GMLAN is not shown due to its exclusivity to GM.

	BUS NAME	
FEATURE	CAN 2.0 ISO 11898 ISO 11519-2 ISO 11992 J2284 J1939	SAE J1939
AFFILIATION	BOSCH/SAE/ISO	TMC - ATA
APPLICATION	CONTROL & DIAGNOSTICS	CONTROL & DIAGNOSTICS
TRANSMISSION MEDIA	TWISTED PAIR	TWISTED PAIR
BIT ENCODING	NRZ-5 MSb first	NRZ-5 MSb first
MEDIA ACCESS	CONTENTION	CONTENTION
ERROR DETECTION	CRC	CRC
HEADER LENGTH	11 or 29 BITS	29 BITS
DATA FIELD LENGTH	0-8 BYTES 11 or 29-bit ID	MOST ARE 8 BYTES 29-bit ID
MESSAGE OVERHEAD	9.9 % - 22 %	9.9 % - 22 %
IN-MESSAGE RESPONSE	NO	NO
BIT RATE	1 Mb/s	250 Kb/s
MAXIMUM BUS LENGTH	Not Specified 40 (Typical)	40 METERS
MAXIMUM NODES	Not Specified 32 (Typical)	30 W/ SHIELDED TWISTED PAIR 10 W/ UNSHIELDED TP
μ NEEDED?	YES	YES
SLEEP/WAKEUP	NO	NO
H/W AVAIL?	YES	YES
COST	MEDIUM	MEDIUM

Table 3a: Comparison of Class C Protocols

NEW CATEGORIES

EMISSIONS DIAGNOSTICS – Usage is to satisfy OBD-II, OBD-III, or E-OBD. Must be a legally-acceptable protocol. Protocols used today (or soon) are listed in Table 4. There is overlap with some of the other categories.

NAME:	USER:	USAGE	MODEL YEARS:	COMMENTS:
J 2480	GM, Ford, DC	OBD-III	2004+	U.S. CAN for passenger cars
ISO 15765-4	Europe	E-OBD	2000+	E-OBD CAN
J 1850	GM, Ford, DC	OBD-II	1994+	Not accepted in Europe
ISO 9141-2	Europe	OBD-II, general	1994+	Old OBD-II UART
ISO 14230-4	Many	OBD-II, OBD-III	2000+	Keyword 2000

Table 4: Some Emission Diagnostics Protocols

Since this data link is only needed between the engine controller and the off-board connector, a simple approach is sufficient. Most automakers and truckmakers are using KW2000 already so this is rapidly becoming the emissions diagnostic standard.

FEATURE	J2480	ISO 15765	J1850 ISO 11519-4			ISO/DIS 9141 ISO/DIS 9141-2	KEYWORD XX
AFFILIATION	SAE	ISO	GM	FORD	CHRYSLER	WORLD	Various
APPLICATION	EMISSIONS DIAGNOSTICS	EMISSIONS DIAGNOSTICS	GENERAL & DIAGNOSTICS	GENERAL & DIAGNOSTICS	GENERAL & DIAGNOSTICS	DIAGNOSTICS ONLY	DIAGNOSTICS
TRANSMISSION MEDIA	TWISTED PAIR	TWISTED PAIR	SINGLE WIRE	TWISTED PAIR	SINGLE WIRE	SINGLE WIRE	1-WIRE
BIT ENCODING	NRZ	NRZ	VPW MSb first	PWM MSb first	VPW MSb first	NRZ (strt, 7D, P, stop) LSb first	NRZ
MEDIA ACCESS	TESTER/ SLAVE	TESTER/ SLAVE	CONTENTION	CONTENTION	CONTENTION	TESTER/SLAVE	MASTER/ SLAVE
ERROR DETECTION	CRC	CRC	CRC	CRC	CRC	PARITY (odd)	x-bit CS
HEADER LENGTH			32 BITS	32 BITS	8 BITS	Not Specified	16 BITS
DATA FIELD LENGTH			0-8 BYTES	0-8 BYTES	0-10 BYTE	Not Specified	0 - 85 BYTES
MESSAGE OVERHEAD			33.3 %	33.3 %	8.3 %	Variable	Variable
IN-MESSAGE RESPONSE			Optional Normally NO	Optional Normally YES	Optional Normally YES	NO	NO
BIT RATE			10.4 K b/s	41.6 K b/s	10.4 K b/s	<10.4 Kb/s	5 b/s - 10.4 Kb/s
MAXIMUM BUS LENGTH			35 METERS (5 Meters for scan tool)	35 METERS (5 Meters for scan tool)	35 METERS (5 Meters for scan tool)	Limited by total impedance to ground	Not Specified
MAXIMUM NODES			32	32	32	Limited by total impedance to ground	10
μ NEEDED?	YES		YES	YES	YES	YES	YES
SLEEP/ WAKEUP	YES		YES	NO	NO	NO	NO
H/W AVAIL?	YES		YES	YES	YES	YES	YES
COST	LOW		LOW	LOW	LOW	LOW	LOW

Table 4a: Comparison of Emission Diagnostics Protocols

MOBILE MEDIA – Usage is for "PC-on-wheels" applications. Three different networks and protocols may be necessary. These sub-categories are low speed, high speed, and wireless. SAE nomenclature is IDB-C, IDB-M, and IDB-Wireless respectively. Bit rate is between 250 Kb/s and 100 Mb/s+.

Low Speed - Usage is for telematics, diagnostics, and general information passing. Cost is around $3 per node. IDB-C has turned out to be a token-passing form of CAN at 250 Kb/s.

NAME:	USER:	USAGE:	MODEL YEARS:	COMMENTS:
IDB-C	many OEMs	Many	2002+	250 Kb/s CAN

Table 5a: Some Low-Speed Mobile Media Bus Protocols

High Speed - Usage is for real-time audio and video streaming. Cost is around $10, mainly due to fiber optics. Fiber optics will be necessary due to high speed. Will probably have to be compatible with industry-standard systems such as Connected Car PC, or AutoPC. D2B has seen the first usage (Mercedes 1999 S-class) but MOST is a contender. The IDB Forum is leaning toward Firewire at this time.

NAME:	USER:	USAGE:	MODEL YEARS:	COMMENTS:
MML	GM	tbd	2004+	Delphi-D only
D2B	Mercedes	tbd	1999+	Has momentum
MOST	Saab, etc.	tbd	2000+	Type of D2B
Firewire	unknown	tbd	2000+	IEEE 1394
USB	unknown	Aftermarket	1998+	Clarion AutoPC product

Table 5b: Some High-Speed Mobile Media Bus Protocols

Wireless - Usage somewhat undetermined at this time. Will be necessary (initially) for cell phones. Cost target is around $5 per node.

NAME:	USER:	USAGE:	MODEL YEARS:	COMMENTS:
Bluetooth	tbd	tbd	2005+	

Table 5c: Some Wireless Mobile Media Bus Protocols

FEATURE	BUS NAME							
	IDB-C	IDB-M	MOST	D^2B Copper	D^2B Optical	MML	USB	IEEE 1394
AFFILIATION	SAE	SAE	Philips	PHILIPS	PHILIPS	DELCO		IEEE
APPLICATION	Aftermarket Entertainment	Aftermarket Entertainment	Stream Data & Control	STREAM DATA & CONTROL	STREAM DATA & CONTROL	STREAM DATA & CONTROL	PC DEVICES	PC DEVICES
TRANSMISSION MEDIA	2-Wire	TBD	Optical	TWISTED PAIR	OPTICAL	OPTICAL FIBER	SHIELDED TWISTED PAIR	SHIELDED TWISTED PAIR
BIT ENCODING	NRZ	TBD	BiPhase	PWM	BiPhase	NRZ	NRZ	NRZ
MEDIA ACCESS	TOKEN-SLOT	TBD	MASTER/ SLAVE	MASTER/ SLAVE	MASTER/ SLAVE	MASTER/ SLAVE	CONTENTION	CONTENTION
ERROR DETECTION	15-bit CRC		CRC	Parity	CRC	CORRECTING (optional)	CRC	CRC
HEADER LENGTH	11 BITS					1 BYTE		
DATA LENGTH	8 BYTES					1 - 200+ BYTES		
MESSAGE OVERHEAD	~ 32 BITS					5 - 10 %	25 %	25 - 30 %
IN-MESSAGE ACK.	1 ACK BIT					NO		
BIT RATE	250 Kb/s	High	25 Mb/s	29.8 kb/s	12 Mb/s	110 Mb/s	12 Mb/s	98 - 393 Mb/s
MAXIMUM BUS LENGTH	TBD	TBD	Infinite	150 METERS	NO LIMIT	10 METERS		72 METERS
MAXIMUM NODES	16		24	50	24	16	127	16
μ NEEDED?	YES	YES	YES	YES	YES	YES	YES	YES
SLEEP/WAKEUP	YES	YES				YES	NO	NO
H/W AVAIL?	NO	NO	YES	YES	YES	NO	YES	YES
COST	LOW	HIGH	HIGH	HIGH	HIGH	HIGH	MEDIUM	MEDIUM

Table 5d: Comparison of Mobile Media Protocols

SAFETYBUS – Usage is for airbag systems. There may be two, or more, buses such as for firing, sensing, etc. Must support at least 64 nodes consisting of squibs, accelerometers, occupant sensors, seatbelt pretensioners, etc. Cost is (hoped to be) $1 to $2 per node. USCAR "SafetyBus" committee is studying the protocols listed in Table 6 and is the source for more information.

NAME:	USER:	PROGRAM(s):	YEARS:	COMMENTS:
SafetyBus	tbd	tbd	2002+	Delphi-D only
BOTE	tbd	tbd	2002+	Bosch-Temic
Planet	tbd	tbd	2002+	Philips
DSI	tbd	tbd	2002+	Motorola/AMP
SI	BMW	tbd	2002+	"ISIS"
BSRS	tbd	tbd	2002+	Breed-Siemens
Surfs	tbd	tbd	2002+	NEW; Siemens
OSCAR	tbd	tbd	2002+	NEW; Bosch-Siemens-Temic

Table 6: Some SafetyBus Protocols

Many issues here involving packaging constraints, existing mechanical envelop, legalities, etc. SafetyBus consortium should have issued the final specification by now. The winning protocol may well be a hybrid of several existing proposals. For now there is no clear industry direction.

FEATURE	BUS NAME					
	SafetyBus	BOTE	PLANET	DSI	SI	BSRS
AFFILIATION	Delphi	Bosch-Temic	Philips	Motorola	BMW	Breed-Siemens
APPLICATION	Airbag	Airbag	Airbag	Airbag	Airbag	Airbag
TRANSMISSION MEDIA	2-Wire	2-WIRE	2-WIRE	2-WIRE	2-WIRE or 3-WIRE	2-WIRE or 3-WIRE
BIT ENCODING	RTZ					
MEDIA ACCESS	MASTER/ SLAVE	MASTER/ SLAVE	MASTER/ SLAVE	MASTER/ SLAVE	MASTER/ SLAVE	MASTER/ SLAVE
ERROR DETECTION	8-bit CRC					
HEADER LENGTH	1 BYTE					
DATA FIELD LENGTH	24 – 39 BITS					
MESSAGE OVERHEAD	20 – 30%					
IN-MESSAGE ACK.	NO	NO	NO	NO	NO	NO
BIT RATE	500 Kb/s	31.25 Kb/s 125 Kb/s	20 Kb/s 250 Kb/s	5 Kb/s +		250 Kb/s
MAXIMUM BUS LENGTH	TBD	TBD	TBD	TBD	TBD	TBD
MAXIMUM NODES	64	12	64	16		
μ NEEDED?	NO	NO	NO	NO	NO	NO
SLEEP/WAKEUP	NO	NO	NO	NO	NO	NO
H/W AVAIL?	YES	YES	YES	YES	YES	YES
COST	LOW	LOW	LOW	LOW	LOW	LOW

Table 6a: Comparison of SafetyBus Protocols

DRIVE-BY-WIRE – Usage is for brake-by-wire, throttle-by-wire, steer-by-wire, etc. applications. Bit rate is between 1 Mb/s and 10 Mb/s. Fiber optics will be necessary due to high speed. Utmost in reliability, performance, and real-time capability is required. Cost is around $15+ per node. Some possible candidate protocols are given in Table 7.

NAME:	USER:	PROGRAM(s):	YEARS:	COMMENTS:
TTP	BMW	tbd	2004+	U-Vienna
FTCA[5]	tbd	tbd	2004+	CAN-based
CANdor[6]	tbd	tbd	2004+	CAN-based
SI	BMW	tbd	tbd	10 Mb/s "Byte Flight"
TTFlex	BMW&DC	tbd	tbd	Still under development

Table 7: Some Drive-by-Wire Protocols

TTP has it's "foot in the door", but work is underway to see if CAN is capable of doing the job. Major issue is how much fault tolerance is really required. TTFlex is a new unproven method. Any scheme will require dual bus interfaces, dual microprocessors, bus watchdogs, timers, etc. Cost is a big problem. The level of fault-tolerance needed requires a lot of silicon and software which, of course, is expensive.

FEATURE	BUS NAME				
	TTP	FTCA	CANdor	SI	TTFlex
AFFILIATION	U-VIENNA			BMW	BMW & DC
APPLICATION	CONTROL & DIAGNOSTICS			X-by WIRE	X-by-WIRE
TRANSMISSION MEDIA	Not Specified			2-WIRE or 3-WIRE	
BIT ENCODING	Not Specified				
MEDIA ACCESS	TIME DIVISION			MASTER/ SLAVE	
ERROR DETECTION	16-bit CRC				
HEADER LENGTH	1 BYTE				
DATA FIELD LENGTH	16 BYTES				
MESSAGE OVERHEAD	18.75 %				
IN-MESSAGE ACK.	YES			NO	
BIT RATE	Not Specified				
MAXIMUM BUS LENGTH	Not Specified			TBD	
MAXIMUM NODES	Not Specified				
µ NEEDED?	YES			NO	
SLEEP/WAKEUP	NO			NO	
H/W AVAIL?	NO			YES	
COST	HIGH			LOW	

Table 7a: Comparison of Drive-by-Wire Protocols

CONCLUSION

Multiple buses per vehicle have been a reality for some time. Primarily this has been because of re-use, and carryover of existing networks and protocols. However, in the near future new functions such as smart connector, drive-by-wire, and mobile media will force the need for additional protocols and networks.

REFERENCES

1. SAE 950293 – "Aspects and Issues of Multiple Vehicle Networks" Emaus
2. SAE 2001-01-0072 "LIN Bus and its Potential for Use in Distributed Multiplex Systems" Ewbank, Lupini, Perisho, DeNuto, Kleja
3. ISO 7498 - Data Processing Systems, Open Systems Interconnection Standard Reference Mode.
4. www.lin-subbus.org
5. "A CAN-Based Architecture for Highly Reliable Communication Systems" Hilmer, Kochs, Dittmar. Proceedings from 5th ICC 1999
6. "A Double CAN Architecture for Fault-Tolerant Control Systems" Ferriol, F. Navio, J. Navio. Pons. Proenza, Julia. Proceedings from 5th ICC 1999

CONTACT

Christopher (Chris) A. Lupini
Senior Project Engineer
Delphi Delco Electronics Systems
Christopher.A.Lupini@delphiauto.com
http://www.delphiauto.com

Mr. Lupini has a BSCompE degree from the University of Michigan and an MSEE degree from Purdue University. He is a registered Professional Engineer and an ASE certified Master Automotive Technician. He has worked in the data communications industry for 13 years, and has been teaching seminars for 8 years. He has spent most of his time assisting in the design and development of the GM Class 2 protocol, as well as the specification and test of numerous J1850 and CAN ICs. He is the lead engineer for the Delphi-Delco Serial Data Center of Expertise and has authored a dozen technical papers and articles along with a textbook and has two patents pending.

PROTOCOLS

Controller Area Network (Can) Protocol

An Automotive Specification of a Time Triggered CAN Implementation: Doubling CAN's Usable Data Throughput

Chris Quigley, Ben Pope, James Finney and Richard T. McLaughlin
Warwick Control Technologies

ABSTRACT

The Controller Area Network (CAN) has seen enormous success in automotive body and powertrain control systems, and in industrial automation systems using higher layer protocols such as DeviceNet and CANopen. Now, the CAN standard ISO11898 are being extended to Time Triggered CAN (TTCAN) to address the safety critical needs of first generation drive-by-wire systems. However, their successful development depends upon the availability of silicon and software support, and appropriate development & analysis tools. This paper outlines the current status of TTCAN technology and describes the implementation of Level 1 TTCAN on the Atmel 89c51cc01/cc02/cc03/cc04 microcontrollers. The descriptions contained show how to implement for different bus speeds, along with suggestion for a user to tailor the drivers for their own application. Level 2 TTCAN is also described for comparison purposes.

Whilst the TTCAN implementation described in this paper is limited to the maximum CAN bit rate of 1Mbit/s, TTCAN can be used to effectively double the bandwidth of a CAN system to the region of 60 to 70% bus loading at 500 Kbit/s and 1Mbit/s. This is approximately twice that of traditional automotive CAN systems. For example, a typical automotive power train control system based on CAN typically runs at maximum of about 35% loading at 500 KBaud to avoid data bus latencies.

INTRODUCTION

Of the currently available open standard network protocols, CAN has become the most prominent across the world's automotive industry. The probable reasons for this have been its huge support from major semiconductor manufacturers, tool suppliers and automotive OEMs. Now, nearly all automotive OEMs have products available with CAN or are intending to develop with CAN in the near future. However, industrial automation applications using DeviceNet and CANopen were amongst the first applications of CAN technology.

This paper focuses on the development of TTCAN control systems based upon the Atmel cc01/cc03 microcontrollers. Other chips with the capability of supporting TTCAN are manufactured by:

- Microchip (8-bit)
- NEC (8-bit)
- Infineon (32-bit)

The *CAN communications protocol* specifies the method by which data is passed between communicating devices on a CAN bus. It conforms to ISO's Open System Interconnection (OSI) model, which is a seven-layer description of a telecommunications network standard.

In fact the CAN protocol can be described by the lowest two layers of the OSI model – the Data Link Layer and the Physical Layer. The Application layer protocols can be proprietary schemes developed by individual CAN users, or one of the emerging standards used within particular industries. A common application layer standard used in the process-control/manufacturing field is DeviceNet, which is especially suited to the networking of PLCs and intelligent transducers. In the automotive industry most manufacturers use their own proprietary standard. In the utility vehicles industry, J1939 has become the popular application layer. This paper assume the reader's prior knowledge of the CAN Physical, Data Link, and Physical Layers.

Time Triggered CAN is standardised under ISO-11898-4 which covers four main parts:

- Time triggered execution of CAN with a central Time Master periodic messages, event triggered messages and Time Master Synchronisation.
- Fault tolerance of this Time Master.
- Drift correction and Global Time.
- Event synchronised - time triggered communication

The main characteristics of TTCAN is that bus access is controlled via a Time Division Multiplexed Access (TDMA) like method using a regularly repeating cycle of time called the *Basic Cycle* (see figure 1). The *Basic Cycle* is divided into a fixed number of time windows (i.e. fixed at design time) which can be a mixture of any one of four types; *Reference Message*, *Exclusive Window*, *Arbitration Window* and *Free Window*.

TTCAN Basic Cycle

Reference Message	Exclusive Window	Exclusive Window	Arbitration Window	Free Window	Reference Message

Figure 1: An example of the Basic Cycle of time triggered CAN communication

Reference Message:

This is sent by the time master control unit (global time master) and controls the timing of the Basic Cycle. The Reference Message signifies the start of the Basic Cycle. CAN communication is initiated by a Reference Message. The global time can be sent in four data bytes leaving the remaining four data bytes available for general data transfer.

- The identifier of the Reference Message is determined by the system designer.
- All but the three least significant bits of this identifier characterise the message as a reference message.
- The three least significant bits distinguish between different potential time masters.
- These three LSBs constitute the time master priority.
- There are up to 8 potential time masters in a TTCAN network.

Reference Message: Format

TTCAN has two levels of implementation, Level 1 and Level 2, Level 2 being an extension of Level 1. The reference message consists of at least one data byte in level 1 and at least four data bytes in level 2. Level 1 or Level 2 will be selectable by the user. The application can increase the length of the reference message up to eight bytes. The MSB in each byte (bit 7) is transmitted first. Figure 2 illustrates the Identifier and data format of the TTCAN Reference message

Figure 3 shows the Byte breakdown of the Reference Message data. Level 1 only provides time-triggered operation using Cycle_Time. The reference message for level 1 consists of at least one data byte containing the

Next_Is_Gap bit and the Cycle count, which is optionally up to the 6 bits.

The other 7 bytes in the message can be used for data. A time triggered FSE (Frame Synchronisation Entity) can deal with Cycle_Count values up to 63 (six bits). In level 1 the NTU (Network Time Unit) is a nominal CAN bit time.

This first byte is present in both level 1 and level 2.

- Cycle_Count: Allows for different basic cycles
- Next_is_Gap: Enables a gap of unspecified length between basic cycles and therefore synchronisation of communication to application events.
- Discontinuity bit: Important for external clock synchronisation.
- MRM (Master_Ref_Mark): Provides global time
- NTU_Res: Important for high synchronisation quality. Three bits in level 2 mandatory.
- Level 2 is an extension of level 1. It is possible to use a level 1 controller in a level 2 network, if the corresponding loss of performance is acceptable.

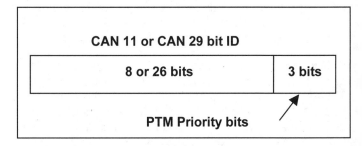

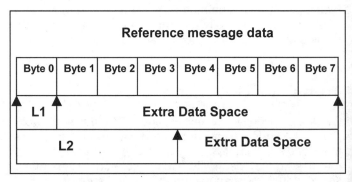

Figure 2. Reference Message Format

Level 2 additionally provides increased synchronisation quality, global time and external clock synchronisation. The reference message consists of at least 4 bytes. The first byte is the same as the byte in level 1. The second byte contains the discontinuity bit (Disc_Bit) and the bits for the NTU's resolution (fractional parts of Master_Ref_Mark). In level 2 at least 3 bits are supported, optionally increasing up to 7 bits. The third byte contains the low byte of Master_Ref_Mark, and the fourth byte the high byte of Master_Ref_Mark (MRM)

(See Figure 3). In level 2 the NTU is a fraction of the physical second.

Byte 0 (Level 1 and Level 2)

Next is Gap	Res.	Cycle Count

Byte 1 (Level 2 only)

NTU Resolution	Disc_Bit

Byte 2 (Level 2 only)

MRM Low Byte

Byte 3 (Level 2 only)

MRM High Byte

Figure 3. Reference Message Data –
Byte break down

EXCLUSIVE WINDOW:

This is a time slice long enough to accommodate the message and data to be transmitted. The *Exclusive Window* is reserved for one particular CAN message only.

ARBITRATION WINDOW:

In an Arbitration Window a number of nodes may attempt to transmit a message. Therefore the nodes that may contend for bus access during the *Arbitration Window* may do so by the usual non-destructive bitwise arbitration method. Thus the message with the lowest CAN identifier will win arbitration. With normal CAN systems, nodes losing arbitration will attempt to retransmit the message. This is disabled in TTCAN since a retransmission would upset the remainder of the operation of the Basic Cycle.

FREE WINDOW:

A Free Window is reserved for future expansion of the TTCAN system. Therefore further nodes can be added at a later date.

All nodes communicating on a bus, which is to be time triggered, must have the retransmission of message that has lost arbitration disabled.

TTCAN MESSAGE TRIGGERING:

It is mandatory that each TTCAN controller is equipped with a local time counter - **Local Time**. The Local Time is incremented each **Network Time Unit (NTU)** - e.g. every CAN bit time. Another timer is necessary to achieve the cycle based sending/receiving of the messages and the synchronisation to the Time Master of the TTCAN network - **Cycle Time**.

THE CYCLE TIME:

Within a Basic Cycle the TTCAN protocol execution is driven by the progression of the Cycle Time based on a Time Master's reference message. The start of an Exclusive and Arbitrating Time Window is defined off-line and is known a priori to all interested nodes of the network. This definition is derived from the System Matrix of the system. The control is based on the matching of so-called **Time Marks** with the current Cycle Time.

The time marks are essential parts of the TTCAN Triggers - **Tx_Ref_Trigger, Tx_Trigger and Rx_Trigger**.

REF_TRIGGER_OUTPUT

The actual time mark, when a potential time master tries to send a reference message is additionally modified by Ref_Trigger_Offset. Ref_Trigger_Offset may take values between -127 and 127. During normal operation Ref_Trigger_Offset = 0 for the current time master and Ref_Trigger_Offset = Initial_Ref_Offset for all other potential time masters. Initial_Ref_Offset is a positive value specified by system configuration for each node. It is possible to use a small Initial_Ref_Offset for a high priority time master. A good choice for Initial_Ref_Offset depends on synchronisation quality. The potential time master with the highest time master priority should have the smallest Initial_Ref_Offset.

SYNCHRONISATION OF CYCLE TIME

Cycle time is resynchronised with every reference message. Accuracy of the starting point depends on width of frame synchronisation pulse and on measurement capability of receiving nodes.

In level 1 the starting point jitter is about one bit time. In level 2 the starting point jitter can be reduced to about one time quanta, if the local time counter has a sufficiently high resolution. The effect of run times is constant and can therefore be neglected for most synchronisation items.

POTENTIAL TIME MASTERS

There can be up to 8 potential time masters in a TTCAN network. At a given time only one of them can be the current time master. Each of the potential time masters has a 3-bit time master priority. Within a TTCAN network the time master priorities of different time masters have to be different. The time master priorities have to be defined offline. If a potential time master sends a reference message the 3 LSBs of the identifier of the reference message are given by the time master priority. In a fault free TTCAN network after initialisation the node with the highest time master priority will be the time master.

TTCAN INITIALISATION PROCEDURE

After a reset or after leaving Config_Mode a TTCAN Slave is not allowed to send a message before receiving two consecutive Reference Messages.

In a Potential Time Master, Ref_Trigger_Offset will be initialised to Initial_Ref_Offset. A Potential Time Master will use the Ref_Trigger for the "Gap" case (modified by Ref_Trigger_Offset) for the first transmission of a reference message.

After successfully transmitting a Reference Message, a node considers itself as time master by adjusting its local time and and setting Ref_Trigger_Offset = 0. Any node receiving a reference message must synchronise to the network before it is allowed to transmit messages (exception: reference messages). An error free node is synchronised to the network after it received at least two successive reference messages (last message did not contain a set Disc_Bit – Level 2 TTCAN only).

After being synchronised any controller sends the messages according to its Tx_Triggers. If, after being synchronised, a potential time master receives a reference message with a lower time master priority than its own, it tries to become the current time master. It sets Ref_Trigger_Offset = 0 and decrements Ref_Trigger_Offset every base cycle by one until it succeeds in transmitting a reference message or it receives a reference message with a higher time master priority.

If two nodes start to transmit a reference message at the same time, the node with the higher time master priority will win arbitration. Once a node transmits a reference message it uses Ref_Trigger_Offset = 0.

Whenever a node receives a reference message with higher time master priority it uses Ref_Trigger_Offset = Initial_Ref_Offset. This procedure guarantees a completely smooth transition to the time master with the highest time master priority.

FAULT TOLERANCE

A missing Reference Message (failure of time master) leads to the following sequence of events:

- The Ref_Trigger of the next best (as defined off-line by Initial_Ref_Offset) Potential Time Master is triggered first.
- This node sends a Reference Message and then is time master.
- After recovery, the previous Time Master can again become time master as during initialisation.
- Depending on the quality of synchronisation, only a gap of a very few NTUs results. This is in fact the difference between consecutive init_ref_offset settings.

Fault tolerance can be achieved to any reasonable level (up to the number of potential time masters with a maximum of 8).

MAIN ADVANTAGES OF TTCAN

- Provides a time deterministic method of communication based on existing technology.
- Since it is based on existing technology, expertise for TTCAN will be based on a company's existing CAN knowledge and experiences.
- Development tools originally aimed at CAN can be used for the analysis of data on TTCAN. Direct monitoring of TTCAN with existing CAN bus analysis tools is possible since TTCAN frames are exactly the same as for CAN. However, this is not necessarily possible if transmitting on the TTCAN bus, since current tools will almost definitely interfere with the message schedule.
- Thus the transition from CAN to TTCAN systems will not require a huge financial outlay.
- Increases data throughput by enabling the bus to be run at high loading.

ATMEL TTCAN TECHNOLOGY SUPPORT IN T89C51CC01/02/03

SILICON FEATURES:

TTCAN is available technology now for the first generation of drive-by-wire systems. The Atmel CANary CAN controller supports the implementation to TTCAN Level 1 currently in their T89C51CC01/02/03/04 8051 based microcontrollers. The main TTCAN Level 1 silicon support features are summarised as:

- **Disable Retransmission**: For any message object that loses arbitration, or is corrupted by an error, an attempt is made to retransmit that message object. The CANary CAN controller allows this retransmission of message objects to be disabled.
- **SOF and EOF Timestamp Capture**: The CANary CAN controller allows for the capturing of a timestamp at the beginning or end of a CAN frame. Capturing of either SOF and EOF allows for network synchronisation.

Due to the limitations of the CANary controller, Exclusive window implementation is described in subsequent sections only. The implementation of Arbitration windows is not recommended. Full reasons are explained in a later section. Appendix C shows the contents and details of these CANary registers specific to TTCAN operation.

SUMMARY OF MEMORY USAGE:

The Memory Requirements Table in Appendix A, illustrates the relationship between the number of cycles, the number of widows in each cycle, and their effect on the memory requirements of the Atmel development unit. For example if 8 Cycles are required, and there are 16 windows per cycle, it can be seen the memory requirements will not be overtaxed. Only 490 Bytes of XRAM are required, and the maximum usage before the need of external memory is 906 Bytes.

The main code base without any messages:

- ~ 2 K bytes code
  - Time Master/Potential Time Master 2100 bytes
  - Slave 1850 bytes
- 66 bytes data
- 65 bytes xdata

SUMMARY OF CANARY CONTROLLER USAGE:

The CANary CAN controller is the on-chip peripheral used for CAN communications. All CAN chips have 15 Message Filtering Objects for on chip message filtering. This removes demand from the microcontroller. For TTCAN communications, each of the CAN communication objects is used as below:

Object 0:
- Transmission of Reference Message

Object 1:
- Reception of Reference Message

Objects 2 to 14:
- Msg *A* to *M* respectively – used as Full CAN Objects
- Can be used for Transmission or Reception of messages

REFERENCE MESSAGE IMPLEMENTATION:

CAN **Object 0** is used for the generation of the TTCAN Reference Message as mentioned previously. The most significant bits of the CAN Identifier field must match. This is the 8 most significant bits of an eleven bit CAN Identifier or the 26 most significant bits of a twenty nine bit CAN Identifier. The 3 least significant bits of the CAN Identifier, are effectively "don't care" bits and are only necessary upon initialisation and re-initialisation situations and required for Time Master Priority.

CAN **Object 1** is used for the Reference Message reception in Potential Time Master and Slave devices. When a Reference Message is received in Object 1, an interrupt is generated. The interrupt service routine then instigates the calculation of Cycle Error so that the node can synchronise to the TTCAN Master.

TIME MASTER FAILURE FAULT CONFINEMENT:

If the current Time Master fails, the next Potential Time Master takes over automatically. This is achieved via the Init_Ref_Offset mechanism described previously. However, for the CANary implementation Init_Ref_Offset must be set to:

$$TM\_Priority * 8$$

for hand over to happen smoothly. This is due to clock drift and internal delays and jitter in the CANary CAN Controller.

Therefore the Init_Ref_Offset settings for different Time Master Priority nodes are shown in the table below.

Time Master Priority	Init_Ref_Offset
0	0
1	8
2	16
3	24
4	32
5	40
6	48
7	56

TTCAN Drivers Configuration:

A TTCAN Application Configuration Tool was developed by Warwick Control to ease the design times of message scheduling. It graphically enables a designer to configure the communication matrix and allow the scheduling of each message window (Reference Message, Exclusive Window, Arbitration Window, Free Window). This tool can configure the Master, Potential Time Masters (PTM) or Slaves of each TTCAN controller. Figure 4 shows an example set-up for a TTCAN node. Once the graphic is completed, C code header files are generated, and then downloaded to the unit under test to complete the matrix table configuration.

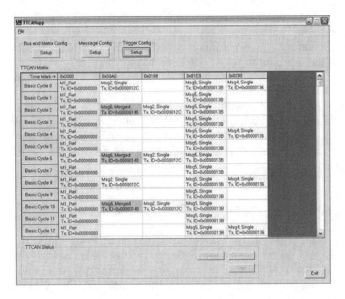

Figure 4: GUI for driver set-up

The tool can also download a configuration to a simulator node, such as the one illustrated in Figure 5. It is also possible to upload the Unit Under Test (UUT) existing configuration into the TTCAN Application Configuration Tool.

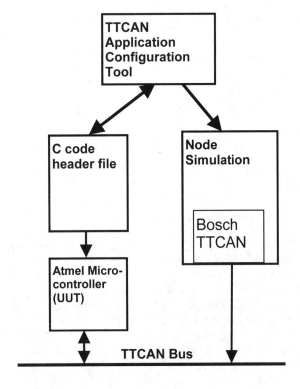

Figure 5: TTCAN Matrix Configuration set-up

ANALYSIS OF TTCAN BUS LOADING OF THE ATMEL DEVELOPMENT BOARD

Referring to the Performance Table in Appendix A, It can be seen that the bus load capabilities of TTCAN is reliant on the internal clock of the microcontroller, and the data bus speed. For example, at the typical clock rate of 16 MHz, and at the typical automotive bus speed of 500Kbps, the bench testing of the Atmel TTCAN development board showed no significant message latencies until the busload reached 67%.

This is a significant improvement over the tradition contention based messaging method of CAN. Past research has shown that contention based CAN messaging experiences significant message latencies at approximately 35%.

INTEGRATION OF OSI LAYER 7 PROTOCOLS

OSI layer 7 protocols such as SAE J1939 can be easily integrated into the Time Triggered CAN protocol and TTCAN drivers such as the ones described in this paper. For a protocol such as SAE J1939, this brings benefits such as:

- Higher achieved bus loading
- Increased message determinism
- Higher system integrity due to this increased determinism
- Plug and play functionality can be implemented by leaving some spare slots

IMPLEMENTATION OF HIGH SPEED AUXILLARY TTCAN SYSTEMS

High Speed Auxillary CAN systems internally to the vehicle are often used to pass signals to one other system and are not usually part of the main vehicle CAN bus. This type of configuration is illustrated in Figure 6. Here the ESP and the ABS are required update every 500 microsecond. This would overtax the bus loading of the Powertrain CAN bus.

The application here needs a closed loop cycle time of down to 500us and this is not possible with CAN. With TTCAN this is possible at 500 Kbaud and 1 Mbaud. By implementing these as TTCAN, the following benefits can be found:

- Increased message determinism
- Which leads to higher system integrity
- Easily implement message watch dogs so that if a message does not arrive in a timely manner, then a failsafe can be activated

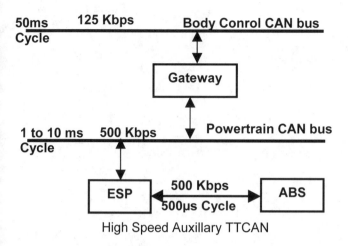

High Speed Auxillary TTCAN

Figure 6: Example of High Speed Auxillary TTCAN

EXAMPLE APPLICATIONS OF TTCAN CANary

Figure 7 shows two possible applications of the CANary TTCAN implementation; *Intelligent TTCAN Transducer* node and *TTCAN Communications Controller* for another host.

In the case of the *Intelligent TTCAN Transducer*, the complete Transducer handling and TTCAN communications is carried out on a single Atmel microcontroller such as the T89c51cc01.

In the case of the *TTCAN Communications Controller*, the Atmel microcontroller is used to solely deal with the

TTCAN communications and application level processing is carried out on a host microcontroller.

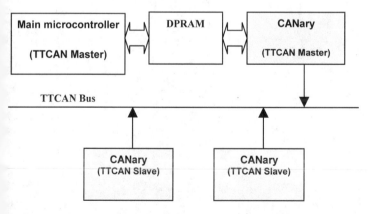

Figure 7: Possible applications of the Atmel CANary TTCAN implementation

developed to address the needs of these applications. Whilst TTCAN is limited in bandwidth and does not provide for all of the needs of such applications, it is very useful as the enabling technology in the first generation of drive-by-wire systems. TTCAN only requires a small amount of additional engineering knowledge on top of that already available from the CAN community and allows the same development tools as used for CAN to be used for message monitoring (e.g. CAN bus analyser software).

It has been shown that the Atmel CANary CAN controller with resides in the T89c51cc01 /02 /03 /04 microcontrollers can be used to implement the TTCAN protocol. The most interesting and useful outcome of this implementation is that it can effectively double the usable bandwidth of CAN to a useable 70% bus loading. Automotive control system designers are limited to 1000 Kbaud by the CAN protocol, but usually limit their usable bit rate to 500 Kbaud for EMC reasons. Use of TTCAN effectively doubles the bandwidth whilst providing reliable deterministic communication latencies.

However, there are limitations to the implementation and therefore the following suggestions to its use are made:

- If the Atmel CANary is to be used in a full TTCAN bus implementation, it is recommended that it is used it as a TTCAN slave and only transmits messages in Exclusive windows.
- If the CANary is to be used as a Potential Time Master (TM0 to TM7) then it is recommended that its failure operation be limited. If when in use it is the Time Master and is therefore responsible for the transmission of the Reference Message, it is recommended that after a failure it should not be allowed to reconnect to the TTCAN bus without a system reset.
- It is recommended that a CANary node should not have the responsibility of transmission in an Arbitration Window. However, Arbitration Window reception is recommended.
- The bit rate of the CANary TTCAN implementation is limited by crystal oscillator used. At 500 Kbaud 70% loading can be achieved.

It is recommended that future work be performed to accomplish two missions:
- Determine the performance of other TTCAN chips as has been done with the Atmel chip here
- Perform interoperability testing where all the different TTCAN chip interoperate on the same data bus, and determine that they all inter-communicate properly.

SUMMARY AND CONCLUSION:

Future automotive control systems will utilise data communication networks in far more safety critical applications such as drive-by-wire. TTCAN has been

REFERENCES

1.   ISO-11898 Part 4
2.   Fuhrer T et al (2000); "Time Triggered Communication on CAN (Time Triggered CAN-TTCAN)", Proceedings of the 7th International CAN Conference, Amsterdam, 24th and 25th October 2000.
3.   Hartwich et al (2000); "CAN Network with Time Triggered Communication", Proceedings of the 7th International CAN Conference, Amsterdam, 24th and 25th October 2000.
4.   McLaughlin & Quigley (2004); "Analysis and Diagnostics of Time Triggered CAN (TTCAN) Systems", SAE Congress paper 2004-01-0201.

CONTACT

All authors are from:
Warwick Control Technologies
Prodrive Warwick
Oldwich Lane East
Fen End, Warwickshire
CV8 1NR
United Kingdom

Richard T. McLaughlin, B.Sc., M.Sc., CEng MIEE
richard@warwickcontrol.com

Chris Quigley, B.Sc., M.Sc.
chris@warwickcontrol.com

Ben Pope, M.Eng.
ben@warwickcontrol.com

James Finney, B.Eng.
james@warwickcontrol.com

Appendix A – T89C51CC01 Time Triggered Communication Performance Metrics Summary

Memory Requirements Table

Absolute xdata requirement for Matrix

		Windows								
		1	2	4	8	16	32	64	128	256
Cycles	1	58	63	73	93	133	213	373	693	1333
	2	64	72	88	120	184	312	568	1080	2104
	4	76	90	118	174	286	510	958	1854	3646
	8	100	126	178	282	490	906	1738	3402	6730
	16	148	198	298	498	898	1698	3298	6498	12898
	32	244	342	538	930	1714	3282	6418	12690	25234
	64	436	630	1018	0	3346	6450	12658	25074	49906

```
          CC03 or External XRAM required          data required
                                                       80
```

Performance Table

The statistics below show the variety of problems associated with Window Length and baud rate. Typically, an increase in Baud Rate requires an increase in Window Length when using the CANary as a Master or Potential Time Master.

Crystal Frequency	Baud rate	Time Master Microcontroller Loading	Slave /Potential Time Master Microcontroller Loading	Max CAN Bus Loading	Suggested Window length (Bit time)	Max. Windows per Basic Cycle
12 MHz	50 Kbps	5.9%	4.8%	79%	165	9
	100 Kbps	11.8%	10.3%	79%	165	19
	125 Kbps	14.5%	12.9%	79%	165	24
	250 Kbps	26.7%	23.6%	73%	180	45
	500 Kbps	45.7%	40.5%	62%	210	78
	1000 Kbps	Not Possible				
16 MHz	50 Kbps	4.2%	4.2%	79%	165	7
	100 Kbps	8.5%	8.5%	79%	165	14
	125 Kbps	10.6%	10.6%	79%	165	18
	250 Kbps	20.0%	20.0%	75%	175	35
	500 Kbps	35.9%	35.9%	67%	195	63
	1000 Kbps	59.6%	59.6%	56%	235	104
20 MHz	50 Kbps	4.8%	4.2%	79%	165	5
	100 Kbps	9.7%	8.5%	79%	165	11
	125 Kbps	12.1%	10.6%	79%	165	14
	250 Kbps	22.9%	20.0%	75%	175	28
	500 Kbps	43.2%	37.8%	71%	185	53
	1000 Kbps	80.0%	70.0%	60%	220	89

Bus loading is calculated by:

It is assumed that 100% loading would be with TTCAN windows of 131 bits per 29 Bit CAN ID with 8 DLC frame.

Therefore bus load = 131bits/Suggested Window Length x 100%

155

Appendix B – **TTCAN GLOSSARY**

Arbitrating time window
Time window assigned to messages that share the same time window

Basic cycle
Row of the system matrix of several consecutive time windows

Cycle_Count
Number of the current basic cycle of the matrix

Cycle_Count_Max
Value of Cycle_Count of the last basic cycle in the given system matrix of the network

Cycle_Error
Difference of Global_Ref_Mark – Ref_Mark, saved at each successful completion of the reference message

Cycle_Offset
Parameter specifying within a matrix cycle the first basic cycle for which an Rx_Trigger or Tx_Trigger is valid

Exclusive time window
Time window assigned to specific message transmitted periodically without competition for the CAN bus

Free time window
Time window free of messages scheduled in the system matrix

Global time
Node view of the global time of the current time master

Initial_Ref_Offset
Initialisation value to load the Ref_Trigger_Offset

Level 1/Level 2
Levels of implementation of ISO 11898-4, where level 2 is an extension of level 1

Local time
Time generated by a cyclic incrementing counter

Master_Ref_Mark
Parameter transmitted be the time master in the reference message

Matrix cycle
Cycle of all basic cycles in the system matrix, consecutive from the first to the last basic cycle

Message status count
Error counter providing means to detect scheduling errors for messages sent in exclusive time windows

Network Time Unit (NTU)
Unit measuring all times, and provides a constant of the whole network

Potential time master
Frame synchronisation entity that is allowed to send a reference message by system configuration

Ref_Trigger_Offset
Parameter used to modify the time mark within a Tx_Ref_Tigger to send a reference message

Reference message
Message (data frame) that starts a basic cycle

Repeat_Factor
Parameter specifying the repetition rate of a message within a transmission column, being part of Tx_Trigger or Rx_Trigger parameters

System matrix
Form containing all messages of all nodes in the network, organised as components, and consisting of time windows, organised in basic cycles (rows of the matrix) and transmission columns (columns of the matrix)

Time mark
Mark within a frame synchronisation entity specifying an instant of Cycle_Time (in NTUs) at which a certain action is expected or planned

Time master
The node sending the Reference message

Time window
Amount of time allocated for a specific transmission column in the system matrix

Tx_Count
Counter that is reset at each start of a matrix cycle, i.e. after identification of the corresponding reference message with Cycle_Count equal to zero

Tx_Ref_Trigger
Special Tx_Trigger parameter referring only to the triggering of reference messages

Tx_Trigger
Parameter specifying when a certain message shall be transmitted and consists of a time mark, the position within the transmission column in respect to the first sending (Cycle_Offset), the repetition rate (Repeat_Factor) within that transmission column, and a reference to a message object for which the Tx_Trigger is valid

Analysis and Diagnostics of Time Triggered CAN (TTCAN) Systems

Richard T. McLaughlin
University of Warwick

Chris Quigley
Warwick Control Technologies

ABSTRACT

The Controller Area Network (CAN) has seen enormous success in automotive body and powertrain control systems, as well as industrial automation systems using higher layer protocols such as CANopen and DeviceNet. Now, the CAN standard ISO11898 is being extended to Time Triggered CAN (TTCAN) to address the safety critical needs of first generation drive-by-wire systems. However, their successful development depends upon the availability of silicon and software support, and appropriate development & analysis tools. Warwick Control Technologies and the University of Warwick are tasked with prototyping a TTCAN analyser within the European Union Media+ project Silicon Systems for Automotive Electronics (SSAE) consortium, and with funding from the British Department of Trade and Industry (DTI). This paper briefly outlines the current status of both CAN & TTCAN technology and describes the requirements of a TTCAN analyser over that of a traditional CAN analyser.

INTRODUCTION

Of the currently available open standard network protocols, CAN has become the most prominent across the world's automotive industry. The reasons for this have been its huge support from major semiconductor manufacturers, tool suppliers and automotive OEMs. Now, nearly all automotive OEMs have products available with CAN or are intending to develop with CAN in the near future. However, industrial automation applications using CANopen and DeviceNet were amongst the first applications of CAN technology.

The Controller Area Network (CAN) is standardised under ISO-11898, which itself is split into 3 parts.

ISO-11898 - Parts 1 to 3

- ISO 11898-1 specifies the data link layer.
- ISO 11898-2 specifies the high-speed physical layer transceiver.

- ISO 11898-3 specifies the fault-tolerant low-speed physical layer transceiver

The main features of CAN are summarised as:-

- Event based communication
- Standard CAN (11 bit identifiers) and Extended CAN (29 bit identifiers)
- Bus access is CSMA/CD with Non-Destructive Bitwise Arbitration, Lowest Value identifier has highest priority for bus access
- Retransmission of messages that lose arbitration
- Silicon available from most semiconductor manufacturers
- Higher layer protocols include CANopen and DeviceNet

There are at least five physical layers available to suit different applications:-

- ISO-11898 High speed CAN – 1Mbaud max.
- ISO-11519-2 Low Speed CAN – 125Kbaud max.
- ISO-11992 - 250Kbaud
- Single Wire CAN – 33.3Kbaud max.
- DC-Bus^{TM} – 250Kbaud max.

There are two main types of silicon available:-

- **Integrated CAN**: The CAN controller is integrated with microcontroller
- **Stand Alone CAN**: Stand alone CAN controller that can be added to any microcontroller

The *CAN communications protocol* specifies the method by which data is passed between communicating devices on a CAN bus. It conforms to ISO's Open System Interconnection (OSI) model, which is a seven-layer description of a telecommunications network standard.

The OSI model describes a layered system of communication between two network nodes, whereby in theory each layer can only communicate with the layers directly above and below it in the local node, and only with the equivalent layer in a remote node.

In fact the CAN protocol can be described by the lowest two layers of the OSI model – the Data Link Layer and the Physical Layer. The Application layer protocols can be proprietary schemes developed by individual CAN users, or one of the emerging standards used within particular industries. A common application layer standard used in the process-control/ manufacturing field is DeviceNet, which is especially suited to the networking of PLCs and intelligent sensors and actuators. In the automotive industry most manufacturers use their own proprietary standard.

fixed at design time) which can be a mixture of any one of four types; *Reference Message*, *Exclusive Window*, *Arbitration Window* and *Free Window*.

REFERENCE MESSAGE:

This is sent by the time master control unit (global time master) and controls the timing of the Basic Cycle. The Reference Message signifies the start of the Basic Cycle. CAN communication is initiated by a Reference Message. The global time can be sent in four data bytes leaving the remaining four data bytes available for general data transfer.

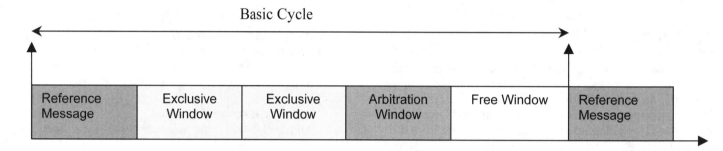

Figure 1: An example of the Basic Cycle of time triggered CAN communication

TTCAN PROTOCOL OVERVIEW - ISO-11898 - PART 4

TTCAN has been developed to address the needs of automotive first generation drive-by-wire systems. Purely time triggered operation of a communication network means that activity is determined by the progression of globally synchronised time. Communication depends upon a pre-defined time schedule, i.e. defined at design time. This is achieved by:-

- Time globally synchronised via a global time master
- Ability to switch off retransmission of messages if arbitration is lost

Time Triggered CAN is standardised under ISO-11898-4 which covers four main parts:-

- Time triggered execution of CAN with a central Time Master periodic messages, event triggered messages and Time Master Synchronisation.
- Fault tolerance of this Time Master.
- Drift correction and Global Time.
- Event synchronised - time triggered communication

The main characteristics of TTCAN is that bus access is controlled via a Time Division Multiplexed Access (TDMA) like method using a regularly repeating cycle of time called the *Basic Cycle* (see figure 1). The *Basic Cycle* is divided into a fixed number of time windows (i.e.

- The identifier of the Reference Message is determined by the system designer.
- All but the three least significant bits of this identifier characterise the message as a reference message.
- The three least significant bits distinguish between different potential time masters.
- These three LSBs constitute the time master priority.
- There are up to 8 potential time masters in a TTCAN network.

Reference Message: Format

TTCAN has two levels of implementation, Level 1 and Level 2, Level 2 being an extension of Level 1. The reference message consists of at least one data byte in level 1 and at least four data bytes in level 2. Level 1 or Level 2 will be selectable by the user. The application can increase the length of the reference message up to eight bytes. The MSB in each byte (bit 7) is transmitted first. Figure 2 illustrates the Identifier and data format of the TTCAN Reference message

Figure 3 shows the Byte breakdown of the Reference Message data. Level 1 only provides time-triggered operation using Cycle_Time. The reference message for level 1 consists of at least one data byte containing the Next_Is_Gap bit and the Cycle count, which is optionally up to the 6 bits.

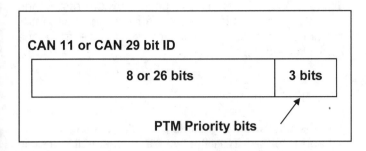

CAN 11 or CAN 29 bit ID

8 or 26 bits	3 bits

PTM Priority bits

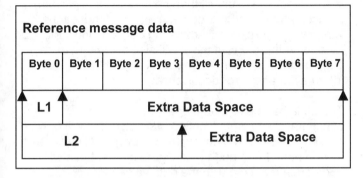

Reference message data

Byte 0	Byte 1	Byte 2	Byte 3	Byte 4	Byte 5	Byte 6	Byte 7

L1 — Extra Data Space

L2 — Extra Data Space

Figure 2. Reference Message Format

Byte 0 (Level 1 and Level 2)

Next is Gap	Res.	Cycle Count

Byte 1 (Level 2 only)

NTU Resolution	Disc_Bit

Byte 2 (Level 2 only)

MRM Low Byte

Byte 3 (Level 2 only)

MRM High Byte

Figure 3. Reference Message Data –
Byte break down

The other 7 bytes in the message can be used for data. A time triggered FSE (Frame Synchronisation Entity) can deal with Cycle_Count values up to 63 (six bits). In level 1 the NTU (Network Time Unit) is a nominal CAN bit time.

This first byte is present in both level 1 and level 2.
- Cycle_Count: Allows for different basic cycles

- Next_is_Gap: Enables a gap of unspecified length between basic cycles and therefore synchronisation of communication to application events.
- Discontinuity bit: Important for external clock synchronisation.
- MRM (Master_Ref_Mark): Provides global time
- NTU_Res: Important for high synchronisation quality. Three bits in level 2 mandatory.
- Level 2 is an extension of level 1. It is possible to use a level 1 controller in a level 2 network, if the corresponding loss of performance is acceptable.

Level 2 additionally provides increased synchronisation quality, global time and external clock synchronisation. The reference message consists of at least 4 bytes. The first byte is the same as the byte in level 1. The second byte contains the discontinuity bit (Disc_Bit) and the bits for the NTU's resolution (fractional parts of Master_Ref_Mark). In level 2 at least 3 bits are supported, optionally increasing up to 7 bits. The third byte contains the low byte of Master_Ref_Mark, and the fourth byte the high byte of Master_Ref_Mark (MRM) (See diagrams above). In level 2 the NTU is a fraction of the physical second.

EXCLUSIVE WINDOW:

This is a time slice long enough to accommodate the message and data to be transmitted. The *Exclusive Window* is reserved for one particular CAN message only.

ARBITRATION WINDOW:

In an Arbitration Window a number of nodes may attempt to transmit a message. Therefore the nodes that may contend for bus access during the *Arbitration Window* may do so by the usual non-destructive bitwise arbitration method. Thus the message with the lowest CAN identifier will win arbitration. With normal CAN systems, nodes losing arbitration will attempt to retransmit the message. This is disabled in TTCAN since a retransmission would upset the remainder of the operation of the Basic Cycle.

FREE WINDOW:

A Free Window is reserved for future expansion of the TTCAN system. Therefore further nodes can be added at a later date.

All nodes communicating on a bus which is to be time triggered, must have the retransmission of message that have lost arbitration disabled.

TTCAN ANALYSIS - REQUIREMENTS

What does a TTCAN analyser need to do that is not provided by a traditional CAN analyser?

A traditional CAN bus analyser is useful for monitoring TTCAN messages, since they are exactly the same as CAN messages.

To bring TTCAN functionality into the X-Analyser a new TTCAN Dynamic Link Library (DLL) will be created which will mimic the Softing API and add a new WCT-TTCAN API. The DLL will communicate with the TTCAN Card, based on the Bosch evaluation chip and with two further configuration data files storing saved schedule data and message ID references. Figure 4 illustrates this functional configuration.

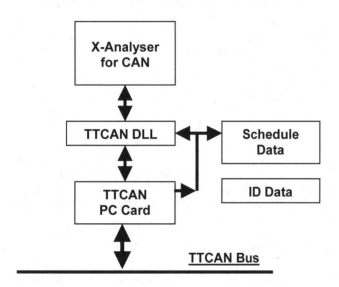

Figure 4. TTCAN Analyser Functional Diagram

The following features are important in a TTCAN analysis tool and are currently being prototyped in the TTCAN X-Analyser within the SSAE project.

Node Simulation
Node simulation in a TTCAN network is when the analysis tool generates all of the messages that a particular node would generate and therefore emulate its operation.

Bus Simulation
Here the analysis tool generates all of the messages required to simulate all the other nodes in order to test a particular node.

Error Injection
A traditional CAN analyser can often inject a bus disturbance such as an Error Frame. However, for TTCAN the main error injection that is useful is *Incorrect Schedule Injection*. This is concerned with the injection of TTCAN frame at the inappropriate time in the schedule; the inappropriate time being when another,

different TTCAN frame, is expected by the rest of the control system. This is useful for testing the control system's response to an incorrect schedule from one particular node. See figure 5.

TTCAN RECEIVING

The X-Analyser will be able to generate a timing schedule for the TTCAN network and present it in a table to the user. The user will then be able to identify when particular time windows occur, the type of these time windows i.e. Exclusive, Arbitrating or Free, and which ID is sending them. From this it will also be able to extract extra information from the network such as timing information and error occurrences.

The table can be set to an update or a static configuration. In static configuration the table will only provide the system matrix with window identification and ID numbers. In an update configuration the table will be updated in real time with the data values.

NB: - The X-Analyser will also provide the standard CAN frame information such as ID, Data, and Error Detection to the TTCAN screen.

The X-Analyser will be able to group messages by ID and buffer them so that the most recent 50,100,etc messages can be viewed after/during capture. The software will also be able to accept all messages or filter particular ID's, much like the current filter in the X-Analyser. Alternatively the user can display the information in a similar format to the raw CAN screen on the current X-Analyser. The displayed data columns would be as follows:-

- **Time Stamp** - Marks the time the message was received
- **T/R** - Shows whether the message was transmitted or Received
- **CH** - Channel the message was Transmitted or received on
- **ID** - The messages ID (8 bytes hex)
- **DLC** - Data Length Code (the amount of data in the message)
- **Message Content** - Displays what type of data the message holds (10 char)
- **Cycle Count** - The value of the cycle count when the message was received (0 – 63)
- **Data** - The data held within the message
- **Window Type** - The type of window the message resided in.

TTCAN TRANSMISSION

Using the received schedule the user will be able to configure messages and inject them into the received schedule, in any of the available windows, and monitor the effects of the valid/invalid message in the system (See section 4.1 – Injection of Invalid Messages).

Inside the X-Analyser the user will be able to create their own schedules in a schedule editor. They will be able to configure the messages, window types, and timing. These schedules can be saved and opened at a later date for further editing. The X-Analyser will also facilitate the creation of ID references which will link the ID number to the user-defined signal names.

INJECTION OF INVALID MESSAGES

The diagram in Figure 5 shows a basic system matrix consisting of several Exclusive, Arbitrating, and Free time windows. Once the X-Analyser has analysed the TTCAN network and created the system matrix table, like the one shown below, the user will then be able to create their own message object and inject it into a selected position within the matrix. The message could be injected during a valid Tx_Trigger or transmitted in the middle of another messages transmission. The software would also facilitate multiple invalid message

POTENTIAL TIME MASTER OR TIME SLAVE SELECTION IN A CAN BUS ANALYSER

The user will be able to choose whether the X-Analyser will be one of the 8 potential time masters or one of many possible time slaves.

A TTCAN network can support up to 8 potential time masters. The priority of each time master is decided by the three LSB of the Reference messages identifier. The lower the value of the 3 bits, the higher the PTM's priority in the system. The PTM with highest priority will become the time master of the network, assuming all nodes are functioning correctly. If the user would like the X-Analyser to be one of the PTM, they can choose their desired priority in the network.

Somewhere on the screen the user will be notified of it's current status in respect to its Time Master Priority, i.e. whether it is the current time master or whether it is currently only a Potential Time Master.

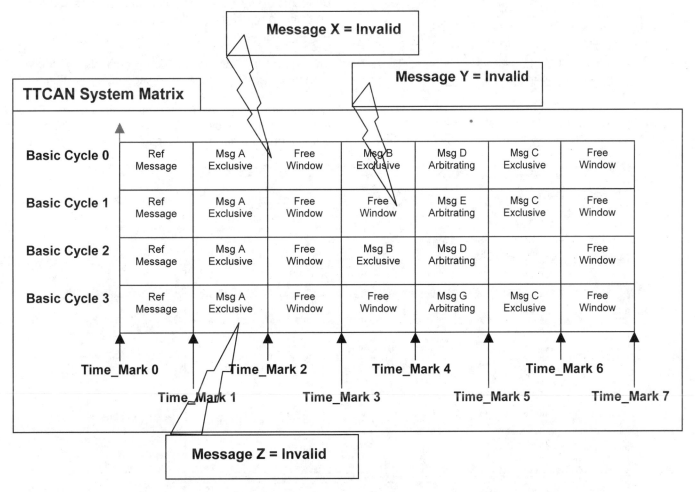

Figure 5 TTCAN system matrix with invalid message injections

injections throughout the matrix if required. The X-Analyser will then be able to monitor the network errors generated, either errors from other nodes in the network or errors from the local controller.

Alternately the user may choose to be a time slave node in the network. A time slave node does not send a reference message or control any of the networks time synchronisation. It can send and receive messages and synchronise itself to the Global_Time of the network (Level 2 only).

TTCAN ANALYSIS

The X-Analyser from Warwick Control Technologies has been extended to support the TTCAN protocol. It is interfaced to the
TTCAN bus via a TTCAN PC interface card as shown in figure 6.

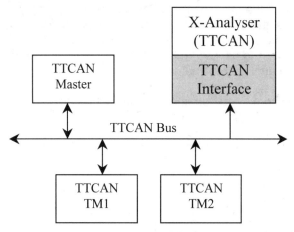

Figure 6: Context of TTCAN Analyser

The software can carry out traditional CAN analysis and can also carry out the aforementioned TTCAN specific analysis. As can be seen in the X-Analyser screen-shot in Figure 7, there are three reference cycles. CAN message ID 000 is Time Master 1. Message ID A11 occurs of all cycles, where message ID A13 occurs on ever other cycle.

User will be able to trigger inter-cycle gaps on/off using a key on the keyboard or clicking a button on the screen. Pressing the key/button once will cause a gap in between the next two windows (only available when the PC is the current time master). Pressing it again will start the next cycle (unless it is outside the maximum gap length and the other PTM have begun to transmit their reference messages.

The TTCAN PC interface card uses the Bosch TTCAN Evaluation chip to carry out the TTCAN protocol.

CONCLUSION

This paper has briefly compared CAN and TTCAN protocols. The need for additional features for TTCAN analysis, beyond those already available in current CAN bus analysers, has been highlighted.

The main support for TTCAN analysis is for error monitoring and Reference Message interpretation.

The largest input of new features concerns TTCAN message transmission for Node Simulation, Bus Simulation and Incorrect Message Schedule Injection.

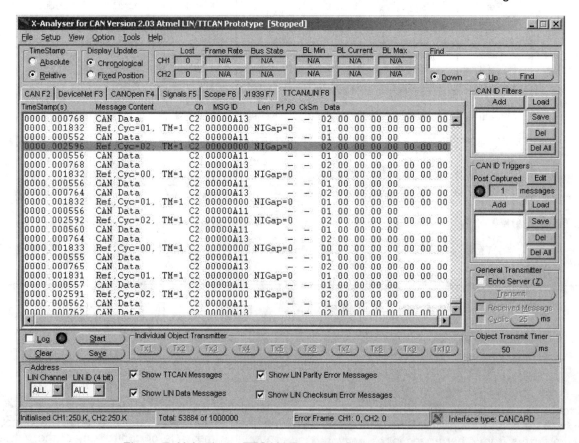

Figure 7: X-Analyser TTCAN Example Screenshot

Future features that may be useful will significantly depend upon which applications TTCAN is used in. For example, redundant networks etc. TTCAN has been developed to address the safety critical needs of first generation drive-by-wire systems, but as the next generation of by wire protocols are implemented, TTCAN will most likely become the replacement to today's conventional CAN in power train systems. Therefore the advancement of CAN analysers to fully diagnose TTCAN is essential.

REFERENCES

1. ISO-11898 Part 4
2. Fuhrer T et al (2000); "Time Triggered Communication on CAN (Time Triggered CAN-TTCAN)", Proceedings of the 7th International CAN Conference, Amsterdam, 24th and 25th October 2000.
3. Hartwich et al (2000); "CAN Network with Time Triggered Communication", Proceedings of the 7th International CAN Conference, Amsterdam, 24th and 25th October 2000.

CONTACT

Richard T. McLaughlin, B.Sc., M.Sc., CEng MIEE
International Manufacturing Centre
University of Warwick
Coventry, CV4 7AL
United Kingdom
r.mclaughlin@warwick.ac.uk

Chris Quigley, B.Sc., M.Sc.
Warwick Control Technologies
Sir William Lyons Road
University of Warwick Science Park
Coventry, CV4 7EZ
United Kingdom
chris@warwickcontrol.com

2003-01-1199

Modeling the Response of an Automotive Event-Based Architecture: A Case Study

Juan R. Pimentel
Kettering University

ABSTRACT

While many current vehicle network systems for body bus applications use event triggered analysis processes, the deterministic point of view raises concerns about system timing due to message latency. This paper studies the latency performance characteristics of a typical body bus vehicle network using event triggered analysis over the CAN bus.

INTRODUCTION

The CAN protocol has been used successfully in many automotive applications for a number of years. One of the most widely used applications involves a vehicle network architecture operating in real time or near real time. A *vehicle network architecture* is a comprehensive protocol-based communication system intended to support vehicle functions. A CAN network involves a number of nodes broadcasting information onto the bus according to the CAN protocol. How often messages are broadcast onto the bus constitutes what is termed the *network traffic (load)* or traffic *(load)*. In the CAN protocol, messages can be from 0 to 8 bytes in length. It is left to the applications to determine how often and how big the messages are to be broadcast.

From the viewpoint of an application, there are three philosophically different ways to generate messages (i.e., traffic) for the network: time triggered, event triggered, or a mix of these two. Some messages (those corresponding to sampled data control systems for example) are generated in a time triggered fashion, for example periodically. Time triggered in this context is used to mean at pre-determined time instants. In this paper, messages generated in a time triggered fashion are referred to as *periodic messages* (although not necessarily with the same repetition rate). As the name implies, event triggered messages are generated whenever an event occurs. Another name used to identify event triggered messages is *aperiodic* messages. We distinguish between the following types of events: simultaneous, concurrent, sporadic, or other events. *Simultaneous events* are events that occur at the same time or are delivered to the network in a manner that the system perceives they have occurred at the same time.

Concurrent events are defined as events that occur in a relatively close time proximity to one another in such a way that the communication system cannot distinguish that they have not occurred simultaneously. When the rate of events is bounded they are referred to as *sporadic events*. The last category of events (i.e., other) includes events that follow a specific event generation model (e.g., following an exponential distribution) or which are totally arbitrary in nature. In practice, it is extremely difficult to have simultaneous events. However, concurrent events are possible. For example, several active children in a car could be closing and opening all windows simultaneously, thus generating concurrent events.

Perhaps the main way to distinguish time triggered messages from event messages is that the latter occur at random times. For the special case of simultaneous and sporadic events, we can assume that a set of N events have occurred at the same time. In this paper we treat simultaneous and concurrent events to be the same and refer to them as just simultaneous events and accordingly simultaneous messages. The frequency and size distributions of periodic and event messages are referred to as periodic and aperiodic (event based) loads. In this paper, we assume that some messages are periodic while others are simultaneous.

Most of the literature on CAN timing analysis assumes that all messages are periodic, even event messages are treated as periodic messages with the minimum inter-arrival time between messages as its period [1-3]. In most studies, event messages are converted to periodic messages prior to performing timing analysis. This analysis technique is referred to as *time triggered analysis*. There is little CAN timing analysis done for treating messages as they are generated by applications, i.e., periodic and event.

In this paper we describe a process for designing CAN-based distributed applications using event triggered analysis involving simultaneous or concurrent messages. The following are the most important features of the new approach. The event triggered process considers both periodic and event messages. The analysis does not assume any values for deadlines, as deadline values are not obvious for some applications. Instead, a set of curves is generated that depict the cumulative probability distribution function of message latencies versus deadlines indicating the probability that a simultaneous message has a latency less than a certain value of deadline. End users then can decide what deadlines are appropriate for their application.

A BODY BUS NETWORK

Today's electrical-related vehicle subsystems can be divided into three major functional areas:

- Engine Area Subsystems
- Driver Area Subsystems
- Entertainment Area Subsystems

Accordingly, there are three major vehicle network architectures, each supporting these functional areas. These vehicle network architectures provide the opportunity to partition the intended application either functionally or geographically. For example, many German car companies have two CAN-based vehicle networks, one to support engine subsystems and another for the driver subsystems. The latter is referred to as the *body bus network*. The functions of the vehicle network architecture are performed at various ECU's (electronic control units) geographically dispersed within the vehicle. Typically, for engine and driver area subsystems, the various ECU's are interconnected by a CAN network. Under such a configuration, the functions are distributed over various nodes on the network. Such functions are referred to as *distributed functions*. Figure 1 shows a typical vehicle network architecture corresponding to a body bus network. It is composed of four nodes, each node generating a combination of event traffic and/or periodic traffic. The latter is referred to as the *background load* in this paper.

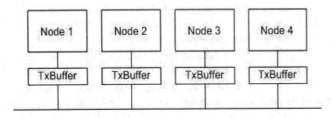

Figure 1. Body bus CAN network

We further assume a CAN network with a data rate of 125 Kbps. Let event messages be represented by e_i where i indicates the priority of the message (highest priority has a lower value). Likewise p_j represents a periodic message with j indicating its priority. We also assume that the highest priority is 1 (in the CAN protocol it is 0).

Application Scenarios

There are many ways to configure all of the message combinations available in a CAN network. For example, in CAN Version 2.0A, with 11 priority bits, there is roughly 2^{11} ways to allocate message priorities, whereas in Version 2.0B, with 29 priority bits, the number is about 2^{29}. There are also many implementation features, and for some studies not all of these features are (e.g., priority inversion). In summary, studying the detailed behavior of messages going through a CAN network the way it is set up in actual applications is extremely complex. An effective way to overcome this complexity is to reduce the scope of the study and consider what is called a *scenario*. Scenarios make assumptions to simplify the analysis, but care must be taken so that the scenarios do not result in a trivial network. Some scenarios focus on answering some critical questions, while others focus on understanding some unknown behavior. By defining appropriate scenarios, models can be simplified and thus they are more manageable. The disadvantage of scenarios is that one scenario alone will not answer all important questions and it is unlikely to yield a complete analysis of an application. However a carefully designed set of scenarios would prove to be adequate for most applications. Defining a set of appropriate scenarios is not trivial, as it requires in-depth familiarity with the applications and a good knowledge of the protocol.

Three scenarios

All scenarios involve the body bus CAN network of Fig. 1 (i.e., with 4 nodes) with 9 messages (4 event messages and 5 periodic messages). It is assumed that the assignment of priorities to event messages and periodic messages is arbitrary and likewise the assignment of messages to nodes is also arbitrary. The scenarios basically result from a specific assignment of message priorities and another specific assignment of messages to nodes. For Scenario 1 depicted in Fig. 2, the priority of event messages per node is higher than the priority of periodic messages assigned to the same node. However, on the overall network basis, it is possible that the priorities of some event messages are lower than the priority of some other periodic messages. Message priorities are also independent from nodes, i.e.; any node can have messages of any priority level. Scenario 2 (Fig. 3) is similar to Scenario 1 with the additional limitation that messages in a certain priority range are collected and assigned to a certain node, thus in effect assigning priority levels to nodes even though the

CAN protocol is based on message priorities rather than node priorities. For Scenario 3 (Fig. 4) we will assume the basic configuration of Scenario 1 except that all event messages have higher priority than the periodic messages.

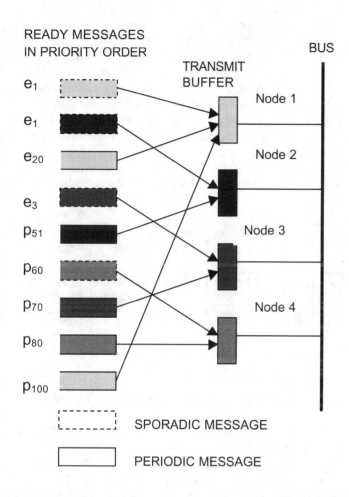

READY MESSAGES
IN PRIORITY ORDER

BUS

TRANSMIT
BUFFER

e_1

Node 1

e_1

e_{20}

Node 2

e_3

p_{51}

Node 3

p_{60}

p_{70}

Node 4

p_{80}

p_{100}

SPORADIC MESSAGE

PERIODIC MESSAGE

Fig. 2. CAN network where message priorities are independent of nodes.

Performance

In this paper we are primarily interested in message latency (or message delay) as the main indicator of performance. For the purpose of this paper we define message latency as the time interval between the instance when a message is ready for transmission and queued at a transmitting node until the message is successfully assembled at a receiving node. According to this definition, message latency includes the wait time at the transmission node, the time for the bus contention resolution, and the transmission time. Obviously, the performance of the configuration of Scenario 1 is different than that of the corresponding configurations of Scenarios 2 and 3. Scenario 1 is unconstrained and the designer is free to assign message priorities as he (or she) sees fit. In

contrast, message priorities in Scenarios 2 and 3 must comply with the order depicted in Figs. 2 and 4. The main advantage (that is not obvious) of some configurations over others is that the message latencies can be optimized if some rules for assigning message priorities and assigning messages to nodes are followed. This is proved later in this paper.

Other scenarios

Other important practical considerations that have an impact on performance have not been considered in this paper because it is beyond its scope. These include the effects of transmission errors due to external interference's, and failures in systems with redundant components.

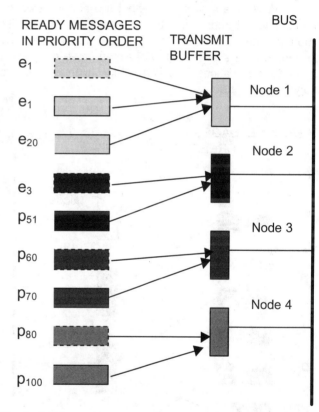

READY MESSAGES
IN PRIORITY ORDER

BUS

TRANSMIT
BUFFER

e_1

Node 1

e_1

e_{20}

Node 2

e_3

p_{51}

Node 3

p_{60}

p_{70}

Node 4

p_{80}

p_{100}

Fig. 3. CAN network where message priorities are defined by nodes.

Priority Inversion

A *priority inversion* situation occurs, if on a certain node, a message with higher priority is ready to be transmitted and another message with lower priority is already pending in the Tx buffer of the same node. In this situation, the higher priority message has to wait until the low priority message is sent. This transmission may be delayed by other, mid-priority messages being sent by other nodes on the bus. In many cases, the Tx buffer register is empty, when a request for a new message transmission arrives. Let us illustrate an example of a bad situation using the scenario depicted in Fig. 4. Let us assume that while the Tx buffer of node one has a pending message p100 (i.e., a periodic message with priority 100), a new event message e_1 arrives at Node 1 followed by e_{20}. Note that e_1 has Priority 1, the highest priority assumed in this paper. All other nodes (2 to 4) have messages with priorities ranging between 10 and 90 in their corresponding transmit buffers. Message p100 in node 1 has to wait until all other higher priority messages have been sent (6 messages). Then p100 is sent and then e_1. This means that e_1 has a delay equivalent to 8 messages times. The delay is the sum of all single transfer times for the corresponding messages. To complicate matters, additional messages with priorities between 1 and 100 may arrive at Nodes 2, 3, or 4 before message p100 has the chance to transmit, thus resulting in additional delays for e_1. In the worst case, e_1 has to wait forever, even though e_1 is the highest priority message in the network!

While the assumption about the already blocked Tx buffer for all nodes may be an *excessive worst case* situation; the possibility that Tx buffers are already blocked only for a subset of nodes may prove dangerous for critical applications involving safety. In this paper we analyze both extremes.

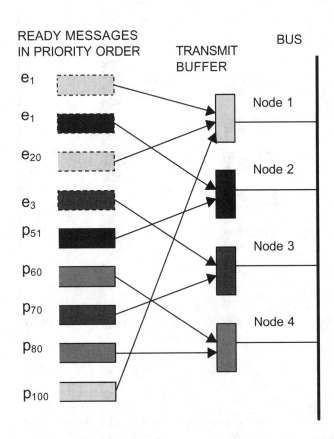

READY MESSAGES IN PRIORITY ORDER   TRANSMIT BUFFER   BUS

Fig. 4. CAN network where message priorities are higher for event messages.

We configure the background load as a load benchmark for the body bus network similar to the "SAE Benchmark" reported in [1]. The benchmark corresponds to many typical signals (53 in the SAE benchmark) in a prototype electric vehicle. Many of the signals have identical periods and have small message sizes ranging from 1 to 8 bits. To increase bus utilization, we assume that we can combine the group of signals into a number of periodic messages making up the background load as listed in Table 1. We control the total bandwidth of the background load by varying the length and period of each of the messages. Table 1 lists the parameters for two example loads. Load A corresponds to a background load of 20.48% (i.e., they generate a total of 25600 Kbps) and a total background load (including overhead and bit stuffing bits) of 52000 Kbps (41.60 % of total bandwidth).

We further assume that all simultaneous messages have a constant message length of 2 bytes. In Section 4, we will analyze the worst-case message latency when a number of event messages arrive simultaneously.

Table 1. Two examples of background loads (Rate = 125 Kbps).

Ld	P	Dm	T (ms)	Cm (ms)	OL Kpbs	TOL Kpbs
A	5	8	12.5	1.0448	25600 20.48%	52000 41.6%
B	8	2	16	0.584	8000 6.4%	36500 29.2%

TIME TRIGGERED AND EVENT TRIGGERED ANALYSIS

As noted, time triggered analysis assumes that all messages are *periodic* and that event messages can be treated as periodic messages with the minimum inter-arrival time between messages as its period. Given this assumption, exact timing analysis can be performed on a CAN network subjected to periodic loads. An example of an analysis technique is that of Tindell et al. [1] where an algorithm is provided for calculating message latencies on CAN networks. Appendix A contains a summary of the algorithm. Provided that the message traffic is schedulable (the latency calculations converge or the network can satisfy the requirements of the offered load) the analysis is said to be *deterministic*. This is so because a specific procedure is known for calculating message latencies for all messages and the latency values are unique and bounded.

Problems with time-triggered analysis

The main problem with the time triggered analysis approach, is that it cannot deal with event messages accurately. Another related problem is that it cannot deal with simultaneous or concurrent events. Theoretically two or more simultaneous events can occur at the same time, but no network can handle a system with an arbitrary number of such simultaneous events. In practice, systems are designed to handle those concurrent events that have a specified minimum inter-arrival time. For some body bus networks a value of a minimum inter-arrival time is typically 20 ms. Thus, time triggered analysis alone is not sufficient to accurately evaluate message latencies involving event messages in applications.

In this paper we use combined time and event triggered analysis, thus overcoming the principal disadvantage of the time triggered analysis approach, that of handling event messages. By definition, event messages occur whenever they are generated by an application which occur on a random basis (e.g., one does not know the precise instant when an event generated by a passenger rolling a window up or down occurs). Thus, the analysis of event messages is *non-deterministic* in that a specific procedure may or may not be known for calculating message latencies for all messages and the latency values are not unique but rather they are random. It follows that the analysis involving a combined network load of periodic and event based messages is also non-deterministic. This is perhaps the main reason why many analysis methods use time triggered analysis instead of a combined approach involving time triggered and event triggered analysis [4].

DESIGN PROCESS USING EVENT TRIGGERED ANALYSIS

We assume a network such as that depicted in Fig. 1 with each node having a Tx buffer holding the message to be transmitted next. Other ready messages are waiting at each node, and whenever the Tx buffer becomes empty, the highest priority ready message will move into the empty Tx buffer and contend for the bus according to the CAN protocol. Thus, priority inversion is possible. Each node can generate periodic and simultaneous messages. We assume that a finite number of n_e event messages are generated simultaneously.

Unlike previous design processes available in the literature, in this paper we propose a design process that is centered on event triggered analysis. The main idea of the process is to begin with the main requirement that we have n_e simultaneous messages in addition to a certain number of n_p periodic messages that constitute the background load. The design process involves solving two related sub-problems:

I. The calculation of the maximum number of simultaneous event messages for a given level of background load, and

II. The calculation of the maximum level of background load for a given number of simultaneous event messages.

The solution of these two problems involve scheduling all the messages (periodic and simultaneous) over the CAN network. Scheduling messages for transmission is done efficiently using message latencies as performance indicators. In addition, network bus utilization needs to be taken into account to detect an upper bound for the number of messages that can be scheduled. In the following we address each problem in turn.

A. Calculating the maximum number of simultaneous messages. The idea is to schedule the given background (periodic) load first followed by simultaneous messages being added one at a time. After each simultaneous message is added, both the maximum (i.e., worst case) message latency and the current bus utilization is calculated. The process is terminated whenever a limit for a desired message latency or the maximum bus utilization (i.e., $B_u = 1$) is reached, whichever occurs first.

B. Calculating the maximum level of background load for a given number of simultaneous event messages. The idea is the same as the previous method in A above, with the exception that the simultaneous messages are scheduled first and the periodic messages are added one at a time.

In both cases, the maximum message latency is calculated by any analytical method (e.g., using the recursive formulas developed by Tindell et.al.). Let us consider the maximum message latency network (i.e., the value of message latency for the message with the highest latency) as the criterion to schedule additional messages. Using Scenario 3 (Fig. 4) presented in Section 2, we analyze the relative order of message transmission by the various nodes. At every step, the CAN contention group is made up of the set of messages currently in the Tx buffer of the respective nodes. For example, at Step 2, the contention group = {p100,e10,p70,p80}. The winner in this group is e10, which is denoted by an * in Table 2. At every step, one message is sent over the CAN bus and deleted from the corresponding row. The last two rows of Table 2 contain the actual order of transmission and a rearranged set of messages for the purpose of worst-case message latency calculations. The rearranged set of messages makes up a rearranged set in such a way that all periodic messages in the set are transmitted first.

It can be noticed that, as a group, the statistical properties of the rearranged set are equivalent than the original set. More specifically, assuming that the message lengths for periodic and event messages are the same, the original set has an actual delay = {2,4,8,9} of units of message transmission times with mean and maximum values of 5.7 and 9 respectively. The rearranged set has delays = {6,7,8,9} with mean and maximum values of 7.5 and 9. There are advantages to reordering messages before making message latency calculations. First, priority inversion is taken into account; second, message latency calculations are simplified; and third, as far as worst case latency calculations is concerned, the statistical properties of the rearranged set is the same as that of the original set. Thus the worst-case calculations on the rearranged set continue to be the worst case. However, the actual message latencies for any message depends on the actual values of priorities assigned to messages and the allocation of messages to nodes.

Table 2. Evolution of messages at each node.

Step	Node 1	Node 2	Node 3	Node 4
1	P_{100}, e_1, e_{20}	p_{51*}, e_{10}	p_{70}, e_{30}	p_{80}, p_{60}
2	P_{100}, e_1, e_{20}	e_{10*}	p_{70}, e_{30}	p_{80}, p_{60}
3	P_{100}, e_1, e_{20}	---	p_{70*}, e_{30}	p_{80}, p_{60}
4	P_{100}, e_1, e_{20}	---	e_{30*}	p_{80}, p_{60}
5	P_{100}, e_1, e_{20}	---	---	p_{80*}, p_{60}
6	P_{100}, e_1, e_{20}	---	---	p_{60*}
7	P_{100*}, e_1, e_{20}	---	---	---
8	e_{1*}, e_{20}	---	---	---
9	e_{20*}	---	---	---
Original order: $p_{51}, e_{10}, p_{70}, e_{30}, p_{80}, p_{60}, p_{100}, e_1, e_{20}$				
Rearranged order: $p_{51}, p_{60}, p_{70}, p_{80}, p_{100}, e_1, e_{10}, e_{20}, e_{30}$				

* Denotes the message with the highest priority in the contention group.

APPLICATIONS TO NETWORK ANALYSIS AND DESIGN

To illustrate the process, let us consider the problem of calculating the maximum number of simultaneous messages that can be added to Load B of Table 1 (8 periodic messages each with 2 bytes, T = 16 ms, 6.4 % offered load, 0.292 bus utilization at 125 Kbps.). The terminating condition for the process is an upper limit to simultaneous message latencies of 16 ms (in this case equal to the period of the periodic messages). As outlined in the approach, we calculate the message latencies for the periodic messages and then the corresponding latencies for the simultaneous event messages. Table 3 shows the results of the calculations assuming that all event messages e_m have a length of 8 bytes. The maximum number of simultaneous messages is 9. Table 4 shows that the maximum number of simultaneous messages that can be sent are 11, 13, and 17 if we change the length of simultaneous messages to 6, 4, and 2 bytes respectively. Fig. 5 depicts the probability that message latency for a simultaneous message is less than a value of deadline.

SUMMARY AND CONCLUSIONS

A process for designing CAN-based distributed applications involving event triggered analysis has been presented. The process takes into account priority inversion, a situation that occurs in practical implementations of the CAN protocol involving a transmit buffer to hold the pending message. For worst-case message latency calculations, message latencies in an actual network are the same as those in an equivalent network where the order of message transmission on the CAN bus has been re-arranged. The main advantage of the rearranged network is that the latency calculations are simplified. Message latencies of simultaneous messages depend on their order of arrival relative to messages in the background load, their priority, and the distribution of priorities across all nodes. It is possible to optimize the worst case message latencies of simultaneous messages by carefully assigning message priorities in such a way that priority inversion is minimized. The process involves the calculation of the maximum number of simultaneous event messages for a given level of background load and the calculation of the maximum level of background load for a given number of simultaneous event messages. Once the maximum number of simultaneous events and the maximum allowed message latency is known for a system, a developer can easily determine the appropriate mix of time-triggered and event triggered traffic that the network can sustain.

Table 3. Message latencies of periodic messages followed by simultaneous messages.

Message	Length dm (bytes)	Period Tm (ms)	Latency (ms)
p_k	2	16	1.6288
p_{k-1}	2	16	2.2128
p_{k-2}	2	16	2.7968
p_{k-3}	2	16	3.3808
p_{k-4}	2	16	3.9648
p_{k-5}	2	16	4.5488
p_{k-6}	2	16	5.1328
p_{k-7}	2	16	5.7168
e_m	8	NA	6.7616
e_{m-1}	8	NA	7.8064
e_{m-2}	8	NA	8.8512
e_{m-3}	8	NA	9.896
e_{m-4}	8	NA	10.9408
e_{m-5}	8	NA	11.9856
e_{m-6}	8	NA	13.0304
e_{m-7}	8	NA	14.0752
e_{m-8}	8	NA	15.12
e_{m-9}	8	NA	16.1648
e_{m-10}	8	NA	17.2096

NA: not applicable.

Table 4. Maximum numbers of simultaneous event messages over a background load of 29.20% on a network with a rate of 125 Kbps.

Message length (bytes)	Transmission time @ 125 Kbps	Max. # of simultaneous event messages (M)
8	1.0448	9

6	0.8912	11
4	0.7376	13
2	0.584	17

REFERENCES

[1] K. Tindell, and A. Burns, Guaranteeing message latencies on controller area network (CAN), 1st International CAN Conference, ICC'94, 1994.

[2] K. Tindell, and J. Clark, Holistic Schedulabiltiy Analysis for Distributed Hard Real-Time Systems, Microprocessors and Microprogramming, March 1994.

[3] K. Tindell, A. Burns, and A.J. Wellings, Calculating controller area network (CAN) message response times, Control Eng. Practice, 3(8): 11-63-1169, 1995.

[4] N. Navet, Y.-Q. Song, and F. Simonot, Worst Case Deadline Failure Probability in Real-Time Applications Distributed over CAN, Journal of Systems Architecture, 46(7), 2000.

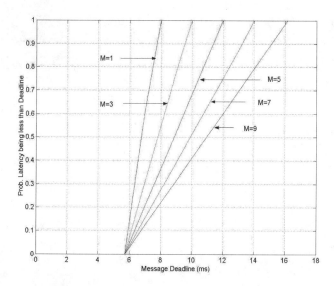

Fig. 5. Probability distribution of simultaneous message latencies in terms of deadlines. M = number of simultaneous messages.

CONTACT

Dr. Juan Pimentel is a Professor of Computer Engineering at Kettering University. He holds a Ph.D. degree in Electrical Engineering from the University of Virginia. Dr. Pimentel has done extensive research in the U.S, Germany, Spain, and Colombia. He is a Fulbright Scholar and an associate editor of the IEEE Transactions on Industrial Electronics and the Transactions on Mobile Computing. His main research areas are: distributed embedded systems, real-time networks and protocols, dependable systems, and electric propulsion systems. He is a member of Tau beta pi, Eta kappa nu, Sigma xi, IEEE, and SAE.

Email: jpimente@kettering.edu

ACRONYMS, ABBREVIATIONS

CAN: Controller area network.

Ld: Load
P: Messages per load
T: Period
Cm: Transmission time
dm: Message size
OL: Offered load
TOL: Total offered load

APPENDIX

The following is a summary of the method given in [3]. A message m can be delayed by higher priority messages and by a lower priority message that has already obtained the bus (this time period denoted as Bm is the transmission time of the largest lower priority message). Thus the worst case message latency Rm is given by,

$$Rm = Cm + Jm + Im \qquad (1)$$

Where Cm is the message transmission time of a message, Jm is the queueing jitter, and Im is the queueing delay referred to as interference time. Im is the longest time that all higher priority messages can occupy the bus plus Bm, thus

$$I_m^n = B_m + \sum_{\forall j \in hp(m)} \left\lceil \frac{I_m^n + J_j + \tau_b}{T_j} \right\rceil C_j \qquad (2)$$

Where hp(m) is the set of messages of higher priority than m and Tj is the message period. The superscript n on Im denotes that it is a difference equation that needs to be solved starting with an initial value of 0 until convergence. The transmission time of a message Cj is given by,

$$Cj = [(34+8dj)/5 + 47 + 8dj]t_b \qquad (3)$$

Where dm is the message size and t_b is the time to transmit one bit on the bus (on a bus running at 1 Mbit/sec this is 1 μs). In (), the term 47 is the size of the fixed-form but fields of the CAN frame and (34 + 8dm)/5 is the maximum number of *stuff bits*.

TTCAN from Applications to Products in Automotive Systems

Patrick Leteinturier, Nico A. Kelling and Ursula Kelling

Infineon Technologies AG

ABSTRACT

This paper outlines the results of a study performed to analyze the mission of TTCAN from applications to products for automotive systems.

As commonly acknowledged communication is one of the key elements for future and even present systems such as an automobile. A dramatically increasing number of busses and gateways even in low- to midrange vehicles is putting significant burden upon the validation scenario as well as the cost. Accordingly, numerous new initiatives have been started worldwide in order to find solutions to this; some of them by the definition of enhanced or new protocols.

This paper shall have a look particular on the new standard of TTCAN (time-triggered communication on CAN).

This protocol is based on the CAN data link layer as specified in ISO 11898-1 and may use standardized CAN physical layers such as specified in ISO 11898-2 (high-speed transceiver) or in ISO 11898-3 (fault-tolerant low-speed transceiver). This particular property is beneficial when migrating towards time-triggered communication approaches.

Furthermore TTCAN provides a mechanism to schedule CAN messages either time-triggered or event-triggered. This feature opens new ways to partition, link and structure systems more efficiently in terms of cost and validation. Examples are implementation of a sensor bus, distributed or split control functionality and increased real-time performance in CAN-based in-vehicle networks without software overhead.

This paper will identify automotive applications and approaches that require or benefit from the TTCAN protocol. Strengths and limits of the solution are addressed for the domains of powertrain and safety vehicle dynamics. Partitioning will be proposed with the key advantages and system benefits for the applications. A new implementation of a level2 TTCAN node providing full TTCAN functionality with very low software overhead is presented as well.

INTRODUCTION

The increasing legal requirements for safety, the reduction of emissions, fuel economy and onboard diagnosis systems are driving the market for more intelligent systems. Therefore, the embedded systems that will have to control automobiles have grown to such an extent that they are now equivalent in scale and complexity to avionics systems, the requirements in terms of computational performance will even further increase by a factor 10 over the next 5 years.

Accordingly the automotive industry has to face a disruptive challenge with respect to validation possibilities as well as effort and – to complicate the matter – an increasing number of software components delivered from different sources to be executed on one common core. The introduction of x-by-wire, enabled by new communication protocols, in Powertrain and Chassis will offer new possibilities, such as:

1. Centralized systems can be distributed to control the right functions at the right places. Provided the right protocol features, this will allow to minimize hardware redundancy whilst at the same time increased system safety, reduce the validation complexity by encapsulation and to reduce cost, weight as well as electromagnetic emission by replacement of analog with digital links.

2. Greater flexibility and scalability at system level achieved by observable and well defined interfaces. New functions can more easily be introduced to satisfy the coming regulations.

3. Harmonization of control strategies. Loosely coupled control systems can become tightly coupled or even integrated. Local control can be migrated towards global control strategies.

4. Increased efficiency in resource management as well for processing performance, as for communication bandwidth and even cost of e.g. processing performance provided within the system implemented with different characteristics and in different technologies.

It is not, that some of these to a certain degree could not be realized with existing and established concepts – But beside the demand for highly sophisticated electronics, the demand on cost reduction (especially for small cars) is probably the number one driving factor for a smart system partitioning. Time-triggered system architecture is one commonly accepted approach as to manage the system complexity with respect to validation, predictability and safety. Hence, upcoming architectures are much supported with a new bus principle such as TTCAN as will be seen in the next chapters.

AUTOMOTIVE APPLICATIONS

Advantages of the application can be elaborated at various different levels. In this case we chose to distinct between design time and validation aspects on the one side and run-time benefits at the other side.

This section will concentrate first on architectural aspects, provide a description of four major categories with respect to control strategies in the context of smart partitioning and give reference to typical application use cases. This section will conclude with an application matrix that illustrates the capability and the limits of the TTCAN solution by exploring the applications where this bus can be used, relative to other emerging high-speed communication solutions on the market such FlexRay or TTP.

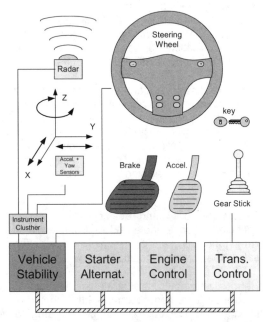

Figure 1: Connection between Powertrain and Vehicle stability

Credit to the runtime advantages will be given in subsequent chapters that exploit implementation aspects of the TTCAN module. Possibilities to improve available (usable) communication bandwidth without complex software strategies, the real time performance, and

determinism of messages, fault detection and contingency action will be outlined.

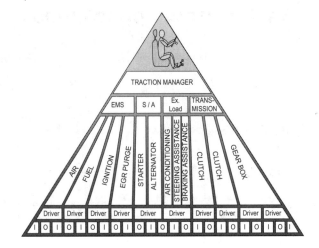

Figure 2: Control Pyramid of a Powertrain System

Powertrain and chassis systems are undergoing a huge evolution. Taking powertrain as an example, with early focus on engine management followed by transmission control and now advanced starter alternator concepts with the capability of energy recovery. Besides fuel efficiency that reaches very high values (40 to 45% with new diesel common rail), regenerative braking systems will even bring kinetic energy back into the system during braking.

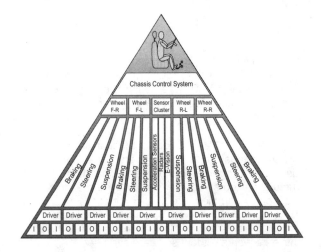

Figure 3: Control Pyramid of a Chassis System

On the other hand road safety is significantly enhanced by vehicle stability and traction control systems (ABS, ASR, ESP) that are further improved by integration of damping, suspension and steering control. More and more systems have to be linked, sharing sensor data and execute harmonized control over various actuators with the common target to optimize overall vehicle stability and wheel to road traction. Further harmonization of powertrain with chassis functions extents from unified chassis control towards unified drivetrain.

This trend is amplified by the introduction of x-by-wire that gradually replaces mechanical and hydraulic functions by electronic control. The technology of by-wire is particularly relevant in drive-by-wire applications. Step-by-step, comfort functions migrate to become warning functions, later driver assistance, emergency react and ultimately (eventually) autonomous driving functions. The revolution in automotive is to assist the driver on the top of a decision process with lots of electronics. The order coming from the driver is no longer transmitted by mechanical elements. This new approach allows to remove some constraints of previous architectures and to focus on the right function in the right place.

Application challenges & requirements

For drive-by-wire functions, powertrain and chassis systems are sharing numerous common information such as: key, acceleration pedal, brake pedal, gear ratio... The challenge of the new combined system will be to establish the right HMI (human machine interface), not only regarding correct interpretation of the driver's intention but also to provide him the right feedback. The second step will be to provide the information to the appropriate control unit in real- and deterministic time. The system has to be safe and failure tolerant. The last challenge but not the least is to ensure complete consistency in between the distributed and embedded systems. Figure 4 displays one possible, fully distributed safety architecture for electrical (dry) braking as it has been implemented within the European research project BRAKE (Delphi Systems, Infineon Technologies, Volvo Car Corporation, Wind River).

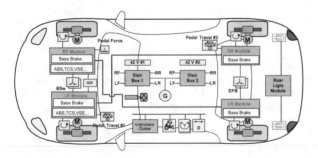

Figure 4: Distributed Safety Architecture for Electro-Mechanic Braking

Such application or architecture certainly marks the upper range in terms of safety- and bandwidth-requirements. In the given application, the used high-speed failure-tolerant, deterministic safety-bus operates as some sort of communication backbone to the entire system and encapsulates the different network nodes from each other.

Due to cost reason it is not desirable that each sub-system like engine management, starter-alternator, and transmission control or vehicle stability is measuring for itself the driver expectation (acceleration pedal, brake pedal...). In fact, this would even jeopardize the harmonization of various control functions towards global vehicle level control. Hence, most available information will have to be shared in between the sub-system.

In addition, not all applications have equivalently high bandwidth requirements compared to a complete and unified, distributed chassis controller. The most common closed-loop control scenarios for TTCAN suited partitioning are addressed in the following chapter.

Control Architectures

A typical, simple closed-loop control for a system is sketched in Figure 5. A sensor (S) measures the system, the controller computes how to modify the system and the actuator (A) performs the modification.

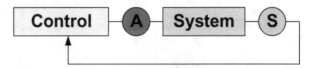

Figure 5: Control loop with (A)ctuator and (S)ensor

TTCAN exhibits multiple capabilities to change the partitioning of closed loop control. These can be clustered in the following four groups:

1   Sensor bus
2   Remote controlled sub-system
3   Split controlled system
4   Distributed System

Sensor bus

A typical situation particularly in chassis applications is a system with multiple sensors at various locations, where some or many of these have to be shared across multiple functions. An efficient approach based on TTCAN communication is as follows: The sensors are "smart" and able to communicate with the controller. A time window will be assigned to each of them.

For instance, for a chassis system with ABS, ASR, ... the sensors could include the lateral and longitudinal accelerometers, wheel speeds, yaw sensor(s), the radar sensor, and more. These sensor data have to be shared across different functions that control wheel traction, spin, braking, and overall vehicle stability such as roll, pitch and spin, that are also influenced by steering and damping.

Once the communication schedule is configured, the TTCAN communication controller will autonomously trigger the sensor if needed and communicate the sensor data at the configured rate and times. There is no need to provide additional software to establish timing correlation in between different sensor values, as all are sampled relative to the common network time. The control algorithm will be also able to run consistency checks, as well as some data fusion to improve the observation of the system. Some algorithms depend on the availability of a fixed number of samples from sensor A along with one or another fixed number of samples from another sensor B (e.g. filters). This can be inherently ensured by means of the communication schedule.

This application is opening new possibilities for a sensor cluster. The TTCAN inherent synchronization features allow for example cost efficient networked sensor nodes that provide their information at pre-defined and synchronized times.

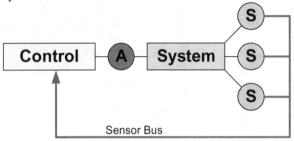

Figure 6: Sensor bus application

Fixed sample ratios between different sensors as may be used e.g. in various filter designs can easily be configured without need for software interaction. In addition, the complexity of a smart sensor can be reduced, as the sample triggering can be managed by means of the bus and no further controller is needed – a significant cost saver when looking at smart sensor modules.

Remote controlled sub-system

In this application the controller and the sub system are not located at the same place, the sub-system outsource the control algorithm to a central unit. As a consequence the sub-system has to communicate the value coming from the sensor and receive the new values for the actuator.

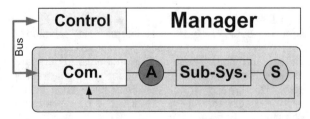

Figure 7: Remote controlled sub-system

This kind of control requires high bi-directional bandwidth. A time window will be reserved for the sensor value, another time window for the actuator value. This

could be used for the electronic throttle in engine management.

Split controlled sub-system

This application is a variant of the remote control loop. Here, two control loops are identified: a fast one and a slow one. For example the fast control loop could be the control of the current and the slow control loop could be the control of the position. The benefit of this partitioning is to remove the need of high bandwidth to control the fast loop. In addition the local control can also guaranty some back-up scenario in case of communication fault. Further more, the complexity of the remote algorithm can take into account some adaptive control and learn the evolution of the sub-system during its life. Many applications can use this principle – for instance electronic turbo-charger, electronic throttle, electronic valves…

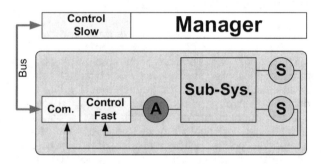

Figure 8: Split-Control sub-system

Distributed System

This application case is by far the most complex one. The control of the system is distributed, the actuators are distributed the sensors are distributed. To compute the values of the local actuators, each local controller requires access to the values of its local sensors and the values of the remote sensors and actuators. The system inherent controller redundancy can easily be used to establish a distributed safety case without adding costly redundancy to a central controller.

Many projects and activities among the automotive society have been analyzing the benefits, drawbacks and mechanisms of distributed versus centralized control architectures. A complete documentation could probably fill at least a book, and can therefore not be handled here. Nevertheless, studies have shown that distributed redundancy is increasingly efficient with higher system complexity. Part of the reason is the validation issue: The growth in system complexity is primarily limited by the ability to validate it, not by technological or physical limits. Distribution supports partitioning to cost, enforces clean and well-defined interfaces and consequently supports encapsulation and modularization. Ultimately, a transparent software design on top of a distributed architecture entirely abstracts control software from underlying hardware and network structures and finally will increase software reusability and scalability.

However, these studies have also shown that additional work has to be performed in the domain of development methodology and appropriate tool support. Model based design and model based system validation is only the first step. Existing design guidelines and interface definitions need thorough review. Besides that, the safety scenario can be greatly improved with communication protocol inherent features for replica determinism, such as configurable mutual exclusion algorithms. These are not part of TTCAN.

In summary, the approach of distributed system architecture is very efficient in cost and validation once fully transparent function design is established and well defined functional API interfaces are defined. On system level, this will initiate the transition from multi gateway architectures to common backbone topologies that enables further cost and flexibility advantages. Such will very soon demand high-speed communication. TTCAN is using the CAN physical layer and is limited to 1Mb/s and accordingly not well suited for this control architecture.

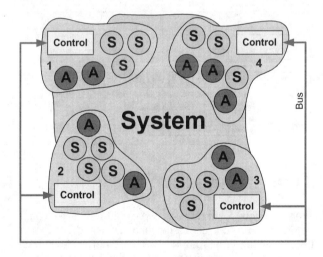

Figure 9: Distributed Control Strategy

Nevertheless TTCAN has very good opportunities in conventional gateway topologies to cover sub-applications like the intercommunication between engine management – transmission control – starter alternator and vehicle stability system. Most by-wire applications except fully distributed electrical braking indeed have relatively relaxed requirements that are well satisfied by the TTCAN protocol. Typical examples are rear electromechanical braking, parking brake, rear steering and some suspension functions. In this aspect, TTCAN can provide a convenient migration path until further

harmonization and unification of control functions will demand high-speed safety communication.

ENHANCED CAN USABLE BANDWIDTH

Another application advantage is the capability of enhanced bandwidth utilization compared to standard CAN applications that do not implement a comparable application layer. Although some different application layers are used today that can feature similarly high bus utilization, these typically are implemented in software and therefore consume precious CPU load and are less efficient than a hardware solution. If no such application layer is used and sensitive messages have to be transmitted on the media, typical planned network load does not exceed somewhere around a third of the capacity. This is, as collisions on the media and automatic retransmissions will eventually load the bus up to a state where messages with lower priority may not come through any more. Typical planned CAN load cannot increase above approximately 40% without further provisions as the inbuilt collision mechanism may otherwise collapse the communication of messages with low priority. Using a software session layer, this planned load maybe extended up to eventually 80%. This is achieved typically by statistic based scheduling optimization for the network communication. Using TTCAN will allow the users to reach around 90% of the network load without further software provisions. Composability and testability can be improved implicitly by firstly deterministic scheduling and secondly by decoupling in time of the network from the CPU activities.

Application Decision Matrix

Summarizing the discussions about design benefits and applications that benefit from TTCAN communication, we present the following table (Figure 10).

Applications	TTCAN	FlexRay or TTP
Unified Engine Management, Transmission, Starter-Alternator, Vehicle dynamic	X	
Electronic Turbo-charger	X	
Electronic Throttle	X	
Sensor Cluster	X	
Parking Brake	X	
Electro-Hydraulic Braking	X	
Hybrid Braking	X	
Full Electro-Mechanical Braking		X
Rear Steering	X	
Basic Suspension System	X	
Full Active Suspension System		X
Full Chassis Control		X

Figure 10: Application's Requirements Matrix

The table compares, which applications can benefit from TTCAN communication compared to those, that have higher requirements e.g. in bandwidth such that a high-speed safety bus becomes mandatory. Certainly, in many cases TTCAN can be classified as a soft requirement depending on the system architecture, meaning that - especially when primarily point-to-point communication is required - other solutions can be applied. However, when targeting a migration towards harmonized control strategies, system unification and standardization in interfaces and protocols, TTCAN is a good option.

Limitations of TTCAN usage has been identified primarily with respect to total available bandwidth needed for distributed, full electro-mechanical braking, and the harmonized, unified global chassis control system.

TTCAN Protocol Implementation

Status

The upcoming TTCAN standard as an hardware layer extension to CAN has been implemented in a new module named "MultiCAN". MultiCAN is a more powerful scalable CAN, which is able to support two up to eight CAN nodes V2.b active. The current implementation of the MultiCAN module supports 4 CAN nodes, whereby one of these nodes is able to run TTCAN level one and two. The module supports 128 message objects flexible assignable to each CAN node. It also features a flexible, double-chained list management to enable list based FIFOs and gateway implementations.

Silicon will be available for open market as integrated part of the 32bit microcontroller devices in June 2003.

Protocol Overview

CAN itself is a completely event triggered system. To provide a possibility of a time-triggered system, TTCAN is specified as an additional layer, using the layers of the CAN standard. The CAN standard describes the conditions for synchronization of a CAN network. If the nodes are synchronized, any message may be transmitted at a specific point in time, without competing with any other message on the bus. By avoiding more than one message on the bus, a loss of arbitration is prevented and latency becomes predictable. This means, that TTCAN gives the possibility of a deterministic schedule on a TTCAN network with hard deadlines. Four different types of time slots are distinguished within TTCAN. One on a complete time triggered base is called exclusive time window, here two triggers define, the earliest starting point of a message, as well as the latest starting point of a message (so called Tx_Enable_Window.). In this window only a predefined message is allowed to be started, which leads to the fact, that no arbitration will take place. In case of a

transmission error in an exclusive time window, no retransmission is allowed. To have the option of arbitration, so called arbitrating windows or merged arbitrating windows exist. In standard arbitrating windows messages are only allowed to be started if the bus is idle and the message does not violate the deadline of this time window, but no retransmission. In case of merged arbitrating windows, which consist out of several consecutive arbitrating windows, a retransmission due to an error or an arbitration loss is possible, as long as the deadline is not violated. The fourth and last window type is "free time windows". In these windows the bus idles. In case future extensions of the bus are needed, these windows can be used.

Scheduling, Matrix and Frame

The transmission schedule is predefined at design time and stored in a system matrix, which consists of transmission columns and basic cycles (rows). The columns of the matrix represent the transmission (time) windows that are sequentially processed and then repeated all over within each basic cycle.

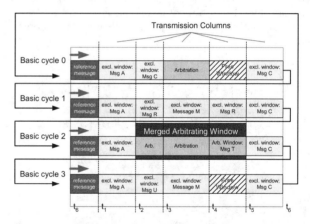

Figure 11: Example of a system matrix

Transmission columns may have different time spans, but in each basic cycle a special n^{th} transmission column has the same time span as in each other cycle. A basic cycle is a row inside the system matrix. Each basic cycle starts with a reference message, which includes global time. Basic cycles are identified by the cycle count number. Cycle count starts at zero and ends at 2^n-1. Within a basic cycle different time windows may exist. The shortest basic cycle includes a reference message. Inside a transmission column periodic messages may be defined by the repeat factor.

Figure 12 introduces a small example how the scheduler of MultiCAN works. MultiCAN has a linear scheduler, which has to be in increasing time mark order. The scheduler gets a start address. Starting from this address, predefined command sequences will be entered. A defined window is per default an exclusive time window. To change the default status an arbitration window has to be defined. The last part of an arbitration window and a single one close the arbitration window and are so called single or short arbitration windows. A window after short arbitration window is again an exclusive time window.

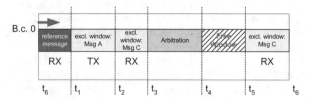

Figure 12 : Single basic cycle of a time slave

The scheduler knows eight different types of events (entries), which can be defined: Time mark, interrupt control, arbitration, transmit control, receive control, reference message, basic cycle end and end of scheduler memory. On every time mark, it is possible to generate an interrupt. A special interrupt is the interrupt on the reference message. This interrupt can trigger actions on other modules or give a regular timer event for the system. In the example of Figure 12, the TTCAN node shall transmit Message A and receive message C.

command	Arbitration mode	Interrupts	Event time
timer mark	No action	Here: none	t1
command	Check enable	Destination RX MO of interest	1st basic cycle / repeat factor
RX control	Yes	MO of ref msg	0 / 1
command	Alternative msg.	Transmission enabled	1st basic cycle / repeat factor
TX control	No msg., if destination object not valid.	Next TX trigger sends MO of msg A.	0 / 1
command	Arbitration mode	Interrupts	Event time
timer mark	No action	Here: none	t2
command	Arbitration mode	Interrupts	Event time
timer mark	No action	Here: none	t3
command	Check enable	Destination RX MO of interest	1st basic cycle / repeat factor
RX control	Yes	MO of msg. C	0 / 1
command	Arbitration mode	Interrupts	Event time
timer mark	short arbitration window	Here: none	t4
command	Arbitration mode	1st basic cycle / repeat factor	
Arbitration	short arbitration window	0 / 2	
command	Arbitration mode	Interrupts	Event time
timer mark	no action: excl time window	Here: none	t5
command	Gap mode	Event time	
basic cycle end	No	t6	
command	Arbitration mode	Interrupts	Event time
timer mark	No action	Here: none	t6
command	Check enable	Destination RX MO of interest	1st basic cycle / repeat factor
RX control	Yes	MO of msg. C	0 / 1

Figure 13: Scheduler example for Figure 12

The scheduler example in Figure 13 shows a way, how a schedule is defined on MultiCAN. Additionally, the scheduler has an error detection mechanism, which helps find, the following configuration errors: At the end of basic cycle, an arbitration window is still open. A slave device has a reference message scheduled. A time mark is missing during instruction collection. In a reference message window, another message window type is scheduled. A reference message is missing.

As all events refer to a time mark this schedule has a deterministic behavior. Only inside arbitration windows, it is not known, which message will be sent on the bus before the complete system is defined. A further feature that is not used in this schedule example is the so-called emergency message. In exclusive time windows, it is possible to transmit a scheduled message or – if that is not available - a message that is triggered by another object. The user has the possibility to configure, the message, which shall have higher priority. For example the message inside scheduler memory has higher priority. The exclusive time window starts and this message is not ready. In this case the user is sending the message of the message object, which has been defined backup. Maybe the user likes to get noticed in case measurements are not ready, here an emergency message will be sent.

Time Synchronization

In TTCAN the time synchronization takes place during synchronization phase and without synchronization losses after a gap. A gap has to be declared inside a reference message and takes place at the end of corresponding basic cycle. In a TTCAN network up to eight potential time masters exist. Every potential time master has a different priority, shown by the identifier, which is differing, in the last three significant bits. In case a potential time master wakes up, it waits for a predefined waiting time, listening to the bus. When no reference message of higher priority comes in, the device starts sending. In this stage arbitration takes place between potential time masters. A slave or a potential time master with lower priority will regard itself as synchronized as soon as it received two valid reference messages. During synchronization all potential time masters have Ref_Trigger_Offset set to Initial_Ref_Trigger as positive value. Ref_Trigger_Offset modifies the time mark, which determines the starting time of a reference message. In case a device with a positive Ref_Trigger_Offset sees a lower priority reference message on the bus, it will reset its Ref_Trigger_Offset to zero. In case a message with lower priority still starts its reference message first, the Ref_Trigger_Offset can be decrement until −127 or until it becomes time master. All potential timer masters who stop sending become backup time master. The new time master sets it Ref_Trigger_Offset to zero.

On MultiCAN the Ref_Trigger_Offset can be written during initialization phase, and will be transferred to a user write protected register. As soon as the synchronization phase starts, the potential time master takes part inside the protocol. The Ref_Trigger_Offset is taken care of, without any interaction with the user. As soon as the device is time master or a time slave (these are also potential time masters as long as the time master sends its reference messages) the scheduler controls the operation.

Outside the initial synchronization phase in strictly time-triggered systems the local time is adjusted to global time. In a level 2 TTCAN network NTUs are the time base of the system. One NTU is fraction of a second. The node view of an NTU usually differ slightly from the network view, which is caused by the local oscillators. After the network synchronization, the global time is known, sent out by the reference message. The time master provides the global NTU, which is valid for the complete system. A time slave get this global NTU by the reference message and adjusts is local time by the shift correction calculated from global time. The TUR is calculated as:

$$TUR = \frac{\text{local length of basic cycle}}{\text{global length of basic cycle}}$$

On MultiCAN the user does not have to take care of the adjustment of local time. The MultiCAN module takes care of this adjustment, so that the user is always sure, that local time and global time are almost identical.

Initialization

After reset the MultiCAN module is in configuration mode. To get the system running, the following steps have to be taken:

- The hardware sets local, global and cycle time to zero.

- Reception filtering is not possible, because it is not enabled.

- A TURADJ value has to be chosen. In case for automatic TUR adjust the corresponding enable has to be set. The TURADJ value will automatically transferred to TUR.

- The user has to setup the TTCAN schedule into scheduler memory.

- Further TTCAN register configurations have to be done.

- The bit for leaving configuration mode has to be set.

- The hardware starts local time.

- Synchronization mode is started automatically.

- The TTCAN protocol starts running.

Error Management

TTCAN contains four levels of error management. The protocol distinguishes between no error, warning, error and severe error. In case of no error, the TTCAN protocol runs without any restrictions. The second level is the warning state. Here a value called MSC is important. This value has the following functionality: Message status count (MSC) (range zero to seven) is an error counter, which provides a possibility to detect scheduling errors for exclusive time windows. It shall be remarked that each message object for exclusive time windows has its own MSC. The overall maximum and minimum MSC inside a matrix cycle is stored inside the MultiCAN module.

Three different root causes for a warning are possible:

- MSCMAX – MSCMIN > 2 at the end of a matrix cycle.

- A receive-message object reached an MSC of seven.

- Tx_Underflow: Not all transmit triggers got active.

A warning exceeds an interrupt flag.

The third stage is the event error. Reaching the error state, no transmit enable window will be opened, additionally the Reference_Trigger_Offset will be set on 127. The root cause of an error is:

- A transmit-message object reached an MSC of seven.

- Tx_Overflow, more than the number of specified Tx_Triggers has been taken.

An error interrupt flag is set.

The forth and last error type is a so-called severe error. Here all CAN bus traffic is stopped. The configuration phase of the MultiCAN module is entered again. There are four different types of root causes for a severe error:

- Application Watchdog, the application failed to service the watchdog. In this case it is not insured that the host is still working properly.

- Bus Off, a CAN bus error leads to bus off

- Config error, a merged arbitration window is not properly closed or a Tx_Trigger occurs in reference message time window.

180

- Watch_Trigger_Event, this trigger occurs if the reference message is missing, which means, that the time master stopped sending reference messages.

The INIT bit is set.

CAN Debug and Safety Concept

To avoid errors and to ensure to use the right error-handling MultiCAN offers a variety of mechanisms.
In case the device has two CAN modules, the following possibilities exist:

a) Two CAN modules, two CAN transceivers

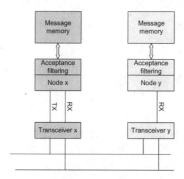

Figure 14: Two separate CAN nodes, two transceivers

Minus: This concept needs two CAN transceivers
Plus: Transceiver errors are detectable.

b) Two CAN modules, one CAN transceiver

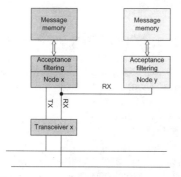

Figure 15: Two separate CAN nodes, one transceiver

Plus: This concept needs one CAN transceiver.
Minus: A transceiver error is not detectable.
Both concepts do have synchronized message access, which basically leads to the fact, that short time disturbances are recognized by both CAN nodes and lead to an error situation.

c) MultiCAN in Analyzer Mode

In Analyzer Mode the device listens to the bus, but it is not actively taking part of the protocol. The nodes are handled asynchronous as the protocol handler is requesting the information one node after another. Message objects are appended to a node. If the application shall test, that a received message is really correct, for each node a message object has to be setup and compared via software. A disturbance on the bus is seen in different states and has a different impact. It is even possible to find a problem between protocol handler and message memory.

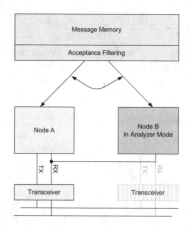

Figure 16: MultiCAN in Analyzer Mode

Beside the Analyzer Mode the MultiCAN module has additional features to avoid bus errors before they occur. The MultiCAN offers a synchronization analysis features as well as driver delay measurement feature, which help to make further extensions of the bus safer and small disturbances on the bus less critical.

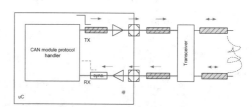

Figure 17: Schematic signal way, TX to RX

The synchronization analysis feature helps to prevent a possible bus error in case of slight changes or impacts on the bus. This feature monitors the time between the first dominant edge and the sample point measured a stored in a frame counter register. By using this feature it is easier to adjust the sample point to the actual sample point of the complete bus system.

The driver delay measurement feature helps to find a circuitry, where a sent edge is received time quanta before the sample point.
Both features help to find a configuration and circuitry, which is not fault-prone to slide changes or disturbances on the bus.

CONCLUSION

The usefulness of TTCAN besides competing high-speed safety communication protocols for automotive has been shown for different applications. Its benefit for more flexible system partitioning at improved cost levels has been outlined for common closed-loop control tasks. Finally, the description of Infineon's implementation of a full TTCAN level 2 protocol controller has demonstrated, how the supported mechanism ensure the positive runtime effects and consistency support without additional software overhead as would be needed with TTCAN level 1 implementations. The described protocol controller has been implemented in silicon and will be available in Infineon's 32bit microcontrollers.

REFERENCES

1   X-by-Wire: Opportunities, Challenges and Trends; Nico Kelling and Patrick Leteinturier; SAE 2003-01-0113

2   Integrated Self-test Technologies for Embedded Automotive; Wilhard Von Wendorff; SAE 2003-01-xxxx

3   New µController links: Micro-Second-Bus and Micro-Link-Interface; Klaus Scheibert and Jens Barrenschen, Hannes Estl; SAE 2003-01-xxxx

4   Real Time Data Fusion: a challenge for distributed intelligence; Patrick Leteinturier, Juergen Hoika and Claus Preuschoff; ATA 2002-01A4011

5   Seamless Solutions for Powertrain Systems; Juergen Hoika, Hannes Estl, Gunther Krall, Joachim Weitzel and Patrick Leteinturier; SAE 2002-01-1303

6   AUDO Architecture. A solution to Automotive µController Requirements; Patrick Leteinturier, Boris Vitorelli; SAE 2000-01-0145

7   TwinCAN A new milestone for inter-network communication; Jens Barrenscheen and Patrick Leteinturier; SAE 2000-01-0144

8   Semiconductor Technologies for the Automotive Electronics of the Future; Eschert, Hoika, Sirch Taeuber, Vennmann; VDA MICRO.tec 2000 EXPO Hannover

9   The Brake Project - Centralized versus Distributed Redundancy for Brake-by-Wire Systems; Nico Kelling, Worthy Heck; SAE 2002-01-0266

10 ISO/WD 11898-4: Road vehicles – Controller area network (CAN) – Part 4: Time triggered communication

11 MultiCAN Specification for AUDO-NG, Infineon Technologies AG, Jens Barrenscheen

CONTACT

Patrick Leteinturier
System Definition Automotive
Tel: +49 (89) 234-83828
Patrick.Leteinturier@infineon.com

Nico A. Kelling
System Marketing Powertrain Worldwide
Tel: +49 (89) 234-82218
Nico.Kelling@infineon.com

Ursula Kelling
Technical Marketing Serial Bus Systems
Tel: +49 (89) 234-83287
Ursula.Kelling@infineon.com

Address:
Infineon Technologies AG
Balanstr. 73
81541 München
Germany
http://www.infineon.com

DEFINITIONS, ACRONYMS, ABBREVIATIONS

ECU: Electronic Control Unit
HMI: Human Machine Interface
IC: Integrated Circuit
PCB: Printed Circuit Board
I/O: Input/Output
ASIC: Application Specific IC
ASSP: Application Specific Standard Product
DSP: Digital Signal Processor

Low-Cost CAN-Based Communication System for High-End Motor-Scooter

F. Baronti, D. Lunardini, R. Roncella and R. Saletti
Dipartimento di Ingegneria dell'Informazione: Elettronica, Informatica, Telecomunicazioni. University of Pisa

R. Hippoliti and S. Mangraviti
PIAGGIO & C. S.p.a

ABSTRACT

A low-cost communication system, based on the Controller Area Network (CAN) protocol, for overall electronics management in a high-end motor-scooter has been designed. The system consists of a single 1 Mb/s CAN network and is designed for PIAGGIO X9 500 vehicle, though it can be easily adapted to any other two-wheel vehicle. The solution presented drastically reduces assembling cost, significantly increases the system reliability and makes the vehicle upgrades very fast and easy. Particular care has been adopted in satisfying safety issues and in minimizing the power dissipation.

INTRODUCTION

Since the 1960s the number of electrical and electronic devices used in a vehicle has gradually increased and the growth of automotive electronics has dramatically accelerated in the last few years. Great advances in semiconductor technology, more stringent environmental regulations and customer increasing demand of enhanced performance, safety, comfort and convenience have driven this trend. As a result, many of the control functions, previously realized electrically or mechanically, are now performed by electronic systems. Besides, a lot of new features have been added: complex Engine and Powertrain Control Systems, Antilock Breaking Systems, Satellite Navigation Systems, on-board computers are only a few examples. Moreover, the growth rate of on-vehicle electronics is expected to increase in the next future, because of the upcoming introduction of new applications, such as X-by-wire control systems, collision avoidance systems, night vision systems, etc [1].

The need of communication among the electronic systems spread into the vehicle and the consequent wiring harness length and weight increase with the complexity of each system and the sophistication of the control algorithms, creating difficulties in the assembling process and making unaffordable the harness cost. Therefore, multiplexed systems, in which the communication among the devices is fully managed by a serial bus, were developed and introduced during the 1980's. Nowadays, multiplexing is widely used in automobiles and heavy trucks in which more than one network is used to connect devices with different bit-rate requirements. In this field, the Controller Area Network (CAN) protocol (ISO 11898) has become a standard at least in Europe for Class B and C networks [2].

Two-wheel vehicle industry comes now to feel the same need that has been first felt by heavier vehicle manufactures. Actually, the comfort and safety demands that are fast increasing, especially in high-end motor-scooter market, can be satisfied only by introducing more and more complex and sophisticated electronic systems on the vehicle. This quickly leads to the introduction of multiplexed systems also in the two-wheel world. However, the problems found in two-wheel vehicles are by far different, because of the more stringent requirements for costs, available space and the different environment in which the system should operate. Besides, the lesser number of electronic systems present in a two-wheel vehicle hardly justifies the realization of the overall electronics management by multiple networks [3]. On the other hand, a limited growth of costs can be tolerated if it is justified by great advances in the satisfaction of the customer demand of quality. The introduction of a multiplexed system based on a single network causes an increase of cost because of the additional electronics needed, increase that is small and affordable even in the present vehicle, particularly in the high-end motor-scooter. However, its use will be more and more convenient, also in terms of costs, in the future vehicles equipped with a large number of electronic devices and control functions.

Aim of this work is to describe the design and realization of a multiplexed system, based on the CAN protocol, for the communication management of all the switches, sensors, actuators and Electronic Control Units (ECUs) that can be found in a high-end motor-scooter. To our knowledge, this is the first application of a CAN network in a two-wheel vehicle for volume production. The system is based on a single CAN network and has been designed keeping always in mind the issues of cost, reliability and safety of the vehicle and portability of the system. While an ECU is a smart device that can be directly connected to a serial bus, other devices such as switches, lights, phonic wheel and other sensors, need additional electronics to be able to share information through the bus. The solution described in the next sections can be viewed as a cost-saving approach to add communication capabilities to the "dumb" devices [4], [5].

SYSTEM ARCHITECTURE

The system consists of 4 CAN nodes, each one controlling the devices located in one of the 4 areas in which the vehicle has been divided. From now on, we will refer to them as *front*, *dashboard*, *seat* and *back* area. Figure 1 shows the positions where each one of the 4 nodes is located into the scooter. Each node is connected by dedicated short wires only to the devices pertaining to its own area, while the longer link between different areas is realized by a single unshielded twisted pair (CAN network communication medium). This architecture eliminates the bottlenecks of the cabling system, reducing the assembling time and improving the system reliability. Indeed, the proposed solution is the result of a trade-off between the manufacturing demands of reduction and simplification of wiring harness, improvement of reliability and minimization of the additional electronics needed for the overall communication management. We also observe that a node can now directly execute some control functions, implemented before by dedicated ECUs.

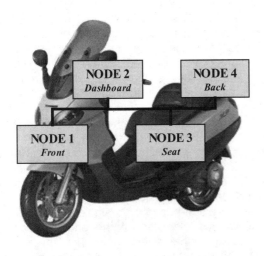

Figure 1. Positions of the 4 nodes into the X9 vehicle.

Let us now briefly describe how the various devices have been assigned to each one of the 4 areas. The first area is controlled by the *front* node that manages the front lights (including the turn lamps), the key switch, the phonic wheel and external temperature sensors, the claxon and the cooling fan. The *dashboard* area is located over the *front* one and the corresponding node controls all the user commands, the digital display and the analog instrumentation. The *seat* node is placed in the middle part of the vehicle and is connected to the engine sensors (fuel

level, engine temperature and oil pressure) and the electrical centerstand circuit. Finally the *back* node controls the back lights.

One of the obvious consequences of the migration from the traditional wiring harness to a multiplexed system is that the control functions, before realized by switches and relays, are now implemented by intelligent power switches and software drivers. Particular care has been adopted to assure a high safety level for the critical controls, such as the engine turn-off. Besides, the 4 nodes must be supplied also when the key is off, in order to maintain the vehicle always under a certain degree of control. However, this need challenges the battery duration and so low battery matters have to be prevented. These and other issues have been faced in the design of the node hardware and software and the adopted solutions will be described in detail in the following sections.

We finally observe that the introduction of a multiplexed system into a two-wheel vehicle opens the way to the realization of new functions based on the vehicle-wide communication, such as distributed powertrain control, diagnostics, X-by-wire, entertainment systems etc. In fact, ECUs, such as the engine control module, the dashboard, information management units, etc., would become additional CAN nodes able to share information through the bus. In the view of these future developments, we have chosen to set the bit-rate of the CAN network to the maximum possible value, i.e. 1 Mb/s.

NODE ARCHITECTURE

A node is a smart unit that essentially controls the devices that are located in its own area and manages the inter-area communication. In particular, it first gets information from on/off inputs, resistive sensors, frequency variable inputs and received CAN messages. Then, the input data are elaborated and the node produces digital and power outputs and transmits CAN messages. In order to achieve the aim of manufacturing cost minimization, the 4 nodes are identical from a hardware point of view. This means that the node hardware has to be designed with general-purpose capabilities, because it must be able to fit in the 4 different environments for both the logical and the physical points of view. This feature also increases the portability of the system. On the contrary, the firmware of each node will be customized according to the required control functions of the area in which the node is physically mounted.

HARDWARE

Due to the hostile environment in which the system should operate, great attention must be played to the hardware design. In particular, robust components should be chosen and the digital control electronics must be carefully protected from out-of-range voltages. Figure 2 shows a simple block diagram of the hardware structure of each node.

The heart of the node is the control block that consists of the T89C51CC01 ATMEL microcontroller. This integrated circuit (IC) is an 8-bit microcontroller with an integrated CAN protocol controller and is equipped with 32 digital I/O lines, 3 timers, UART, PCA module, 32K bytes of on-chip Flash memory and other useful features [6]. The other blocks are necessary to make the microcontroller able to operate and communicate with the external environment.

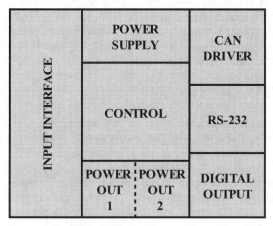

Figure 2. Block diagram of the node.

As far as power management is concerned, the microcontroller is always supplied. However, it can disable the power supply of the other circuitry and in particular of the CAN driver, that is a very demanding component for the global power budget. This feature has been realized adopting a robust dual voltage regulator with a disable logical input.

In order to make the node input signals suitable for the microcontroller, they are conditioned by the input interface block. Made by low-cost elemental electronics, this block electrically isolates the sensitive electronics from the external harsh environment by means of a set of photo-couplers. Moreover, the input interface contains the circuitry that is necessary to read the resistive sensors. As a result, every input and output of the microcontroller is electrically isolated from the external environment, thus preventing to exceed its absolute maximum ratings.

Another important part of the node is the power output block that makes the microcontroller able to drive high current loads. Aiming to the target of minimizing the manufacturing costs, this block has been designed as an external board with modular structure. Each module provides 4 power outputs, hence the total number of them can be customized depending on the requirements of the area in which the node will be mounted. This solution permits to maintain an identical manufacturing process for all the nodes, limiting the on-board unused circuitry at the same time. The power output module is based on a 4-channel high-side driver, which is able to provide up to 6 A and a load fault-detection capability for each channel [7]. It is worth noting that the driving capability of the node allows the elimination from the traditional cabling system of several relays, thus avoiding the problems relevant to their short lifetime.

Also, an RS-232 interface is provided to increase the node communication capability. As an example, this block can be used to link a node to a digital display with the same standard interface.

FIRMWARE

As explained above, the firmware of a node depends on the area in which it is inserted. Nonetheless, the firmware of each node has the same structure. The regular tasks are managed by a cyclic scheduling with an activation period equal to 10 ms. The block diagram reported in Figure 3 shows the tasks sequence; only the *elaborate* block is different in each node, because it contains the specific control algorithms proper of each area.

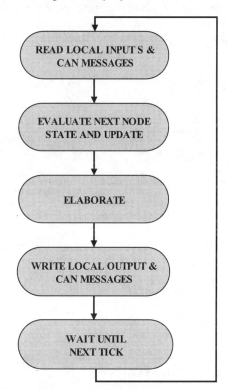

Figure 3. Flow diagram of a node firmware.

It is worth noting that not all the processes can be included in the cyclic scheduling, because some critical control functions cannot wait for the 10 ms periodical tick to be executed. An example of this kind of tasks is the stoplights turn on, where the value of the activation latency is not negligible with respect to the human reaction times. Therefore, asynchronous activation of some tasks has been implemented by means of interrupts.

POWER ISSUES

One of the most challenging issues that has to be faced during the design of an automotive electronic system is power dissipation. Actually, the power consumption of the circuitry is not important when the engine is running, but it becomes critical when the vehicle is parked. However, in that state the system has just to recognize the key insertion, so it is not necessary to maintain the whole functionality of each node as in the active state. In particular, we can adopt a control strategy that allows us to turn off the communication circuits of all the nodes for most of the time. Indeed, if these circuits were always supplied, the power dissipation would lead to a quick discharge of the battery. When the vehicle is off, all the nodes are usually in the *sleep* state in which only the microcontroller, that stays in the idle mode, is supplied. Hence, the power dissipation is negligible in this state. In order to detect the key insertion, the node that controls the key switch is waken by this event and it starts to send the *key-on* message through the CAN bus. On the other side, the other nodes wake up periodically, monitor the bus activity and, if they do not receive the *key-on* message, they come back to the *sleep* state. Otherwise, they go to the *wake-up* state where they answer to the key node sending their own *wake-up* message. Figures 4 and 5 show with more detail the state sequence of the key node (node 1) and of the other nodes respectively.

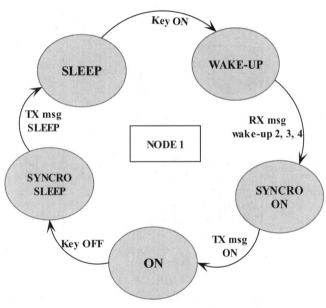

Figure 4. State diagram of the key node.

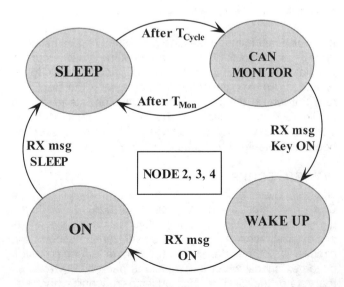

Figure 5. State diagram of the other nodes.

As we can note in Figure 4, node 1 leaves the *wake-up* state after it has received the *wake-up* messages from all the other nodes. Then it synchronizes the passage towards the *on* state, where the overall system is active.

It is worth noting that the monitoring time T_{Mon} is very low with respect to the cycle time T_{Cycle}, so that the power dissipation is drastically reduced. At the same time, the system is able to detect the key insertion with a delay that is not appreciable to the user. Finally, we observe that node 1 determines the passage from the *on* state to the *sleep* state of the whole system once it has detected the key turn-off.

SAFETY ISSUES

Automotive engineers must always adopt particular care in safety issues. Indeed, the design of any vehicle control system should prevent all the events that can cause accidents or at least reduce their probability. Dangerous situations can be due to wrong human behaviors or hardware or software failures [8]. Presently, safety issues are usually faced by adopting a strategy that prevents or neutralizes the wrong system behaviors, by conditioning the electro-mechanical activation of a critical resource. Analyzing the status of one or more sensors generates the relevant enable. As an example, on high-end scooters like the X9, the engine control module supply, necessary for the engine to run, is conditioned to the correct position of key switch, side stand and emergency button. Moreover, the key switch must be on and at least one brake hand lever must be pulled to activate the starter motor.

In a multiplexed approach, the nearest node collects the output of each sensor and the relevant information is shared among all the nodes through the bus. Therefore, the node that generates the activation consent for a critical resource knows the overall vehicle status. Thus, it is obviously possible to adopt the same logical control strategy of the traditional approach, but more sophisticated activation rules, based on the whole vehicle status, can also be followed.

Unfortunately, a correct logical strategy is not sufficient to satisfy the safety requirements. A proper implementation that avoids any activation, due to hardware or software failures, is also necessary. First of all, the circuits devoted to the sensors reading must be designed in a fail-safe way: any hardware fault must be interpreted as a condition in which the consent is denied. Then, more in general, system fault detection must immediately lead to the critical resource disabling. This goal can be reached by continuously renewing the consent only if all the system is running correctly. At this point, a false activation of the critical resource can only be due either to an incorrect execution of the software or to a hardware fault of the power driver. In order to reduce the probability of these undesirable situations, two design strategies have been followed. First, the power supply of the critical resource is obtained by an ac coupled square wave, generated by the node microcontroller. In this way, each stuck-at type fault leads to a condition in which the resource is not supplied.

Then, the software generation of the square wave is conditioned to the correct execution of the overall software, avoiding the use of a peripheral that automatically generates square waves, and inserting the relevant instructions in the cycle that executes the regular tasks.

IN-SYSTEM-PROGRAMMING AND PORTABILITY

The hardware design of the node and the flexibility of its interfaces guarantee a large portability of the system. In fact, the same hardware can be mounted on different types of two-wheel vehicles, once a dedicated firmware has been written for each model. As we have already observed, the 4 nodes are physically identical, but they must be customized by means of a dedicated firmware for each area. In a traditional approach, this means that the microcontroller has to be mounted on the node board through a socket, so that it can be removed and programmed by a dedicated hardware. Therefore, it is necessary to maintain a different storage item for each kind of node or a qualified and expensive assistance must be available. In addition, this kind of approach requires disassembling the system, whenever a software upgrade is needed.

In order to reduce the considerable costs coming from all of these difficulties, we can exploit the in-system programming feature of the chosen microcontroller [6]. As first consequence, this solution permits to eliminate the on board microcontroller socket, saving manufacturing costs. Then, the CAN bootloader of the microcontroller [9] allows to update the firmware of each node by easily connecting a host (PC) directly to the CAN bus, also when the system is already mounted on the vehicle. This drastically simplifies the updating operations that can become necessary for bug corrections or vehicle upgrades, because the disassembling process is no longer necessary. Besides, only one storage item can be enough and skillful assistance is no longer required. Moreover, this technique opens the way to sophisticated maintenance and diagnostics strategies. As an example, it is possible to write dedicated software for diagnostics to be downloaded on the vehicle during the maintenance or repair operations, with the aim of simplifying troubleshooting.

EXPERIMENTAL RESULTS

At the time of this paper, the node has been completely designed and realized using low cost electronics. In a first prototype not-optimized realization, both the node control part and the power module fit in a 7.5 cm x 12.5 cm board, so the physical size of the node is largely suitable for the space available on the vehicle. We also note that the node is already equipped with the protections needed to face the electrical disturbances that can be found in a vehicle and that could threaten the reliability of the system. The functionality of a single node has been extensively tested and in particular it has been proven that the node is able to correctly read the local inputs (switches, resistive sensors, frequency variable signals) and drive the local loads (relays and lights) through the modular power block.

Then, two nodes has been linked through an unshielded twisted pair that realizes the CAN physical layer. The correct communication between the nodes has been verified and experiments of remote activations of loads and reading of sensors have been successfully carried out. These experiments also show that the total power consumption constraint has been satisfied. The system is presently going to be mounted on the vehicle. Only at the end of this step the full functionality will be checked and it will also be possible to quantitatively evaluate the benefits obtained in terms of harness volume reduction and assembling process simplification. In any case, however, the advantages described in the above sections strongly support for the utilization of this system. Actually, a coarse evaluation of the overall system manufacturing cost is less than 50 €, an interesting number and business-compatible, especially if we take into account other cost-savings due to the elimination of all low-medium power relays and the electrical central stand ECU. In conclusion, we can state that the limited costs of the system together with the great advantages introduced, can lead to the industrial volume production of this kind of communication system in PIAGGIO X9 500 motor-scooter.

CONCLUSIONS

A communication system based on the CAN protocol for the management of all the electronic devices in a two-wheel vehicle has been presented in this paper. The system allows the replacement of the global traditional wiring, real bottleneck as far as weight, complexity, reliability and allocation on the vehicle are concerned, with a single CAN network composed by 4 physically identical nodes. The vehicle sensors, actuators and electronics, in general, are associated to well defined local areas connected to the nodes with short wires. The experimental results demonstrate the feasibility of the proposed technique that gives great advantages in terms of harness reduction, reliability improvement, assembling and upgrading simplification, portability of the system and diagnostics opportunities with a cost that can be afforded particularly in high-end motor-scooter.. The experience described in this work shows that the benefits of multiplexed systems can be enjoyed not only in four-wheel vehicles, but also in two-wheel ones in a cost-effective way. Indeed, a first evaluation of the system cost shows the good business sense of our solution, therefore we expect that this kind of architecture will soon be adopted by the two-wheel vehicles industry.

REFERENCES

1. D. K. Ward and H. L. Fields, "A Vision of the Future Automotive Electronics," SAE Paper 2000-01-1358
2. C. A. Lupini, "Multiplex Bus progression," SAE Paper 2001-01-0060
3. B. D. Emaus, "Current Vehicle Network Architectures Trends – 2000," SAE Paper 200-01-0146
4. D. Sparks, T. Noll, D. Agrotis, T. Betzener and K. Gschwend, "Multi-Sensor Modules with Data Bus Communication Capability," SAE Paper 1999-01-1277

5. R. Hadeler and H.J. Mathony, "Design of Intelligent Body Networks," SAE Paper 200-01-0152
6. "T89C51CC01 Enhanced 8-bit MCU with CAN controller and Flash technical data sheet," Atmel Wireless & Microcontrollers, January 2001
7. "VNQ830 quad channel high-side driver technical data sheet," STMicroelectronics, May 2001
8. N. G. Leveson, "System Safety in Computer-Controlled Automotive Systems," SAE Paper 2000-01-1048
9. "In-system Programming – T89C51CC01 CAN Bootloader Description technical data sheet," Atmel Wireless & Microcontrollers, November 2001

CONTACT

Contact author: Prof. Roberto Saletti, Dipartimento di Ingegneria dell'Informazione: Elettronica, Informatica, Telecomunicazioni, University of Pisa, Via Diotisalvi 2, I-56122 Pisa, Italy. Tel. +39-050-568648, Fax. +39-050-568522, E-mail: r.saletti@iet.unipi.it.

2002-01-1606

Controller Area Network (CAN) Simulation of Alternative Vehicle Electronic Architectures

Setha Pan-ngum

Department of Computer Engineering, Faculty of Engineering, Chulalongkorn University

R. J. Ball

Advanced Technology Centre, University of Warwick

ABSTRACT

The increasing amount of vehicle wiring harness, due to the rise of vehicle electronic contents, results in reliability, assembly and EMC concerns. In order to help ease the problems as well as gaining improved vehicle dynamics, the authors study an alternative networking solution, namely a Vehicle Electronic Architecture with Individual Wheel Control. The concept proposes having an ECU on each wheel, responsible for its combined ABS, suspension and four wheel steering control.

As part of a feasibility study of the proposed architecture, CAN message simulation was made to investigate its associated delay time. The simulation was done on software *Simul8*, using message load data based on a modern luxury car [1]. Comparison was made between the proposed and several other existing or future vehicle electronic architectures. The results show that under the current level of message load, the message delay time of the proposed architecture, whilst longer than that of the others, is still within limits, which is defined as the data update rate. And the CAN network still has considerable spare capacity.

INTRODUCTION

Modern cars contain a large amount of electronics to improve performance, ride, comfort and provide entertainment. Its apparent benefits and the fall of electronic component price will further increase its applications in future vehicles. This trend will cause vehicle wiring problems, since modern luxury car wiring harness has already reached several kilometres in length and weighs more than twenty kilograms [2,3]. Large amount of wiring harness induces assembly, EMC, reliability, as well as vehicle space limitation concerns. In order to help ease the problems as well as gaining improved vehicle dynamics, the authors study an alternative networking solution, namely a Distributed Wheel Controller Architecture. The concept proposes introducing an ECU on each wheel, responsible for its

combined ABS and active suspension control, and a Central ECU which controls four wheel steering.

The preliminary study of the proposed architecture (will be called *Architecture 4* in this paper when compared to others) was executed to determine its feasibility for implementation, as well as learning its inherent benefits and weaknesses in comparison to other vehicle electronic architectures. One of the three other networked architectures is the existing Conventional Centralised Architecture (*Architecture 1*), which contains a centralised ECU for each of the ABS, active suspension and 4WS system. Its incoming successor is the Conventional Centralised Architecture with Functional Integration (*Architecture 3*), which enhance vehicle control by integrating the functions of the three systems, and hence more data sharing among the ECUs. Finally, the future Total Centralised Architecture (*Architecture 2*), which combined the three ECUs into a single centralised ECU. Since Controller Area Network (CAN) is the widely used high speed vehicle network at present, it is useful to learn if CAN could satisfy the timing requirements of the proposed architecture. Its associated message delay was also compared between different architectures.

TIMING ASPECTS OF CAN

CAN uses *non-destructive bitwise arbitration* process to select a message when message collision occurs. The simultaneously transmitted messages are allowed to enter the bus, while the process monitors their identifiers bit by bit. Message transmissions can go on, as long as the messages' identifier bits are the same. Once the difference is detected, the message with a passive bit (1) will have to stop, while the message with dominant bit (0) continues. The message which is backed off will be transmitted again as soon as the CAN bus is free.

By giving each data a unique identification number, same priority message collision where all the messages have to be taken off line, can be avoided. In practice, more important or urgent messages are given lower identification numbers, which give them higher priorities to be transmitted. Under the CAN protocol, the message with the highest priority always get the first access to the bus, which effectively guarantees its delivery time. It can be seen that a problem can arise when a large number of high priority messages are transmitted, and lower priority messages will have to keep backing off transmission. This may cause long delays before lower priority messages can be sent.

The vehicle dynamic control systems considered here are safety related. Hence it is vital that their control related messages, for example a wheel speed data, arrive in time. It is, therefore, important that the CAN message delays of the above vehicle electronic architectures are examined. An excessive delay of signal transmission is a delay longer than the period of the transmission, causing the delayed data to be obsolete. The results of message missing or late arrival, forcing ECUs to use previous, out-of-date data could be hazardous. Thus, a simulation of CAN data transfer between ECUs had been carried out.

SIMULATION DATA

In order to realistically simulate this operation, real information on data signalling in the target vehicle is needed, including all the other messages that would be using the CAN bus. The work in [1], which also studies the timing characteristics of CAN, provides this information. The information contains common messages shared between ECUs of a modern luxury car [4]. This data was primarily used, together with the signal information from individual ABS, active suspension and 4WS systems found in literature [5-7], to form a vehicle signal model.

The signal information of the individual systems was employed in place of those of the identical systems from the complete vehicle model. This was done for consistency, because this information was also used in other vehicle modelling works by the authors.

The combined vehicle model arranged in the Conventional Vehicle Electronic Architecture with Functional Integration (*Architecture 3*) comprises 117 periodic signals that are shared between 10 ECUs. The signals include all classes (Class A, B and C), characterised as low, medium and high speed messages, respectively. The list of signals is shown in Table A1 in Appendix.

From the data, class C signals have transmission frequencies of 200 and 100 Hz. Those frequencies of class B signals are 50, 10 and 5 Hz. Class A signals are transmitted at 1 and 0.1 Hz.

The number and the combined bandwidths of the signals of different frequencies are shown in Figures 1 and 2.

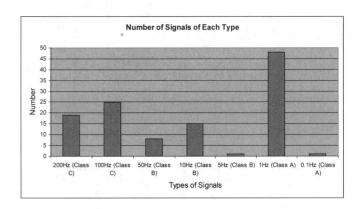

Figure 1 Number of signals of each type

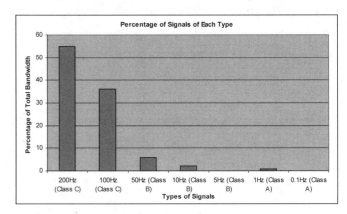

Figure 2 Percentage of bandwidth of different types of signals per total bandwidth

Note that the percentage of class A signal bandwidth to the overall bandwidth is very small as seen from Figure 2. Due to this reason, some of the class A signals such as horn or L/R indicator signals controlled by the driver, whose nature is not periodic, are assumed periodic for simplicity. Since their bandwidth is very small, the effect of the assumption is minimal.

SIMULATION PACKAGE AND METHODS

The author has carried out a discrete event simulation in a similar fashion to the referred work [1], but using different software, due to its availability.

The simulation in this paper was done on *Simul8* software, created by Visual Thinking International Ltd. *Simul8* is primarily used to simulate step by step process, such as factory automation or a hospital receiving patients. It runs under *Microsoft Windows* and is readily available, and can be modified to simulate the CAN bus.

Simul8 provides a source item, which can be used to represent a CAN node. Its function is to arrange its messages and attempt to send the highest priority one first when more than one messages are ready to be transmitted. From these functions, the timing aspect of CAN bus access can be simulated accurately.

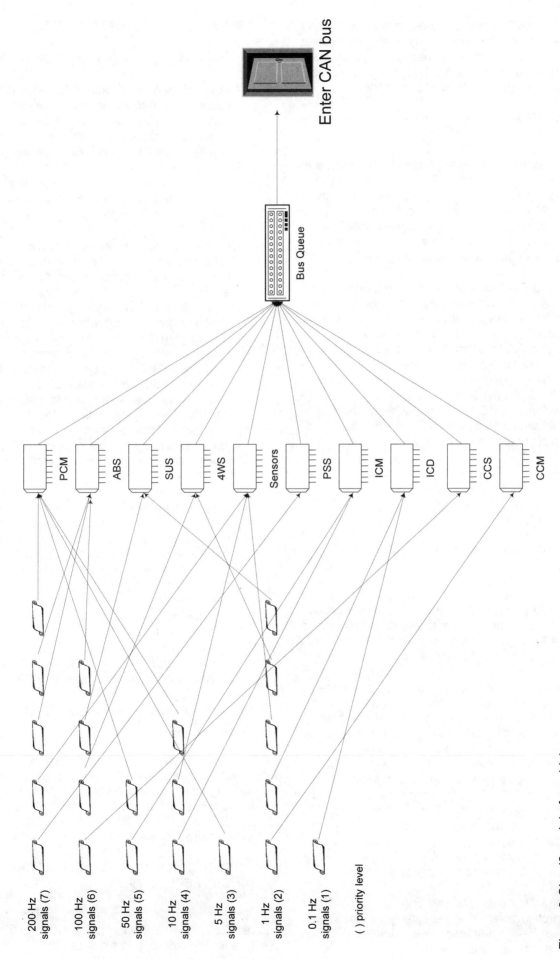

200 Hz
signals (7)

100 Hz
signals (6)

50 Hz
signals (5)

10 Hz
signals (4)

5 Hz
signals (3)

1 Hz
signals (2)

0.1 Hz
signals (1)

( ) priority level

PCM

ABS

SUS

4WS

Sensors

PSS

ICM

ICD

CCS

CCM

Bus Queue

Enter CAN bus

Figure 3 *Simul8* model of a vehicle

Furthermore, *Simul8* contains a queuing item, which can act like a CAN contention resolver by letting the highest priority one among the waiting messages go on CAN bus first. As soon as the CAN bus is free, the next highest priority message is allowed to be sent.

The *Simul8* vehicle simulation model is as shown in Figure 3

Small scrolls on the left represent individual signals, that are from the list in Appendix. The model is arranged such that the high priority signals are above the lower priority ones. 200 Hz signals have priority level 7, 100 Hz signals have priority level 6, 50 Hz signals have priority level 5, and so on. Note that for clarity, most of the 117 signals from the list are left out of the diagrams.

The scrolls and ECU images are for signal generation purpose in the simulation. The initial occurrence of each signal is firstly generated at a random start time by *Microsoft Excel*. Once the first signal is created, the subsequent signals of the same type follow according to their specified period. For example, as shown in Figure 4, the first vehicle speed signal is created at the random time of 2.52 ms from the start of simulation. Vehicle speed has a period of 10 ms, so the subsequent vehicle speed signals would arrive at 12.52, 22.52, 32.52 ms and so on from the starting time.

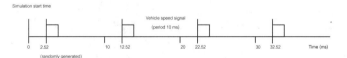

Figure 4 Timing diagram of a 10 ms period signal generation

As indicated by arrows, these signals go to the originating ECU to which they belong, to be transmitted onto the CAN bus. The whole process described above is equivalent to ECUs creating signals to be sent onto the CAN bus, in real vehicle applications.

From ECUs linked by arrows to Bus Queue represents data queuing to be transmitted from ECUs onto the CAN bus. An ECU will let its signals join a queue one by one, equivalent to an ECU attempting to transmit one signal at a time. More than one signals in the Bus Queue at a time symbolises message collision. The Bus Queue arranges incoming signals in order according to their priority, equivalent to CAN message contention. The highest priority signal is put in front of the queue. The Bus Queue then let the highest priority signal onto the CAN bus (displayed as a door image) once the bus is free. Each signal occupies the CAN bus for 64 ms, equal to the time taken to transmit the 8 byte message.

The simulation models representing the 4 different electronic architectures are essentially of the same format, except for the difference in the number of ECUs and signals, and their signal routing.

ASSUMPTIONS AND DATA SETTINGS

In this simulation, the following assumptions are made:

- all the signals are assumed to be short (2 bytes) and periodic which is the nature of real-time data such as sensor and control signals. Each message, therefore, is eight byte long, containing the two bytes of data and six bytes of CAN overhead as in [1].

- all the messages are given priorities according to their transmission frequencies. Ones with higher frequencies have higher priorities.

- messages with the same transmission frequencies are given the same priority. There are 7 different frequencies and hence 7 priority levels. This is not exactly like actual CAN application, in which each message has its unique priority level. The messages were not given unique priority levels here because it was preferred to study the transmission delay of a group of messages with the same frequency, rather than individual messages. Besides, this practice does not affect the timing of CAN bus access. Same priority level messages would gain access to the CAN bus arbitrarily. It could be assumed that the one that misses out has lower priority. In terms of timing, one message gets access and the other gets a delay, just like in real application, hence no effect.

- as a discrete event simulation, all the messages initial occurrence time is randomly generated for each simulation run. However, the following message arrivals are consistent with their periods. Two messages in the same node will not be allowed to be simultaneously generated. This is because in reality a node does not try to send two messages at the same time. It would organise all the ready to transmit messages in order in its buffer before sending them.

- CAN bus transmission speed is 1 Mbits/s which is the highest bit rate for automotive CAN use [8].

VEHICLE ELECTRONIC ARCHITECTURE MODELS

The signal Table A1 in Appendix represents the simulation data for *Architecture 3* as an example. Although similar in simulation model structure, some types and sources of data vary among architectures. In this section, the four different vehicle electronic architecture arrangements for simulation are explained.

CONVENTIONAL CENTRALISED ARCHITECTURE (*ARCHITECTURE 1*)

Architecture 1 is intended to represent the electronic architecture of current vehicles, which utilises CAN but to a limited extent. It is expected that CAN will be under more message load in the future, as more electronic systems and sensors/actuators are introduced or put on to the network. Since ECU connections to sensors and actuators of current vehicles are by hard wires, the sensor node and its related signals were taken off the vehicle signal list for *Architecture 1*. ABS, active suspension and 4WS control signals to their actuators were also taken off for the same reason.

The signals taken off from Table A1 are number 2-11, 22-42, 57-64, 79-82.

TOTAL CENTRALISED ARCHITECTURE (*ARCHITECTURE 2*)

The architecture contains one centralised ECU in place of the ABS, active suspension and 4WS ECUs. The number of signals in the vehicle signal list is the same, but the three dynamic control ECU nodes are replaced with the centralised ECU node.

CONVENTIONAL CENTRALISED ARCHITECTURE WITH FUNCTIONAL INTEGRATION (*ARCHITECTURE 3*)

The simulation model of *Architecture 3* was constructed according to the vehicle signal list in Table A1.

DISTRIBUTED WHEEL CONTROLLER ARCHITECTURE (*ARCHITECTURE 4*)

In *Architecture 4*, the ABS, active suspension and 4WS ECUs are replaced by four Distributed Wheel ECUs and the Central ECU. Sensor signals which are closely located to the distributed wheel ECUs were assigned to be transmitted by the ECUs. The distributed wheel controllers also took over the controls of ABS and suspension wheel actuators. The modified signal list is shown in Table A2 in Appendix.

SIMULATION RUN

A simulation was run for an equivalent of 1 real-time second at time. One second covers the periods of all the signals except for signal no. 117, whose period is 10s and hence of little significant to the CAN bus load. Since all the signals are assumed periodic, any longer simulation run would give a repetitive result to the 1 second run.

For each architecture, the simulation was run 100 times with different sets of random numbers. Each simulation was run for 1 real-time second. A number of simulations were run in order to simulate different possibility of messages arriving on the CAN bus at different times. Each simulation involves 4800-6800 messages getting access to the CAN bus. The time which the two groups of class C signals (of period 5 and 10ms), which are for real-time control, wait in the CAN Bus Queue plus the transmission time was collected. This is equivalent to the signal time delay associated with CAN in real applications.

RESULTS AND ANALYSIS

The frequency distribution of the CAN delay of the two groups of class C signals are shown in Figures 5 and 6.

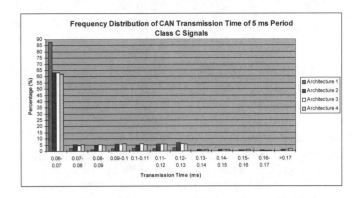

Figure 5 Frequency Distribution of CAN Transmission Time of 5 ms Period Class C Signals

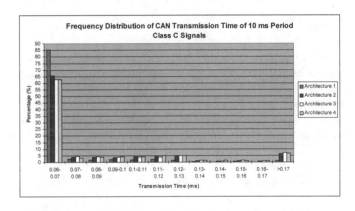

Figure 6 Frequency Distribution of CAN Transmission Time of 10 ms Period Class C Signals

From Figures 5-6, it can be seen that the majority of the two groups of class C signals are transmitted within 0.7 ms. The minimum possible CAN transmission time (no collisions) for each signal is 0.64 ms. This indicates that those signals are transmitted virtually without delay. *Architecture 1* has the highest percentage of signals sent without delay, because it has the lowest utilised message bandwidth.

The percentage of messages transmitted with increasing delay then falls drastically for all the architectures. However, a larger number of 10 ms period signals have experienced long delays than the 5 ms periods signals. This can be seen from the higher percentage of signals with transmission time longer than 17 ms (the last bars on the chart) in Figure 6 than those in Figure 5. Also the 10 ms period signals have experienced longer worst case delay than the 5 ms signals. This could be expected, since the 10 ms period signals have lower priority than the 5 ms signals. The longest delays of all the architectures are shown in Table 1.

Architecture	Longest delay (ms)	
	5 ms Period Class C Signals (% of delay per period)	10 ms Period Class C Signals (% of delay per period)
Architecture 1	0.18 (3.69%)	0.26 (2.55%)
Architecture 2	0.27 (5.34%)	0.54 (5.37%)
Architecture 3	0.25 (5.07%)	0.67 (6.67%)
Architecture 4	0.73 (14.66%)	1.00 (9.97%)

Table 1 Longest Delay of Class C Signals of all the Architectures

From Table 1, the worst case delay of the class C signal transmission for all the architectures are of low percentage to their periods. A signal delay of longer than its period would cause a problem to control systems involved. This indicates that under the current level of message load, CAN is applicable in terms of speed to all the electronic architectures.

CONCLUSION

The simulation results suggest that the proposed Distributed Wheel Controller Architecture has a larger longest CAN message delay times for highest priority messages than other architectures. This is primarily due to its larger number of ECUs, since the amount of message load is similar to that of the Architectures 2 and 3, which is higher than that of the Architecture 1. However, its maximum delay time was still found to be well within the system requirement, leaving CAN with considerable capacity for message addition, and means that the architecture's other benefits (safety, additional functionality) could be obtained at no cost.

REFERENCES

1. B. Upender, "Analysing the Real-Time Characteristics of Class C Communications in CAN Through Discrete Event Simulation", SAE 940133, 1994
2. J. Fenton, "Focus on Networking of On-Board Vehicle Electronic Systems", Automotive Engineer, June/July 1996
3. R. Schreffler, "Wiring Harness of the Future", Automotive Industry, February 1997
4. Ford Motor, "Electrical and Vacuum Trouble Shooting Manual FPS-12119-93: 1993 Town Car", Ford Motor, 1993
5. Matsutomi S., et al., "Development of ABS and Traction Control Computer", SAE 901707, 1990
6. Aoyama Y., et al., "Development of the Full Active Suspension by Nissan", SAE 901747, 1990
7. Sato H., et al., "Development of Four Wheel Steering System Using Yaw Rate Feedback Control", SAE 911922, 1991
8. Bosch, "CAN Specification Version 2.0", Robert Bosch GmbH, 1991

CONTACT

Setha Pan-ngum, Setha.P@chula.ac.th

APPENDIX

Signal No.	Name	Period (ms)	Class	Source ECUs
1	spark output timing signal	5	C	PCM
2	front left wheel brake demand	5	C	ABS
3	front right wheel brake demand	5	C	ABS
4	rear left wheel brake demand	5	C	ABS
5	rear right wheel brake demand	5	C	ABS
6	ABS solenoid control	5	C	ABS
7	ABS motor control	5	C	ABS
8	front left wheel speed sensor	5	C	SEN
9	front right wheel speed sensor	5	C	SEN
10	rear left wheel speed sensor	5	C	SEN
11	rear right wheel speed sensor	5	C	SEN
12	crash sensor 1	5	C	PSS
13	crash sensor 2	5	C	PSS
14	crash sensor 3	5	C	PSS
15	brake position sensor	5	C	PSS
16	clutch position sensor	5	C	ICM
17	crankshaft position sensor	5	C	ICM
18	profile ignition pickup	5	C	ICM
19	speed control signal	5	C	CCS
20	transmission speed sensor	10	C	PCM
21	vehicle speed	10	C	ABS
22	front left wheel height demand	10	C	AS
23	front right wheel height demand	10	C	AS
24	rear left wheel height demand	10	C	AS
25	rear right wheel height demand	10	C	AS
26	fail-safe mode demand	10	C	AS
27	flow control valve control	10	C	AS
28	fan motor control	10	C	AS
29	variable timing pump control	10	C	AS
30	rear wheel steering demand	10	C	4WS
31	front left wheel height sensor	10	C	SEN
32	front right wheel height sensor	10	C	SEN
33	rear left wheel height sensor	10	C	SEN
34	rear right wheel height sensor	10	C	SEN
35	vertical acceleration sensor 1	10	C	SEN
36	vertical acceleration sensor 2	10	C	SEN
37	vertical acceleration sensor 3	10	C	SEN
38	lateral acceleration sensor 1	10	C	SEN
39	lateral acceleration sensor 2	10	C	SEN
40	longitudinal acceleration sensor	10	C	SEN
41	steering angle sensor	10	C	SEN
42	yaw rate sensor	10	C	SEN
43	rear wheel spin sensor 1	10	C	PSS
44	rear wheel spin sensor 2	10	C	PSS
45	cylinder Id sensor	20	B	PCM
46	manual level position	20	B	PCM
47	delta pressure feedback electronic	20	B	PCM
48	heated exhaust gas oxygen sensor	20	B	PCM
49	mass air flow sensor	20	B	PCM
50	throttle position sensor	20	B	PCM
51	engine RPM	20	B	PCM
52	ignition diagnostic monitor	20	B	ICM
53	transmission oil temperature	100	B	PCM

54	A/C compressor clutch	100	B	PCM
55	engine coolant temperature	100	B	PCM
56	transmission lubricant pressure	100	B	PCM
57	ABS solenoid status	100	B	SEN
58	ABS sensor status	100	B	SEN
59	4WS motor status	100	B	SEN
60	brake status	100	B	SEN
61	reverse switch	100	B	SEN
62	height switch	100	B	SEN
63	door switch	100	B	SEN
64	active suspension pressure status	100	B	SEN
65	battery current	100	B	ICM
66	battery voltage	100	B	ICM
67	shift sensor	100	B	ICM
68	intake air temperature	200	B	PCM
69	octane adjust plug	1s	A	PCM
70	transmission control switch	1s	A	PCM
71	engine idle speed	1s	A	PCM
72	engine status	1s	A	PCM
73	fuel flow	1s	A	PCM
74	transmission control indicator	1s	A	PCM
75	EGR vacuum regulator	1s	A	PCM
76	check engine indicator	1s	A	PCM
77	ABS warning lamp	1s	A	ABS
78	active suspension warning lamp	1s	A	AS
79	steering oil level	1s	A	SEN
80	suspension hydraulic oil level	1s	A	SEN
81	thermistor	1s	A	SEN
82	brake fluid	1s	A	SEN
83	power locks	1s	A	PSS
84	power seats	1s	A	PSS
85	power windows	1s	A	PSS
86	shift inhibit signal	1s	A	PSS
87	shift in progress	1s	A	PSS
88	seatbelt sensor	1s	A	PSS
89	door sensor 1	1s	A	PSS
90	door sensor 2	1s	A	PSS
91	door sensor 3	1s	A	PSS
92	door sensor 4	1s	A	PSS
93	door sensor 5	1s	A	PSS
94	anti-theft sensor	1s	A	PSS
95	airbag indicator lamp	1s	A	PSS
96	seatbelt lamp	1s	A	PSS
97	door lamps status	1s	A	PSS
98	airbag status	1s	A	PSS
99	fuel level sensor	1s	A	ICM
100	alternator warning indicator	1s	A	ICM
101	auto headlamp sensor	1s	A	ICD
102	ignition switch position	1s	A	ICD
103	horn sensor	1s	A	ICD
104	hazard sensor	1s	A	ICD
105	L/R signal	1s	A	ICD
106	control to tone maker	1s	A	ICD
107	oil pressure	1s	A	ICD
108	SET/ACCEL/RESUME	1s	A	CCS
109	cruise control indicator	1s	A	CCS
110	outside temperature	1s	A	CCM
111	desired temperature	1s	A	CCM

112	cabin temperature	1s	A	CCM
113	rear window defrost	1s	A	CCM
114	blower speed control	1s	A	CCM
115	damper control	1s	A	CCM
116	hear/cool control	1s	A	CCM
117	washer fluid sensor	10s	A	ICD

Table A1 Complete List of Signals for Conventional Centralised with Functional Integration CAN Simulation (*Architecture 3*)

Source ECU codes:

PCM — Powertrain Control Module
ABS — Antilock Braking System
AS — Active Suspension
4WS — 4 Wheel Steering
SEN — Sensors
PSS — Passenger Safety Systems
ICM — Ignition Control Module
ICD — Instrument Cluster Display
CCS — Cruise Control System
CCM — Climate Control Module

Signal No.	Name	Period (ms)	Class	Source ECUs
2	front left wheel brake demand	5	C	FLC
3	front right wheel brake demand	5	C	FRC
4	rear left wheel brake demand	5	C	RLC
5	rear right wheel brake demand	5	C	RRC
6	ABS solenoid control	5	C	DCC
7	ABS motor control	5	C	DCC
8	front left wheel speed	5	C	FLC
9	front right wheel speed	5	C	FRC
10	rear left wheel speed	5	C	RLC
11	rear right wheel speed	5	C	RRC
21	vehicle speed	10	C	DDC
22	front left wheel height demand	10	C	FLC
23	front right wheel height demand	10	C	FRC
24	rear left wheel height demand	10	C	RLC
25	rear right wheel height demand	10	C	RRC
26	fail-safe mode demand	10	C	DCC
27	flow control valve control	10	C	DCC
28	fan motor control	10	C	DCC
29	variable timing pump control	10	C	DCC
30	rear wheel steering demand	10	C	DCC
31	front left wheel height	10	C	FLC
32	front right wheel height	10	C	FRC
33	rear left wheel height	10	C	RLC
34	rear right wheel height	10	C	RRC
77	ABS warning lamp	1s	A	DCC
78	active suspension warning lamp	1s	A	DCC

Table A2 Changes in Signals from Table B1 for Distributed Wheel Architecture (*Architecture 4*) CAN Simulation

ECU codes:
FLC — Front Left Wheel Controller
FRC — Front Right Wheel Controller
RLC — Rear Left Wheel Controller
RRC — Rear Right Wheel Controller
DCC — Distributed Central Controller

2002-01-1470

CANopen in Industrial Vehicles

Thilo Schumann
CAN in Automation

ABSTRACT

Industrial vehicles cover a broad range of different applications including fork-lifts, agriculture and forestry machines, as well as road construction machines and truck-based cranes, garbage trucks and even truck-based aircraft washing robots. All of these vehicle applications are characterized by low-volume to mid-volume production. There is also a high variety in different versions. That is the reason why in these applications embedded networks should be standardized. In addition, the vehicle manufacturers like to use off-the-shelf products, even if they are manufactured for other market fields, in order to get better prices.

INTRODUCTION

Most of the industrial trucks are using CAN (Controller Area Network) for embedded network purposes. Because CAN provides only data link layer services, most of the companies have developed their proprietary application layer. The device and module suppliers have to support all these different solutions. This causes additional costs for the software adaptation. It would be much more cost-effective, if they will accept a standardized higher-layer protocol.

The Industrial Truck Association (ITA) is the first non-profit trade organization that has decided to recommend CANopen profiles for use in forklifts. CANopen is a CAN-based application layer and communication profile. The CANopen specification is submitted for European standardization (prEN50325 Part 4). The CANopen-based device, interface, gateway, and application profiles are published by CAN in Automation (CiA) non-profit users and manufacturers group.

BASIC CANOPEN COMMUNICATION SERVICE

The CANopen application layer and communication profile (CiA DS-301) supports both direct access to device parameters and time-critical process data communication. The CANopen network management services simplify project design, system integration, and diagnostics.

OBJECT DICTIONARY - In each decentralized control application there are different communication objects required. CANopen defines all these communication objects and describes them in the Object Dictionary. This dictionary is unique for each device and accessible by a 16-bit index and, in the case of arrays and structures, additionally by an 8-bit sub-index. It also describes all application objects of the device.

Index (hex)	Object
0000	not used
0001-025F	Data Types
0260-0FFF	Reserved for further use
1000-1FFF	Communication Profile Area
2000-5FFF	Manufacturer Specific Profile Area
6000-A8FF	Standardized Device Profile Area
A900-FFFF	Reserved for further use

Table 1: Object dictionary structure

PROCESS DATA OBJECT (PDO) - Process Data Objects are mapped to a single CAN frame using all 8 bytes of the data field to transmit application objects. Each PDO has a unique identifier and shall be transmitted by only one node, but it can be received by more than one (producer/consumer communication). PDOs may be transmitted in modes that are driven by an internal event, by an internal timer, by remote requests, and by a sync message received from a specific node. The default mapping of application objects as well as the supported transmission mode is described for each PDO in the Object Dictionary. PDO identifiers should be high priority to guarantee a good real-time performance. PDOs are transmitted with no confirmation. Which application objects are transmitted within a PDO is defined in the PDO Mapping Object. It describes the sequence and length of the mapped application objects. A

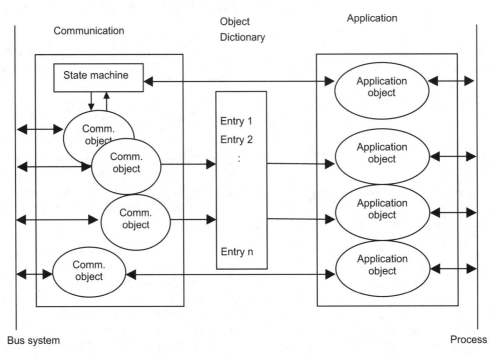

Figure 1: Interaction between communication and application
handled by the object dictionary

device that supports variable mapping of PDOs must support this during the Pre-Operational state. If dynamic mapping during Operational state is supported, the SDO Client is responsible for data consistency.

SERVICE DATA OBJECT (SDO) - Service Data Object reads or writes to entries in the object dictionary. The SDO transport protocol allows transmitting objects of any size. The first byte of the first segment contains the necessary flow control information including a toggle bit to overcome the well-known problem of double received CAN frames. The next three bytes of the first segment contain index and sub-index of the object dictionary entry to be read or written. The last four bytes of the first segment is available for user data. The second and the following segments (using the very same CAN identifier) contain the control byte and up to seven bytes of user data. The receiver confirms each segment or a block of segments, so that there is a peer-to-peer communication (client/server).

NETWORK MANAGEMENT OBJECTS - The network management (NMT) objects include Boot-up object, Node/Life-guarding object, Heartbeat object, and NMT object. Boot-up, Node/Life-guarding, and Heartbeat objects are implemented as CAN frames with 1-byte data field. The NMT object is mapped to a single CAN

frame with a data length of 2 bytes. It has the identifier 0. The first byte contains the command specifier and the second contains the node ID of the device that must perform the command (in the case of node ID 0 all nodes have to perform the command). The NMT object transmitted by the NMT master forces the nodes to transit to another NMT state. The CANopen state machine specifies the Initialization state, the Pre-Operational state, the Operational state and the Stopped state. After power on each CANopen node is in the Initialization state and transitions automatically to the pre-operational state. In this state, transmission of SDOs is allowed. If the NMT master has set one or more nodes to the Operational state, they are allowed to transmit and to receive PDOs. In the Stopped state no communication is allowed except NMT objects.

The Initialization state is divided in three sub-states in order to enable a complete or partial reset of a node. In the Reset_Application sub-state the parameters of the manufacturer-specific profile area and the standardized device profile area are set to their default values. In the Reset_Communication sub-state the parameters of the communication profile area are set to their power-on values. The third sub-state is the Initialization state, which a node enters automatically after power on or after reset communication or reset application. Power-on values are the last stored parameters.

SPECIFIC FUNCTIONALITY OBJECTS - CANopen defines also three specific objects for synchronization, emergency indication, and time stamp transmission.

defines several Emergency error codes to be transmitted in the Emergency object, which is a single CAN frame with 8 data bytes.

Time Stamp object - By means of the Time Stamp object a common time frame reference is provided to application devices. It contains a

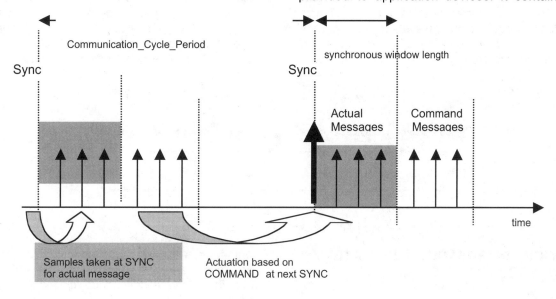

Figure 2: Synchronization cycle

Synchronization object - The Sync Object is broadcasted periodically by the Sync Producer. This object provides the basic network clock. The time period between Sync messages is defined by the Communication Cycle Period object, which may be written by a configuration tool to the application devices during the boot-up process. There can be a time jitter in transmission by the Sync Producer due to some other objects with higher prior identifiers or by one frame being transmitted just before the Sync object. The Sync object is mapped to a single CAN frame with the identifier 128. By default, the Sync object does not carry any data, but it can have up to 8 bytes of user-specific data.

Emergency object - Emergency objects are triggered by the occurrence of a device internal fatal error situation and are transmitted from an emergency client on the concerned application device. This makes them suitable for interrupt type error alerts. Emergency objects may be transmitted only once per 'error event'. As long as no new errors occur, on a device no further Emergency object must be transmitted. Zero or more Emergency object consumers may receive these. The reaction of the emergency consumers is application-specific. CANopen

value of the type Time-of-Day. This object transmission follows the producer/-consumer push model. The associated CAN frame has the pre-defined identifier 256 and a data field of 6 bytes length.

IDENTIFIER ALLOCATION

As CAN is a communication object (COB) oriented network, where each COB has one or more associated identifiers which specifies its priority implicitly, the allocation of identifiers to the COBs is an essential issue in the system design. In order to reduce configuration effort for simple CANopen networks a mandatory default identifier allocation scheme is defined. These identifiers are available in the Pre-Operational state and may be modified by means of dynamic distribution. A CANopen device has to provide the corresponding identifiers only for the supported communication objects. The default profile COB-ID allocation scheme consists of a functional part, which determines the object priority and a node ID part, which allows distinguishing between devices of the same functionality. The COB-ID allocation scheme corresponds to a pre-defined master/slave connection set and allows peer-to-peer

201

communication between a single-master device and up to 127 slave devices. It also supports the broad casting of non-confirmed NMT, Sync and Time-Stamp objects. The pre-defined master/slave connection set supports one Emergency object, one SDO as well as up to 4 Receive-PDOs and up to 4 Transmit-PDOs, and the Error Control object. In order to optimize the identifier allocation, the system designer may change the allocation of identifiers to communication objects.

PDOs. The recommendation also prefers Heartbeat as error control service instead of Node Guarding. Node Guarding is based on remote frames.

The Recommended Practice manual includes an example for a CANopen implementation as shown in the figure. The Vehicle Manager is connected via CANopen to a Steer Manager, two Motor Control Units and the Display/keypad Control Unit. The communication with the Steer

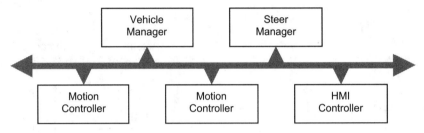

Figure 3: Vehicle design

DEVICE, INTERFACE, AND GATEWAY PROFILES

CiA has already published CANopen device profiles for generic I/O modules (DSP-401), drives and motion control (DSP-402), human-machine interfaces (DSP-403), and encoders (DSP-406). There is also the interface profile for IEC 61131-compatible controllers published. Under development are device profiles for hydraulics, sensors and closed-loop controllers as well as an interface profile for truck gateway. This gateway specification is suitable for devices connecting J1939/ISO 11992 networks to CANopen networks.

The University of Magdeburg has pre-developed CANopen device profiles for diesel engine controllers, for electronic gear controllers, joysticks and other vehicle-specific devices. The German VDMA ("Verband Deutscher Maschinenbauer"; engl: non-profit trade association for machine builders) supports this research project.

RECOMMENDED PRACTICES FOR FORKLIFT

The Recommended Practices for forklifts specified by ITA is based on the CiA DS-301 Version 4.0 application layer and the CiA DSP-402 Version 1.1 device profile for drives and motion control. The ITA does not recommend to use remote frames to trigger the transmission of

Manager is manufacturer-specific. The motion controllers are compliant to CiA DSP-402 using only the profile velocity mode (closed-loop speed control) or velocity mode (open-loop speed control). If a manufacturer-specific mode is used, it is recommended that the manufacturer describes the objects associated with that profile. For the display/keypad controller ITA recommends the use of the CiA DSP-403 device profile.

REFERENCES

[CiA DS-301] CANopen application layer and communication. Erlangen 2000.
[CiA DSP-401] CANopen device profile for generic I/O modules. Erlangen 2001.
[CiA DSP-402] device profile for drives and motion control. Erlangen 2001.
[iCC proceedings] Erlangen 1994-2000.
[CAN Newsletter] Volume 1 to 8. Pz marketing, Nuremberg.
[ITA Recommend] Practice for forklifts.

AUTHOR

Thilo Schumann
CAN in Automation (CiA)
Am Weichselgarten 26
91058 Erlangen
Germany
p: +49-9131-69086-0
f: +49-9131-69086-79
e-mail: headquarters@can-cia.org
web: www.can-cia.org

In-Vehicle Communication Network Modeling and Simulation Environment – A Tool to Study the Vehicle Network Architecture and Vehicle EE System Architecture

Yibing Dong and Salim Momin
Virtual Garage of Detroit, Motorola, Inc.

ABSTRACT

A methodology of modeling and simulating a whole vehicle CAN network will be presented in this article. Concentrating on the distributed characteristics of the in-vehicle communication network, this methodology not only provides all detailed information on traditional communication parameters such as communication bus bandwidth, message latency time, bus arbitration behavior and so on, but also provides detailed information for each node in the communication network. The detailed information on each node includes the behaviors of hardware (in terms of delay time), CPU performance (in terms of band width and throughput), and software tasks (in terms of execution time and deployed operation system). All these dynamic linked behaviors of each node and behaviors of network give engineers a deeper understanding of their network communication strategy and system impact of chosen hardware and software implementations.

INTRODUCTION

With more and more electrical components and computation power integrated into vehicles, the needs to have a communication network arises in order to reduce the complexity of vehicle wire harness, increase reliability of overall system, and enhance the flexibility of entire system layout. Controller Area Network (CAN) is the one type of in vehicle network standard developed by European OEM's in past 10 years and has been adapted by US OEM in past two years. But, due to the lack of understanding of architecture impact of vehicle networking, and mentality of "adds on" concepts when introducing newer functions to a vehicle, the integration of a cost effective, reliable CAN network becomes a very difficult task.

Most OEM's have developed their CAN message strategy based on their past experiences on J1850 and helps from their European counter parts. How the CAN message strategy performs, how efficient of CAN bus running, and how those CAN communication impact ECU performance still remain as a mystery waiting to be validated. The engineers have to wait for the prototype vehicle to validate the message strategy, which could be too late to discover any problems.

To discover system level communication network problems at early development stages, to identify the ECU impacts at earlier development stages, and to gain more confidence of the CAN network behaviors are becoming very urgent goals. Vehicle level CAN network modeling and simulation is one way to start answering these problems.

SIMULATION ENVIRONMENT SET UP

We selected "eArchitect" from Innoveda to set up our vehicle network simulation environment. For demonstration purposes, we constructed a vehicle level CAN communication strategy, which is captured in the Vector's CAN database format. This communication strategy includes three CAN sub-networks (CAN b, CAN c and CAN diagnostic) and requires gateway function in order to share signals between different CAN sub-networks.

The models used in the simulation were developed using the "EML" language, a language used in eArchitect, and captured in the library format. The libraries used for the CAN network demonstration simulation include:

- Hardware block model of MSCAN based CAN specification 2.0-B.
- CAN bus model to instrument the bit level arbitration at message level.

- ECU performance model, priority-based preemptive scheduler
- Software task models for HC12 vector low-level drivers. This software task model was developed based on software profiling works done at Motorola, Virtual Garage of Detroit.
- Generic software tasks for CAN TX and RX tasks.

The top-level architecture of the simulation environment model consists of two parts, hardware architecture and software architecture. All relationship between each node has been defined in the hardware architecture and each node will operate independently during simulation.

Both hardware architecture and software architecture can be set up at node level (ECU level). The software architecture for each node is build up through software tasks. Those tasks are mapped to the CPU mode based on their execution cycle time and priority of the execution. At any time, the penalty of the hardware is represented as time delay and the penalty of the software tasks is represented as execution time. The CPU model under each node is a preemptive priority-based scheduler, with which the higher priority task can interrupt the lower priority tasks.

Modeling vehicle communication protocols is the key for vehicle network simulation. The capability of modeling CAN, LIN, MOST, J1850, TTP, etc. is the heart of vehicle network simulation. In this simulation, we implemented CAN communication protocols based on CAN specification 2.0-B. The features included in this CAN protocol model are as follows:

- Bus arbitration
- Bus load monitor
- Trace of message behavior
- Random error frame injection (passive and active)
- Random overload frame injection

In conjunction with the hardware mode of CAN physical layer (MSCAN in this case), the behavior of the communication protocol can be dynamically implemented even with error and data eruptions in vehicle network.

CPU models were created based on a priority-based pre-emptive scheduler. The CPU architecture was mapped to each node in terms of execution time of each instructions/tasks. The fine-tuned execution time due to different kind of methods of access memory and penalty of interrupts can also be mapped into model. With this mapped CPU model, one can create software tasking in this CPU, assign priority of each tasking to manipulate Operating System (OS) for the node. One can specify variety of details for CPU performances, such as cycle counts for a particular instruction, a subroutine, or a task.

Whole vehicle CAN network model were put together by hardware mapping and software task mapping. Hardware mapping has been done mainly through CPU architecture mapping. And software mapping were done through priority mapping on associated CPU. Each node will have at least one CPU model, at least one pair of receiving block and transmitting block for CAN physical layer (two pairs for dual CAN application and three pairs for gateway application). All models, including bus model, were automatically generated to an EML module.

The outputs of simulation provide rich sets of information regarding vehicle network behaviors (bus traffic, message latency, average bus load, etc.), node behaviors (CPU load, hardware delays) and communication task behaviors (tasking time, software filtering, interrupt, etc.). All those information would be very valuable for vehicle communication network architects to debug and optimize the network architecture strategy before vehicle ever been built.

In additional to the analysis tool provided by *eArchitect*, the post process tool developed by Motorola Virtual Garage of Detroit, also provides a more convenience method to analyze the simulation results such as:

1. Information of CAN bus:
   - Overall bus utilization
   - Average bus utilization over a sliding time window
2. Information of node on the bus:
   - Overall CPU utilization.
   - Average CPU utilization over sliding time window
   - Task penalty
   - Interrupt behavior and how a regular scheduled task has been effected by interrupts.
   - CPU load due to unwanted messages
   - CPU load for all communication tasks
3. Information of a message
   - Message TX latency: The time between message due to transmit to message get into transmit buffer of CAN physical layer.
   - Message Bus arbitration latency: The time between the message getting into the transmit buffer and the time the message wins arbitration.
   - Message propagation latency time.
   - Message RX latency: The time between the message getting into receiving buffers to the time the RX interrupt has been generated.
   - Message RX process latency: The time between the message received by software to the time of message has been unpacked.

SIMULATION RESULT ANALYSIS

Network simulation can be run at any time period depending on the information requested. The flowing graphics in Figures 1-4 are derived from a 500ms simulation run. Figure 1 is the hot spot analyzer of the network. The color-coded CPU loads illustrate the dynamic performance of each node in the network at any given time and any given average time period. It also can be used to look into how the token (some of containing messages) traveled in it in the debug mode.

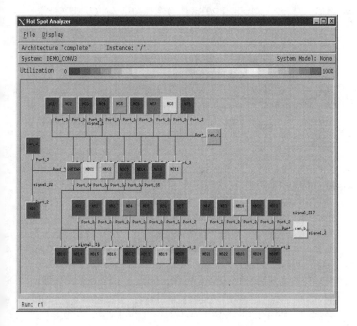

Figure 1 Hot Spot Analyzer At Top Level

The figure 2 is the message window, which can be used to monitor activities of each message. These activities include all critical time information for a given information and the debug information coded in the simulation model to detect any error conditions that may occur, such as buffer over flow, message dropping, message missing and message overwriting. In the example simulation, there were several instances that the designed message was not sent out due to several reasons, and then overwritten by new messages.

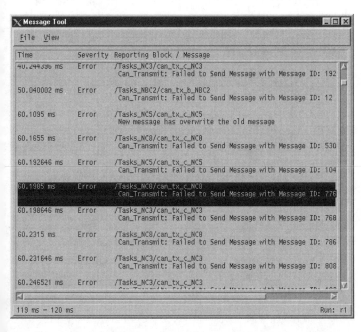

Figure 2 Detailed Message Window

The figure 3 shows a real example of the bus arbitration. In this instance, one message was initiated from the TX task, put into the transmit buffer and did not win the bus arbitration until 19 milliseconds later. This will help system engineers to identify if the given message strategy can meet system level requirements.

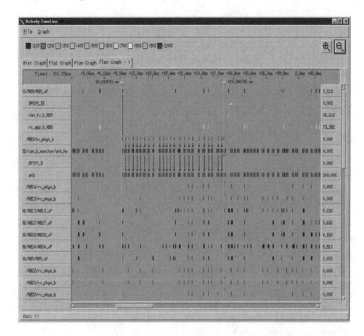

Figure 3 An Example of Message Arbitration on the Bus

The figure 4 shows the average CPU load for each CPU node, execution time and behavior of each task on a given CPU, and hardware penalty of CAN physical layers. It also shows the interrupt impact for a given application task.

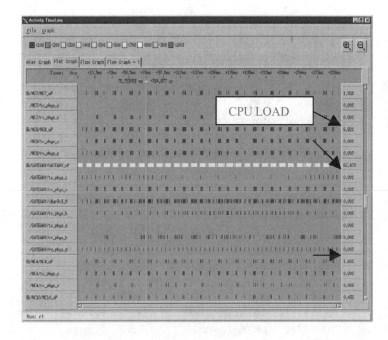

Figure 4 Average CPU Load and Utilization

CAN protocol is not deterministic communication protocol. Message latency cannot be statically predicted and will be dynamic during runtime. What is the upper or lower boundary of the latency time of a given message and a given message strategy is still the most difficult question to be answered by the network design engineers. The vehicle network simulation and analysis tool developed by Virtual Garage of Detroit, Motorola, is able to answer those type of questions based on above simulation environment. Using this tool, engineers can perform analysis on every message, every node in the network in terms of the statistic behaviors of CPU load, bus utilization, and message latency time. It can also generate report on these analysis results (Please contact Motorola Virtual Garage for detailed information).

CONCLUSIONS

The in-vehicle communication network simulation is the first step to the complete simulation of a distributed computation system of a vehicle. Using this simulation environment, engineers can have better understanding on what are the interactions between each node on a vehicle.

The utilization of CPU and throughputs of the communication bus are very important for a message strategy design. With this simulation environment and the analysis tool, simulation results can provide detailed insight before the actual vehicle network is built. The system engineers will be able to use this information to intelligently define a CAN message strategy and be able control system integration and system latency time better.

The CPU performance of each node also plays very important roles to maximize the CPU utilization and meanwhile guaranteeing the required performance. This simulation environment provides the first hand information on the CPU load due to the tasks perform communication. With proper OS model (like OSEK) and task scheduling running on the OS, this set up can also be used to analyze the performance of an OS and the efficiency of the scheduler. Using "budgetary" information, engineering can play what-if game for the software architecture of the ECU.

At the vehicle level, this simulation environment has the capabilities to simulate the entire network(s) at varying degrees of detail, which enables the OEM to study network behaviors very early in the design stage long before the vehicle is built such as:

1. Network issues.
   a. Network efficiencies
   b. Message latencies,

   c. Efficiencies of frame packing,
   d. Message prioritization
   e. Bus bandwidth utilization
2. Understand integration and interaction of the different Networks at a vehicle level.
3. Gateway placement issues
   a. CPU bandwidth requirements
   b. Task time requirements
   c. Scheduler timing requirements
4. Validation of network management strategy.
5. Potential use in design/validation of distributed Systems.

At each node, this setup allows engineers to explore the trade offs between hardware and software tasks, evaluate the performance of CPU, OS, and scheduler under the "real traffic" of the network.

This simulation environment has been used to conduct several studies of CAN network, LIN network and mixed CAN-LIN networks to validate message strategies. The simulation environment also has been validated through hardware setup with very good correlations. Using this validated simulation environment we were able to understand the message strategy, the system latency time associated with different stages of a message transmitting and the CPU performance at each node (include scheduler).

FUTURE WORKS

The future works for this simulation environment will be introducing the OS into CPU model and more communication protocol models. Introducing "OS" will help better modeling of CPU performance, which will lay down the foundations for the simulations of a distributed system.

REFERENCES

1. *eArchitect Reference Manual*, July 2000, Innoveda, Inc.
2. *eArchitect User's Manual*, July 2000, Innoveda, Inc.
3. *68HC12 Reference Manual*, Rev.1 Motorola, Inc.
4. *MC68HC912DG128 Advanced Information*, December, 1999, Motorola, Inc.
5. *CAN 2.0: Robert Bosch CAN Specification*, Revision 2.0

CONTACT

For more detailed information, please contact Dr. Yibing Dong at 248-3477952 or send him an email at Yibing.dong@motorola.com

2002-01-0444

A CAN Based Real-Time Embedded System for DC Motor Control

**Mohsin M. Jamali, Mark M. Brown, C. C. Sheh,
C. Suriyakamol and M. Y. Niamat**
Department of Electrical Engineering and Computer Science,
The University of Toledo

ABSTRACT

This paper describes a design of a system that controls the speed and direction of a small DC motor though a network system. This work is important as large numbers of motors are employed inside a modern vehicle. Moreover most of the electronic devices inside a vehicle are networked through a Controller Area Network (CAN). The system uses commands from a CAN node and sends this message through a CAN bus to another CAN node which is controlling the DC motor. The system uses two Phytecs boards (Infineon C505C Microcontrollers based) and communicates through a CAN bus. The system bus is monitored with a Dearborn Protocol Adapter II (DPA-II). The system is broken into four major parts: two CAN nodes, the driver circuitry, and a feedback sensor. This paper provides detailed design, flow chart of the programs and testing of the system.

INTRODUCTION

With the explosion of electronic technology more and more electronics are being introduced in today's cars and trucks [1-2]. There are a number of Electronic Control Units (ECUs) in a vehicle. These ECUs may be used to control engine, transmission, Anti-lock Brake Systems (ABS), cruise, steering, air bag system, vehicle,

traction, Heating Ventilation Air Conditioning (HVAC) systems, entertainment, doors, and locks etc. There is large number of motors inside the vehicle and this number is growing as more options are being incorporated. Some examples of use of motors inside vehicle are window rolling functions, wiper control, seat & mirror movements and others. There is a need of digital control of these motors.

Auto manufacturers are using multiplex buses within their vehicles for integration of all their vehicle electronics. In the past manufacturers used proprietary buses for the transfer of data within their vehicle or to satisfy the regulatory agencies for access to their emission data [3-5]. The Controller Area Network proposed by Robert Bosch [6] has become or approaching to become an industry standard. The CAN protocol provides specifications of data link layer only and does not specify the physical layer. There are number of physical layers currently proposed by various organizations such as ISO, SAE and others [7-9].

A generalized version of the system for controlling the speed and direction of the motor is shown in Figure 1. The system uses two Phytecs boards (Infineon C505C Microcontrollers based) and communicates through a CAN bus [10]. The system is broken into four major parts: two CAN nodes, the driver circuitry, and a feedback sensor.

207

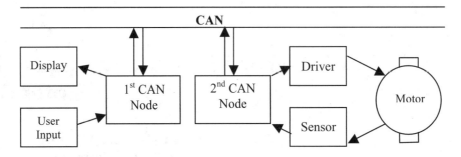

Figure 1: Generalized System Schematic

C505C Microcontroller

The C505C microcontroller is a member of the Infineon C500 family of 8-bit microcontrollers [11]. It is fully compatible to the standard 8051 microcontroller. The C505C is an enhanced version of the C505 and incorporates a Full CAN Module with a top transfer rate of 1 M Baud. Additional on-chip peripherals are three 16-bit timers/counters, full duplex serial interface with programmable baud rate generator (USART) and an 8-bit A/D Converter.

DC Motor Control

Control of the speed of DC motors below and above the base speed can easily be achieved. The technology of speed control of DC motors has evolved considerably over the past quarter-century. In the classical method a Ward-Leonard system with rotating machines is used for speed control of DC motors [12]. Recently, solid-state converters have been used for this purpose. These solid-state converters are mainly controlled rectifiers or choppers.

Choppers convert a fixed-voltage DC supply into a variable-voltage DC supply. It acts mainly like a high-speed switch turning the voltage on and off. The waveform of the chopper's output is shown in Figure 2. The average output voltage, V_t, of this waveform can be determined using the following equation:

$$V_t = \frac{t_{on}}{T} V$$
$$= \alpha V$$

Where t_{on} is the time the voltage is on (high)
T is the period of the waveform
α is the duty ratio of the chopper

It is obvious from the previous equation that the average voltage varies linearly with the duty cycle. The voltage across a dc motor's lead will determine its corresponding speed. So it follows, by changing the duty cycle of the chopper, one can change the motor's speed.

The C505C's timer 2 with additional compare/capture/reload features can be used for all kinds of digital signal generation. One of the more powerful of these digital signals is Pulse Width Modulation (PWM). PWM is the modulation of a digital signal's duty cycle. In

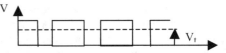

Figure 2: Chopper output

fact, PWM in the case for motor control can be called a chopper as they have similar output and function.

There are two types of motor control systems: open loop and closed loop. In the open loop system, the controller's output is based on an input or inputs that do not come from the motor. For our application, a potentiometer and two switches are the inputs to the controller, which in turns varies the input to the motor. While it allows the motor speed to be set, it does not allow for changes in load. As the load on the motor increases the motor's speed will decrease. The closed loop system incorporates real time speed as a feedback signal. By making continuous adjustments to its speed error, the controller can maintain desired speed under different loads.

Number of vehicular applications has fixed load this project uses an open loop system. This work can be extended to closed loop applications. An optical sensor has been added for feedback to display the motors speed, so the system can be modified for a closed loop system.

FIRST CAN NODE

The first CAN node acts like a user control panel. It monitors all user inputs and displays the real time motor speed to a hyper terminal. There are three user inputs to the system: speed control, forward/reverse, and run/stop. Speed control is implemented using a 50kΩ potentiometer connected between V_{cc} and the ground. As the user turns the potentiometer, the voltage level is read continuously by channel 0 (Port 1_0) of the on-chip 8-bit A/D converter. Run/stop and forward/ reverse are both digital inputs, implemented using dip-switches and read in through ports Port 1_1 and Port 1_2 respectively.

Every 250ms the first CAN node reads the output of the A/D converter and the two digital inputs. It then constructs a CAN data frame. The data frame consisted of only two bytes of data and an identifier of 555H. Contents of two data bytes are shown in Figure 3. The first byte contains encoded input from two digital switches. The last byte is for the speed from the ADC. The data frame is then transmitted over the CAN bus to the second CAN node.

7	4	3	0	7	0
Run/Stop		Forward/Reverse		Speed	

Figure 3: Two Data Bytes

Every second, the first CAN node detects a message on the CAN bus with an identifier of 666H. This data frame was sent by the second CAN node and contains one byte of data. The data is the actual motor speed in terms of revolutions per second. Since speed is commonly represented in RPM, the first CAN node multiplies the data by sixty. The result is then sent to the on-chip USART using the printf ( ) function. Actual display shown as "Current motor speed is ### RPM". A flow chart of various operations performed by the first CAN node is given in Figure 4.

SECOND CAN NODE

The second CAN node is responsible for generating various signals for motor control. This includes PWM signal generation, output

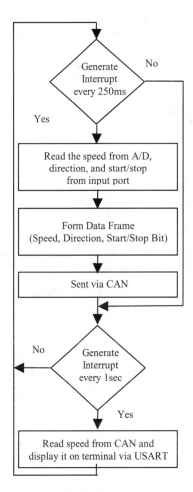

Figure 4: First CAN Node Operations

port location, and signal reproduction. It is also responsible for monitoring the feedback signal and relays its frequency to the first CAN node.

Every 250 ms, the second CAN node detects a message on the CAN bus with an identifier of 555H. The first CAN node sent this data frame with its contents as described in the last section. Depending on the value of the direction half byte, registers CCH and CCL of Timer2 are updated for either channel 1 or channel 2. Channel 1 corresponds to the forward direction and channel 2 to the reverse direction. The other channel's registers are set with the value of 0x0000, which will produce a five-volt DC output. Port 1_0 is also updated with the value of the start/stop half byte.

Every second Timer 1 is halted and the number of produced pulses at the feedback sensor in the previous second is recorded. It then constructs a CAN Data Frame with only one byte of data containing the recorded value. Timer 1 is then

209

reinitialized. A flow chart of various operations performed by the first CAN node is given in Figure 5.

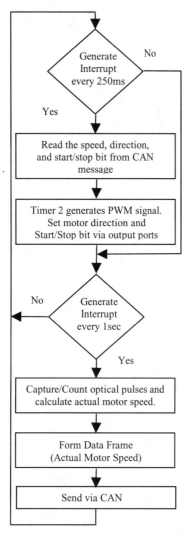

Figure 5: Second CAN Node Operations

DRIVER CIRCUITRY

The motor for this application was chosen for its small price and that it can be found at any local hobby shop. It has a voltage range of 1.5V to 6V and a current range of 115mA and 209mA. The chosen motor can operate at 5 volts. The C505C cannot supply the necessary current to drive a motor directly. In fact, current requirements of any practical motor are higher than any microcontroller can supply. Therefore a suitable driver must be selected to take the control signal from the C505C and deliver the necessary current to the motor. A motor draws its maximum current when it is fully loaded and starts from a stand still condition. As the motor's speed increases, its power consumption decreases. Once the speed of a motor reaches a steady state, the current depends on the load and the voltage across the motor. However, since this application requires reversing the motor, the current demand will even be higher. These factors must be taken into account when choosing a driver.

Standard motor drivers are available in many current and voltage ratings. The L293D was chosen for this application. It can output up to 600mA per channel with a supply voltage of 36 V. There are four channels per device and it has a separate logic supply and takes logical input (0 or 1) to enable or disable all channels. The main reason the L293D was chosen over the other drivers in the L293 series is that it also includes clamping diodes needed for protecting the driver against the back EMF generated during the reversing of motor and eliminates the need of the additional circuitry. Figure 6 and Table 1 show various connections and their functions.

Input		Function
VINH=H	C=High; D=Low	Turn Right
	C=Low; D=High	Turn Left
	C=D	Motor Stop
VINH=L	C=Don't Care	Free Run
	D=Don't Care	Motor Stop

Table 1: L293D Function Table

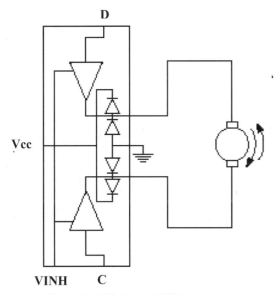

Figure 6: Half of the L293D

FEEDBACK SENSOR

A feedback sensor was added for displaying the real time motor speed. Different types of sensors such as optical encoders, infrared detectors, Hall effect sensors, etc. can be used for this motor-tachometer application. The selection of sensor should take into consideration environment and its intended use. For example, if the sensing device were to be used in oil-laden air environment, an optical sensor would not be prudent. Also, if an optical sensor is to be used, it should be encapsulated to underscore any unexpected factors such as ambient light, dust, and dirt that maybe normally uncommon t o that environment.

An optical infrared interrupter switch (H21LOB) was selected as this project is for demonstration purposes and for laboratory environment. The H21LOB is a member of the OPTOLOGIC® H21L series of optical switches by Fairchild Semiconductor™[13]. It features an open-collector, buffered digital output that is compatible with TTL logic. Its infrared emitter and detector are built into a single package for easy installation. Since its output is TTL logic

compatible, an additional 1KΩ pull-up resister was added to its output to make it compatible with the CMOS based C505C microcontroller.

The sensor is at no time physically connected to the motor shaft as shown in Figure 7. It will not add a load to the motor and there will be no effect on motor speed and its current requirement. A disk made up of a material, which allows the infrared light to propagate through, was painted black over its whole surface except for a small notch. In short, the painted surface will not let the infrared beam pass but the unpainted portion will. The disk

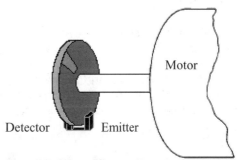

Figure 7: Motor/Sensor Interaction

was then attached to the motor shaft. The sensor was mounted to a stationary part of the system with the disk located between its emitter and detector as shown in Figure 7.

As the disk rotates with the motor, the unpainted portion in the disk passes between the emitter and detector, allowing infrared beam to pass. The detector is turned ON for each rotation producing one pulse per revolution. One should also note that the width of the unpainted portion would determine the duty cycle of the output signal. By determining the number of times a pulse is produced each minute, one can calculate the motor's speed (RPM).

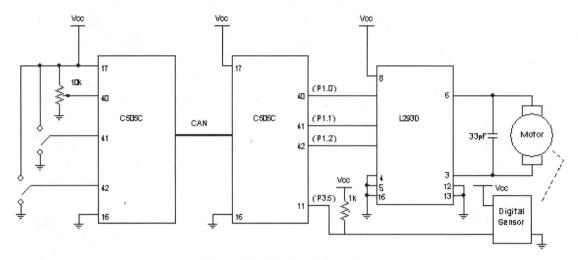

Figure 8: Full System Schematic

TEST RESULTS

A schematic of the system is shown in Figure 8. The serial connections from the USART to the PC as well as a 1 µF capacitor across the V_{cc} and ground terminals to clean the power supply are not shown in Figure 8. To test the system, the potentiometer was turned to its halfway mark.

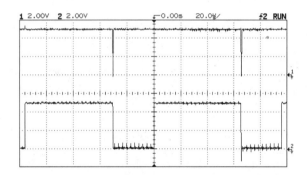

Figure 9: Output From the Second CAN Node

The corresponding value produced by the 8-bit A/D converter is 7FH, which produces a duty cycle of fifty percent. The change of direction was tested by flipping the input directional switch on the first CAN node and detecting the change in the motor direction. Figure 9 shows the actual output from the second CAN node. Its waveform does have a fifty percent duty cycle. The motor speed displayed to the hyper terminal. The CAN frames were monitored using the DPA system available from the Dearborn Group [15]. The C-programs were written and loaded in the flash memories of two CAN nodes. C-program code was generated using the Digital Application Virtual Engineer (DAVE) CD available from Infenion [16]. Programs were tested using Keil software [14].

CONCLUSIONS

This paper presented an example of the DC motor control via CAN system. One of the CAN node specifies the speed and direction of the motor and sends a message on the CAN bus. The other CAN node receives this message and appropriately runs the DC motor with the specified speed and direction. The system also has a feedback sensor, which captures the DC motor speeds and sends to the first CAN node. The system can be implemented in just one CAN node and can interact with other nodes on the network.

This work was done as part of the Real Time Embedded Course offered at the University of Toledo. The motivation of presenting this work is to show engineers in the industry how to build an embedded system in a laboratory setting. This example would be useful in controlling various motors in modern vehicles and other applications. It is just an example to show operation of CAN based systems and would require further consideration and modifications in terms of its appropriateness and cost for practical automotive usage. One should also need to consider other protocols and mechatronics.

REFRENCES

1. William B. Ribbens, "Understanding Automotive Electronics," SAE Publications 1998.

2. R. K. Jurgen,"Automotive Electronics Handbook" McGraw Hill, ISBN 0-07-033189-8

3. Y. Inoue, S. Uehara, K. Maguro, Y. Hirabayashi, "Multiplex Systems for Automotive Integrated Control," SAE Paper # 930002.

4. K. Furuichi, K. Ishida, K. Enomoto, K. Akashi, "An Implementation of Class A Multiplex Application," SAE Paper #920230.

5. SAE J 1850 SAE Standard Class B Data Communication Network Interface.

6. Robert Bosch GmbH, "CAN Specification" Version 2.0 ISO/TC22/SC3/WG1 document 1992.

7. SAE J 2411 SAE Recommended Practice-Single Wire CAN Network Vehicle Applications.

8. SAE J 2284-500 SAE Recommended Practice-High Speed CAN (HSC) for Vehicle Applications at 500 K bits/s.

9. ISO 11898, International Standard Road Vehicles-Interchange of digital information-Controller area network (CAN) for high-speed communication 1993.

10. KitCON-505C   Hardware-Manual www.phytec.de/

11. C505 and C505C 8-bit CMOS Microcontroller, User's Manual www.infineon.com/

12. Sen P.C. "Principles of Electric Machines and Power Electronics" second edition, John wily & Sons Inc., 1997, pp. 125, 455, 493, 533.

13. OPTOLOGIC® H21L series of optical switches by Fairchild Semiconductor™ www.fairchildsemi.com

14. 8051/251 Evaluation Kit and C-compilers, www.keil.com

15. Dearborn Protocol Adapter II (DPA II) www.dgtech.com

16. Digital Application Virtual Engineer CD available from www.infenion.com/dave.

CONTACT

Mohsin Jamali received the B.Sc. from the Aligarh Muslim University, India in 1975, M.S. From the University of Saskatchewan, Canada in 1979 and Ph.D. degree from the University of Windsor, Windsor, Canada in 1984. He is a professor in the department of electrical engineering and computer science at the University of Toledo, Ohio. He teaches courses in the areas of microcontroller based digital systems designs, VHDLs, PLCs, and automotive electronics. His research interests are in application specific hardware design for real time applications. His e-mail address is mjamali@uoft02.utoledo.edu.

Mohammad Yaqzan Niamat is an Associate Professor in the Department of Electronic Engineering Technology at the University of Toledo. He teaches embedded systems and works in the area of testing of FPGAs. Email: mniamat@utnet.utoledo.edu

2002-01-0438

Development of Car Intranet Infrastructure

Rami Baroody and Nizar Al-Holou
University of Detroit Mercy

Salim Hariri
University of Arizona

ABSTRACT

The automobile currently has a number of processors to control different subsystems such as engine controller, transmission controller, ABS controller, lighting controller, entertainment controller, and airbag controller. These subsystems are connected as a single vehicle network. Different vehicle networks can run under different protocols such as CAN, VAN, SCP, DLC, ACP, and J1939. Controller Area Network (CAN) has become the standard for the automotive industry. However, CAN has limited speed to incorporate new applications, such as in-vehicle multimedia, entertainment, navigation, and computing. A new technology, Media Oriented Systems Transport (MOST), provides high bandwidth to accommodate such applications. Moreover, a trend exists to include other functionality, such as access to information in the car through wireless technology to provide Internet access that enables remote diagnostics, status update, roadside assistance, navigation, global positioning, infotainment, traffic monitoring, emergency, safety, and security service.

A system architecture approach to integrate CAN, MOST and wireless communication system into the car environment has been presented in this paper. Moreover, an open and universally accessible architecture of a Car Intranet infrastructure that includes three network technologies CAN, MOST, and Wireless communication has been proposed. This infrastructure can provide an entirely new type of service in addition to current services can be delivered to cars in an efficient and secure way through wireless data connectivity over the Internet.

INTRODUCTION

Electronics were introduced in automobiles with the advent of integrated circuits. The main application of electronics was found in the automotive engine's electrical system, with the next application of electronics found in the fuel ignition control, cruise control, ABS, and transmission control unit. Further, use of electronics was extended to the instrument panel, suspension control, air-conditioning system, and other body control systems.

With advances in (VLSI) and computers more electronics are being introduced in today's cars and trucks, providing many new features such as: navigation system, intelligent highway system support, automated collision avoidance system, etc [1]. Furthermore automotive electronics have been used to introduce other new features (e.g., accessing the internet, read newspaper, send e-email, search the net, listen to music, make phone calls, participate in video conferencing, and conduct remote diagnostics) with all services that are available to people through the Internet extended to the automobile [2].

As the application of automotive electronics increases to enhance exiting features and introduce new features in automobiles, the need for more electronic control units (ECU) becomes mandatory, fast communication links are needed [3]. As the number of node increase, associated costs also increase. To overcome this difficulty, multiplexing is used in automotive applications, with many signals transmitted over a single medium using a time-sharing basis. In multiplexing, a serial bus connects nodes inside a vehicle; allowing data transfer between sensors, actuators, switches, and processors over a common bus. A Controller Area Network (CAN) can be used as an embedded communication system for micro controllers, as well as an open communication system for intelligent devices. Also the extended version of CAN protocol (J1939) also can be used.

As electronics are used inside the car for multimedia applications, such as CD quality audio, surround sound, and high quality video, a high-speed network running protocols that satisfy the data rate requirements for those applications is needed. CAN have a limit speed, which is 1 Mbps [4], which is not sufficient for some Multimedia applications. A high-speed network, fiber-optic bus, known as Media Oriented Systems Transfer (MOST), is a high performance, low-cost multimedia network technology, created by the MOST Cooperation (Karlsruhe, Germany) that uses a 25-Mbit/second fiber-optic bus and is best suited for these applications.

While initial vehicle multiplex applications concentrated on single vehicle-wide networks running certain protocols, additional factors are causing a shift toward the use of multiple vehicle networks. Speed, application differences, costs, new emerging interconnection trends, and other influences are impacting the single network philosophy. The application of multiple vehicle networks is growing and methods for supporting this direction are required. Multiple vehicle networks may result in major changes in the meaning of vehicle diagnostics that may require the use of gateways. These changes may require modification of the original single network solutions to handle new requirements. From the industry perspective, a variety of vehicle network protocols exist, including CAN, MOST, VAN, LIN, etc. Each of these robust, powerful solutions represents a major technological investment in a single vehicle-wide network [5].

In this paper, an open, and universally accessible, architecture along with development of a car Intranet infrastructure that enables each vehicle to access and use Internet technologies for communication (voice, video), entertainment, monitoring traffic, emergencies, etc has been proposed. In Section II, classification of automotive communications, multiplexing protocols, CAN protocols and J1939 are presented. Section III, MOST is presented. In section IV the proposed System architecture and networks Gateways between CAN, MOST are presented. Finally, conclusions are discussed in section V.

CLASSIFICATION OF AUTOMOTIVE COMMUNICATION

The Society of Automotive Engineers (SAE) has defined three distinct protocol classifications, Class A, Class B, and Class C [6]. Class A is the first SAE classification and maintains the lowest data rate, a rate that peaks as high as 1Kbps and supports up to 100 nodes. Class A devices typically support convenience operations like actuators and "smart" sensors. The second SAE classification is Class B protocol. Class B supports data rates as high as 100Kbps and supports up to 50 nodes. Class B typically supports non-real time control and communications. Utilization of Class B can eliminate redundant sensors and other system elements by providing a means to transfer data between different

nodes. Class C protocol is used for real time control applications with critical speed and accuracy and supports a data rate of up to 1Mbps.

MUTIPLEXING PROTOCOLS CONTROLLER AREA NETWORK (CAN)

Communication signals are transmitted and received by various network nodes according to a set of rules called protocols. Protocols set rules for coding, address structure, transmission sequence, error detection, and handling. The protocols also define the medium (copper wires or optical fiber), transmission speed, and electrical signal requirements. Protocols for automotive networking cover most of the Open System Interconnection (OSI) model [7]. In the harsh, noisy automotive environment, a multiplexing protocol should be optimized to meet technical and functional requirements of the application. Various network protocols have been developed for automotive multiplexing (e.g., controller area network (CAN), SAE J1939, vehicle area network (VAN), class2, etc.).

CONTROLLER AREA NETWORK (CAN)

CAN is an advanced serial bus system that efficiently supports distributed control systems. Robert Boost of GmbH, Germany initially developed CAN for the use in motor vehicles in the late 1980s. CAN is internationally standardized by the International Standardization Organization (ISO) and the Society of Automotive Engineers (SAE). The CAN protocol uses the Data Link Layer and the Physical Layer in the ISO - OSI model. CAN is most widely used in automotive and industrial market segments. Typical applications for CAN are motor vehicles, utility vehicles, and industrial automation. Other applications for CAN are trains, medical equipment, building automation, household appliances, and office automation. CAN is a protocol for short messages, with each transmission carrying 0 to 8 bytes of data, making it suitable for transmission of trigger signals and measurement values. CAN is a CSMA/AMP (Carrier Sense Multiple Access / Arbitration by Message Priority) type of protocol. The protocol is message oriented and each message has a specific priority according to which message gains access to the bus in case of simultaneous transmission. An ongoing transmission is never interrupted. Any node that wants to transmit a message waits until the bus is free and then starts to send the identifier of its message bit by bit. A zero is dominant over a one and a node loses the arbitration when it has written a one but reads a zero on the bus. As soon as a node has lost the arbitration, it stops transmitting but continues reading bus signals. When the bus is free again the CAN Controller automatically makes a new attempt to transmit its message. The CAN Controller chips complete this procedure, as well as error checking and retransmission of corrupted messages. The arbitration procedure requires that a limited number of identifiers (more than 500 million in Extended CAN) and that a specific identifier be sent only by one node. The only exception from this rule is when a message carries no data. As the amount of data that can be sent in one transmission is limited to eight bytes, the maximum latency time of the highest priority message can be calculated if the nodes are restricted to the use of the same message identifier, once transmitted, until a specified time has elapsed. Every CAN Controller in a network receives any message transmitted on the bus. Each node has to check whether a message is for him or not. Any CAN controller on the market offers some filtering capacity to reduce the processor capacity needed for this activity; with some more elaborate than others DO [8]. Figure 1 shows an example for CAN network within the car.

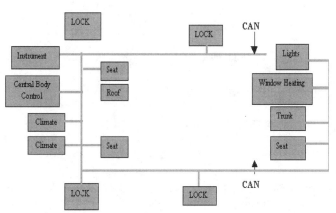

Figure 1 shows an example for CAN network within the car.

SAE J1939

CAN, introduced by Robert Boost, is a network protocol designed specially for in-vehicle networking [9]. The previous version of CAN version 1.2 has been successfully implemented in passenger cars, trains, and factory automation [3]. CAN protocol has been extended with a second message format called CAN V 2.0 [9]. The ASE has accepted this extended version of CAN protocol as the basis for in-vehicle networks in trucks and buses. The extended version protocol is called SAE J1939 [10].

J 1939 is a high speed, Class C type communications network designed to support real-time closed loop control functions between electronic control devices that may be physically distributed throughout the vehicle. SAE J1939 defines all relevant communication layers to support automotive multiplexing. Within J1939, a message is called the protocol data unit (PDU). There are two existing formats for the protocol data unit called PDU1 and PDU2. Format PDU1 includes source and destination address. Format PDU2 is used to send the data frames that are not destination specific.

MEDIA ORIENTED SYSTEMS TRANSFER (MOST)

Electronically controlled systems in automobiles are becoming increasingly dependent on rapid and reliable exchange of significant quantities of data. Within the communication and infotainment field, the secure transfer of high data volumes becomes more and more important. Apart from the data transfer for radio, CD players, telephone and navigation systems, the multi-channel usage of video data for TV-images, video games, DVD, voice-activated control for many functions, Internet accesses from the car, text-to-speech, and e-mail. These functions require a high bandwidth and reliability. These requirements can be met by the newly developed optical bus systems, which allow signal transfer via Plastic Optical Fiber (POF), combining advantages of a high bandwidth and insensitivity to electromagnetic interference with the weight-saving that can be compared to copper technology. This high-speed network, fiber-optic bus known as Media Oriented Systems Transfer (MOST), created by the MOST cooperation (Karlsruhe, Germany), uses a 25-Mbit/second fiber-optic bus is best-suited [11]. The topological structure of the MOST networks is a ring. Light signals are transported via POF from one connected apparatus to the next. In each apparatus, received light signals are transformed into electrical signals, which are then further processed by the MOST-processor. Generated signals are subsequently sent to an LED, which in turn converts this back into optical signals again. Figure 2 shows the MOST networks within the car.

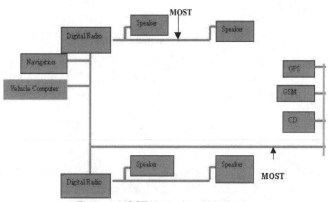

Figure 2 MOST Networks within the car

SYSTEM ARCHITECTURE

In automobiles, many subsystems need extensive interactions for information exchange. For example, an engine management system and transmission controller may work closely together for proper gear change. These kinds of interactions are becoming so numerous that the traditional view of vehicle electronics as independent sub-systems my not be adequate. The concept of local area network (LAN) has been introduced in automobiles to improve communications and interactions among different electronics systems. In applying a LAN to automotive electronics systems, an optimal protocol has been adopted for each system and each system may run different protocols that satisfy the data rate for their applications.

Multiple sub-networks are proposed in this section. Different groups of an electronics system may need different bus speeds. To accommodate this need, multiple buses can be introduced in automobiles, with these buses connected through Gateways. Figure 3 shows the proposed multiple networks scheme for automotive applications.

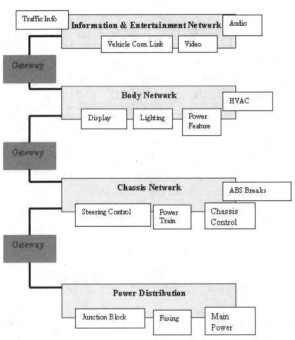

Figure 3. In-car network four Layers architecture

Each sub network consists of necessary electronics devices. Electronics devices within a sub network communicate with each through the local bus. For communications among various sub-networks, electronics devices can use Gateway. Because of the different requirement speed for applications running in these sub-networks, individual sub networks can contain individual protocols. In this scheme, no global bus has been used. This feature eliminates the bottleneck of global bus failure, and provides more reliability and bandwidth for multi-channel usage of video data for TV-

images, video games, DVD; Voice activated control for many functions etc. Because the general trend today in vehicle development is to carry an increasing number of information and control messages via serial busses, we proposed system architecture inside the vehicle as we mentioned before in this section and as shown in Figure 3. For technical information and machine control, the CAN bus is dominating. For radio, CD player, telephone and navigation system, the multi-channel usage of video data for TV-images, video games, DVD, voice activated, control for many functions, Internet accesses, MOST Technology dominates. The proposed four basic sub-networks are:

- Information Network. Digital High Speed "Source Data" (MOST,)
- Body Network. Low / Medium Speed "Control Data" (J1939, CAN-B/C, LIN, etc. Multiplex).
- Chassis Network. Control of Critical Power train, Chassis & Safety Systems (CAN C)
- Power Distribution.

NETWORK GATEWAYS

Gateways between CAN, MOST

The Gateway between CAN and MOST provides connectivity from an automotive control bus (such as CAN or J1939) to MOST and is the emerging networking standard for in-vehicle multimedia and entertainment. The basic function of a data gateway is data translation and data routing. Figure 4 shows the network Gateway between Controller Area Network and Media Oriented Systems Transport. The Gateway functions are:

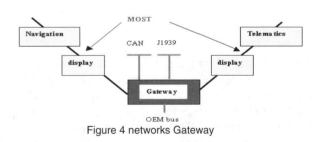

Figure 4 networks Gateway

Connecting two of the same or two different networks. Data from one or more input messages feeds in one or many output messages. The Gateway has the ability to connect networks with different baud rates and can provide a firewall for security of data between networks.

MOBILE IP

To provide access from automobiles to the Internet we consider the wireless communication aspect and the idea of Mobile Internet Protocol (IP), first, at the Home LAN (home, office), and then during travel time (foreign LAN). Standardization efforts have been made in this area already in the Internet community (e.g., the Mobile IP along with the IPv6)[2]. In summary, Mobile IP assigns temporary IP addresses to visiting hosts and directs Internet packets addressed to them at home locations to be forwarded to the point of attachment when a mobile IP (Car) moves from one sub-net to another. Figure 5 shows a mobile vehicle with services available on the Internet.

Networking can be carried out using a wireless technology, with satellite one of the most attractive technologies in USA for connectivity between the Internets from the road

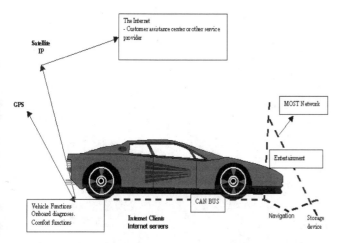

Figure 5 shows a mobile vehicle with services available on the Internet

CONCLUSION

As people continue to demand additional functions and features in their automobiles and technology continue to expand to meet these demands, developing the appropriate networks and connectivity protocols has become a priority for the automotive companies. Engineers and computer specialists need to work collaboratively to produce the types of systems and networks that will allow automobiles to be able to perform these functions at a reasonable cost and with good reliability.

REFERENCES

1. Ronal K.Jurgen, " The electronic Motorist," IEEE Spectrum, March 1995, pp.37-48.

2. Aktar Jameel, Axel Fuchs, and Matthlas Stuempfle " Internet multimedia on Wheels: Connecting Car to Cyberspace. " IEEE 1998, pp 637-642.

3. Tom Denton, "Automobiles electrical and electronics systems," 1995,SAE International, Warrendale, PA.

4. http://www.microcontroller.com/Learn Embedded/CAN1_sie/CAN1_files/frame.htm

5. Bruce D. Emaus " Aspects and Issues of Multiple Vehicle Networks. SAE 1995.

6. Friedrick H.Phail and David arnet, "In-Vehicle Networking-Serial Communication requirements and directions, SAE paper 8603901998.

7. http://www.nsi.fr/can2.html

8. http://www.kvaser.se/can/index.htm

9. Bosch, "CAN specification, ver2.0, Robert Bosch GmbH, Stuttgart 1991.

10. http://www.kvaser.se/can/hlps/index.htm

11. http://www.oasis.com/technology/technology.htm

12. Video and audio applications in Vehicles Enabled by Networked systems IEEE 1999.

13. Akihiro Tanaka, Masachika kamiya and kazunori Sakai " Gateway Application for Automotive Network System" "BEAN" SAE 1999.

CONTACT

Rami Baroody

Rbaroody@ecs.syr.edu

Professor: Nizar Alholou

Alholoun@udmercy.edu

Time-Triggered Communication on CAN

Holger Zeltwanger
CAN in Automation

ABSTRACT

The TTCAN (time-triggered communication on CAN) protocol is based on CAN (Controller Area Network). It may use standardized CAN physical layers such as specified in ISO 11898-2 (high-speed transceiver) or in ISO 11898-3 (fault-tolerant low-speed transceiver). TTCAN provides mechanism to scheduled CAN messages time-triggered as well as event-triggered. This will allow using CAN-based networks to close control loops. Another benefit is the increase of real-time performance in CAN-based in-vehicle networks.

INTRODUCTION

Controller Area Network (CAN) is data link layer protocol featuring event-triggered communication. The event may be of any kind occurring inside or outside of the CAN node. Event-triggered communication is very efficient regarding network bandwidth. It is suitable for a broad range of applications in in-vehicle networks as well as in decentralized or distributed general-purpose control systems. However, the real-time performance is limited. Due to this limitation, it is not generally suitable to implement CAN-based distributed closed-loop control systems. The closed-loop control function should be handled locally in one CAN node. Closed-loop control applications require time-equidistant message transmission. This can be achieved in networked systems only if a time-triggered communication is implemented. Event-triggered CAN communication is not sufficient even if the event is the elapsing of the local timer, because different drifts of local timers have to be considered as well as automatic retransmitted faulty messages. If time-equidistant transmission is demanded, a global time has to be distributed. This means, the scheduling of frame transmissions is triggered only by the progress of global time. In a time-triggered system, all message transmission and reception schedules have to be known in advance. Time-triggered communication is suitable in any kind of periodical transmission.

In automotive and industrial applications, event-triggered as well as time-triggered communication may be requested by the application. In order to implement both message-scheduling methods, time-triggered communi-

cation shall be added to the already existing event-triggered CAN communication. In some industrial and off-highway vehicle applications, a simplified time-triggered communication is implemented in software. For example in CANopen (recommended by ITA for electrical-powered truck lifts) there is the Sync Message available which allows a synchronized sampling of inputs and a synchronized validation of outputs. If a higher time-precision is required, there is also a Time-stamp Message available, which can be used in conjunction with the high-resolution protocol. However, this doesn't avoid the automatic retransmission of faulty CAN data and remote frames. The real-time performance of this software solution is still limited.

TIME-TRIGGERED COMMUNICATION

In state-of-the-art and future passenger cars you will have multiple in-vehicle networks. There will be networks for different purposes connecting not only ECUs (electronic control unit) but also an increasing number of sensors and actuators. Event-triggered communication will dominate the entertainment network and the body electronics network. Powertrain networks require event-triggered as well as time-triggered communication. For example, the vehicle engine is not running with constant revolution therefore, the injection time is not periodically. However the synchronization with the crankshaft saves gas and reduces exhaust. This application requires both event-triggered and time-triggered communication.

ISO (International Standardization Organization) has specified in ISO 11898-4 the TTCAN (time-triggered communication on CAN) protocol. TTCAN is based on the CAN data link layer protocol (ISO 11898-1) and does not infringe it at all. In TTCAN protocol, there are three different types of time windows: exclusive time window free time window, and arbitrating time window. Exclusive time windows are assigned to a specific message that is transmitted periodically without competition for the CAN network access. For the support of the time-triggered communication, any node needs to provide a time base which is provided either by an internal or by an externa clock. Any message received or transmitted invokes a capture of the time base taken at the respective message's reference point. After successful message completion, the capture value is provided to the control unit for at least one message and it is readable until the next message is completed. It has to be possible to generate at least one programmable event-trigger from the above mentioned time base. The trigger should be freely programmable by the control unit. The transmission schedule's time slots are divided by trigger events. The automatic retransmission of messages that could not be transmitted successfully must be disabled. The basic cycle defines the length of windows transmitted between two reference messages, which represent the global time synchronizing the local timers. The matrix cycle specifies the sequence of messages transmitted in each basic cycle.

In the arbitrating windows, all nodes with pending event-triggered transmission requests try to get bus access. The node with the highest priority message will get bus access. In order to allow longer periods for arbitrating messages, consecutive arbitrating time windows may be merged. The Tx_Enable window can be extended to the point of time when the Tx_Enable of the last arbitrating time window will be closed. This gives existing CAN applications a migration path to time-triggered communication without changing the bus system. The network

paradigm has not to be changed. All the experiences made with CAN physical layer options can be reused.

Between two reference messages, the drift of different local timers may be too high for a specific application requiring very high network bandwidth. For such applications, the global time producer may send the global time within the Reference message. The Reference message may be used to synchronize the cycle times. Cycle time restarts in each node at zero when a Reference message occurs. Cycle time is incremented whenever local time is incremented. When cycle time reaches a certain value indicating for instance the beginning of a time window, a protocol action is initiated. The global time unit implemented compensates the drift between the local clock and the global time by calibrating the local time. The calibration process is on hold when the node is not synchronized to the system and is (re-) started when it (re-) gains synchronization. The accuracy of the calibration is specified by the requirement that the remaining drift shal be less than three nominal bit times in one basic cycle The synchronization accuracy is about 1 µs, which should be sufficient for most of the automotive and industrial applications

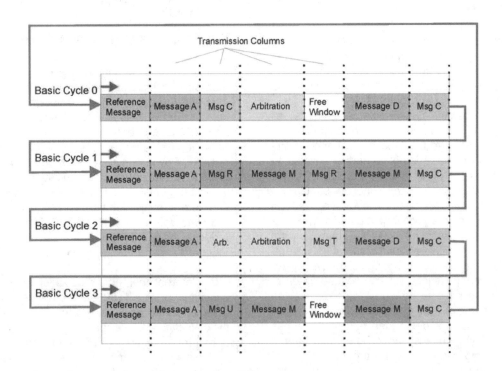

Figure 1: System matrix

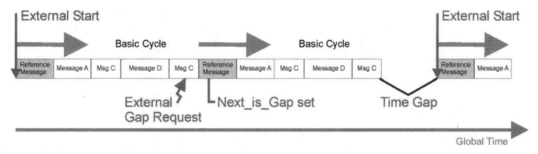

Figure 2: Event-synchronized Basic cycle

SYNCHRONIZATION METHOD AND GLOBAL CLOCK

A synchronized operation of the tasks within a network requires a local counter at each node and a synchronization of these. Once all local counters are synchronized, any task execution on any node in the network can be triggered by the elapse of the time provided by the current time master. The CAN protocol offers an event that is present at all nodes in the network at the same time, the start of frame pulse (SOF). Neglecting signal propagation the SOF can be used to generate a network-wide synchronization. In any node the SOF will initiate the capturing of a free running counter. This capture process will only be performed when the message identifier indicates a reference message that was sent by the time master of the network.

The reference message transfers the captured value of the time base of the time master in the data part. The receiving node is able generate a drift and an offset correction out off subsequent receptions of a reference message. While a TTCAN level 1 implementation just needs a 16-bit time base, level 2 needs at least 3 more bits representing fractional parts not visible to the scheduling. In TTCAN level 2 the drift correction is done in hardware only influencing the fractional part of the time base.

The TTCAN hardware implementation will perform tasks that currently still bother the CPU. Calculating the globa time, applying drift corrections and error confinement wil be integral services of the TTCAN module. As well, the scheduling will be completely autonomously handled by the CAN interface. No auxiliary timers taken from the device are needed anymore.

7	6	5	4	3	2	1	0
Next_is_Gap	Reserved	Optional: Cycle_Count (5)	Optional: Cycle_Count (4)	Optional: Cycle_Count (3)	Optional: Cycle_Count (2)	Optional: Cycle_Count (1)	Optional: Cycle_Count (0)

Figure 3: Format of reference message in Level 1

7	6	5	4	3	2	1	0
Next_is_Gap	Reserved	Optional: Cycle_Count (5)	Optional: Cycle_Count (4)	Optional: Cycle_Count (3)	Optional: Cycle_Count (2)	Optional: Cycle_Count (1)	Optional: Cycle_Count (0)

7	6	5	4	3	2	1	0
NTU_Res (6)	NTU_Res (5)	NTU_Res (4)	Optional: NTU_Res (3)	Optional: NTU_Res (2)	Optional: NTU_Res (1)	Optional: NTU_Res (0)	Disc_Bit

7	6	5	4	3	2	1	0
MRM (7)	MRM (6)	MRM (5)	MRM (4)	MRM (3)	MRM (2)	MRM (1)	MRM (0)

7	6	5	4	3	2	1	0
MRM (15)	MRM (14)	MRM (13)	MRM (12)	MRM (11)	MRM (10)	MRM (9)	MRM (8)

Figure 4: Format of reference message in Level 2; byte 3 and 4 contains the Master_Ref_Mark (RMR)

SCHEDULING

TTCAN requires at least a 16-bit counter that is used to schedule messages. This time base delivers the cycle time that gets restarted every basic cycle. The system designer typically assigns the LSB of the cycle time to be the duration of a bit time. All messages in the network are scheduled by the cycle time. On top of the maximum value of the cycle time the TTCAN session layer offers up to 64 basic cycles. In result the user may set up a system matrix that may schedule as much as 600 times 64 different messages assuming an average message length of approximately 100 bits. The length of the rows and columns of the system matrix can be scaled to the actual requirements of the system. The system designer uses a system matrix to define transmit triggers. These Tx-triggers specify time marks at which a particular message object is sent. As well the width of a column is set up which designates the maximum message length that can be applied.

Further the Tx-trigger designates the window type, exclusive or arbitrating. While exclusive windows are referencing to one specific message object, arbitrating windows allow the transmission of any message object. Subsequent arbitrating windows are merged which uses the bandwidth most efficiently. The arbitrating windows in TTCAN offer a migration path from a pure event-driven communication towards a strict time triggered data exchange. As well as messages can be sent at fixed points in time, the application can make use of this and even synchronize the task execution on different nodes. This is of course out of the scope of the TTCAN specification and may therefore be implemented be proprietary software.

TTCAN IMPLEMENTATION

The main components of a TTCAN implementation includes global time unit (GTU), time-base builder (TBB) and cycle time controller (CTC). The time base serves on the one hand the scheduling by the time scheduling organizer (TSO) and the trigger memory and on the other hand it is presented to the CPU that may even synchronize its task execution to the global time of the network. The master state administrator (MSA) governs the operation of the TTCAN-module. Dependent on the error status the MSA partially or completely limits the activity of the node.

The master state administrator (MSA) governs the operation of the TTCAN-module. Dependent on the error status the MSA partially or completely limits the activity of the node. In order to avoid that a TTCAN protocol controller is still transmitting data while the CPU is not working properly anymore, an application operation monitor (AOM) is installed. According to the watchdog principle the TTCAN-node will cease to operate once the CPU did not serve the AOM.

While effort on the implementation of the blocks in the frame synchronization entity (FSE) is quite clear, the design of the trigger memory will be key parameter of a TTCAN implementation. Since memory is a cost driving factor the trigger memory must be kept small. Imagining a system matrix with 100 or more messages the RAM quickly becomes larger than 1 Kbytes. As a major benefit of TTCAN only that part of the transmission matrix must be provided that is directly assigned to messages that are processed at this node. The complete system matrix is distributed over all nodes of the network.

FUTURE DEVELOPMENTS

Networks connecting safety-critical ECUs will need time-triggered communication and additional redundancy due to availability requirement [3]. This may include x-by-wire applications. Event-triggered CAN networks are already used for x-by-wire applications in off-highway vehicles (e.g. wheelchairs), agriculture and forestry machinery as well as in industrial control systems. In particular, CAN-connected joysticks can be regarded as x-wire-application. Other applications demanding time-triggered communication include flight- and mission-critical contro systems. First hardware and software implementations from different companies have been tested successfully Prototype chips for evaluations are already available. Off-the-shelf TTCAN semiconductors will be shipped by end of this year and beginning of 2003.

In order to achieve interoperability between different TTCAN implementations, ISO is going to standardize a conformance test plan. An appropriate new work item proposal has been already started. The standardized conformance test plan will allow implementing certification tools. However, certification of TTCAN implementations is not in the scope of ISO.

REFERENCES

1. ISO WD 11898-4 (unpublished)
2. TTCAN workshop proceedings 2001. CiA, Germany.
3. B. Mueller: Fault tolerant TTCAN networks. In: iCC 2002 proceedings, CiA, Las Vegas.

CONTACT

headquarters@can-cia.org

Integration of Time Triggered CAN (TTCAN_TC)

Florian Hartwich, Thomas Führer, Bernd Müller and Robert Hugel
Robert Bosch GmbH

ABSTRACT

Time Triggered CAN (TTCAN) is an extension of the well-known CAN protocol, introducing to CAN networks time triggered communication and a system wide global network time with high precision. Time Triggered CAN has been accepted as international standard ISO CD 11898-4.

The time triggered communication is built upon the unchanged standard CAN protocol. This allows a software implementation of the time triggered function of TTCAN, based on existing CAN ICs. The high precision global time however requires a hardware implementation. A hardware implementation also offers additional functions like time mark interrupts, a stopwatch, and a synchronization to external events, all independent of software latency times.

The TTCAN testchip (TTCAN_TC) is a standalone TTCAN controller and has been produced as a solution to the hen/egg problem of hardware availability versus tool support and research. The TTCAN_TC supports both TTCAN level 1 and TTCAN level 2; its time triggered communication is - apart from the configuration and the watchdog function - not depending on software control. The package has been designed to be pin-compatible to existing standalone CAN controllers.

The first application of the TTCAN_TC is the evaluation of TTCAN networks, both for research and for system development. TTCAN network monitoring tools (TTCANanalyzer), based on the TTCAN_TC, will soon be available.

INTRODUCTION

CAN is the dominating network for automotive applications. New concepts in automotive control systems (x-by-wire systems) require a time triggered communication. This is provided by TTCAN. The main features of TTCAN are the synchronization of the communication schedules of all CAN nodes in a network, the possibility to synchronize the communication schedule to an external time base and the global system time.

TTCAN is fully compatible with the existing CAN controllers, both in the data link layer as well as in the physical layer, it uses the same bus line and bus transceivers. Dedicated bus guardians are not needed in TTCAN nodes, bus conflicts between nodes are prevented by CAN's non-destructive bitwise arbitration mechanism and by CAN's fault confinement (error-passive, bus-off).

Existing CAN controllers can receive every message in a TTCAN network, TTCAN controllers can operate in existing CAN networks. TTCAN controllers can be seen as existing CAN controllers enhanced with a Frame Synchronization Entity. A gradual migration from CAN to TTCAN is possible.

The minimum additional hardware that is required to enhance an existing CAN controller to time triggered operation is a local time base and a mechanism to capture the time base, the capturing triggered by bus traffic. Based on this hardware, which is already existent in some CAN controllers, it is possible to implement in software a TTCAN controller capable of TTCAN level 1.

A TTCAN controller capable of TTCAN level 2, providing the full range of TTCAN features like global time, time mark interrupts, and time base synchronization, has to be implemented in silicon.

TTCAN PROTOCOL

TTCAN, specified in 2000 by ISO TC22/SC3/WG1/TF6, is accepted as ISO CD 11898-4, the time triggered extension to the internationally standardized CAN protocol.

This chapter gives only a brief description of the main features of the TTCAN protocol and of time triggered system requirements, a more detailed description can be found in [2].

SYSTEM MATRIX

In a time triggered system, all messages of all nodes in the network have to be arranged in a fixed time schedule, specifying a time window for each message. In time triggered CAN, the correlation between the messages and the time windows in which they shall be sent is specified by the system matrix. The components of the system matrix are the time windows, it is organized in basic cycles (rows of the matrix) and transmission columns

(columns of the matrix). Each basic cycle starts with a specially characterized message, the Reference Message. Inside the basic cycle, the boundaries of the time windows are defined by time marks relating to the Cycle_Time. The cycle of all basic cycles in the system matrix is the matrix cycle. All basic cycles of the matrix cycle have the same length, the matrix cycle is continuously repeated.

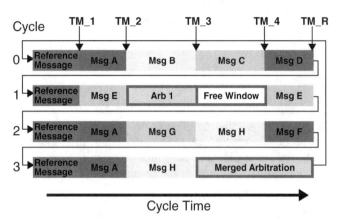

Figure 1: Matrix Cycle

Time windows

Each message is transmitted in a particular time window, the transmission may only be started at the beginning of the time window.

A basic cycle may consist of time windows of different type and length. All time windows of a transmission column shall have the same length but may have different types.

Three different types of time windows exist: Exclusive time windows, free time windows, and arbitrating time windows.

Exclusive time windows are assigned to a specific message transmitted periodically without competition for the CAN bus. Arbitrating time windows are assigned to messages that share the same time window; consecutive arbitrating time windows may be merged to one single window. Several nodes in the network may start a transmission in the arbitrating time window, bus conflicts are resolved by CAN identifier arbitration. Free time windows are reserved for future extensions of the network.

Messages in the Matrix Cycle

It is not required that each node knows all components of the system matrix, a node's view of the system matrix may be limited to those messages that it needs to receive and those that it transmits.

Within a basic cycle a message may be assigned to more than one time window, i.e. a specific message may belong to more than one transmission column. Within a matrix cycle the different basic cycles are identified by Cycle_Count, the current value of Cycle_Count (cyclically incremented each basic cycle) is transmitted by the actual time master as part of the Reference Message.

Within a transmission column (column of the matrix cycle), a specific message may be transmitted periodically. A specific message may belong to more than one transmission column and may be transmitted in more than one time window of a transmission column.

Figure 1 and figure 2 show an example for three nodes, where Node_1 and Node_2 are potential time masters. Figure 2 specifies for three nodes at which time mark which message is to be transmitted in which basic cycle.

Time mark TM_W is reached when the reference message is missing. Additional time marks may specify when the reception of a message is to be checked.

	Node_1	Node_2	Node_3
TM_1	0,2,3: Msg A	———	1: Msg E
TM_2	1: Msg x	0: Msg B 1: Msg x 3: Msg H	1: Msg x 2: Msg G
TM_3	3: Msg x	2: Msg H 3: Msg x	0: Msg C 3: Msg x
TM_4	0: Msg D 2: Msg F	———	1: Msg E
TM_R	Reference Msg	Reference Msg	———
TM_W	Error Handler	Error Handler	Error Handler

Figure 2: Local Views of System Matrix

TIME BASES

Local Time

Each node has its own time base, Local_Time, which is a counter that is incremented each Network Time Unit NTU. In TTCAN level 1, Local_Time has a resolution of 16 bit and the NTU is identical to the CAN bit time. In TTCAN level 2, Local_Time has an integer resolution of 16 bit and an additional fractional part of at least 3 bit. The length of the NTU is defined by the TTCAN network configuration, it is the same for all nodes. It is generated locally, based on the local system clock and the Time Unit Ratio TUR. Different system clocks in the nodes are compensated by different (non-integer) TUR values.

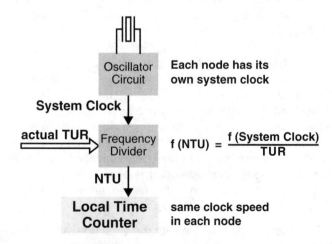

Figure 3: Local Time

Cycle Time

In the TTCAN network, the synchronization of the nodes is maintained by so-called Reference Messages that are transmitted periodically by a specific node, the time master. The Reference Message is a CAN data frame, characterized by its identifier. Valid Reference Messages are recognized synchronously (disregarding signal propagation time) by all nodes. Each valid Reference Message starts a new basic cycle and causes a reset of each node's Cycle_Time.

The value of Local_Time is captured as Sync_Mark at the start of frame (SOF) bit of each message. When a message is recognized as a valid Reference Message, this message's Sync_Mark becomes the new Ref_Mark; Cycle_Time is the actual difference between Local_Time and Ref_Mark.

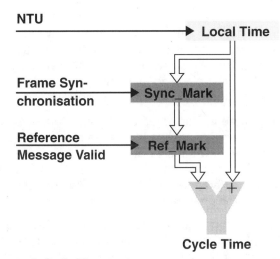

Figure 4: Cycle Time

Global Time

There are two levels of implementation in TTCAN, level 1 and level 2. In TTCAN level 1, the common time base is the Cycle_Time which is restarted at the beginning of each basic cycle and is based on each node's Local_Time. In TTCAN level 2, there is additionally the Global_Time which is a continuos value for the whole network and which is the reference for the calibration of all local time bases.

The time master captures its view of Global_Time at each Sync_Mark and transmits that value in the Reference Message, as Master_Ref_Mark. For all nodes, Global_Time is the sum of their Local_Time and their Local_Offset, Local_Offset being the difference between their Ref_Mark in Local_Time and the Master_Ref_Mark in Global_Time, received (or transmitted) as part of the Reference Message. The Local_Offset of the current time master is zero if no other node has been the current time master since network initialization.

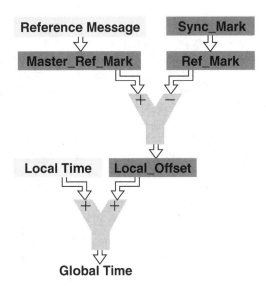

Figure 5: Global Time

The TTCAN nodes compensate the phase drift between their Local_Time and the Global_Time at each Reference Message. They monitor the Master_Ref_Mark and their local Ref_Marks, which are all captured at the same time, and calculate their clock speed drift by dividing the differences between two consecutive Master_Ref_Marks and two consecutive Ref_Marks (see figure 9).

The clock speed drift is compensated by adapting the prescaler (TUR) that generates the NTU from the local system clock. This calibration, together with the (at least) three bits of additional resolution for the NTU, provides a high precision time base.

TTCAN COMMUNICATION

Time Master Redundancy

Since the Reference Messages transmitted by a time master are vital for the TTCAN network, any TTCAN network may have up to eight potential time masters. One of them is the current time master, the other operate as backup time masters. Which of the potential time masters becomes the actual time master is defined by their time master priorities.

The most significant bits of the Reference Message's identifier are defined by the network configuration, the three least significant bits define the time master priority. Each potential time master transmits a different Reference Message identifier, the CAN bit arbitration decides which node becomes the actual time master.

When the actual time master fails to transmit a Reference Message, the node with the next highest time master priority instantly becomes the new actual time master.

Time master arbitration is done on system start-up or on the (transient) failure of the actual time master, otherwise backup time masters recognize the actual time master's priority and do not transmit Reference Messages.

Start-up of the Network

When the TTCAN network is powered up, all nodes need to be synchronized and the actual time master needs to be established before the time triggered communication is possible. The initialization procedure has to take into account that the nodes may have different setup-times before being able to take part in bus communication. The communication is initiated by the potential time masters. The actual time master establishes itself among the other potential time masters by winning the CAN bit arbitration when transmitting the Reference Message, using the same algorithm that in normal operation allows the reintegration of transiently failed time masters.

The transmission of the other messages of the system matrix is enabled after two consecutive valid Reference Messages have been seen.

Synchronization to external events

TTCAN has the option to synchronize the communication schedule to specific events in the time masters' nodes. When the communication is to be synchronized, the cyclic message transfer is discontinued after the end of a basic cycle and a time gap may appear between the end of the last periodic basic cycle and the beginning of the next, event synchronized basic cycle. The time gap is announced by the current time master in the last basic cycle's Reference Message. The time gap ends as soon as the current time master or one of the potential time masters sends a Reference Message to start the following basic cycle of the matrix cycle. The transmission of the Reference Message will be triggered by the occurrence of a specific event or after a maximum waiting time.

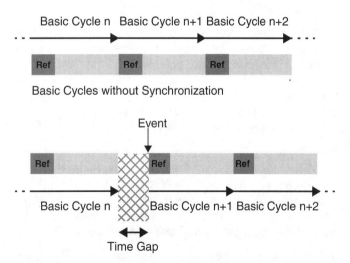

Basic Cycles without Synchronization

Basic Cycle n+1 synchronized to event

Figure 6: Synchronization of Basic Cycles

FAULT TOLERANCE

Networks in the automotive environment require powerful measures for error detection, error signalling, and fault confinement.

The measures that are implemented in every CAN node (bit monitoring, bit stuffing, CRC and format check for error detection – error frames, error-passive and bus-off for error signalling and fault confinement) give TTCAN networks the same data consistency as CAN networks.

TTCAN provides additional measures to detect violations of the timing restrictions, e.g. when a message is transmitted outside its dedicated time window or when it is not transmitted at all.

A Message Status Count (MSC) for each periodic message shows whether that message was transmitted or received on time.

Multiple TTCAN busses may be combined to build a fault tolerant multi channel TTCAN network [3].

MIGRATION FROM CAN TO TTCAN

The introduction of a new protocol and a new bus line in automotive applications takes a major effort in the regard of the development of new protocol controller and transceiver ICs, new planning and analysis tools, and in acquiring the know-how for system integration.

TTCAN can be implemented on the same physical layer as any existing CAN network, with the maximum bit rate of 1 Mbps. No new transceivers or bus lines are needed; whether shielding is required depends on the electromagnetic interference conditions. An independent bus guardian for the prevention of bus conflicts is not needed.

The bit rate cannot be increased, neither for CAN nor for TTCAN [4]. CAN's non-destructive bitwise arbitration mechanism for the resolution of bus conflicts requires a lower limit (in automotive applications 1000 ns) for the length of the CAN bit time.

All messages in a TTCAN network are transmitted in the CAN format, so the wide range of existing analysis tools for CAN bus traffic can also be used for TTCAN. New tools are needed for the temporal configuration of the message transmission schedule.

TTCAN is downward compatible with CAN, TTCAN nodes can be configured to operate as existing CAN nodes; existing CAN nodes can operate as listen-only receivers in a TTCAN network. A CAN node transmitting in a TTCAN network would disturb the transmission schedule.

A gradual migration from a CAN network to a TTCAN network is possible. Starting with a CAN network, all CAN nodes can be replaced step-by-step with TTCAN nodes before switching the communication and the application programs to TTCAN.

TTCAN PROTOCOL IMPLEMENTATION

An ISO 11898-4 TTCAN [2] controller can be seen as an existing ISO 11898-1 CAN [1] controller (e.g. Bosch's C_CAN module) enhanced with a Frame Synchronization Entity FSE and with a trigger memory containing the node's view of the system matrix. An analysis of the structure of an FSE allows an estimation of the effort to implement TTCAN in software or in hardware.

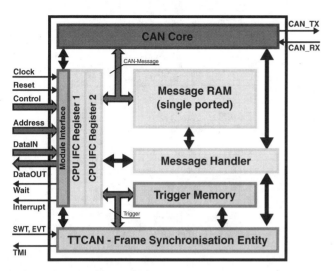

Figure 7: TTCAN Controller Module

To allow a more detailed analysis, a TTCAN protocol FSE can be divided into several functional blocks (implementation in hardware) or processes (implementation in software): TBB, CTC, TSO, MSA, AOM, and GTU. In the following, each part is considered separately.

Time Base Builder – TBB

This block generates NTU and Local_Time, based on system clock and TUR. The Local_Time is the output of a counter that is incremented each NTU. The resolution of Local_Time is different for TTCAN level 1 and 2.

In TTCAN level 1, TUR is a constant and Local_Time is a 16 bit integer value, incremented once each NTU. The NTU is the CAN bit time. In a software implementation, Local_Time is generated by one of the μC's timer-counters.

In TTCAN level 2, Local_Time consists of a 16 bit integer value extended by a fractional part of N (at least three) bit. Local_Time is incremented 2^N times each NTU, providing a higher time resolution than in level 1. TUR is a non-integer value and may be adapted to compensate clock drift or to synchronize to an external time base. A software implementation in usual μCs could not provide an adequate time resolution in TTCAN level 2.

Cycle Time Controller – CTC

This block controls the sequence of time windows (Cycle_Time) in the basic cycle and the sequence of basic cycles (Cycle_Count) in the matrix cycle.

The CAN's protocol controller signals the occurrence of each start of frame (SOF) bit and each valid Reference Message seen (transmitted or received) on the CAN bus. The Local_Time is captured each SOF as Sync_Mark. At each valid Reference Message, the Sync_Mark register is copied into the Ref_Mark register.

Even in a software implementation of TTCAN, the capturing of Local_Time into Sync_Mark at each SOF must be done in hardware (see figure 4). ISO 11898-1 [1] specifies the necessary hardware interface as an optional feature, it is already implemented in some CAN controllers.

The Cycle_Time is the actual difference of Local_Time and Ref_Mark, restarting at the beginning of each basic cycle when Ref_Mark is reloaded. Cycle_Time is the reference for the time marks that define the time schedule. In a hardware implementation, Cycle_Time is calculated by logic and is compared to the time mark registers. In software, all time marks of the basic cycle are adjusted by the amount of Ref_Mark and are then compared to Local_Time.

Cycle_Count identifies the actual basic cycle inside the matrix cycle. Its actual value is defined by the value received (or transmitted) in the last valid Reference Message. In potential time masters, the actual value of Cycle_Count is cyclically incremented inside the CTC in preparation for the next Reference Message.

Time Schedule Organizer – TSO

This block is a state machine that maintains the message schedule inside a basic cycle and checks for scheduling errors. The TSO gets its view of the message schedule from an array of time triggers in the trigger memory. Each time trigger has a time mark that defines at which Cycle_Time the trigger becomes active. Tx_Triggers and Rx_Triggers also point to specific messages and specify for which basic cycles they are valid.

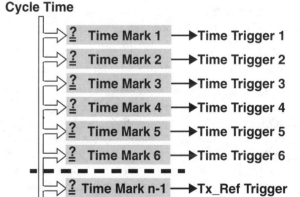

Figure 8: Time Trigger

A Tx_Trigger specifies when a certain message shall be transmitted. Depending on the success of the transmission, that message's MSC is incremented or decremented.

An Rx_Trigger specifies when the reception of a message shall be checked and that message's MSC shall be updated.

A Tx_Ref_Trigger triggers the transmission of a Reference Message, it finishes the current basic cycle and starts a new cycle.

A Watch_Trigger has a Time_Mark with a higher value than the Tx_Ref_Trigger and checks if the time since the last valid Reference Message has been too long.

This state machine is evaluated at the time marks that define the time windows; an implementation in software has the advantage of flexibility in the size of the system matrix.

Master State Administrator – MSA

This block controls tasks like operating mode, initialisation, time master redundancy, and error responses.

Application Operation Monitor – AOM

The function of the application program is checked to avoid the transmission of invalid data. The AOM is implemented as a watchdog that has to be served regularly. If the application fails, the MSA disables all CAN bus activities. Even in a software implementation, the AOM has to be supported by a hardware watchdog.

Global Time Unit – GTU

The GTU block only exists in TTCAN level 2. It has two functions, it generates the node's view of Global_Time and it calculates the necessary drift correction when the length of the NTU of the local time base and of the actual time master's time base differ.

Global_Time is not generated by a separate counter, the node's view of Global_Time is its Local_Time adjusted by an offset, the Local_Offset. The node's Local_Offset is calculated new at each Reference Message's SOF, it is the difference between the node's Ref_Mark and the received Master_Ref_Mark.

Changes in Local_Offset show differences in local node's NTU and the actual time master's NTU.

After initialisation, before synchronizing to the network, each node sees its own Local_Time as Global_Time, the Local_Offset is zero. The actual time master establishes its own Global_Time as the network's Global_Time by transmitting its own Sync_Marks in the Reference Message, as Master_Ref_Marks. When a backup time master becomes the actual time master, it keeps its Local_Offset value constant, avoiding a discontinuity of Global_Time.

The actual time master may adjust the phase of Global_Time by changing its Local_Offset value, e.g. to synchronize to an external clock. Any such intended discontinuity of Global_Time is signalled in the Reference Message, by setting the Disc_Bit.

The local NTU is built from the local system clock period t_{sys} and the TUR in the TBB, $NTU = TUR \cdot t_{sys}$. TUR is usually a prescaler value dividing the system clock; the NTU is adjusted by modifying TUR. A different concept using a voltage controlled oscillator is also possible.

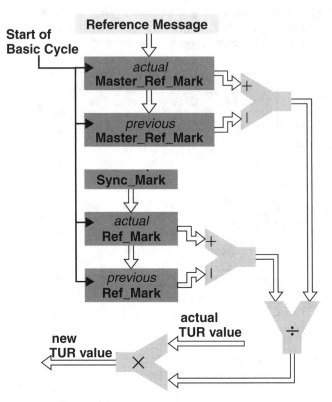

Figure 9: Drift Compensation

When the TTCAN communication is initialized, any two consecutive Reference Messages are used to compare the clock speeds of local clock and time master's clock when in the second Reference Messages the Disc_Bit is not set. The factor df by which the local NTU has to be adjusted is calculated from the Ref_Marks:

$$df = \frac{Ref\_Mark - Ref\_Mark_{previous}}{Master\_Ref\_Mark - Master\_Ref\_Mark_{previous}}$$

$$TUR = df \cdot TUR_{previous}$$

The calibration process is on hold when the node is not synchronized to the system and it is (re-)started when it (re-)gains synchronization. The necessary accuracy of the calibration is defined by the system's requirement e.g. that the remaining drift must be less than three nominal CAN bit times in one basic cycle.

Comparison of the Implementations

Advantages of the software implementation are that it can be done based on existing silicon and that it allows flexibility of trigger memory and TSO; it can be adapted to any size of system matrix. Disadvantages are that only TTCAN level 1 is possible and that the CPU load rises with the bus traffic. The TTCAN protocol processes require a short interrupt response time to be able to maintain the timing conditions.

Disadvantages of the hardware implementation are the increased silicon area (about 50% more that existing CAN implementation) and that the size has to be scaled for large trigger arrays. Scaled-down "basic"-TTCANs for small applications may be designed in future. Advantages are the improved timing capabilities (TTCAN level 2, drift correction) and that all protocol functions are handled independent of the application software. The application just has to provide the data to be transmitted.

The implementation of TTCAN in hardware allows to implement some additional features (not required by TTCAN protocol) that cannot be provided in software.

FEATURES OF HARDWARE IMPLEMENTATION

The TTCAN hardware implementation gives a better time resolution, does not depend on interrupt response time and can provide additional interfaces to the application. This makes it possible to implement functions like automatic clock calibration, bus-time-based interrupts, a stopwatch function, and a direct event trigger.

Clock Calibration

In TTCAN level 2, the time master provides a global time to all nodes in the network. All nodes calibrate their own local time to the same clock speed as the global time. The time resolution is at least NTU/8, meaning that all time marks are measured with a fractional part of at least three bit. That means that for reasonable system clock speeds TUR may be a fractional, non-integer number. An approximation algorithm calculates df and the adjusted TUR, a non-integer clock divider provides the calibrated Local_Time. Cycle_Time and Global_Time are available as continuous values by adding offsets to Local_Time.

Time Mark Interrupt

Local_Time, Cycle_Time, and Global_Time can be compared to a time mark interrupt register. When the selected time value matches the register value, an interrupt is generated. This event may trigger the CPU's interrupt line or may be directly connected to an output port.

Stopwatch

An input port may be used to trigger the capturing of Local_Time, Cycle_Time, or Global_Time to a stopwatch register. This allows the clocking of events in a TTCAN network time base without interrupt response time jitter.

Event Trigger

The TTCAN protocol allows to synchronize the message schedule by inserting time gaps between basic cycles. The time gap is announced by the transmission of a Reference Message with the bit Next_is_Gap = 1. The gap starts at the end of the basic cycle initiated by that Reference Message. The gap is finished when the actual time master (or one of the backup time masters) starts a new Reference Message. This new Reference Message may be triggered by the application program. In a hardware implementation, it is possible to provide the trigger via an input pin, linking the message schedule directly to an external time reference.

Synchronization of several TTCAN Networks

Time mark interrupt, stopwatch, and event trigger can be used for the synchronization between application and TTCAN network as well as for the synchronization between different TTCAN networks.

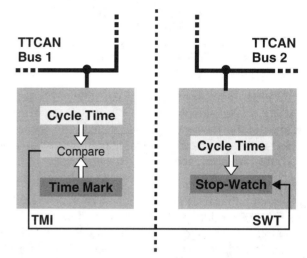

Figure 10: Synchronization of TTCAN Networks

When a node is connected to more than one TTCAN network, as a gateway node, it can measure the differences in the clock speed and in the phases of Cycle_Time and Global_Time by connecting the time mark interrupt output (TMI) of one TTCAN controller to the stopwatch input (SWT) of the other TTCAN controller (see figure 10). When this node is time master in one of the TTCAN networks, it can adjust clock speed and clock phase of that network until the synchronization is achieved. When the gateway node is not time master, it transmits the results of the measurements to the time master that will perform the synchronization.

A time master can adjust the network's clock speed by modifying its actual TUR value. In TTCAN level 2, the other nodes will synchronize their local clock speeds to the time master's clock speed.

The phase of Global_Time is adjusted by the time master by adding an offset to the actual value of Global_Time. This causes a discontinuity in Global_Time that is signalled to the other nodes in the network by setting the Disc_Bit = 1 in the Reference Message. This Reference Message will not be used for clock calibration.

The Cycle_Time's phases of two TTCAN networks are adjusted by inserting a time gap into the message schedule of one of the networks. The time gap is started by the time master sending a specific Reference Message. The message schedule is restarted at the occurrence of a specific event. This event may be an edge at the event trigger input of the TTCAN controller connected to the network to be synchronized that is driven by the time mark interrupt output of the other TTCAN controller.

Any number of TTCAN networks that are connected by (a chain of) gateway nodes can be synchronized that way, providing a common Global_Time for the whole system.

CONCLUSION

This paper describes the requirements for the implementation of the TTCAN protocol in hardware and in software, in both cases based on existing CAN controllers.

The TTCAN protocol provides several synchronization features, not only the calibration of the clock speed inside the TTCAN network, but also synchronization between application and TTCAN network and synchronization between different TTCAN networks.

TTCAN level 1 can be implemented in software based on a CAN controller supporting ISO 11898-1. The implementation of the full range of TTCAN features, including the global time and clock calibration, requires a dedicated hardware approach.

In parallel to the ISO standardization process, Bosch has implemented the TTCAN protocol (level 1 and level 2) into a CAN IP module [5] that performs all protocol functions in logic, not depending on software control and that provides time mark interrupts, a stopwatch, and an event trigger input. The module, synthesized into FPGAs, was used for research applications in support of the standardization.

The TTCAN IP module has also been implemented in silicon. The IC is available as samples in PLCC44 package and is called TTCAN_TC, the TTCAN testchip.

The TTCAN_TC supports both TTCAN level 1 and TTCAN level 2, its package is pin-compatible to existing standalone CAN controllers (Intel 82527 / Bosch CC770). It is currently used in the design of the first TTCAN network planning and analysis tools.

REFERENCES

1. ISO/DIS 11898-1; Road vehicles – Controller area network (CAN) – Part 1: Controller area network data link layer and medium access control.

2. ISO/CD 11898-4; Road vehicles – Controller area network (CAN) – Part 4: Time triggered communication.

3. Fault Tolerant TTCAN Networks; B. Müller, T. Führer, F. Hartwich, R. Hugel, H. Weiler, Robert Bosch GmbH; Proceedings 8th International CAN Conference; 2002; Las Vegas.

4. The Configuration of the CAN Bit Timing; F. Hartwich, A. Bassemir, Robert Bosch GmbH; Proceedings 6th International CAN Conference; 1999; Turin.

5. TTCAN User's Manual; Robert Bosch GmbH; 2001; http://www.can.bosch.com/docu/Users_Manual_TTCAN.pdf.

CONTACT

Florian Hartwich
Robert Bosch GmbH, AE/EIS
P.O.Box 13 42
72703 Reutlingen
Germany
Phone: +49 7121 35-2594
Fax: +49 7121 35-1746
E-mail: Florian.Hartwich@de.bosch.com

Thomas Führer
Robert Bosch GmbH, FV/FLI
P.O.Box 10 60 50
70049 Stuttgart
Germany
Phone: +49 711 811-7597
Fax: +49 711 811-7136
E-mail: Thomas-Peter.Fuehrer@de.bosch.com

Dr. Bernd Müller
Robert Bosch GmbH, FV/FLI
P.O.Box 10 60 50
70049 Stuttgart
Germany
Phone: +49 711 811-7053
Fax: +49 711 811-7136
Mail: Mueller.Bernd@de.bosch.com

Robert Hugel
Robert Bosch GmbH, FV/SLN
P.O. Box 30 02 40
70442 Stuttgart
Germany
Tel: +49 711 811 8517
Fax: +49 711 811 1052
E-mail: Robert.Hugel@de.bosch.com

ADDITIONAL SOURCES

http://www.can.bosch.com/content/TT_CAN.html

DEFINITIONS, ACRONYMS, ABBREVIATIONS

CAN: Controller Area Network
CTC: Cycle Time Controller
FSE: Frame Synchronization Entity
FPGA: Field Programmable Gate Array
GTU: Global Time Unit
IC: Integrated Circuit
IP: Intellectual Property
MSA: Master State Administrator
NTU: Network Time Unit
PLCC: Plastic Leaded Chip Carrier
SOF: Start of Frame
TBB: Time Base Builder
TM: Time Mark
TSO: Time Schedule Organizer
TTCAN: Time Triggered CAN
TUR: Time Unit Ratio

Sophisticated CAN on Embedded Microcontrollers for Smart In-Vehicle Real-Time Control Systems

Shigeo Uno, Naoyuki Hirayama, Akihiko Watanabe and Fumio Tsuchiya
Semiconductor & Integrated Circuits Group Hitachi Ltd.

ABSTRACT

In the past 10 years there have been a drastic evolution in car electronics. Most mechanically operated systems as like an engine control for example has switched to an electronic system to have more precise method to control the system and in some cases more cost effective. Due to the implementation of electronically controlled systems more and more information could be handled within these systems, and with more and more information there is a turning point that there is a large benefit to have a common use of the information. Now in-vehicle network is playing a major part in car electronics. CAN(Controller Area Network) is one of the most popular candidate for real time communications between the systems

This paper will describes the development concept of our new CAN module for embedded microcontrollers targeting automotive applications. Currently CAN have been used as one of the most popular control network systems. Due to expansion of data transactions via CAN, data handling by the module or by the controller has been a major subject. We have developed a method to handle the large amount of data with the least CPU(Central Processing Unit) overhead to provide maximum performance of the CPU. This will be shown with a simulation result in comparison of the new CAN module and the conventional CAN module, along with an example to show the efficiency in a application such as ACC(Adaptive Cruise Control) system. And furthermore this new CAN module will have the capability to cooperate with the latest requirements of the CAN specification such as the time triggered CAN of the ISO(International Organization for Standardization) standard.

INTRODUCTION

In the past decade there has been great progress in automotive control systems. The biggest motivation of the progress is the close-up on environmental preservation. And the biggest step forward was the introduction of electronic controlled systems. Sensors and actuators that are operated by ECU(Electronic Control Unit)s give the big advantage of more precise control and cost efficiency compared to the conventional mechanical control systems.

In today's high class passenger vehicles there are more than 50 ECUs existing to control each system. Each of these ECUs handles a certain amount of information to control the system. Although each ECU has a different role in the control system, there is lots of common information handled within the system. Therefore now in the current systems, data communication between the ECUs is playing a major role for advanced vehicle control systems. For data communication CAN has become the standard real time network system which supports the data transactions up to the speed of 1Mbits/s. Many real time network system have adapted the CAN because of its high speed capability and more over the high reliability of the data communication.

In the early stages of the CAN implementation, very few nodes (for example 2 to 3 nodes) were connected to the CAN bus, and also during these phases communication data that was transferred between the nodes were very few and not very frequent. But as the control systems grows and the control precision required in the systems gets higher, the data information increases and the data transaction speed will expand.

To solve these problems and cooperate with the expanding requirements the controller of the ECU needs to higher performance and in cases more intelligent data handling methods.

SYSTEM INTEGRATION TRENDS

As the vehicle control systems enhance, the information that is handled by the ECUs will automatically expand to achieve higher performance and higher precision. Introduction of additional sensors and actuators will expand the system and a large amount of common information to operate the additions above will exist within several systems. This common information could be utilized by centralizing the system. Figure 1 shows an example of the system integration for a powertrain control module. This is consisted by the engine management, automatic transmission and electronic throttle control.

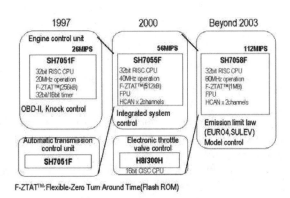

Figure 1 System integration of a powertrain control module

Integration of ECUs has been the major trend due to handling common information from the sensors and actuators in the 1990's. This integration will allow reducing the number of parts used in the total solution and also the reduction of the physical mounting area, which is also another critical factor in the limited space of an automobile.

But as the control systems and information expands, to harmonize lots of control applications such as engine management, AT control, ABS, ACC, etc. there is a point where system integration may not be the most effective solution. Figure 2 shows the trend of system integration and system distribution in control network. While common information continues to increase, the system integration advances and due to the help of advanced technology of semiconductors and peripheral components. One of the largest factors for the transition of system integration to distribution is the introduction of the now standardized network named CAN. CAN has been in the market for more than 15years and it's reliability has been proven on the market. CAN have become one of the major real time network systems. Figure 3 shows the trend of network nodes existing in one vehicle. As shown in this figure the increase of control application and information leads to more node connection for more interactive communication.

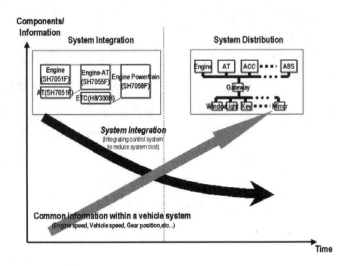

Figure 2 System integration & distribution

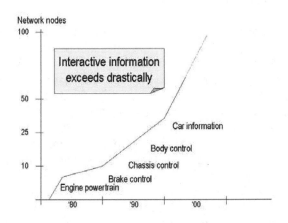

Figure 3 Number of network nodes in 1 vehicle

IMPLEMENTATION REQUIREMENTS

In recent real time applications, as mentioned above CAN is now a standard for car networking systems. Large numbers of information transactions are done via the CAN and currently in high class vehicles may have up to several hundreds of messages to obtain the high precision control of the vehicle and comfort. From this vast expansion, more and more performance is required for the CAN controller as well as the microcontroller. There are 3 key issues to drastically increase the performance of the microcontroller to gives the performance to concentrate on other control methods.

The 3 key implementations are :

- Sophisticated receive message handling

- Time triggered message transmission

- Time stamp functionality

Our newly developed HITCAN module will have the functionality of the 3 key items mentioned above. This will allow the microcontroller to handle a large amount of data with the least CPU overhead to provide the maximum performance of the CPU. Furthermore several other implementations have been done to give the user the opportunity to have the optimized functionality for real time applications. Table 1 shows the summary of the CAN implemented features and Figure 4 shows the hierarchical structure of our new HITCAN module.

Other key implementations:

- Thirty-two transmit/receive message mailboxes

- Non-read message overwrite function

Automatic re-transmission stop functionality

Table 1 HCAN implementation features

	Functions	HCAN	HITCAN(New CAN)
Mailbox	Numbers	16 mailboxes	32 mailboxes
	Transmit/Receive programable	Available	Available
Transmit functions	Message completion flag	Available(Each mailbox)	Available(Each mailbox)
	Message abort	Available(Each mailbox)	Available(Each mailbox)
	Time triggered data transmission	-	Available(Each mailbox)
	Time stamp at data transmission	-	Available(Each mailbox)
	Automatic remote frame respons	-	Available(Each mailbox)
	Automatic re-transmission stop	-	Available(Each mailbox)
Receive function	Filtering	1 LAFM	32 LAFM(Each mailbox)
		16 bit wise filtering	32 bit wise filtering
	Message overwrite detection	Available(Each mailbox)	Available(Each mailbox)
	Message overwrite abort	-	Available(Each mailbox)
	Time stamp at data reception	-	Available(Each mailbox)
Others	Time base counter	-	16bit free running counter
	Global sychronization	-	ID compare match register to clear timer counter

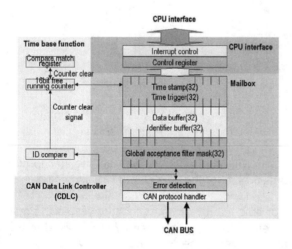

Figure 4 Hierarchical structure of the new HITCAN

SOPHISTICATED RECEIVE MESSAGE HANDLING

As number of messages handled within the CAN bus system increases, It is necessary to reduce the CPU process power to the minimum by having a sophisticated method to handle the receiving messages. Figure 5 shows the block diagram of how the HITCAN receives messages and stores it into the appropriate mailbox. Our HITCAN controller will provide 32 mailboxes which is capable to program receive or transmit on the fly. Each mailbox contains an independent identifier(standard & extended) area and data area for maximum 8 bytes. Furthermore each mailbox has its independent receive message filtering function with the full identifier range.

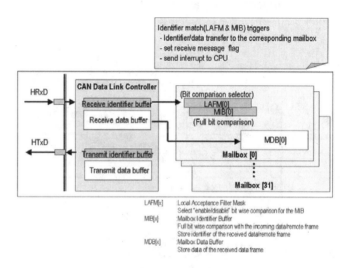

Figure 5 Block diagram of receive messages

For message receive a global mask filter and a bit wise filter is common. This HITCAN gives 2 filtering functionality for each of the 32 mailboxes. The first filter is called the Local acceptance filter and this allows setting the corresponding bit to be matched with the identifier buffer or not. For example if this is set to "1" this would mean a "must match" for the corresponding identifier buffer bit. And if "0" was set the receive identifier does not have to match the identifier buffer bit. This allows to receive a group of messages in certain mailboxes.

Once the identifier matches the condition of the LAFM and the identifier buffer this will trigger the transfer of the data value which is in the receive data buffer into the data buffer automatically. If the receive identifier structure is set to be optimized this will allow to have the minimum CPU load which could be used for other control which could be very important for real-time applications.

TIME TRIGGERED DATA TRANSMISSION

In the past, data transmission from the CAN has been realized by software tasks via the CPU. The bus arbitration will go into affective as soon as there is a request by the CPU. This method is very useful when the transferred message amount is fairly low and the data transaction system structure is an event driven system. But since the message transfer needs to be handled by the CPU, when the message amount increases and the data transaction system structure shifts to time base driven, this will directly effect the performance of the CPU. The performance will reduce drastically. Also as the side effect, if there is a necessity to transmit message in a critical time, on an event request and the CPU has other critical tasks or if the CPU load were heavy there would be a possibility to miss the critical time slot.

One solution to solve this problem is the introduction of the time trigger data transmission. Figure 6 shows the block diagram of the time triggered functionality.

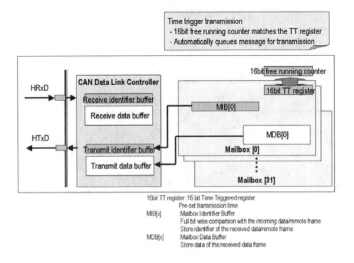

Figure 6 Time triggered function block diagram

This function consists of a independent 16bit free running counter which counts up on a internal clock which creates the time base, and in addition each individual mailbox has a register that consist of a compare match register. This compare match register determines the time when the corresponding mailbox is set for transmission. If the counter matches the compare match register the corresponding mailbox will be queued for transmission. At this point the information(identifier, DLC, data, etc.) which is stored in the mailbox previously is automatically queued for transmission. Therefore to update data which is set for periodic transmission at a pre-determined time slot, the CPU only needs to update the new data (DLC + data only) based on a task which does not have to relate to a CAN transmission task. This will allow the CPU to be free from a CAN transmission task and could perform other important tasks.

Also to secure the complete update of the new message content (to prevent half-new data and half-old data transmission) there is a secure bit which will indicated if the message is currently being updated or not. If this indication bit is inactive, if the compare match occurs this

postpones the transmission queue and will be delayed until the next compare match occurs.

Along with this feature, this module has a register, which contains an identifier compare register. This register is dedicated to allow a global synchronization. When a message with a specific identifier which matches the identifier compare match register this will send a reset signal to the free running counter. This will also allow a global base time and is possible to synchronize without any CPU intervention.

By this methodology, It Is able to achieve data transmission request with the least CPU overhead as well as an accurate periodical message transmission via the CAN bus.

TIME STAMP FUNCTIONALITY

Time trigger functionality is previously known functionality as a timekeeper to identify when the message has been transmitted or when it has been received.

This functionality could also be implemented using the 16bit free running counter. Figure 7 shows the methods of the time stamp functionality.

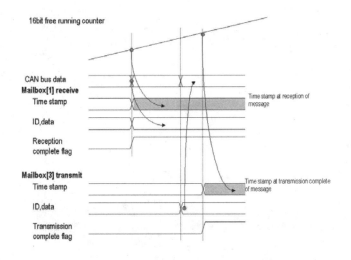

Figure 7 Time stamp functionality

Each individual mailbox has a 16bit register to store the counter value for the time stamp functionality.

As for time stamp for data reception, when the message is received the timer counter value is stored in the time stamp register of the corresponding mailbox. This allows to identify if the received data stored in the mailbox is up to date or not. This functionality would be useful when the CPU has a critical task other than the control of the CAN and polls the received data in the mailbox.

For data transmission, when the message is actually transmitted the timer counter value is stored in the time stamp register of the corresponding mailbox. This will also be useful to identify when the data actually has been sent on the CAN bus. In addition to this feature, if the data length of the message(data frame) is less than 6 bytes you have the possibility to capture the 16bit timer counter value to the last 2 bytes when the message is set for transmission queue. If the CAN system has been globally synchronized the receiver could identify when the data was queued for transmission and could possibly calculate the latency of the CAN. This functionality could be very useful in time critical systems.

IMPLEMENTATION EXAMPLES ON ACC

Here is an example of our HITCAN(Hitachi Intelligent Time triggered CAN) simulated in a real time application called ACC(Adaptive Cruise Control) which is a system that requires real time control. Figure 8 shows the diagram of the ACC control system.

ACC is a system to advance the conventional cruise control which controls the vehicle at a constant vehicle speed. In addition to the conventional cruise control, this new system detect the distance to the object ahead using a radar sensor system and gives command to the powertrain system (engine control, AT control and electrical throttle valve control) as well as the ABS system to control the vehicle to avoid collision with the object ahead or even acceleration to keep a certain distance with the obstacle ahead. The control information within this system is all transmitted or received via the CAN network which is the current trend of system distribution. This ACC system main processing unit runs on a 32bit RISC SH-2 microcontroller(40MHz) with a CAN interface with our new HITCAN.

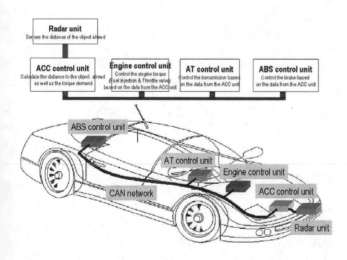

Figure 8 ACC control system diagram

The sequence of the control is based on every 20ms event to control the system. The ACC unit measures the desired driving speed via the radar by calculating the distance of the obstacle ahead which is located at the front of the vehicle. After the application program inside

ACC calculates the desired drive force and the ACC sends desired drive force to the engine control unit via the CAN network. The engine control unit invokes the corresponding controls to the brake control system or the throttle control system via the CAN network again. The time the brake system or the throttle system will start the execution every 20ms.

PERFORMANCE EVALUATION

Since the system needs to use a various number of messages within the CAN system, the CAN will need to handle data with the least CPU overhead. One of the most efficient methods to handle the incoming messages would be to use the message filtering method mention previously. As the new HITCAN has an individual Local Acceptance Filter Mask, it is capable to compare the identifiers of the receive messages without CPU intervention. Once the message has passed the LAFM it will be stored in the corresponding mailbox which is to reduce the times of compare to identify which message has been received.

For example, if it is able to group the identifiers to distribute the identifier comparison equally this could give the minimum average comparison time.

LAFM settings = message numbers / receive mailbox numbers

Using 16 of the 32 mailbox as receive, and group the messages of 128 to the 16 mailboxes, this would give approximately 90% reduction compared to the conventional HCAN method which you needed to compare the identifiers with a single LAFM mailbox. Figure 9 shows the simulated performance difference with the number of LAFMs.

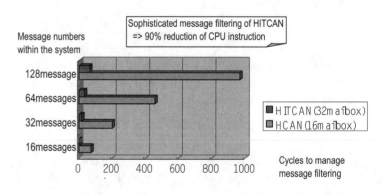

Figure 9 CPU load for message reception

As well as data reception, data transmission is also critical. To control the system accurately and give the driver immediate response of the transactions, the data transmission to the control systems via the CAN needs to time triggered rather than event driven. Figure 10 shows the periodic data transmission using the time triggered function. Using this functionality this will allow the CPU to run tasks independent to the 20ms data transmission to control the ACC system. To update a mailbox would take roughly 1.5us/message (using extended ID with 8 bytes of data) which the CPU load for would be approximately be 0.008%(Based on a 40MHz operation device).

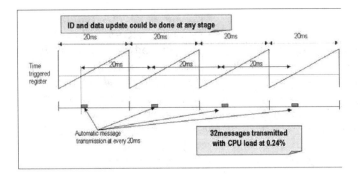

Figure 10 Time triggered function timing chart

This shows that the HITCAN will be capable to handle up to 32 messages keeping the CPU load down to 0.24%. This will allow the CPU to be 99.76% free and more than sufficient to handle the main application and other tasks.

Other than this control application, the CPU load is an inverse proportion to the task control frequency. As for a system with 5ms, 10ms task the CPU load would be 0.48% and 0.96%. As this shows a real-time network system could be operated without a large CPU overhead.

CONCLUSION

In vehicle real time application more and more electronic controlled systems will appear to contribute to the technology innovation. Networking is and will remain the key factor for control systems and CAN will remain state of the art.

The new developed HITCAN module offers a high functionality because of the 32 mailboxes functionality and also with the implemented of a timer gives you the capability to manage the new required time triggered CAN applications.

As simulation results using HITCAN in a real-time application, the CPU load for data reception was reduced 90% to the conventional HCAN and for data transmission the CPU load was held down to 0.24% in 32 message/20ms task application. This shows a very low CPU load of handling messages both data receive and data transmission and is suitable for real-time control network applications.

ACKNOWLEDGMENTS

This HITCAN module is planned to be lined up with our 32bit high performance SH microcontroller along with the 16bit H8S microcontrollers and provides a wide range of products to meet vehicle real-time control network application requirements.

REFERENCES

1. S. Suzuki, W. Nagaura, T. Imai, S. Kuragaki, T. Yokoyama, "A Distributed Control System Framework for Automotive Powertrain Control with OSEK Standard and CAN Network" in SAE 1999-0-1276
2. ISO 11898-1:Road vehicles – Interchange of digital information – Part 1: Controller area network data link layer and medium access control
3. ISO 11898-2:Road vehicles – Interchange of digital information – Part 2: High-speed medium access unit and medium dependent interface
4. CAN specification V2.0, 1991 Robert Bosch GmbH
5. S. Tanaka, M. Kaneyasu, A. Kodama, K. Matsubara "A New RISC Microcontroller with On-chip FPU to Introduce Adaptive Control into Powertrain Management" in SAE 1999-01-0865

CONTACT

Shigeo Uno
Hitachi Ltd.,
Automotive Semiconductor Marketing Department
Semiconductor & Integrated Circuits Group
Nippon Bldg., 6-2 Otemachi 2-chome
Chiyoda-ku, Tokyo Japan
Tel : +81-3-5201-5056
Fax : +81-3-3270-6254
E-mail : suno@denshi.head.hitachi.co.jp

DEFINITIONS, ACRONYMS, ABBREVIATIONS

ABS : Antilock Brake System
ACC : Adaptive Cruise Control
AT : Automatic Transmission
CAN : Controller Area Network
DLC : Data Length Code
HCAN : Hitachi CAN
HITCAN : Hitachi Intelligent Time triggered CAN
ID : Identifier
ISO : International Organization for Standardization
LAFM : Local Acceptance Filter Mask
RISC : Reduced Instruction Set Computer

Time Triggered CAN (TTCAN)

Thomas Fuehrer, Bernd Mueller, Florian Hartwich and Robert Hugel
Robert Bosch GmbH

ABSTRACT

Connecting microcontrollers, sensors and actuators by several communication systems is state of the art within the electronic architectures of modern vehicles. The communication among these components is widely based on the event triggered communication on the Controller-Area-Network (CAN) protocol. The arbitrating mechanism of this protocol ensures that all messages are transferred according to the priority of their identifiers and that the message with the highest priority will not be disturbed. In the future some mission critical subnetworks within the upcoming generations of vehicle systems, e.g. x-by-wire systems (xbws), will additionally require deterministic behavior in communication during service. Even at maximum bus load, the transmission of all safety related messages must be guaranteed. Moreover it must be possible to determine the point of time when the message will be transmitted with high precision.

One way to solve this issue using CAN is the extension of the standard CAN protocol to a time triggered protocol TTCAN. The communication is based on the periodic transmission of a reference message by a time master. This allows to introduce a system wide global network time with high precision. Based on this time the different messages are assigned to time windows within a basic cycle. A big advantage of TTCAN compared to classic scheduled systems is the possibility to transmit event triggered messages in certain "arbitrating" time windows as well.

INTRODUCTION

TTCAN is a higher layer protocol above the unchanged standard CAN protocol. It synchronizes the communication schedules of all CAN nodes in a network and it provides a global system time. When the nodes are synchronized, any message can be transmitted at a specific time slot, without competing with other messages for the bus. Thus the loss of arbitration is avoided, the latency time becomes predictable. Apart from the synchronized communication schedule the TTCAN nodes operate according to the standard ISO 11898 [6, 7]. The time triggered extension "TTCAN" is

accepted in TC22/SC3/WG1/TF6 (ISO11898-4) [8] after standardization work in 2000. In parallel to the standardization process, Bosch has implemented the time triggered communication function into a CAN IP module [9] that maintains the cyclic transmission schedule autonomously, not depending on software control.

SYSTEM REQUIREMENTS

Due to the present system structure where microcontrollers, sensors and actuators are connected by a communication system, a careful design becomes more and more necessary. To overcome the increasing complexity of these systems, a deterministic behavior of the communication network must be provided, in particular when dealing with distributed functionality or redundant realization of certain nodes. This can be achieved by using the design philosophy of time triggered operation [1] at the communication network (time triggered protocol activities [2]) and at the application level (time triggered task activation [3]). The optimum is reached if a globally synchronized time base (global time) is available at all nodes of the network with a precision which fulfills the real-time requirements of the application.

Several systems will require this behavior of the communication system, e.g. x-by-wire systems (xbws), motor management systems and sensor subnetworks. Within the automotive domain, xbws will be one of the most mission critical systems. They control the vehicle and its dynamics while the input from the driver, e.g. steering wheel, to the system is mechanically and/or hydraulically de-coupled from the physical transaction on the road, e.g. road wheels. The communication network is the back-bone of these by-wire applications [4] and often has to be redundant as well. The first generations of xbws will still have mechanical/hydraulic backup, but determinism of the message transfer, the service of a global network time and redundancy at the timing services will already be important as well as for various systems supporting active safety in vehicles, e.g. electronic stability program (ESP).

But also several designers from other industry domains, e.g. automation industry or medical service industry build many systems based on CAN and will welcome the additional features of TTCAN improving the determinism of their networks.

TIME TRIGGERED OPERATION AND THE STANDARD CAN

One of the most powerful features of the CAN protocol [6, 7] is the bitwise arbitration to control the media access among the controllers of the network. The bitwise arbitration guarantees a controller with a high priority message to access the bus even if other controllers try to access the media, without destroying any message. The access may be delayed if some other message is already in the process of transmission or if another message with higher priority also competes for the bus. This means that even the temporal behavior of the message with the highest priority may show a small latency. The lower the priority of a message is, the higher the latency jitter for the media access may be [10].

If the media access should be controlled by a time triggered mechanism, any protocol activity of the communication system is determined by the progression of a (globally synchronized) time. Sending, receiving, or any other activity depends on a predefined time schedule and on the current state of the clock as it is shown in Figure 1. Message "a" is sent if the system clock reaches 3 and 6 while message "b" is sent at 5. If the whole communication traffic is summarized in such a time table, a deterministic and predictable communication matrix results. The necessary information can be mapped into each node within the network. This results in a highly composable system in the domain of time and value [5]. In principle, no arbitration mechanism would be necessary.

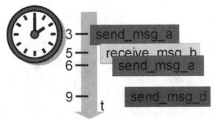

Figure 1 - Principle of time triggered activation

If time triggered operation on CAN is supported the latency jitters can be reduced and a deterministic communication pattern on the bus can be guaranteed. Moreover, this allows to use the physical bandwidth of a CAN network much more efficiently (under the constraint of determinism).

TTCAN LEVEL 1 AND LEVEL 2

Within the extension work for time triggered execution on CAN within ISO 11898-4, the listed features are going to be realized in two levels. Extension level 1 guarantees the time triggered operation of CAN based on the ref-

erence message of a time master. Fault-tolerance of that functionality is established by redundant time masters – the so called potential time masters. In extension level 2, a globally synchronized time base is established and a continuous drift correction among the CAN controllers is realized.

TIME TRIGGERED CAN - TECHNICAL DESCRIPTION

THE REFERENCE MESSAGE

TTCAN is based on a time triggered and periodic communication which is clocked by a time master's reference message. The reference message can be easily recognized by its identifier. Within TTCAN's level 1 the reference message only holds some control information of one byte, the rest of a CAN message can be used for data transfer. In extension level 2, the reference message holds additional control information, e.g. the global time information of the current TTCAN time master. The reference message of level 2 covers 4 bytes while downwards compatibility is guaranteed. The remaining 4 bytes are open for data communication as well.

THE BASIC CYCLE AND ITS TIME WINDOWS

The period between two consecutive reference messages is called the basic cycle (see Figure 2). A basic cycle consists of several time windows of different size and offers the necessary space for the messages to be transmitted.

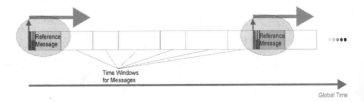

Figure 2 - The Reference Message starts the TTCAN basic cycle.

The time windows of a basic cycle can be used for periodic state messages and for spontaneous state and event messages. Any message that is sent has the CAN data format and is a standard CAN message. A time window for periodic messages is called an exclusive time window. Within exclusive time windows the beginning of the time window determines the sending point of a predefined message of a node. If the system was properly specified, e.g. with the help of an off-line design tool analyzing the communication pattern, no conflicts will happen. However, even in the error case of a conflict, the CAN protocol properties (bit arbitration, only sending when the bus is idle) are valid. The system engineer has to decide off-line which message must be sent at which exclusive time window. To provide higher flexibility to the system designer, an exclusive time window may be repeated more than once within a basic cycle. The automatic retransmission of CAN messages is not allowed in exclusive time windows.

A time window for spontaneous messages is called an arbitrating time window. Within an arbitrating time window, the bitwise arbitration decides which message of which node in the TTCAN network will succeed (see Figure 3) on the bus. At design time it is allowed to schedule more than one message for an arbitrating time window. So the application can decide at runtime if it would like to use an arbitrating window for a message to be sent and which message should be sent in a certain arbitrating window. The automatic retransmission of CAN messages is not allowed within an arbitrating time window.

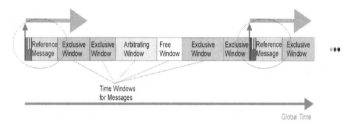

Figure 3 - Exclusive time windows and arbitrating time windows of a TTCAN basic cycle.

During the design phase it is also possible to reserve free time windows for further extensions of the network. They can be changed to arbitrating or exclusive time windows if new nodes need further space for communication or the bandwidth has to be extended for existing nodes.

NODE SPECIFIC KNOWLEDGE OF A TTCAN NODE

It is important for any hardware realization of a protocol, that the necessary information is kept at a minimum. In TTCAN a network controller does not have to know the complete time schedule of all messages. The controller only gets the necessary information it needs for time triggered sending and receiving of messages as well as for sending of spontaneous messages. E.g., if the controller is only interested on the reception of message A in a certain exclusive time window, then it only gets this specific knowledge as well as it gets only the time information for sending its own messages in exclusive and/or arbitrating time windows. This design principle allows highly optimized memory utilization in hardware but offers still enough information for network management, e.g. within OSEKTime's FTCOM [3]. Moreover, this provides very high flexibility during development as changes in the schedule and the change of participating controllers imply a new download of the schedule only for the affected controllers.

THE SYSTEM MATRIX

The ongoing system designs of future embedded and distributed systems try to cover several control applications. The associated control loop design requires different timings and different message latencies – several time schedules result. TTCAN covers this topic by assembling these time schedules within the mechanism of the basic cycle and the system matrix

(matrix cycle). Several basic cycles are connected to build the matrix cycle. This allows to combine multiple sending patterns, e.g. sending every basic cycle, sending every second basic cycle, or sending only once within the whole system matrix. An example is shown in Figure 4.

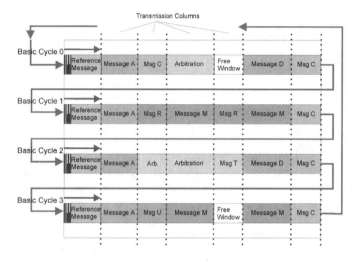

Figure 4 - TTCAN's System Matrix Mechanism

MERGED ARBITRATING WINDOWS

TTCAN specification allows another useful mechanism – merged arbitrating windows. It often makes sense to ignore the columns of the system matrix in the case of two or more arbitrating time windows in series (see Figure 5).

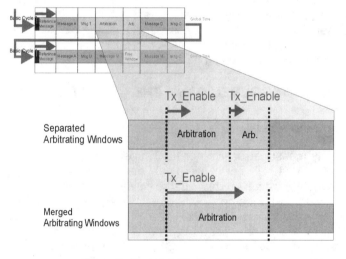

Figure 5 - Merged Arbitrating Windows

The arbitrating time windows are merged to a longer time window where bitwise arbitration can take place for several spontaneous messages. The most important constraint for this construct is that it is not allowed to start an spontaneous message within this merged arbitrating window if it will not fit in the rest of the time window. The start of the next periodic time window must be guaranteed. The automatic retransmission within a merged arbitrating time window is allowed as long as the constraint already described above is satisfied.

TIME MASTER AND POTENTIAL TIME MASTERS

As the time master plays a vital role within the TTCAN approach, fault-tolerance of this functionality must be established. This is done by predefining more than one TTCAN controller to be a potential time master. They get corresponding identifiers which can be used to control the startup behavior. After reset, a potential time master checks if there is already traffic on the bus and if there is already a reference message sent. If not, the potential time master sends a reference message with its identifier and, in TTCAN level 2, with its local time as the very first global time of the network. This TTCAN controller assumes to be time master. Whenever a reference message with a higher priority is received, the potential time master stops sending the reference message and synchronizes to the basic cycle given by the higher priority time master. Whenever a reference message with a lower priority is received, the potential time master first synchronizes to the existing basic cycle and then tries to become time master by sending its own reference message at the start of the next basic cycle. Due to higher priority it will win the arbitration. Hence the protocol mechanisms ensure that out of all error free potential time masters the one with the highest priority eventually becomes active time master without violating the structure of the basic cycles. This state then is stable as long as no errors occur.

During operation a missing reference message is recognized by all potential time masters within short latency. The latency is realized by a timeout. After this timeout is reached all other potential time masters start sending the reference message. Again, the bitwise arbitration of the standard CAN protocol decides among competing potential time masters.

TIME TRIGGERED CAN – THE OPERATION

TTCAN CYCLE TIME

Any message received or transmitted invokes a capture of the local time taken at the message's frame synchronization. This frame synchronization event occurs at the sample point of each start of frame (SOF) bit and causes the local time to be loaded into the synchronization mark. Whenever a valid reference message was detected, the synchronization mark becomes the valid reference mark. The difference between this reference mark and the local time is the cycle time of TTCAN which drives the protocol execution within a basic cycle. Figure 6 describes the generation of the cycle time in a more implementation specific view [12]. The cycle time results out of the difference between the current local time and the last valid frame synchronization (Sync_Mark) of a reference message (Ref_Mark).

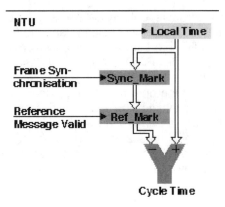

Figure 6 - TTCAN Cycle Time generation

TTCAN TIME MARKS - SENDING AND RECEIVING OF MESSAGES

The necessary link between the cycle time and the system matrix to make the protocol execution run are the so called time marks. They specify the beginning of the exclusive and arbitrating time windows. Time marks for sending periodic or spontaneous messages are called TxTriggers. RxTriggers have to be defined to check the reception of a periodic message.

A time mark furthermore consists of the cycle offset and the repeat factor information. The cycle offset determines the number of the first basic cycle after the beginning of the matrix cycle in which the message must be sent/received. The repeat factor determines the number of basic cycles between two successive transmissions/receptions of the message (see Figure 4).

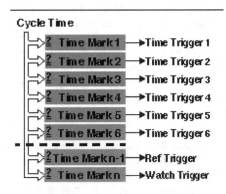

Figure 7 - Scheduling based on the cycle time

Figure 7 shows the principle of the time triggered protocol execution based on the cycle time. If the cycle time equals a certain predefined value of a time mark, sending or receiving of a message happens (time mark 1 to 6). In the figure above, two additional time marks are defined. The time mark which activates the reference trigger is used to start the next basic cycle at the time master. The time mark which activates the watch trigger is used to supervise the correct functionality of the actual time master. If this time mark is reached, any potential time master will have started sending its reference message to keep TTCAN alive.

GENERATION OF THE NETWORK TIME UNIT (NTU) IN TTCAN

An important property of the cycle time is its granularity. The granularity of any timing information within TTCAN is the network time unit (NTU). So the cycle time is measured in NTU and is based on the nominal CAN bit time in TTCAN level 1 (see chapter about time triggered communication and TTCAN's Level 1 and Level 2) and on the physical second in level 2. In level 2, to establish a system wide NTU, the node local relation between the physical oscillator of a TTCAN controller and the system wide NTU has to be established.

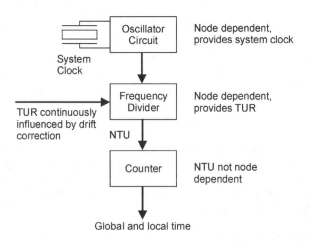

Figure 8 - Network Time Unit and Time Unit Ratio

Figure 8 demonstrates one possible principle for NTU generation. The node dependent oscillator circuit provides the system clock to a frequency divider. This frequency divider generates the system wide NTU while a node local time unit ratio (TUR) takes care for the correct relation between the system clock and NTU. NTU now can be used to build a local time (see also Figure 6) and to build the global time.

GLOBAL TIME IN TTCAN

The node sending the reference message is the time master of the TTCAN network. In TTCAN level 2 all nodes take a snapshot of their time values at the frame synchronization pulse (see Figure 6). The time master sends its (by definition correct) local time value for this frame synchronization pulse as part of the reference message to become the global time. After reception, each node can build its local offset as the difference between the master global time snapshot value and its own local time snapshot value. During the next basic cycle the node can compute the global time by adding the measured local offset to the current local time. Figure 9 shows the generation of the global time in a more implementation specific way. The local offset is derived from the measured values of the master reference mark in Ref_Mark and the transmitted reference mark by the master in Master_Ref_Mark.

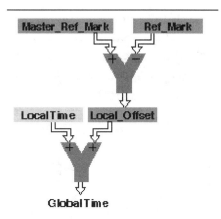

Figure 9 - Generation of the global time

DRIFT CORRECTION MECHANISM

If local time and global time have the same speed this ensures that all nodes have a consistent view on the global time. Due to slightly different clock drifts of the different nodes, a mechanism has to be introduced to guarantee that local and global time have in fact the same speed. This mechanism is the continuous update of TUR (see also Figure 8). An initial value of TUR, depending on the local oscillator specification, is a priori known by each node. During operation, to adapt this value to the correct value determined by the master clock speed, the node measures the length between two successive frame synchronization pulses both locally (number of oscillator periods in this interval) and in global time (difference between the two master snapshot values). The quotient of these two values gives the actual TUR (limited only by the precision of the measurement). The achievable precision determines a reasonable [11] choice of the NTU-value in physical seconds. In level 2 the global time values of two nodes will then not differ by more than one NTU. The NTU typically will be in the order of a CAN bit-time.

TTCAN IMPLEMENTATION

In parallel to the standardization process and the functional specification of TTCAN, Bosch develops an IP module that implements the TTCAN protocol including level 2. This TT_CAN module is based on the existing C_CAN IP Module [7] and will be available as VHDL code to be synthesized in FPGAs, supporting the development of CAN based time triggered communication networks.

TTCAN EXTENSIONS

For a TTCAN realization the C_CAN IP module with the already existing components (CAN Core, Message RAM, Message Handler, Control Registers, and Module Interface) are extended by two functional blocks (see Figure 10) – the Trigger Memory and the Frame Synchronization Entity (FSE).

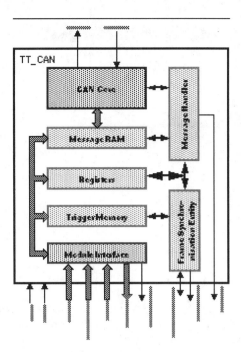

Figure 10 - Block Diagram of the TT_CAN IP Module

TRIGGER MEMORY

The Trigger Memory stores the time marks of the system matrix that are linked to the messages in the Message RAM; the data is provided to the Frame Synchronization Entity.

FRAME SYNCHRONIZATION ENTITY

The Frame Synchronization Entity is the state machine that controls the time triggered communication. It synchronizes itself to the reference messages on the bus, controls the cycle time, and generates time triggers. It is divided into five blocks, the Time Base Builder (TBB), the Cycle Time Controller (CTC), the Time Schedule Organizer (TSO), the Master State Administrator (MSA), the Application Operation Monitor (AOM), and the Global Time Unit (GTU).

Time Base Builder

The Time Base Builder generates the local time from the node's local clock and the time unit ratio (TUR). In TTCAN level 1, the TUR is defined at configuration, in level 2, it is continuously adapted by the drift correction mechanism in the Global Time Unit (GTU).

Cycle Time Controller

The Cycle Time Controller gets the local time from the TBB, the frame synchronization events from the CAN Core, and the reference messages from the CAN Message Handler. The task of CTC is furthermore to generate the cycle time and to control the sequence of basic cycles.

Time Schedule Organizer

The Time Schedule Organizer maintains the message schedule inside a basic cycle and checks for scheduling errors. The schedule is defined by the data in the Trigger Memory. The data consists of the time mark and function information (trigger for transmission or for checking reception of messages) and is linked to a message in the CAN Message RAM.

Master State Administrator

The Master State Administrator controls the state machine within the Frame Synchronization Entity. The operating state depends on whether the node is synchronized to the network, whether it is time master or whether it is a backup time master. In case of errors, transmissions are disabled and the master state is resigned.

Application Operation Monitor

The Application Operation Monitor checks the function of the application program. The application controller has to serve a watchdog regularly. If the application fails, the MSA disables the transmissions, preventing invalid data to disturb the system.

Global Time Unit

The Global Time Unit only exists in TTCAN level 2 and generates the node's view of the global time (GT) and controls the drift correction of the local time. Time base errors are signaled to the MSA, causing it to stop all TTCAN operations.

CONCLUSION

This document describes the extension of the CAN protocol by a time triggered execution for periodic messages while the possibility for spontaneous messages still remains. The time triggered communication is based on a time master. TTCAN furthermore provides fault tolerance of this time master functionality and even offers a network wide synchronized global time.

Due to the time triggered approach the communication structure in TTCAN is deterministic and hence seems more suitable to fulfill requirements of future applications and system architectures in vehicles while still maintaining flexibility during development as well as in use. TTCAN also supports the needs of other industries, where the time triggered approach, determinism and global network time is required, e.g. medical services and automation industry. Furthermore, TTCAN still allows the use of the CAN based monitoring and analyzing tools. The technical knowledge of the engineers about CAN is still vaild and only has to be updated for the TTCAN extensions. The current specification of TTCAN does not cover all requirements of safety related distributed systems (redundancy or data transmission rate), e.g. for dry xbws. Further development steps concerning these characteristics are checked in the context of the ISO TC22.

REFERENCES

1. The Time-Triggered Approach to Real-Time System Design; H. Kopetz, TU-Wien;
2. TTP – A Protocol for Fault-Tolerant Real-Time Systems; H. Kopetz, G. Grünsteidl; IEEE Computer; January 1994, pp. 14-23.
3. OSEKTime: A dependable Real-Time Fault-Tolerant Operating System and Communication Layer as an enabling Technology for By-Wire Applications; St. Poledna, M. Glück, Ch. Tanzer, TTTech; S. Boutin, Renault; E. Dilger, Th. Führer, Robert Bosch GmbH; Ch. Ebner, BMW Technik; E. Fuchs, DeComSys; R. Belschner, B. Hedenetz, DaimlerChrysler; B. Holzmann, A. Schedl, BMW AG; R. Nossal, B. Pfaffeneder, Siemens AG; Th. Ringler, TU-Stuttgart; Y. Domaratsky, A. Krüger, Motorola; A. Zahir, ETAS GmbH; SAE 2000, Detroit, Michigan, U.S.A.
4. The Steer-By-Wire prototype implementation: Realizing time triggered system design, fail silence behavior and active replication with fault-tolerance support; Th.Führer, Robert Bosch GmbH; A. Schedl, DaimlerChrysler; SAE 1999, Detroit, Michigan, U.S.A.
5. Composability in the Time-Triggered Architecture; H. Kopetz, TU-Wien; SAE 2000, Detroit, Michigan, U.S.A.
6. Road vehicles – Controller area network (CAN) – Part 1: Controller area network data link layer and medium access control; ISO 11898-1.
7. Road vehicles – Controller area network (CAN) – Part 2: High-speed medium access unit; ISO 11898-2.
8. Road vehicles – Controller area network (CAN) – Part 4: Time triggered communication; Working Draft ISO 11898-4.
9. C_CAN User's Manual; Robert Bosch GmbH; 2000; http://www.bosch.de/de_e/productworld/k/products/prod/can/docu/Users_Manual_C_CAN.pdf.
10. Guaranteeing Message Latencies on Controller Area Network (CAN); K. Tindell, A. Burns; Proceedings 1st International CAN Conference; 1994; pp 2-11.
11. Real-Time Systems: Design Principles for Distributed Embedded Applications; H. Kopetz; Kluwer Academic Publishers; 1997; ISBN 0-7923-9894-7.
12. CAN Network with Time Triggered Communication; F. Hartwich, B. Müller, Th. Führer, R. Hugel, Robert Bosch GmbH; Proceedings 7th International CAN Conference; 2000; Amsterdam.

CONTACT

Dipl.-Ing. Thomas Führer
Robert Bosch GmbH
P.O.Box 10 60 50
70049 Stuttgart
Germany
Phone: +49 711 811-7597
Fax: +49 711 811-7136
Email: Thomas-Peter.Fuehrer@de.bosch.com

Dr. Bernd Müller
Robert Bosch GmbH
P.O.Box 10 60 50
70049 Stuttgart
Germany
Phone: +49 711 811-7053
Fax: +49 711 811-7136
Email: mueller.bernd@de.bosch.com

DEFINITIONS, ACRONYMS, ABBREVIATIONS

CAN: Controller Area Network
CTC: Cycle Time Controller
FPGA: Free Programmable Gate Array
GTU: Global Time Unit
IP: Intellectual Property
LT: Local Time
MSA: Master State Administrator
NTU: Network Time Unit
SOF: Start of Frame
TBB: Time Base Builder
TSO: Time Schedule Organizer
TTCAN: Time Triggered CAN
TUR: Time Unit Ratio
XBWS: X-By-Wire System

TwinCAN – A New Milestone for Inter-Network Communication

Jens Barrenscheen and Patrick Leteinturier
Infineon Technologies

ABSTRACT

Recent trends in field bus applications, especially in the automotive section, show a very high demand for data exchange between decentralised, intelligent functional units and modules. These functional units can be grouped together to power train applications or body/convenience applications.

In many cases, the coupling of local modules is done with one or more independent bus systems. The actual design and the partitioning of the modules strongly depend on application-specific requirements, such as the total amount of data to be transferred or the maximum of the tolerated latency in data delivery.

A very powerful and fast field bus is the CAN bus (Controller Area Network), which supports transfers with data rates up to 1 Mbits/s. Due to the higher transmission speed and the standardized functionality, CAN is a very interesting alternative to and improvement on bus systems based on other protocols.

In order to meet the specific requirements of the large variety of applications, the CAN module's internal structure has to have wide ranging flexibility as well as be adapted to this special communication task. An optimised integrated device, such as the Infineon TwinCAN module, can be seen as a cheap and efficient solution. Microcontrollers with integrated peripherals supporting the CAN protocol can be found on the market from most semiconductor manufacturers.

INTRODUCTION

Commonly, automotive applications dealing with a large amount of coupled functional units with different tasks contain more than one independent bus system. This structure allows the manufacturer to regroup the functional units and to optimize each communication channel according to application-specific requirements. Most manufacturers in the automotive field offer systems with two or more independent CAN buses, especially in the mid-range and high-class range.

One of the buses is typically built as a high-speed bus (500kbits/s to 1Mbits/s) to control the power train modules, such as the engine management unit, the injection system or the ignition system. Typical examples for slower modules (50kbits/s to 250kbits/s) can be found in body/convenience applications, such as electrical seat positioning systems, heating/ventilation, air conditioning or door modules.

CAN bus systems allow for easy interconnection of the required functional units (ECU) due to the built-in multi-master capability. One important advantage of CAN-based systems is the easy communication between the modules, because all CAN chips on the market can communicate with each other thanks to the standardized CAN protocol. An example from the automotive section is shown in figure 1.

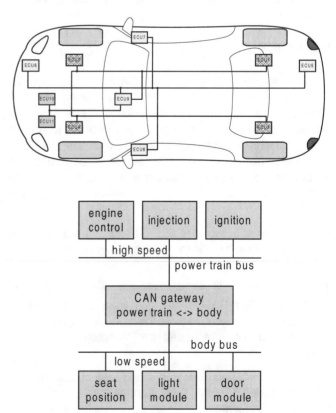

Figure 1.   CAN bus in automotive applications

The modules, which are participating in the data exchange via the CAN bus are called CAN nodes. The total number of CAN nodes on each bus depends on their implementation and the connected functional units and is not limited by the protocol.

MAIN TASK

In most automotive applications, the separate bus systems are not totally decoupled. Some functional units might need application-specific parts of the data from another bus, such as reference values or values actually measured. This data, which may be generated or stored in other modules, may then be transferred via the different buses.

Therefore, interconnection and data exchange between the buses are required. In order to reduce the overall bus load, only selected data has to be transferred from one bus system to another (selection as a function of the identifier). This data transfer can be done between two CAN nodes, one on each bus, see figure 1. In this example, a gateway functionality allows for data transfer between one node on the power train bus and another node on the body bus.

The gateway module itself can be implemented as a standalone module without any additional functionality or can be an integrated part of a functional unit. In the latter case, the module can be easily used for data transfer between both buses and for the normal communication task of the functional unit, for example the dashboard controller.

HIERARCHICAL APPROACH

In the internal structure of CAN nodes, several hierarchical levels (referring to OSI levels) can be distinguished, see figure 2. The physical bus line itself represents the lowest of these levels. In order to ensure the transfer characteristics with low error rates, the bus implementation can be adapted to the application requirements and to the environment. In general, copper wires can be used as well as fiber optics to connect the modules.

In many cases, a two-wire bus is used in the automotive field. The wires can be a twisted pair and can, as an option, be shielded.

The connection of the CAN controller devices, such as the Infineon TwinCAN to the bus lines is achieved via transceiver chips, converting the voltage levels on the bus into signals, which can be interpreted as logical levels by the control logic.

The communication between the protocol part of the CAN controller and the CAN transceiver is done via a read line and a write line.

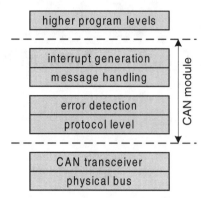

Figure 2.   Hierarchical structure of a CAN node

The low-level functionality of the protocol level (such as bit timing and bit stuffing) and the error detection and handling are defined in the standardized CAN protocol, which is common for all CAN devices.

The data link layer, which is responsible for message handling, is one of the key features of a CAN controller. This layer provides functionality such as the acceptance filtering in Full-CAN devices and message buffering. The higher levels are then normally handled by software, based on the status information delivered by the CAN controller hardware.

On the level of the data link layer, incoming messages on the bus are automatically checked by hardware to determine whether they have to be received by the node and whether the data has to be passed to the higher program levels.

So called message objects are set up inside the CAN device to receive and to send the desired data. As a consequence, the overall CPU load significantly decreases, because the message handling is done by hardware.

In applications with a low data flow, a basic automatic message handling capability of the CAN controller is sufficient to ensure communication without data loss.

In highly sophisticated CAN bus systems, the requirements for the message handling are greater. The Infineon TwinCAN device has been designed to meet the requirements of communication systems based on multiple buses. In order to support two independent CAN buses, the protocol level and the error handling have been duplicated for bus A and for bus B.

The message data link layer comprises the message handling capability (acceptance filtering), an automatic buffer management unit to deal with bursts of messages, a special gateway functionality to exchange selected messages between the nodes and a sophisticated interrupt generation unit.

The hierarchical structure of the TwinCAN device is shown in figure 3.

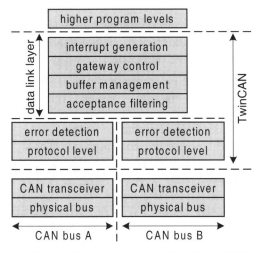

Figure 3.   Hierarchical structure of the TwinCAN

TWINCAN: ONE MODULE FOR TWO CAN NODES

The TwinCAN module controls two independent CAN nodes, which have been designed to reduce the amount of additional hardware and software required for the interconnection of two independently running CAN buses. The optimized, built-in data processing capabilities of the module result in an improved functionality. Furthermore, the CPU load in complex applications is lower than in conventional systems due to the specific hardware support. Particular emphasis is placed on the time-critical processes of transmitting and receiving messages on both buses with the associated message handling, as well as on their analysis.

In applications with two separate CAN buses exchanging data, the built-in, freely programmable gateway function between two buses is a decisive improvement. The automatic forwarding of selectable CAN messages from one bus to the other significantly reduces the CPU load and improves the real time capability of the entire system.

All data, control and status information related to a CAN message are grouped together inside the CAN module to so-called message objects. The number of usable message objects has been increased to 32, which can be independently configured, comprising the identifier, the data and all control information. Standard 11 bit long, as well as the extended 29 bit long identifiers are supported by the module (according to CAN specification V2.0B active) and can be used without constraints on both buses.

Furthermore, to improve the acceptance filtering and the allocation of the message objects, an individually programmable acceptance mask for each message object has been introduced. By means of this structure, complex groups of message identifiers can be received in the message object buffers.

The acceptance filtering is described in figure 4. Parts of the received messages are compared to the contents of the message objects with matching node information ("NODE"), identifier length ("XTD") and data transfer direction ("DIR").

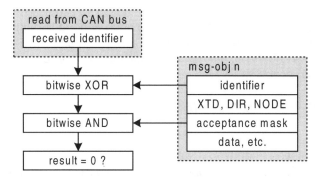

Figure 4.   Acceptance Filtering

The received identifier is compared bitwisely to the programmed identifier in the message object and is then combined by a bitwise logical AND with the programmed acceptance mask ("0"= don't care, "1"= take into account). Only if all bits of the result are zero has the matching message object been found and is selected for reception. If the result is not zero, the next message object is checked.

If all message objects have been checked without finding a match, the received CAN frame was not targeting this CAN module.

Each message object contains an independent "NODE" bit, selecting which CAN bus the message object is assigned to, see figure 5. In a system with two separate CAN buses connected to the TwinCAN module, each message object can be individually assigned to one of the buses, e.g. a "0" selects node A and a "1" selects node B. The selection can be reprogrammed during runtime, while the corresponding message object is disabled.

As a consequence, the number of objects connected to the buses can be easily changed by software allowing flexible and application-specific repartitioning of the system while it is running. The number of message objects assigned to the nodes is programmable and not fixed by hardware. Variable system partitionings, such as 16/16, 10/22 or 28/4 can be supported by the device.

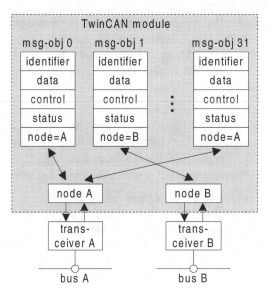

Figure 5.   TwinCAN node selection

249

Both CAN protocol layers (called node A, node B in fig. 5) can be programmed individually and as a result, the baud rates and the bit timing can be adjusted independently for both buses.

BUFFER MANAGEMENT

One of the key features in CAN controller devices is the message handling capability. In order to increase functionality and flexibility and to increase the tolerated latency for CPU actions, the module provides the possibility to concatenate a programmable number of message objects (out of the 32 available) to one FIFO buffer (First In First Out). This leads to structures with up to sixteen independent FIFO buffers (if it is always the case that two message objects are combined as two-stage FIFOs), or one 32-stage FIFO buffer.

Each FIFO buffer is organised as a circular buffer, which is automatically controlled by the module for CAN-related actions, whereas the main CPU (in the system) can always access the message objects individually.

The complete data handling and interrupt features of each message object remain available. In general, all message objects can be used as elements of FIFOs. Each FIFO may contain 2, 4, 8, 16 or all 32 objects. Figure 6 shows a 4-stage FIFO, which is connected to a CAN bus via the corresponding protocol layer.

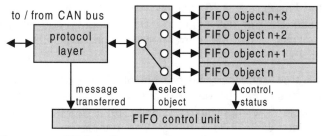

Figure 6. FIFO buffer control structure

The FIFO control unit is charged with providing the access to the message object by the CAN protocol layer, according to the circular buffer rules and the programmed FIFO length. The corresponding control and status information is held inside the message object. The interrupt generation works independently from the FIFO structure.

As an example, a four-stage FIFO is initialised with receive message objects, starting with the message object number n. The first received CAN message with matching identifier will be stored in this object and then the next message object (number n+1) will be targeted.

In the case that the CPU has to be informed that three CAN messages have been correctly received, the interrupt generation for a correct message reception has to be enabled for the message object n+2. After the correct reception of the third CAN message and the automatic storage in message object n+2, the CPU will

be interrupted and can deal with the block of messages already received instead of reading the messages one by one directly after their reception.

If the system CPU is not capable of dealing with each independent CAN message transfer due to real-time requirements, a receive FIFO allows the system to work even with bursty message transfers. The FIFO size has to be adapted correctly to the message transfer characteristics (like bursts) on the CAN bus in order to avoid data loss.

DATA TRANSFER BETWEEN THE NODES

A quite common task in this application field is the transfer of messages from one bus (source bus) to a second bus (destination bus). In order to minimise the required CPU load for the transfer task, the TwinCAN module supports a specifically designed hardware gateway mode.

A message object configured as a receive object on the source bus and another message object configured as a transmit object on the destination bus, see figure 7, are required to fulfil the gateway functionality with full hardware support.

The receive message object on the source bus receives the CAN frame (CAN message) and stores it. If the gateway functionality is selected, the data bytes and, optionally, the identifier and the data length code are copied to another message object. This second message object should be configured as a transmit object for the destination bus.

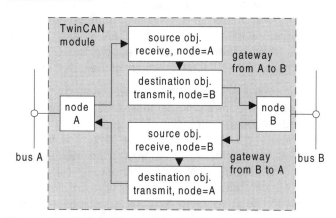

Figure 7. TwinCAN gateway functionality

These copy actions are entirely controlled by the TwinCAN module's hardware and lead to a lower CPU load as it is no longer responsable for the moving of all the data bytes. This hardware copy support can take place from CAN bus A to CAN bus B or vice versa, depending only on the programmable NODE bits in the corresponding message objects. Bidirectional data transfer with the TwinCAN module is ensured by free configuration of each message object for both the source and the destination bus.

An automatic transmission of the copied CAN frame on the destination bus is possible, as is remote handling. Remote frames on the destination bus can not only be detected, but can also, optionally, be transferred to the source bus. This allows a node on one bus to ask for data from the other bus.

The data transfer functionality is freely programmable by the user and permits selective data exchange between two buses without any CPU load (selection criteria: message identifier).

GATEWAY BETWEEN BUSES WITH DIFFERENT BAUD RATES

The asynchronous data flow and bus timings of both buses due to different baud rates and bit timings represent the major problems for data exchange between the nodes. The synchronization problem has been solved by implementing two independently running protocol layers, one for each bus.

As a result, the bit timing (e.g. sample point and baud rate) of each CAN node can be programmed to adapt to bus-specific requirements.

In many automotive applications, data has to be exchanged between the different CAN buses. In spite of their different baud rates and bit timings, they have to be coupled by means of a gateway. In this case, there is a risk of data corruption or loss due to message bursts or unbalanced bus loads. This can only be avoided by an intelligent message handling system. The FIFO structure discussed above is a reliable solution to this problem.

In order to avoid the above mentioned problem, the TwinCAN module permits the configuration of FIFO structures for gateway functionality. A receive object on the source bus delivers the selected information to a transmit message FIFO on the destination side.

Figure 8 describes a structure for transferring messages from bus A to bus B with a 4-stage gateway FIFO. The complete handling of the FIFO is done automatically by the TwinCAN. Since no CPU load is required for this action, the real-time capability of the entire system is improved significantly.

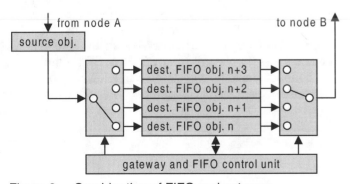

Figure 8.   Combination of FIFO and gateway

The problem of data communication between two buses running with different speeds can of course only be solved if the worst case conditions are taken into account. An average data flow is in itself not sufficient to see, whether data can be transferred without any corruption. It is only by being aware of the maximum length of a burst of messages that the size of the required data buffer can be determined.

Figure 9 shows a burst of four messages on the source bus (with a higher baud rate), which have to be transferred to the destination bus (with the lower baud rate).

The incoming messages are stored one by one in a FIFO structure. As soon as the first message has been completely received by object n, it is transmitted by the destination bus (according to CAN rules), while the next incoming message will then be stored in FIFO object n+1.

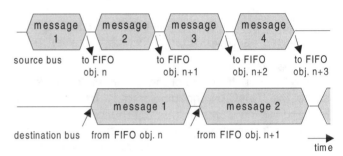

Figure 9.   Data flow in a gateway

The complete interrupt capability of the message objects (independent receive and transmit interrupt requests for each message object) are also available in gateway mode. This enables the user to monitor the gateway data transfer and to react in the case of a risk of data loss or corruption.

INTERRUPT GENERATION

The programmable interrupt generation unit of the TwinCAN module provides the possibility of running independent interrupt routines for different tasks.

The complete TwinCAN module is able to generate 72 different, independent interrupt requests, one receive and one transmit interrupt request for each of the 32 available message objects and four status or error interrupt requests for each CAN node. This large amount of request sources requires powerful and flexible interrupt handling in order to ensure a simple and lean program structure without long "case" or "if..then..else" statements.

The most flexible way to deal with this problem is the introduction of an "interrupt compressor", which combines several individual interrupt request sources and connects them to the outside world via eight interrupt output lines (interrupt nodes).

Each interrupt source can be individually enabled or disabled and contains an interrupt request flag with an associated interrupt node pointer (e.g. INP_r1 for the interrupt node pointer assigned to the interrupt request flag msg.1_received of message object 1 for correct reception), see figure 10.

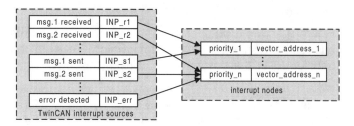

Figure 10. Flexible interrupt generation

The interrupt node pointers allow for the regrouping of the interrupt requests and their assignment to the available interrupt nodes. The interrupt node itself, which is not part of the CAN module is characterised by its priority and its vector address, which are given by the system.

Due to this structure, two or more independent tasks can share the TwinCAN module for data transfer but are handled via independent interrupt routines. This permits the user to separate the tasks, something which is becoming more and more important for creating safer program design and improved testability.

Furthermore, application-specific requirements, such as the number of the message objects used for one task or the interrupt latency tolerated can be easily taken into account in the program flow. The interrupt nodes in the system can be programmed to different interrupt priorities, which leads to a very flexible and easy program structure to control the data transfer via CAN.

TWINCAN IN EMBEDDED CONTROL SYSTEMS

Regarding the large variety of requirements for control applications in the automotive field, a carefully conceived system design becomes increasingly important. The system's hardware resources have to be partioned according to the priority of the control task and the tolerated latency of execution. The different applications and their needs lead to designs based on flexible and powerful numerical control systems.

The use of microcontrollers with embedded memory and a peripheral set dedicated to automotive applications significantly simplifies the system design and reduces the overall system costs. The available computing power and the peripheral functions can be easily adapted for a wide range of application-specific requirements. A simplified example for a microcontroller suited to the automotive world is shown in figure 11.

As the physical CAN bus layer may differ depending on the application, CAN transceiver devices have to be connected to the microcontroller in order to drive the desired CAN bus voltage levels.

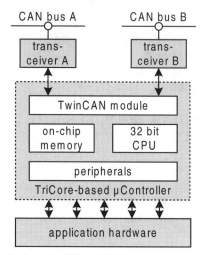

Figure 11. Embedded TwinCAN

The new TriCore microcontroller family from Infineon contains a powerful 32-bit CPU, on-chip memories and a specialised peripheral set, including the TwinCAN module. Thanks to the integration of a large variety of control peripherals into one chip, a very compact system design can be achieved.

Automotive control systems with a high demand for computing power, internal memory and specific and precise peripheral functions, such as flywheel position detection, ignition or injection control, can be efficiently built with very few components.

Another application field for the TwinCAN module is the extension of the CAN functionality in already existing design platforms or mid-range control systems. Particularly in systems with smaller (and cheaper) microcontrollers, the possibility of connecting an external CAN device might be an alternative solution to a change of the complete environment. The Standalone TwinCAN device has therefore been developed. This device can be easily connected to a control unit, see figure 12. Where this is the case, the communication between the host system and the standalone device can be handled via an 8-bit multiplexed parallel bus. In order to be able to communicate with a large variety of host chips, the Intel/ Infineon protocol (/CS, /RD, /WR) can be selected, as well as the Motorola protocol (/CS, R/W, E). Furthermore, the connections to host systems with a low pin count can also be achieved by an SPI-compatible serial channel.

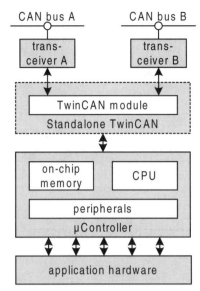

Figure 12. TwinCAN Standalone and μC

Another very interesting application field is the connection of small actuators or sensors to a common field bus structure. Most commonly, these systems require no specific computing power, but only the capability of monitoring external events or of switching on/off loads depending on the messages on the bus.

Therefore, the initialization of the CAN device has to be done without a host system. The Standalone TwinCAN device can be easily initialized by merely connecting a serial EEPROM (with an SPI channel). All internal registers required can be loaded, based on the contents of the EEPROM. What is more, the device enables additional modifications to the contents of internal registers via CAN messages. As a result, the eight available I/O ports of the device can be read by or written to by CAN messages. This permits the design of extremely small systems with input/output functionality, e.g. to read the status of control buttons or to switch on/off electric drives or parts of the light system. A typical example is shown in figure 13.

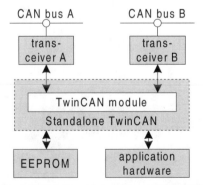

Figure 13. TwinCAN Standalone and EEPROM

CONCLUSION

The large variety of control applications in the automotive field requires sophisticated and adapted control devices in order to build solutions with a minimum of overall system costs. Under these circumstances, the use of microcontroller-based designs becomes more and more important. Thanks to the integration of an optimized peripheral set combined with on-chip memories and powerful CPUs, these devices permit the design of efficient and reliable control systems with low design effort.

In almost all automotive applications, decentralised, intelligent functional units have to be connected together by bus systems to allow for easy data exchange. Since the CAN protocol offers many advantages compared to other solutions, more and more microcontrollers with integrated CAN modules are available on the market.

Many of the well-known Infineon (formerly Siemens Semiconductors) microcontrollers provide CAN functionality and meet all the specific automotive requirements, ranging from the low-cost 8-bit devices (8051-compatible C500 Family) and the 16-bit devices (C166 Family) to the powerful 32-bit devices based on TriCore.

The new TwinCAN module offers a highly efficient and flexible CAN functionality because of the hardware FIFO and gateway support and other new CAN-related features. This enables the user to improve the system's real-time behaviour by reducing drastically the CPU load for data handling. A cost-efficient CAN extension of existing control structures can be easily implemented with the Standalone TwinCAN device. Due to its flexible bus interface, many applications can be covered.

The complete product portfolio of Infineon Technologies covers the entire range of components for automotive applications, from control devices (microcontrollers, Standalone TwinCAN, etc.) to memories and sensors and to power devices (single switches, bridges, voltage regulators, etc.) and bus transceivers. This provides the basis for building cost-efficient, reliable and homogenous control systems.

REFERENCES

1. CAN Specification V2.0, 1991 Robert Bosch GmbH

2. W. Lawrenz, CAN Controller Area Network, Huethig, ISBN 3-7785-2263-9

3. 4th..6th International CAN Conference Proceedings, CAN in Automation (CiA), Am Weichselgarten 26, D-91058 Erlangen, Germany

4. J. Barrenscheen, Flexible Control of AC-Drives using CAN, Embedded Systems 1997, San Jose, USA

CONTACT

Dr. Jens Barrenscheen
Infineon Technologies
Product Definition Microcontrollers
St.-Martin.-Str. 76
81541 Munich, Germany
Tel.: (49) 89 234 83295
Email: jens.barrenscheen@infineon.com

Patrick Leteinturier
Infineon Technologies
Automotive System Definition
St.-Martin.-Str. 76
81541 Munich, Germany
Tel.: (49) 89 234 83828
Fax: (49) 89 234 81029
Email: patrick.leteinturier@infineon.com

Local Interconnect Network (LIN)

2004-01-1742

Cost Effective LIN Bus Automotive Networking Microcontroller

Michael Bender
Melexis GmbH

ABSTRACT

The Can bus has been the default standard vehicle networking bus for more than a decade. Unfortunately it suffers from being relatively expensive and unnecessarily complicated for many non-critical control and communication functions. The LIN (Local Interconnect Network) bus represents a more appropriate engineering solution for many functions in vehicles, appliances and equipment.

To be able to develop optimized LIN applications a new 4-bit Dual-Task RISC Microcontroller was developed, which enables the capability to integrate all functions required for a LIN slave on a single IC. The complete integration of the LIN-transceiver, a voltage regulator, a microcontroller and various application specific blocks equivalent to the cost level of optimized state machines is now possible. This LIN 4-bit Dual-Task MCU combines the advantages of the state machine and classical MCU solution, which is efficiency, robustness and flexibility at a very low cost level.

INTRODUCTION

The increase of electronic modules used in vehicles, appliances, robotics and equipment provide additional comfort, safety and performance features. The coordination of the functions of these modules makes the use of networking protocols a necessity to maximize the advantage of the electronic content. Manufacturers prefer to use open, industry wide network standards in order to reduce costs, increase available design options and leverage the developed skills of their engineering software and hardware providers. A recent networking standard is called LIN (Local Interconnect Network). It is the communication protocol for a low-speed network to which many comfort and convenience applications in a car might be connected.

LIN provides a cost efficient bus communication where the bandwidth and versatility of CAN are not required. It will be the enabling factor for the implementation of a hierarchical network in order to gain further quality enhancements and cost reduction in vehicles, appliances and equipment.

Presently it is common for every function like the air conditioning vent, electrical seat motor and sunroof to be individually wired point-to-point to its own switches, sensors and actuators. Applying the LIN networking protocol makes it possible for one wire to connect up to 16 separate points..

LIN can be used for most every application which has no demand for high data rates. Currently three different main application areas exist in automobiles, the native environment for which LIN has been developed. The three vehicle applications are:

- Intelligent Human Interface Devices for window lifter, sunroof, mirror- and seat position control, steering wheel modules, …

- Actuators for climate flaps, wiper motor, positioning motors, blower, …

- Sensors for light, emission, rain, intrusion, tow away protection,   …

It is also practical to apply the LIN protocol to similar low speed applications in appliances. As an example a modern washer has many sensors and actuators that might benefit from simplification of the wiring by networking. A heating system in a modern building might also benefit from applying such networking protocols.

The LIN protocol enables development of distributed networks within the car. LIN has the potential to cover all applications unlikely to be directly connected to CAN because of high costs. In order to further amplify the economic advantages it is necessary to bring the costs of LIN modules to a level far below CAN. Up to now this was only possible with single purpose state machines. That solution is efficient, robust and cheap, but it has the

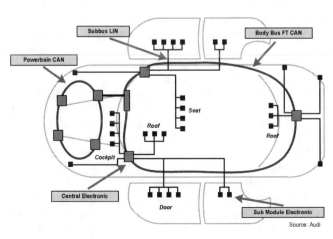

Figure 1 - Future In Vehicle Network

disadvantage that it's only proven for one application and therefore not adaptable to other applications. Changes are difficult to implement and can only be envisaged with a high effort and long development and validation time. Conventional MCU architectures are flexible, but the LIN protocol and application function influence each other. These solutions don't cover most applications at the cost levels that are required for LIN.

A new MCU concept, which is optimized for LIN slave applications, combines the advantages of both solutions: On one side the efficiency, robustness and cost level of a state machine and on the other side the flexibility of a conventional MCU solution.

CURRENT SITUATION ON LIN INTEGRATION

Current implementations within master or slaves are realized with standard MCU's together with a separate voltage regulator and separate LIN transceiver. This realization in a 3 chip solution doesn't meet the target from the cost point of view, but it is an initially flexible and secure solution. For further developments it is necessary to develop more integrated products to bring down the cost. This optimization can be done in different steps. The first step is to put the LIN transceiver, voltage regulator and support functions for the MCU together on one chip. This kind of ICs – the LIN System Basic Chip (SBC) – makes it possible to lower the cost and improve the development time due to decreasing the LIN slave implementation to a 2 chip solution. An example is shown in Figure 2.

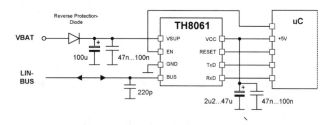

Figure 2 - LIN slave 2 chip solution with SBC TH8061

For further improvements we have to integrate all of the necessary functions of a LIN node within one chip. Melexis currently has developed for different applications completely integrated LIN slave ICs with a new CPU-core. Possible applications areas in vehicles are climate control, door applications, dashboard, seat functions. The devices might be similarly applied to other environments such as appliances, equipment and anywhere a collection of sensors and actuators might be used in a control system.

This integration concept is based on a LIN optimized 4-bit dual task microcontroller (MLX4) together with the possibility of high voltage integration (12V and 42V boardnet). This architecture makes it possible to integrate a complete LIN slave (LIN transceiver, voltage regulator, CPU and application) on one chip. This kind of IC can be directly connected to the network with a minimum of passive components.

MLX4D – THE OPTIMIZED LIN CPU CORE

The MLX4D LIN CPU core is based on RISC architecture and has a calculation power of 4Mips@12MHz. It has a guaranteed timing of one CPU cycle for each instruction and has special instructions for direct command of periphery and direct moves between RAM and I/O are possible. The CPU can address 4k instructions per task, 256 nibbles RAM and 128 nibbles I/O. In addition, special instructions are available to directly address peripherals and non-volatile memory (DMA – Direct Memory Access).

LIN operations are processed very efficient, e.g. a LIN parity calculation needs only 5 cycles (1.25us@12MHz) and 11 ROM words and a 16-bit CRC of a byte needs 36 cycles (9us@12MHz) and 32 ROM words.

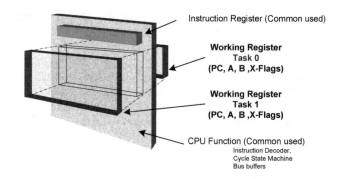

Figure 3 - MLX4D dual task CPU principle

Why is this CPU specifically suited for LIN processing? First of all, the MLX4 is not just an ordinary microcontroller. This CPU-core consists of 2 independent register sets, which gives the possibility of handling 2 different software tasks at the same time. With this system it is possible to run on one task the whole LIN protocol while the other task runs the application. Both tasks run in complete independence of each other and therefore there is no influence of one software part to the other. LIN and Application software are processed in complete independence of each other.

Each task of this CPU owns a private set of peripherals like timer and UART, as well as a private memory part. The data transfer from one task to the other will be done via a protected memory area. This protection prevents that both tasks write to the same RAM address at the same time.

The switching of the register sets occurs after every instruction. Therefore every task has a guaranteed processing power of 50% of the whole MCU power (2 Mips@12MHz). The available processing power can also be shared dynamically. If one task is put into a WAIT mode, the entire processing power is available to the other task.

Register-bank-switching after every instruction

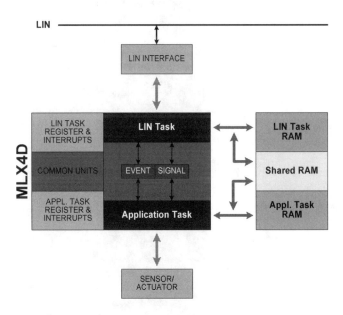

Figure 4 - MLX4D switching mechanism

The synchronization between both tasks is performed via different flags (time, events, and shared resources). For a typical LIN application this is achieved via events. It means that if the LIN task has received data, the application task receives a signal that there is data from LIN ready for processing.

The working principle of completely integrated LIN slaves shows the following picture:

Figure 5 - MLX4D working principle

This CPU is a full dual task implementation. It has integrated functions of an RtOS (Real Time Operating System) in hardware, of course limited to two tasks. The MLX4 CPU core was developed for integration into completely integrated LIN slave ICs in order to provide a very flexible, robust and yet low-cost way to implement LIN applications.

The Lin protocol is running on task0 (called LIN task) of the MLX4 and is supported from a Class A network interface realized in hardware. This interface is directly connected on one side to the LIN physical layer and on

the other side to task0 of MLX4D. It is responsible for bit-level processing of the LIN protocol. The LIN task has access to all bus states, bus events and errors and is responsible for the LIN protocol processing.

The application software runs on Task1 of the MLX4D. This part is directly connected to the application I/O. The data transfer between both tasks is done via a shared RAM area. The used RAM area for data exchange is protected via mutex flags to avoid that both tasks write to the same RAM address. The general communication between both tasks is done via an API.

Both tasks run completely independently from each other. It means that there are no influences from the LIN to the application and vice versa. The application task transfers data to the shared RAM area and the LIN task transfers this data to the LIN bus. From the application point of view the transferring of the data via LIN is done automatically. Data coming from the LIN is processed in the same way.

MLX4 AND LIN SOFTWARE

The LIN software running on task0 of MLX4D processes the serialization / de-serialization of the LIN bit stream, the LIN frame itself and frame filtering, makes the data- and error handling and is responsible for the network monitoring and status reporting. In other words the LIN handler does everything regarding LIN. Therefore the application software must only transfer the data to or from the LIN to the application via shared memory buffer or provide the data to be transmitted in the transmit buffer. The message filtering ensures that the application will only get the data which are valid for these slave nodes.

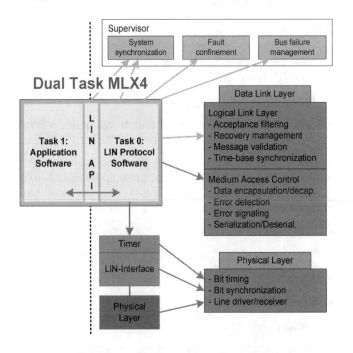

Figure 6 - MLX4 and LIN software acc. to OSI model

The software also supports protocol extensions like auto baud rate detection and auto configuration via dynamic ID assignment.

Because of the realization of the protocol processing via software it is very easy to adapt this software to future protocol extensions or specification changes (SAE J2602).

The necessary LIN protocol software for MLX4 based products will be supplied by Melexis and can be used for every MLX4 product. That means that customers only have to focus on the development of the application software itself for which a software library is also available. Implementation of protocol upgrades is made easy thanks to this separated processing of LIN protocol and application software. Currently LIN specification 1.3 is supported and the upgrade to 2.0 -respectively SAEJ2602- is in development.

MLX4 INTEGRATION EXAMPLES

IC FOR INTELLIGENT SWITCH MODULES - TH8100

The first member of our MLX4 completely integrated LIN slaves is the TH8100. This IC can be used for development of distributed intelligent switch modules, which can be linked together via LIN.

The combination of physical layer LIN transceiver and LIN protocol controller in conjunction with switch matrix or single switch inputs and PWM outputs on one chip makes it possible to develop simple, but powerful and cheap switch slave nodes in LIN Bus systems.

The following functions are implemented:

- LIN transceiver and protocol controller based on MLX4D, wake-up capable
- Integrated voltage regulator
- 17 Switch inputs to connect switch matrix up to 5 x 5 or single switches and encoder reading, wake-up capable
- Three high current PWM outputs up to 50mA with configurable PWM frequency and slew rate limitation, constant current source mode 2 to 30mA
- 3 ADC channels, one with high voltage option
- Very low standby current, 10uA in sleep mode
- Internal RC-Oscillator
- EEPROM with internal charge pump
- Watchdog
- Directly usable in a 12V, automotive environment

The TH8100 is a universal MCU for distributed intelligent LIN modules. It is well suited for applications like switch modules for window lifter, mirror adjustment, seat adjustment as well as switch modules within the steering wheel or the dashboard. It might be applied to a membrane switch control panel on an appliance or other equipment It is also possible to use this IC as LIN gateway for exiting non LIN solutions via the SPI interface.

The external passive support circuitry is very simple. Just a few capacitors and a reverse protection diode are needed. The switch inputs of this IC can be directly connected to the I/Os. These pins are short circuit proof against battery voltage as well as load dump protected for using in automotive environment. The PWM outputs can be used for realization of a search and function lighting. These outputs are slew-rate controlled in order to ensure good EMC performance. These outputs can also be used in different modes that make it possible to connect LED's directly to battery voltage as well as conventional realizations via pre resistor. It is also possible to connect a rotary encoder. The integrated 8-bit ADC can be used for reading analogue values.

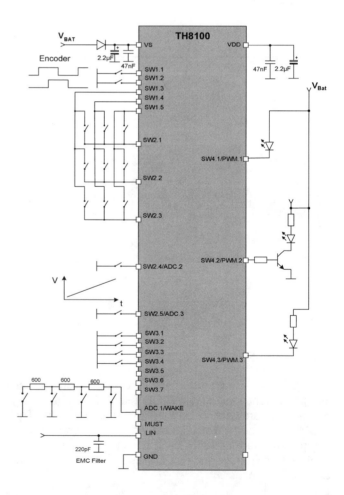

Figure 7 - Application circuitry sample of TH8100

260

INTELLIGENT CONTROLS FOR MIRROR MOTORS AND AIR CONDITION VALVES – MLX90402/04

Similar to the TH8100, the MLX9040x utilizes the MLX4 for communicating with the LIN bus and for controlling DC motors and steppers. Analogous to the TH8100, there is a separation between the LIN bus and the application. The LIN software used for the LIN task of the MLX90402/04 is identical to the software used in the TH8100.

The MLX90402 is a Stepper Motor Controller with an on-chip dual H-bridge for independent control of two motor axes. The IC can be used for LIN-based mirror controls. Apart from the mirror actuators, it can also be used to control mirror heating and lighting features. An 8-bit micro-stepping PWM stage is included for controlling applications like the adaptive curve headlamp.

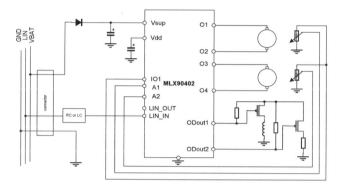

Figure 8 - Mirror control using the MLX90402

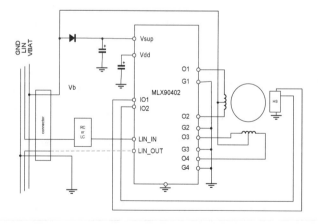

Figure 9 - Air condition valve control using the MLX90402 for stepper-motors

Contrary to the MLX90402, the MLX90404 motor driver is specifically optimized for DC-motor controls in air condition valves. This application only requires a single half-bridge with a nominal output current of 500mA.

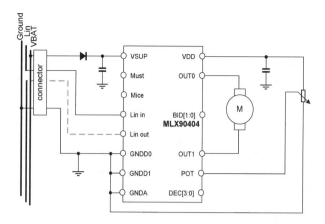

Figure 10 - Air condition valve control using the MLX90404 for DC-motors

As a special feature, these ICs have the capability to automatically assign the LIN ID. This is shown in both figures 9 and 10 where the optional 'LIN out' connection may be made. (For standard BUS configurations the usual LIN or LIN in pin is used). This auto-configuration feature is based on the daisy-chain principle, but also other principles like plug coding and others are possible. It allows the LIN bus to be re-configured at any time without affecting any standard LIN slaves already connected. This feature only requires a minimum amount of hardware. It is secured within the protocol (compatible with LIN 1.3) by checksums and error flags.

CONCLUSION

With the shown 4-bit MLX4 LIN solution LIN bus architects are supported with a very flexible, low cost LIN MCU for slave applications that enable the LIN bus to be a success story.

It is a very safe and robust system with inter task protection and makes it possible to test the LIN software independently from the application software. This LIN software is supported from the IC supplier and therefore the customer has only to develop the application software. This low cost 4-bit CPU is optimized for event processing and is best suited for the LIN protocol. Thanks to the use of a CPU for processing the LIN protocol, it is a very flexible solution that can be adapted to every protocol evolution or extension.

The future of LIN has now started with a new progress of integration of LIN applications. It offers a very low cost solution with an extremely high level of flexibility.

REFERENCES

1. LIN Specification 1.3 and 2.0.
2. Datasheet TH8100
3. Datasheet MLX90402/04

CONTACT

 Michael Bender holds a degree in electronic circuitry from the Technische Universität Ilmenau. He has worked for Melexis since 1996. Within the technical marketing department, Michael is responsible for automotive bus products, including the definition of future automotive products together with customers, customer support for these products and assistance in bus-system standardization. Michael participates as a member of the LIN Committee.

E-mail: michael.bender@melexis.com

Web: http://www.melexis.com

262

In-vehicle Network Verification from Application to Physical Layer

Georg Pelz, Juergen Schaefer, Dieter Metzner, Magnus Hell and Adam Opielka
Infineon Technologies AG

ABSTRACT

The verification of an in-vehicle network often requires to look at more than one level of abstraction at a time. At the moment, this is not addressed by existing methods, which are dedicated either to physical or application layer, but not both. This paper fills this gap by introducing a methodology to insert the protocol related software execution as well as the motor behavior into the physical layer mixed-signal (i.e. analog/digital) simulation. Electronics and mechanics are covered by the hardware description language VHDL-AMS, while the software is given in C.

INTRODUCTION

The verification of an in-vehicle network often requires to look at more than one level of abstraction at a time. The LIN (local interconnect network) bus is a good example for that, especially because being a low-cost solution, the system design and setup requires a lot of careful considerations to guarantee functionality. For instance, some LIN message may switch on the electric motor of a window lifter, which causes a voltage-drop and in turn a shifting of the clock frequency. This may result in a loss of messages. Later on, the window runs into its seal, which again may cause a voltage drop resulting in a loss of the message to switch off the motor. Here – as in many other cases – we have a complex set of cause-and-effect chains, which spread over application and physical layer. This calls for an assessment of physical and application layer at a time. At the moment, this is not addressed by existing methods, which are dedicated either to physical or application layer, but not both. For example, a tool is available from the Vector Group, see [10], which covers the functional simulation of LIN networks. This tool is well suited for purely digital considerations, but if the problem stems from the analog domain and interacts with the application layer, this tool will not be suitable.

This paper fills this gap by introducing a methodology to insert the protocol related software execution as well as the motor mechanics into the physical layer mixed-signal simulation. Electronics and mechanics are covered by the hardware description language VHDL-AMS, while the software is given in C. It will be shown that the transfer of one message over a larger LIN network in a door with window lifter can be simulated in roughly one CPU minute on a SUN Sparc Ultra 80 workstation. Moreover, the network's behavior is made transparent through using a software debugger illustrating and controlling the software execution on the nodes in question.

In contrast to a breadboard solution for the verification, it is also possible to control all parameters of silicon and system, e.g. it is possible to assess fastest as well as slowest silicon in its environment.

MODELING LIN-SYSTEM COMPONENTS

In the following, the modeling of the components of a LIN-system will be detailed. Like CAN (controller area network), LIN is an automotive serial bus, see also [9]. As compared to CAN, it is lower in cost and ideally suited for body applications. It requires just three wires, i.e. the bidirectional data line and two supply lines. Bytes are sent using the standard UART data format in RS232 communication at a rate of up to 20 Kbaud. The clock tolerance of the LIN slaves is +/-15% to allow for low-cost solutions without crystal or resonator. In the sync phase, the slaves calibrate to the master which leads to a precision of +/-2%, as required by the LIN specification.

A LIN network consists of a number of LIN nodes between which the data is transferred. A LIN node contains two essential parts. The first is the LIN protocol engine and the second is the LIN-Transceiver. The protocol engine may be implemented in hardware or software. The latter holds in our case – the respective software is run on a microcontroller, see Figure 1.

The transceiver works as interface between the protocol engine and the physical bus and in this way forms a connection between the physical Layer and the application Layer according to the ISO/OSI-reference model. In addition to this elementary function, the transceiver protects the bus topology with regard to errors like over-current, over-load and so on.

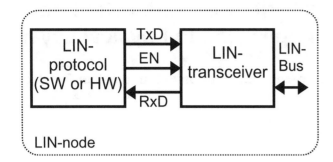

Figure 1: LIN-Node

Apart from the LIN nodes, we have the cable and an electric motor for window lifting, which causes a substantial supply shift.

LIN-TRANSCEIVER AND CABLE

For the investigations in this paper, Infineon's LIN-transceiver (TLE 6259-2G) has been chosen. It comprises analog and digital behavior, which is reflected by using the mixed-signal hardware description language VHDL-AMS. The model is optimized for simulations in a system environment, which requires a careful look at the tradeoff between accuracy, simulation time and numerical stability. To cope with the system simulation requirements (e.g. many components, large numerical equation sets and pieces from different physical domains) Infineon developed a modeling approach, which considers these requirements, see [2][3]. By this method all inputs and outputs of the transceiver are modeled as physics-based compact models while the internal signal processing is described as a mixed signal behavioral model, see Figure 2.

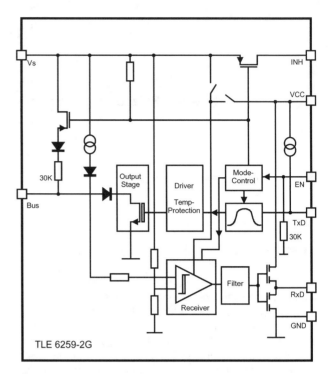

Figure 2: LIN-Transceiver (TLE 6259_2G)

The physics-based compact models are described using DAE[1] systems. These models contain all effects which determine the interaction between transceiver and system environment like cable and PCB board. The most important physics-based compact model is the output stage with its parasitic effects like the MOS-capacitors, see Figure 3. In a transceiver, this effect has a major impact on the bus signal timing characteristics like rise time, fall time and delay times.

---

[1] Differential and algebraic equations set

```
begin

if (vdg>0.0) use
vdepl == vdg +2.0*(kgd_c/cox_c)**2*
(1.0sqrt(abs(1.0+(vdg/(kgd_c/cox_c)**2))));
else
vdepl==0.0;
end use;

break on vdg'above(0.0);
break on vref'above(vtepi);
break on vref'above(vtepi+0.1);

vref == vdi-vdepl;

if (vref<vtepi) use
    cgs1 == cgs_c+cox_c;
elsif (vref < vtepi+0.1) use
    cgs1 == cgs_c+(vtepi+0.1-vref)*10.0*cox_c;
else
    cgs1 == cgs_c;
end use;

vmill == vdg-vdepl;
imill == cox_c*vmill'dot;
icgs == cgs1*vgs'dot;
```

Figure 3: VHDL-AMS model of parasitic MOS-capacitor

```
p1:process
   begin
     if vlin 'above(vbus_dr*v_vs) then
       lin_h<='1';
     else
       lin_h<='0';
     end if;
     wait on vlin 'above(vbus_dr*v_vs);
end process;

p2:process
   begin
     if not vlin 'above(vbus_rd*v_vs) then
       lin_l<='1';
     else
       lin_l<='0';
     end if;
     wait on vlin 'above(vbus_rd*v_vs);
end process;

p3:process(lin_h,lin_l,en,plb)
   begin
     if lin_h='1' and lin_l='0' and en='1' and plb='0' then
       out_p<='1';
       ctrl<=1.0;
     elsif lin_h='0' and lin_l='1' and en='1' and plb='0'
     then
       out_p<='0';
       ctrl<=2.0;
     elsif en='0' or plb='1' then
       out_p<='0';
       ctrl<=0.0;
     end if;
end process;
break on ctrl;
```

Figure 4: Mixed-Signal VHDL-AMS of the bus receiver

The internal signal processing like the bus receiver function or the bus mode control is modeled as a mixed-signal behavioral model, see Figure 4. Here, the modeling uses simple analog threshold functions, controlled switches, controlled voltage- and current-sources and digital events and processes.

LIN-PROTOCOL

In our example, the LIN-protocol is implemented using software to be run on a microcontroller. This calls for some kind of hardware/software co-simulation, while still sticking to the mixed-signal simulation, which is necessary for transceiver and cable.

As we have mixed-signal simulators available for transceiver and cable, two further points have to be addressed in hardware/software co-simulation: simulating the software looking at functionality/execution time and the synchronization between hardware simulation and software simulation. In general, a couple of methods are available for software simulation:

1. **Running software by simulating hardware**
   Here the controller hardware is modeled on register-transfer level and in a hardware description language like Verilog or VHDL. After filling the memory with the program to be executed, the controller hardware is

simulated. While simulating the hardware, the software is executed. On the one hand, the synchronization to the rest of the hardware is trivial as the transceiver interface to the controller also is digital. Moreover, functionality and timing are perfectly met by the simulation. On the other hand, the simulation time is by far too high to allow for reasonable system assessments.

2. **Binding-in models of micro-controller cores**
   These micro-controller core models typically are formulated in C. They execute much faster than register-transfer models, as they only care for the IO-behavior, which is reasonably met in functionality and cycle-accurate timing.

3. **Binding-in instruction set simulators**
   Further compromising the timing inherent to the simulation results leads us to the use of instruction set simulators which can be bound into a mixed-signal simulator using the respective C-interface. Here the cycles for the instructions can be counted, which works pretty well for simple architectures. If more sophisticated features, e.g. pipelining, are used, the simulated timing will not be cycle-accurate any more.

4. **Binding-in the system software directly**
   Finally, if the system software is written in C, it can be directly included through the mixed-signal simulator's C-interface. In this case, the functionality but not the respective timing can be simulated. In many cases, this may not be a serious drawback. E.g. in the early stages of concept engineering and design, no information on the timing will be available anyway. Moreover, many digital control systems work with an underlying timing grid and provide new regulator values after equal time steps. Here it has to be made sure, that the controller is fast enough to provide for these values in time. If this is taken for granted, the timing can easily be taken into account. The big advantages of the direct inclusion of software are on the one hand that we do not need a controller model here and on the other that the overall simulation runtime is virtually just determined by the hardware simulation.

In this paper, we will apply the fourth approach. The central point here – as in the other methods – is the synchronization between hardware and software. First of all, why do we need synchronization? In general, hardware and software may be simulated independently, but it has to be made sure, that they show the same simulated point in time, when data is exchanged between the two. A reasonable scheme for the synchronization may be to work on bit or byte level. In practice, this means that the software part as formulated in a C-routine is called for every bit or every byte of the transferred LIN data. Please note that the synchronization on byte level needs some extra feature, e.g. to cut off the own drivers on the bus in case of transmission errors or lost priority decisions.

ELECTRICAL MOTOR

As presented in detail below, the current consumption (and respective supply shift) of an electrical motor may cause serious problems for the overall functionality of a LIN system. Here it turned out, that looking at the mere electrical behavior is sufficient to assess the system as a whole. The mechanical properties may be neglected as long as measurements of the typical current consumption shapes are available. These may easily be accounted for by introducing respective piece-wise linear current sources. If the mechanical behavior also plays a role, models of the electric motor and its load may easily attached, see [5]. For the simulations below, the purely electrical approach was chosen.

MODELING A LIN-SYSTEM

In the following, a complete LIN system is built out of the previously described components. Specific attention is put upon the connector and lead resistances, e.g. 100mOhm, of the master's and slave's supply. In the configuration as shown in Figure 5, the electric motor causes a shift down of roughly 2V in the master's power and a shift-up in the same range of the slave's ground[2]. This in turn results in a couple of effects:

- As the supply of the master is shifted down, the (pull-up driven) positive edge of the LIN bus line is substantially slower. Result: the RxD signal lags.
- As the slave's GND is shifted up, its input receiver thresholds are likewise shifted up, which again leads to a lagging of its RxD signal.
- In addition, the clocking of master and slave may substantially be affected by these power and ground shifts, as crystals for clock control often are too expensive for LIN systems.

Typically, the slave synchronizes to the master by using the SYNC field's edges. The effects above may pop up after that, when the motor is switched on. This in turn may lead to a mismatch when the slave samples the data read from the bus line.

---

[2] This is higher than the 10% Vbat or Vgnd, which is specified in the LIN specification.

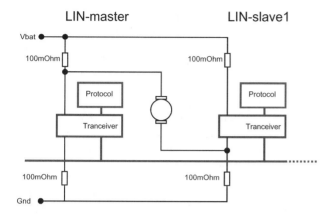

Figure 5: LIN system

SIMULATION

The simulation of the LIN system is based on a commercial mixed-signal simulator supporting the analog/digital hardware description language VHDL-AMS and a C interface for the software. Several tools are on the market for that, we chose AMS-Designer[3].

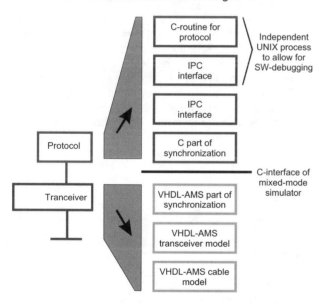

Figure 6: Mapping the LIN system to the simulation

One important aspect in the development and assessment of embedded software is the visualization and control of it within the simulation process. Ideally, the handling of embedded software would be as easy for regular software by using a software debugger. In our case, this is also possible by using the approach as illustrated in Figure 6. Here an independent UNIX process is initiated which comprises the C-routine implementing the LIN protocol. This process virtually is an independent program which can be run in a standard software debugger. In our case this is the xxgdb debugger which comes with the GNU[4] software, see Figure 7. The IPC (inter process communication) interface binds the mixed-signal simulator and the protocol process together.

The simulation in Figure 8 shows the transfer of three messages over the LIN bus. For each message, the various phases (synch break, sync field, id, data, checksum and inter frame spacing) can be identified. The LIN slave interprets the message and may start some action, e.g. switching on some electric motor. According to the sub-optimal configuration, as shown in Figure 5, this leads to a supply shift at the master, i.e. sender, and a ground shift at the slave, i.e. receiver. As pointed out above, this leads to a time shift of the RxD signal. For worst case conditions, this time shift is in the range of 11 us, see Figure 9. This encloses the first two effects as listed above, i.e. the lower supply and the receiver thresholds shifted up. In addition, the clock generation is affected with its 2% precision within a byte. Together with the bit length of 50 us, we end up with an additional time shift of 8 us. Moreover, it is very difficult to maintain this 2% precision with the large supply/ground shifts as seen above. All in all, we have a total time shift of at least 19 us, in case the electric motor is switched on. EMC issues may further add to the delay, so that depending on the LIN protocol and UART implementation, this leads to a situation in which the wrong bit is sampled. There are a number of measures to cope with this situation:

1. Change the connectors and lead layout of the motor's supply, to bring down the transfer resistance
2. Change the implementation of the LIN protocol and the related UART in that it resynchronizes with every edge it experiences
3. Devise a schedule of motor activity and LIN messages, which does not exhibit the named problems

The simulation as shown in Figure 8 took three CPU minutes on a SUN Sparc Ultra 80 workstation. This includes the mixed-signal simulation, the software execution and the software debugging. Please note that the software execution accounts for just a very minor part on the runtime.

---

[3] Cadence Design Systems

[4] Free Software Foundation

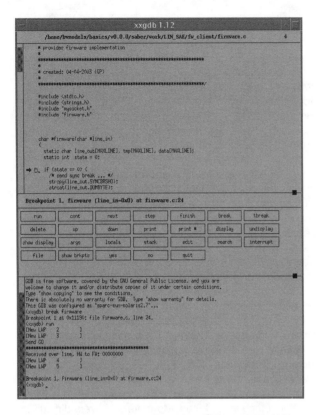

Figure 7: Standard software debugger (xxgdb) for embedded LIN protocol software

CONCLUSION

This paper proposes an approach for in-vehicle network verification, which allows for assessing physical and application layer at a time. In this way, errors and problems can be detected and analyzed which spread over more than one of these layers. In the example shown above, this is definitely the case. An electric motor is switched on and off by LIN messages. On the other hand, the current it draws can disturb these messages. This circular dependency is analyzed in a dynamic way and within one tool, i.e. a circuit simulator, which is enhanced by a model for software implementing the LIN protocol. This HW/SW-cosimulation is accomplished without compromising simulation speed. Even more, measures are taken for a reasonable visualization of the executed software.

REFERENCES

[1] G. Pelz: "Mechatronic Systems – Modeling and Simulation with HDLs", John Wiley & Sons, 2003.

[2] Dieter Metzner, Juergen Schaefer, Chihao Xu: "Mixed Signal, Multi Domain Behavior Models of Smart Power ICs for Design Integration in Automotive Applications", IEEE International Workshop on Behavioral Modeling and Simulation (BMAS), Santa Rosa, California

[3] Dieter Metzner, Juergen Schaefer, Chihao Xu: "HDL Based System Engeneering for Automotive Power Applications", IEEE International Workshop on Behavioral Modeling and Simulation (BMAS), San Jose, California

[4] A. Luedecke, G. Pelz: "Top-Down Design of a Mechatronic System", Proceedings: Third Forum on Design Languages (FDL 2000), Tuebingen, Germany, 151-158.

[5] D. Dammers, P. Binet, G. Pelz und L. Vosskaemper: "Motor modeling based on physical effect models", IEEE International Workshop on Behavioral Modeling and Simulation (BMAS) 2001.

[6] L. Voßkämper, R. Schmid und G. Pelz: "Modeling Micro-Mechanical Structures for System Simulations", Forum on Design Languages (FDL 2001).

[7] G. Pelz, J. Bielefeld und G. Zimmer: "Virtual Prototyping for a Camera Winder: a Case Study", IEEE/VIUF Workshop on Behavioral Modeling and Simulation (BMAS), Orlando, FL, 1998.

[8] G. Pelz, T. Kowalewski, N. Pohlmann und G. Zimmer: "Modeling of a Combustion Engine with Hardware Description Languages", IEEE/VIUF Workshop on Behavioral Modeling and Simulation (BMAS), Orlando, FL, 1998.

[9] P. Topping, "Automotive Keypad in a Distributed LIN Network", AutoTechnology 2/2003, 56-59.

[10] http://www.vector-informatik.de/english

CONTACT

Main author:

Dr. Georg Pelz
Infineon Technologies
Automotive Power
Balanstr. 73, 81541 Munich, Germany
Fax: +49-89-234-9552249
Phone: +49-89-234-83171
E-Mail: Georg.Pelz@infineon.com

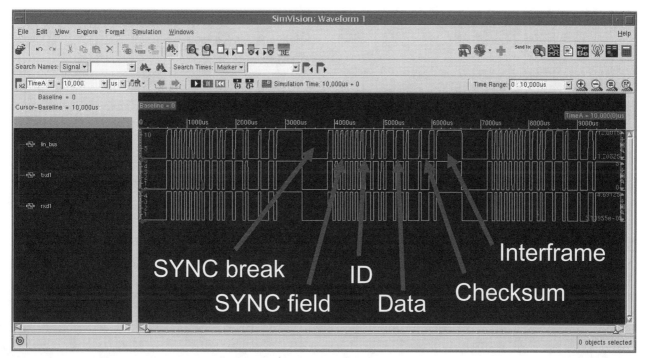

Figure 8: Simulating three LIN messages. Traces shown from top to bottom: LIN busline (analog), TxD (master, digital), RxD (master, digital); the simulation takes roughly three CPU minutes on a SUN Sparc Ultra 80 workstation

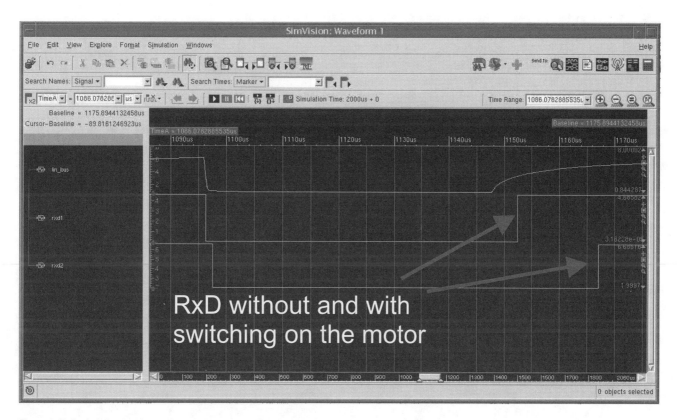

Figure 9: RxD signal at the slave with and without switching-on the motor. Traces shown from top to bottom: LIN bus line (analog), RxD (without switching-on the motor, digital), RxD (with switching-on the motor, digital)

Implementing Local Interconnect Network (LIN) Slave Nodes

Matthew Ruff
Motorola, Inc.

ABSTRACT

As the Local Interconnect Network (LIN) standard gathers more automotive industry interest, it is gaining strength as a likely standard for SAE Class A networking needs. European auto makers are embracing LIN in ever increasing numbers, even abandoning many other formerly competing Class A solutions, and the manufacturers in the United States are also beginning to take notice of this standard.

The driving thrust of LIN is its ability to create low-cost Class A system solutions. In this master-slave based system, the majority of nodes are slave nodes. It is therefore critical to reduce, wherever possible and appropriate, the cost of slave node implementations. Lower cost slave nodes should yield the greatest reduction in system cost due to the number of slave nodes in the system.

This paper explores the different ways a LIN slave node can be realized in an automotive LIN network. The concepts and principles, however, stretch beyond the automotive space and could include any applications in which LIN might be used to communicate between devices. As the options are analyzed, the issues of relative cost, application categorization, application complexity, manufacturability, and design trade-offs will be discussed. In the end, there is a need for all types of LIN slave node implementations and the purpose of this paper is to present the possibilities, advantages and problems associated with each.

INTRODUCTION

Over the last decade, embedded electronics have taken over most major functions in the car. With the advent of multiplexing, these formerly isolated control systems have been given the chance to interact and communicate in increasingly complex ways. These trends continue to pervade more deeply into the vehicle as technology allows for less expensive implementations.

LIN represents the next major step forward in this evolution. It is designed to allow inexpensive connectivity through lower complexity protocol features, use of plentiful existing hardware solutions, and protocol features which allow for the elimination or cost-reduction of components in each node of the network. The LIN concept also provides cost-reduction through standardization of physical layer components and tool interfaces. This standardization allows silicon manufacturers and tool vendors to create interoperable products, fueling competition amongst suppliers.

Upon first examination of the LIN specification, the application programming interface (API) specification would make a designer immediately think that LIN is designed for microcontroller based systems. There are, however, other ways to implement electronics that communicate on a LIN bus which will be explored in this paper. In fact, the rest of the LIN specification, covering the protocol specifics, physical layer, and tools interfaces, does not restrict the implementer in choosing the components for creating electronic control units (ECU) for use on a LIN bus, only in how that ECU interacts with the bus itself.

Microcontroller based solutions are the traditional approach to creating ECUs on a multiplexing network. In the case of a LIN network, however, these designs can still take on several different major variations. Microcontrollers which have no dedicated serial communications interface (SCI), also called a universal asynchronous receiver/transmitter (UART), are amongst the simplest and least expensive microcontrollers available. These devices can be used to communicate on a LIN bus, generally by utilizing an on-chip timer. The next step up would be microcontrollers with dedicated SCI hardware on-chip, which are very common, not very expensive, and widely used for many years in the embedded electronics community. Another feature of the LIN protocol is that it allows microcontroller slave node implementations which depend on Resistor/Capacitor (RC) oscillators rather than crystals for their clock source. RC oscillators are considerably less expensive than crystals, but have wider tolerances on accuracy.

Microcontrollers are not the only method of implementation of LIN slave nodes. The protocol was designed to be simple enough that a fairly straightforward logic network can be realized to decode LIN messages. This logic could be incorporated in application specific integrated circuits (ASIC) or other similar dedicated hardware, potentially reducing the cost of implementation.

LIN BASICS

To understand specifics of how LIN slave nodes operate, it is necessary to understand the basics of how a LIN network operates.

LIN is a low-cost single wire bus, designed to be a sub-bus architecture. This sub-bus concept means that a LIN network may connect to a higher level bus, like the main body network in a vehicle, but the LIN network itself is an independent domain. The LIN architecture is a single-master, multiple-slave arrangement, where the master node drives all communications on the bus based on the standard UART format. For reference, Figure 1 shows the format of a LIN message frame.

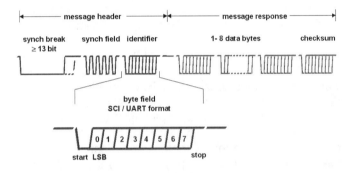

Figure 1: LIN Message Frame (LIN Revision 1.2)

To allow slave nodes to base their timing on less expensive RC oscillators, the LIN protocol transmits synchronization information (0x55 character) from the master node in every message frame on the bus. Since the master node is based on a much more stable crystal, its timing is considered the standard to which slave nodes may synchronize themselves.

The message header is always transmitted by the master node, so that synchronization data and all message traffic initiation is generated from the master node. The data bytes, which can be further sub-divided into signals, can be transmitted by the master node (as a command from master to slave) or by the slave node (in response to recognition of the identifier as a status update or the like). The data integrity of the data field is protected by a checksum.

The LIN protocol dictates that message identifiers are functionally addressed, meaning that the identifiers denote the function or meaning of the message rather than a physical address of a node. In this way, a message can be sent from the master to any number of nodes in the sub-network simultaneously (multi-casting).

When a slave receives a message header, it synchronizes to the bus using the synchronization byte, then decodes the identifier. From the value of the identifier, the slave determines whether the master is going to supply data in the data field (a command-type message), if this slave is supposed to supply data in the data field (a status request-type message) or whether it is an identifier that this slave does not care about. In the first case, the slave would receive the data and proceed with whatever operation was appropriate based on this new data. If the message is a status request-type message, for which the slave is expected to supply data, the slave would begin to transmit the data after the identifier. In the case that the identifier indicates that the message is not intended for the given slave, the rest of the message is ignored. All of this information about what identifiers correspond to which messages, and so forth, is pre-defined by the system designer and is captured in the LIN configuration description language file for the system.

Since the master node transmits all message headers on the bus, the message traffic is also very predictable. The master node uses a scheduling table to determine the order and frequency of message transmissions in a time triggered manner. This scheduling table can be changed and modified during execution to allow for varying requirements of the system at different times.

The full LIN specification covers the physical layer, protocol, and interfaces to the tools and software application layer of embedded programs.

THE LIN APPLICATION SPACE

The range of applications being considered for LIN is wide and growing wider each day. Traditionally, many automotive manufacturers have had proprietary solutions to meet the needs of a Class A network, but with the rising popularity of LIN as a global standard, many are looking to LIN as a replacement for these systems.

In the automotive industry, the birthplace of LIN, the majority of applications are in the body electronics area. Due to the relatively low speed of LIN (20 kbit/s maximum), it is firmly considered a Class A network. This generally means control and command of devices which operate at "human interaction" speeds. Some possible applications of LIN in the automotive world are:

- Interior and exterior lighting systems
- Seat systems (seat motors, heater, fan)
- Door systems (window lifter, latch, switch panel, mirror)
- Heating, ventilation, and air conditioning (HVAC) systems (damper door actuators, fan, sensors)
- Power sliding door, sunroof, and convertible top actuators
- Power door and trunk lock mechanisms
- Windshield wiper motors
- Headlight positioning and leveling actuators

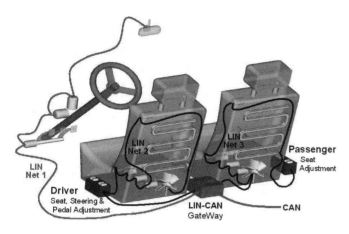

Figure 2: Automotive LIN applications

Applications of LIN are not limited to the automotive industry, however. Industrial systems often have similar requirements to automotive systems, and designers are looking with increased interest at LIN as a potential standardized networking approach for low-speed communications. For example, a partial list of applications for industrial and consumer applications includes:

- Proportioning valves for gasses and liquids
- Paper and materials handling equipment
- HVAC actuators and sensors
- Entertainment equipment (powered, remotely controlled volume switches for audio receivers and mixers)
- Motor control, switch panels, and sensors for white goods (washing machines, dish washers, etc.)

This list primarily covers motor applications, not counting the myriad of switch monitoring applications, remote sensors, and other remote actuator functions that various industrial systems and consumer products contain.

Whether in automotive applications, industrial systems, or consumer products, all of these uses of LIN have varying levels of complexity and differing requirements on performance and cost. For this reason, it is necessary to consider different ways to implement LIN nodes in these systems.

TRADITIONAL EMBEDDED MICROCONTROLLER SOLUTIONS

Historically, ECUs are very similar in some regards. Most consist of an embedded microcontroller with its firmware, power supply conditioning circuitry, and sometimes external memory if the application calls for it, and input conditioning and load control circuitry. All of these components are generally located on a printed circuit board (PCB) and placed in a protective housing with connectors to provide input and output. This method works very effectively, though not always the

most economically, for LIN slave node applications. In the context of LIN slave applications, a few distinctions must be made beyond the previous description.

Microcontrollers are not all alike, even in their most basic functions. The differences in basic functionality which affect LIN slave node implementations will be looked into in this section. Some of those differences include the presence or absence of a UART/SCI, a UART/SCI with enhancements for LIN, and internal clock generation capability. Each of these features plays a role in the implementation of a LIN slave node application.

BASIC MICROCONTROLLER

In the context of this paper, a basic microcontroller is assumed not to have a UART. This requires the application to employ a software based UART, often called "bit-banging," in order to communicate with the LIN bus. There are several ways to accomplish a software UART, which simplify to using a timer channel to measure bit timings or using software timing loops to measure bit timings. These two methods are, respectively, the interrupt driven and polling based methods for timing in embedded programming.

Using the timer module of the microcontroller is an interrupt based solution, where the incoming signal is measured against the main clock of the microcontroller. By capturing the incoming signal and measuring against the local time base, the ECU can be synchronized by measuring the synchronization byte of the LIN message. Another timer channel could be used to generate transitions on the output side, generating outgoing bits based on the measurements from the synchronization byte.

It is also possible to measure incoming portions of the messages and regulate the outgoing portions of the message using software polling loops. These loops rely on a fixed number of microcontroller instructions taking an exact number of clock cycles to execute. This provides a reference time base from which the received and transmitted bits can be measured. The primary difference with software polling loops is that they rely on the loops taking an exact amount of time, which means that the microcontroller cannot execute other instructions, such as servicing interrupts. Software polling loops eliminate the need for timer hardware on the microcontroller, but also restrict the use of interrupts due to the probability of interrupt servicing code changing the length of those timing loops.

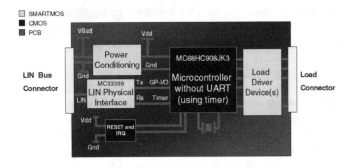

Figure 3: Microcontroller LIN slave using General Purpose Input/Output and Timer Channel instead of UART

Figure 3 shows a microcontroller device using a timer channel on the LIN receive line and a general purpose input /output pin for the transmit line. This is an example of both the timer based method and the software timing loop method. In this scenario, the timer channel, set to capture the input signal, is used to measure the incoming message header, synchronize on the synchronization byte, capture the LIN message identifier, and capture any incoming data bytes. The application then uses the output pin to transmit any responses necessary using the software timing loop method. In this configuration, the impact of interrupts on software timing loops is minimized and localized only to the times when the slave device is sending data bytes.

The hardware configuration shown in Figure 3 could also be used when using a timer for both receive and transmit. The application code must control the output level of the port pin in the section of code which services the timer interrupts.

Using timer modules or software loops for synchronization are both viable options, but certain considerations must be made. For example, if the application that the LIN slave node is designed for is very interrupt intensive, software timing loops might get interrupted in an irregular or frequent fashion, resulting in unreliable and potentially out-of-specification LIN messaging. Software timing loops also depend upon the accuracy of the base clock of the microcontroller and must be properly adjusted and scaled accordingly, any time this frequency is changed in the design cycle. This can be managed reasonably efficiently through proper software design techniques.

The solutions described in this section are useful in cases where a very inexpensive microcontroller is desired or perhaps where a special microcontroller containing peripherals unique to a given application is required, but does not have a UART.

Software drivers which perform the "bit-banging" operations can be crafted which can then interface to the LIN API, thus creating a re-usable and standardized driver block for use with various similar microcontrollers.

MICROCONTROLLER WITH UART

The majority of microcontrollers on the market now have some form of UART. Due to this availability and the resulting cost reductions due to high volumes of devices manufactured, many Class A protocols including LIN have based their messaging on that hardware platform.

In Figure 4, an implementation using a microcontroller with a UART can be seen. This hardware implementation is the most common implementation at present, and is similar to most ECU designs in the automotive industry.

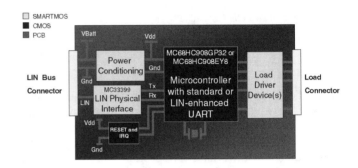

Figure 4: Microcontroller with UART LIN slave implementation

One of the problems with implementing LIN with the current revision (revision 1.2) or older with a standard SCI is dealing with the 13-bit break symbol at the beginning of every message frame. When designing slave nodes, particularly with a crystal or resonator being used, detecting a break symbol is not a serious issue. Crystals or resonators are generally well within the clock tolerance required by LIN to ensure detection of break symbols.

As was stated earlier, the LIN sub-bus is an autonomous network. It is operated at a fixed speed between 1 kbit/s and 20 kbit/s, defined at system design time. This means that an implementer may pre-define the baud rate of the LIN bus into the LIN driver code of the microcontroller. Knowing this baud rate and having a fixed and stable time base (crystal or resonator) means that the UART can detect the break using its standard break detection circuitry. As long as the clock tolerance of the clock driving the UART is within 15% of the target frequency, then a 13-bit length break symbol will appear as a minimum of 11 bits long. Since a standard UART is designed to detect a 10 or 11-bit break symbol (depending on settings), an 11-bit break will always be detected properly as a 0x00 data byte with a framing error (definition of a break symbol). This is handled automatically by the UART. The challenge of generating this 13-bit break symbol is left as an exercise for the master node and is beyond the scope of this paper.

Since the UART can detect the break, a great deal of reuse of existing UART software libraries can be utilized and modified to incorporate the LIN API.

MICROCONTROLLER WITH LIN-ENHANCED UART

As LIN gains momentum, microcontrollers are becoming available with specialized UART implementations containing special features to facilitate ease of use with LIN protocol.

Special features vary by manufacturer, but can include items such as the ability of the UART to generate a 13-bit length break character (used only by the master node) or additional adjustment of the baud rate of the UART to more closely synchronize to the message traffic. The latter is useful in cases where the base clock of the microcontroller is not as accurate as a crystal or resonator, as will be described in the next section. Fine adjustment of the baud rate of a UART might also be needed in cases where non-standard baud rates are desired due to frequency compatibility at the master node between LIN and a higher level network such as CAN.

Other features may include the ability of the SCI to automatically measure the length of an incoming break symbol or bit time, to provide for adjusting of the baud rate from the original target speed. This measurement capability, combined with fine adjustment capability of the UART, is useful in compensating for RC oscillator drift which can occur as a result of temperature changes or aging components.

MICROCONTROLLER WITH INTERNAL CLOCK GENERATION

Microcontrollers with an internal, trimmable RC oscillator are available to eliminate the external crystal or resonator. These products offer some advantages and some challenges to the LIN slave node designer.

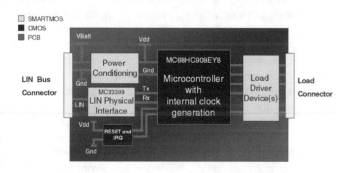

Figure 5: Microcontroller with internal clock generation capability LIN slave implementation

First, in order to ensure reliable communication, it is likely that these devices will need the oscillators to be trimmed during the manufacturing flow of the modules in order to tighten the oscillator tolerances to acceptable levels. Untrimmed, an oscillator's tolerance may be as loose as ±30-40%. After trimming, this may drop to ±5% or often less, which is well within the ±15% required to detect break symbols. Generally trimming brings the tolerance within ±2% which is the tolerance required by the current LIN specification during the synchronized portion of the messaging. Exact numbers here vary by manufacturer.

Oscillator trimming does add a step to the manufacturing flow, however. Also required is a method to perform the trimming. Most methods center around providing a pulse of known length to either a timer channel input or the UART and measuring that pulse in software on the device being trimmed relative to the untrimmed clock. The difference between the number of clock periods measured and that which is expected gives a trimming correction factor. This method requires coordination between the testing/programming development team which creates the pulse and the software development team which writes the firmware needed on the microcontroller to measure the pulse to ensure the pulse width is known and accurate. The trim value must be stored in non-volatile memory (NVM) to ensure that it is not lost when power is removed from the microcontroller.

Once the trimming of the internal oscillator has been accomplished, the use of a microcontroller with an internal oscillator in a LIN slave node is very similar to that of a microcontroller with an external crystal or resonator. The only significant difference is that the RC oscillator node must resynchronize on each message.

If the designer uses a microcontroller with an internal clock generation module, but which does not have a UART present, it adds another dimension to the method mentioned earlier in the BASIC MICROCONTROLLER section. Depending on the accuracy of the trimmed clocks, software timing loops might be less accurate methods of generating and receiving LIN messages, because they are based on a fixed number of clock periods corresponding to a given amount of real time. In these cases, interrupt based methods, such as the use of a timer, are more reliable.

DEDICATED LOGIC HARDWARE SOLUTIONS

Up to this point, all of the solutions explored in this paper deal with embedded microcontrollers. Those solutions are very flexible, but can also be relatively expensive. One of the primary goals of the LIN standard is to provide low-cost communications.

One of the ways to accomplish the goal of low-cost communications for smart sensors and actuators is to create dedicated hardware logic devices which can communicate on the bus. Until a standard protocol is agreed upon, this goal remains elusive and is one of the reasons why these types of devices have not been created in significant numbers.

LIN is poised to become the industry standard for Class A communications, making dedicated logic devices more feasible. At this point, application control logic can be added to create specific implementations to do specific jobs, such as a mirror motor controller.

These types of node implementations would likely be prototyped using field programmable gate array (FPGA) devices. FPGAs are far too expensive to be practical for automotive production volumes of ECUs. Once the application's logic design is tested and finalized, an ASIC can be created for the production volumes. Prototypes would also require the building up of all external components needed to interface to the network, power supply, and load.

Physical implementations might vary significantly, but one of the first probable ECU implementations would look very similar to a microcontroller implementation with the ASIC in place of the microcontroller. One major advantage to these types of devices is that they perform better in harsh temperature and electrical interference environments.

Dedicated logic hardware solutions are not without their challenges. For example, in a power seat application, it is necessary to differentiate between each of the motors in the seat. The motors themselves might be of identical type and function, but the headrest tilt motor cannot be mistaken for the front-to-back seat travel motor. In a microcontroller approach, the software can simply be changed to recognize a different set of LIN identifiers. In a dedicated hardware approach, some method must be employed to make this differentiation. Multiple versions of the ASIC are a possibility, but reduce the volume of a single ASIC design and require keeping separate inventories of parts. Both of these effects result in increased system implementation costs. Other possibilities, are the use of external components connected to port pins, and internal or external NVM to uniquely identify different nodes.

Another challenge for the dedicated hardware logic approach is the issue of changing application requirements. Once a design is tested and begins volume production, it cannot be changed quickly. This means that if insufficient forethought was put into the design, changing application requirements could result in large numbers of useless devices and expensive redesign stages.

In order to truly consider dedicated logic hardware solutions, as much complexity as possible must be forced back to the master node. The simpler a slave node application can be made, the more likely that a dedicated logic device can be created to serve that need. Abridging and standardizing the messaging strategy needed by the slave node will also serve to streamline the application's design.

Many challenges face the development of dedicated logic hardware solutions for LIN slave nodes, but the potential cost reduction for the devices continues to provide incentive and hope for further development.

THE MECHATRONIC APPROACH

With the standardization of the Class A protocol to LIN, additional possibilities are created for higher volume production of much more integrated solutions, sometimes called intelligent connectors or mechatronics. These solutions combine the electronic components needed to implement the ECU with the wiring harness connectors in the vehicle. This approach offers many advantages over the options mentioned before.

Figure 6: Example of mechatronic connector, showing internal components

One of the most apparent advantages to the mechatronic method is the reduction in form factor over traditional microcontroller PCB implementations. One method used in a mechatronic solution is the use of an insulated metal substrate (IMS) in place of a PCB. This is a material which is electrically insulated from the electronic components, but thermally conductive. This feature allows the IMS to dissipate heat generated by the devices connected to it, allowing more components to be placed in a smaller area, including power handling semiconductor devices. Figure 6 shows an example of a mechatronic device with the casing removed to reveal the IMS and components which are mounted to it. The example shown here is designed for prototyping, and as such does not use automotive grade connectors, but the concept extends easily to any style of connector.

As semiconductor providers are able to offer more integrated solutions such as the mechatronics concept, they can provide more compact and complete hardware solutions. Another advantage of this approach is that suppliers do not have the expense of creating and testing PCBs for their application. The PCB has been replaced by an IMS, provided by the semiconductor manufacturer, with components already bonded to it. The IMS solution will also be considerably smaller due to the fact that the silicon components placed on it do not have to be packaged separately for manufacturing purposes.

Removing the need for a PCB design also speeds up the design time needed for a given LIN slave node implementation.

REDUCING DISCRETE COMPONENTS

Cost reduction can be accomplished through many means and consolidation of components in a design is often one effective way to cut ECU costs.

Semiconductor manufacturers offer many components which integrate many features which were formerly implemented in discrete components.

Figure 7 shows one possibility of component consolidation can be seen. Integration of the LIN physical interface, power management and conditioning circuitry, and microcontroller management components results in a single device to perform these tasks.

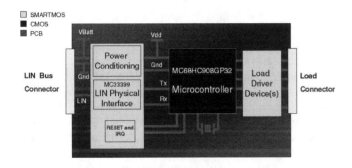

Figure 7: Microcontroller LIN slave implementation with consolidated power, communication, and microcontroller management components

In semiconductor manufacturing, analog components and digital logic components are generally manufactured using different manufacturing processes. Since the analog and power handling components use the same process, however, it is possible for the semiconductor manufacturer to integrate many of them into the same package for a relatively low cost.

The benefits of this integration of components are a reduction of the number of components to inventory, reduced manufacturing steps, increased reliability due to fewer board connections, and smaller PCB layouts resulting in smaller packaging options for the ECU. For these reasons, cost reductions usually result from increased integration.

CHOOSING THE PROPER IMPLEMENTATION

This section is not intended to serve as a design guide for LIN slave node designers, but to raise issues that must be considered when choosing the proper implementation for a given ECU design.

Some of the key considerations when deciding which type of implementation is most appropriate for a given LIN slave node application are:

- Application complexity
- ECU and system cost targets
- ECU form factor
- Reuse of existing resources
- Vehicle optioning levels for the ECU
- Flexibility for design changes
- Flexibility for field upgradability

First, the complexity of the application is a major factor in determining what implementation to choose. Initially, application complexity can quickly determine whether a microcontroller or dedicated logic hardware solution is most appropriate. For example, a simple application such as a switch panel, where the ECU monitors a series of switches and reports status when requested might be simple enough to consider a dedicated hardware logic device approach. In contrast, a window lift motor controller node which must support anti-pinch features is complicated enough to require a microcontroller. The application might also require memory functions or capturing of run-time calibration data which would necessitate the presence of NVM.

Cost is always a driving factor for ECU designs. If application complexity does not indicate an obvious choice, then other factors such as cost must be taken into account. It is impossible to accurately represent all factors of cost within the scope of this paper, but trends are important to note. Increased integration of components and reduction in cost of those components are critical to reducing cost. Cost must, however, be analyzed not just at the ECU level, but also at the system level. Integration can take the form of replacing multiple ECUs with a single, more complex ECU in certain cases, for a net reduction in system cost.

Another factor to consider is the form factor of the ECU in question. If the module is intended to be placed into a small space, a mechatronic solution might provide the smallest form factor. If physical size is not a driving issue for the ECU designer, then any implementation might be appropriate.

Reuse of resources is an important factor for a supplier to consider. Investments in tools, training, testing, and manufacturing equipment can be very substantial and should be reused for as long as is practical.

Vehicle manufacturers wish to tailor their vehicles to suit the needs of many customers with as little added cost as possible. LIN allows for simpler interchangeability of components to support these optioning levels, as the wiring harnesses in the vehicle do not have to change when adding new features. With various optioning levels come various ECU implementations to do a similar job. For example, a vehicle might have a simple power mirror with only a two axis movement or it might have a fully featured mirror adding power folding, heater, puddle lamp, turn signal indicator, and electrochromic reflectivity. The simple mirror ECU could easily be a dedicated logic hardware solution, where the complex mirror ECU might be better served by a microcontroller implementation. Vehicle option levels might best be viewed simply as multiple ECU designs.

Design change flexibility is a significant factor for many manufacturers. Requirements change rapidly and often, resulting in late term design changes. These changes are much more easily accommodated in microcontroller based designs. The tradeoff for this flexibility is the increased cost of a microcontroller over a dedicated logic hardware solution. Another aspect of flexibility is the ability to make changes and upgrades to the ECU once it is deployed in the field. This requires not only a microcontroller, but a microcontroller with NVM code storage. This also increases cost over a dedicated logic hardware solution.

OTHER DESIGN CONSIDERATIONS

There are a few other considerations that designers need to keep in mind. Many of these issues relate to the requirement mentioned above that manufacturers should, wherever possible, conserve and reuse resources. These factors are listed here for illustrative purposes:

- Hardware and PCB design expertise
- Software design expertise
- Development tools availability
- Design time-to-market
- Existing code libraries
- Component cost and availability
- Last minute design change probability
- Field programmability

MANUFACTURING CONSIDERATIONS

In addition to the design considerations mentioned above, there are also factors related to the manufacturing process which should be analyzed. Among these issues are:

- Assembly costs
- Factory programming
- Factory testing
- Maintenance and field reprogrammability

Assembly costs of ECUs, as well as the cost of the assembly of the entire system, should be taken into account when deciding upon slave node implementations. Admittedly, many suppliers designing a particular ECU may not have control over the entire system and therefore will largely be concerned with assembly costs of that ECU.

An additional manufacturing aspect is the step of factory programming. Dedicated logic hardware solutions might require little or no programming at all, whereas microcontroller solutions will require full programming of

firmware. The factory programming step requires manufacturing time and planning, both of which come at a measurable price.

Factory testing of ECUs can vary depending on their implementation. Like the factory programming stage, testing takes time in the manufacturing flow and may require significantly different and new techniques for new ECU implementations.

Maintenance and field reprogrammability or upgradability also figure in to manufacturing costs. Dealing with customer returns and upgrades will change from one implementation to another. It is likely that this might require changes to the manufacturing process, particularly in reusability of testing phases to diagnose customer returns.

CONCLUSIONS

There is a wide range of options for implementing LIN slave nodes. Each option comes with benefits and costs, and ultimately it is up to the designer to weigh these factors in determining the most appropriate choice to implement a particular design. Microcontroller based solutions come with a variety of options for making LIN slave node implementations easier to realize. Although microcontroller solutions offer flexibility, they are more expensive than dedicated logic hardware solutions, given sufficient volumes. Cost is not the only factor in deciding upon an implementation choice for LIN slave nodes. Design considerations such as availability of design expertise in hardware and software, development tools, code libraries, and the like also figure into the decision process. Finally, manufacturing issues such as programming and testing of the finished products must also be taken into account before an informed decision can be reached about how to create a LIN slave node device for a particular application. The LIN standard creates many opportunities to devise new and less expensive ways to network vehicle components, particularly because of the plethora of choices available for implementing LIN slave nodes.

ACKNOWLEDGMENTS

Special thanks to Alan Devine and Hans-Christian von der Wense for their assistance in drafting this paper.

CONTACT

Matthew Ruff is a systems engineer with the Systems Engineering team of the 8/16-bit Microcontroller Division of Motorola. He specializes in automotive multiplexing systems, especially CAN and LIN, and is a member of the Vehicle Architecture For Data Communications Standards Committee of SAE.

Email: mailto:matt.ruff@motorola.com

REFERENCES

1. LIN Consortium, "LIN Specification, Version 1.2", www.lin-subbus.org, November 2000.
2. Will Specks, Antal Rajnák, "LIN - Protocol, Development Tools, and Software Interfaces for Local Interconnect Networks in Vehicles", 9th International Conference on Electronic Systems for Vehicles, Baden-Baden, Oct. 5/6, 2000.
3. John V. DeNuto, Stephen Ewbank, Francis Kleja, Christopher A. Lupini and Robert A. Perisho, Jr., "LIN Bus and its Potential for use in Distributed Multiplex Applications", SAE 2001 World Congress, March 5-8, 2001.
4. Hans-Chr. v. d. Wense, A. J. Pohlmeyer, "Building LIN Applications", SAE 2001 World Congress, March 5-8, 2001.
5. George, Scott, "HC05 MCU Software-Driven Asynchronous Serial Communication Techniques Using the MC68HC705J1A", Motorola Semiconductor Application Note, AN1240, 1995.
6. Biershenk, Brad, "Software SCI Routines with the 16-Bit Timer Module", Motorola Semiconductor Application Note, AN1818, 1999.

ADDITIONAL SOURCES

The LIN-Specification Package and further background information about LIN and the LIN-Consortium is available via the URL: http://www.lin-subbus.org

Motorola application notes referenced can be found at: http://www.motorola.com/semiconductors

DEFINITIONS, ACRONYMS, ABBREVIATIONS

ASIC:
Application Specific Integrated Circuit

CAN:
Controller Area Network

ECU:
Electronic Control Unit

FPGA:
Field Programmable Gate Array

HVAC:
Heating, Ventilation, and Air Conditioning

IMS:
Insulated Metal Substrate

LIN:
Local Interconnect Network

NVM:
Non-Volatile Memory

PCB:
Printed Circuit Board

SCI:
Serial Communication Interface - see also UART

UART:
Universal Asynchronous Receiver Transmitter

Cycle-Accurate LIN Network Modeling and Simulation

Qian Chen, Yibing Dong and Salim Momin
Virtual Garage, Motorola Inc.

ABSTRACT

LIN (Local Interconnected Network) is a serial communications protocol that supports the control of mechatronic nodes in distributed automotive applications. This paper discusses LIN network modeling and simulation based on a token-based and event driven simulation platform. The complete LIN network features are modeled in the behavior level. The simulation is time-accurate and it provides system information, such as CPU load, bus utilization and message latency time. It can also simulate the scenarios such as network sleep and wakeup, switch event and error message. This LIN network simulation model can be integrated with CAN network simulation model for a complete vehicle network simulation.

INTRODUCTION

LIN is a low cost network which complements the existing portfolio of automotive multiplex networks. It is starting to become popular in the automotive industry. One major problem in the LIN network development is that before the hardware network is built, there is no way for the network designer to collect enough accurate system information of the network, such as CPU utilization, bus utilization and system response time. The system response time is especially a problem in the case that we need to consider some system issues, such as network wakeup, error message handling and gateway operation. Building the hardware network is costly and time consuming and any change in the design may require a complete rebuilding of the hardware network. Moreover, it is often too late to find the design flaws after the hardware network has been built and completely tested. The situation can be worse for the LIN network design in the sense that normally LIN network need to be connected to the CAN networks. It means that the thorough tests of the LIN network design may not be performed until both the LIN network and CAN network are built. Obviously, a simulation environment that can provide accurate information of the system performance for the LIN network design before any hardware is available will be a great help in the LIN network development.

LIN is a sub-network with speed up to 20k bit/s, wire length less or equal 40 m and recommended maximum 16 nodes. The cost for LIN is low since it uses single wire, its silicon implementation is based on common UART/SCI interface and it does not require quartz or ceramic resonator in the slave nodes. LIN uses a single-master / multiple-slave concept which guarantees the latency times for signal transmission (assume that there is no error). The LIN network structure can be described briefly with the following Figure 1.1:

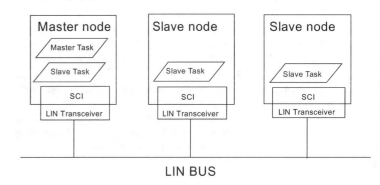

Figure 1.1 LIN network structure

LIN message has a fixed frame format with configurable data length (2, 4 or 8 bytes). A message frame includes two parts: header and response. The header consists of three fields: synchronization break, synchronization field and identifier fields and the response consists of several (2, 4 or 8 bytes) data fields and one checksum field. The header carries synchronization and identifier information from the master task to the slave tasks. Only the master task can send out a header. The response carries the data information. It is sent by the slave tasks from either a master node or a slave node. If the slave task fails to respond to a message header, the master task will transmit a new message header after a maximum time-out. The format of LIN message frame can be described with the following Figure 1.2:

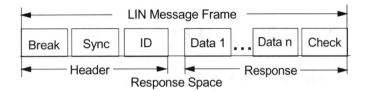

Figure 1.2 The LIN Message Frame

LIN NETWORK MODEL

The LIN network modeling is based on a simulation platform that is token based and event driven. The hierarchy of the LIN network modeling is as follows: The network model consists of the models of master node, slave nodes and LIN bus. The model of each node consists of the partition of hardware architecture and software architecture, which model the hardware and software behavior of the node respectively. The hardware architecture includes the modeling of the hardware blocks that are related to LIN message transmission, such as CPU, SCI and timer. The software architecture includes the modeling of the software tasks and the relationship among various software and hardware tasks. Furthermore the flowchart and cycles information of each software task are profiled and mapped into the software task model so that our LIN network simulation is cycle accurate. The LIN bus is modeled as a monitor of the message traffic in the bus. The hardware and software architectures of the master node and the slave nodes are different and they will be discussed separately in the following sections. Following is an example of the LIN network simulation model.

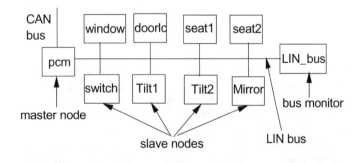

Figure 2.1 LIN network simulation model

HARDWARE ARCHITECTURE MODELING

The hardware architecture is one basic component in our hierarchical model structure. The LIN hardware architecture modeling include the modeling of the master node and the modeling of the slave node.

MASTER NODE HARDWARE MODELING

In the master node model, we have the hardware models of CPU, SCI and timer. These are the basic hardware blocks required for the operation of a LIN master node. The SCI is the physical layer for the LIN master node and a timer is required for the scheduling of message transmission and the detection of slave-no-response time-out. More hardware blocks can be added to the model if required. The hardware architecture for the master node model is described with the following Figure 3.1.

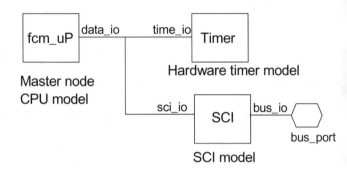

Figure 3.1. Master node hardware modeling

SLAVE NODE HARDWARE MODELING

The hardware architecture for the slave node model is described with the Figure 3.2. Here, compared to the master node model, we have a model block for the switch/activator in addition to the CPU and SCI models but we do not have a timer model block. That is because the switch and activator are normal for the functionality of a slave node, while a timer is not so critical for a slave node except for the detecting of bus idle time out. (This functionality is not simulated in our model since the normal simulation time is not long enough to observe this time-out.) The switch/activator block will generate the event stimulus for the simulation and the response to the switch stimulus, which enable us to check the latency time for the switch response.

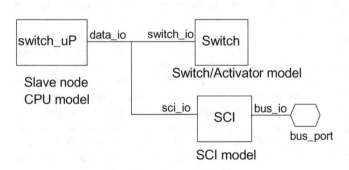

Figure 3.2 Slave node hardware modeling

SOFTWARE ARCHITECTURE MODELING

Besides the hardware architecture, the model also needs the software architecture. The software architecture defines the relationship among various software and hardware tasks that simulate the functions of the LIN protocol. As mentioned before, our LIN network modeling is based on a simulation platform that is token based and event driven. Therefore, each software task block and hardware block will communicate with the transmission of tokens. The software architecture defines the type of the message to be transmitted and the route of the transmission. By doing so, the software architecture describes the relationships among different software tasks and hardware blocks. The software architecture modeling also includes the master node modeling and slave node modeling.

MASTER NODE SOFTWARE MODELING

The software architecture for the master node is described with the following Figure 4.1. As stated before, the master node includes a master task and a slave task. The modeling describes the type of messages to be transmitted (not shown in the Figure 4.1) and the route of transmission among the slave task, master task, timer hardware block and SCI block. Hence, the relationship among these two tasks, the timer model and the SCI model is clearly defined in the software architecture.

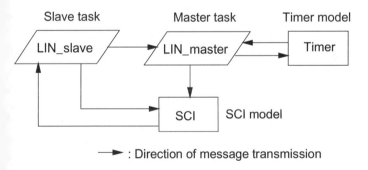

Figure 4.1 Master node software modeling

SLAVE NODE SOFTWARE MODELING

The software architecture for the slave node is described with the following Figure 4.2. The basic concept is the same as that of the master node. The software architecture defines the relationships among the slave task, switch/activator block and SCI block. The difference is that it consists of different component blocks. For example, the slave node only has slave task. In our model, it does not consist of the timer block while it includes a block for the switch/activator (refer to the slave node hardware architecture model). Which block is included in the architecture is determined by the specific application need.

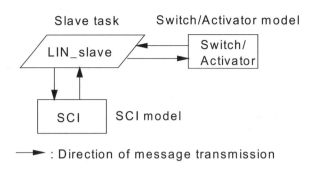

Figure 4.2 Slave node software modeling

SIMULATION AND RESULTS

In order to become a time-accurate simulation model, this LIN network model needs to include not only the hardware architecture and software architecture but also to simulate the software processing with accurate timing information. Therefore, in the simulation model we incorporate the timing information from the Motorola LIN drivers running in Cosmic M68HC12 simulator to make it cycle-accurate. In order to do so, first we need to profile the flow chart of the Motorola LIN drivers and run the LIN drivers in the simulator to profile the desired timing information. This timing information is incorporated in the programming of the software modeling of the LIN network to provide cycle-accurate timing penalty for the CPU. So the simulation result can show the CPU running time for each task and total CPU utilization. Furthermore, each software task can be assigned a priority such that high priority task can interrupt low priority task just as the how the real software works. Therefore, this simulation not only provides the CPU timing information as accurate as we can, but also provides the software processing delay for the message transmission.

The following Figure 5.1 shows an example activity chart for the LIN network simulation result. Each small mark represents an activity for the corresponding hardware or software block. Either it is processing a task or transmitting a message. This activity chart clearly shows the operation of the LIN network. For example, from this chart we can see that our simulation model simulates the following LIN network scenarios:

- How the LIN network wakes up from the sleep mode when a switch is pressed.

- How the LIN master node sends the message header and the slave node sends the response.

- How the slave-no-response time out works when the slave node fails to respond to message header.

- How the switch command is transmitted to the activator and how long is the latency time.

Also the simulation result can show us the information of CPU bandwidth and bus utilization which will be very useful when we analyze the penalty of LIN network operation in a LIN-CAN gateway.

Besides this activity chart, the simulation also provides the detailed information about each node and each message. For example, when does an error occur and what kind of error it is. Also, with a convenient GUI, the user can get the information such as: what is the maximum latency time for each message, how many messages are transmitted and received by each node and what is the dynamic moving average CPU load of each node. Finally, by clicking a button, the user can get an automatically generated report of the statistical data of each node and each message. Figure 5.2 gives a brief screenshot of these results and analysis.

As mentioned before, LIN network is usually used as a sub-network of CAN network. Several LIN networks and CAN networks will work together and messages and signals are exchanged among these networks. One important advantage of our LIN network simulation model is that it can be connected with our CAN network simulation model to perform the complete simulation of several networks as desired.

CONCLUSION

This paper presents the modeling and cycle-accurate simulation of LIN network based on a token based and event driven simulation platform. The simulation can provide accurate system information of the network to the network designer far before any hardware is built. It can also provide the module builder information of the network impact on the performance of each module. Furthermore this LIN network simulation model can be connected with our CAN network simulation model to perform a complete multi-network simulation.

REFERENCES

[1] LIN consortium, "LIN Protocol Specification, Revision 1.2", November 17, 2000.

[2] LIN consortium, "LIN API Recommended Practice, Revision 1.2", November 17, 2000.

[3] Wense, H., "Introduction to Local Interconnect Network", SAE World Congress, Detroit, March 2000. Document No. 2000-01-0145, SAE press.

[4] Motorola Inc., "LIN12 Driver User's Manual, Rev. 1.2", January 12, 2001.

CONTACT

Dr. Qian Chen
Virtual Garage, Motorola Inc.
41700 Six Mile Road
Northville, MI 48167
Email: Qian.Chen@motorola.com

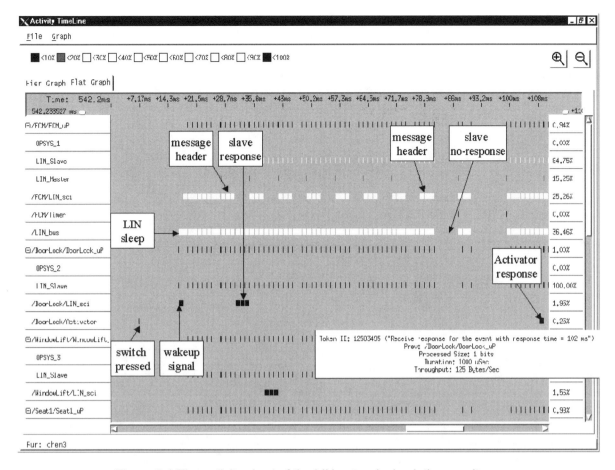

Figure 5.1 The activity chart of the LIN network simulation result

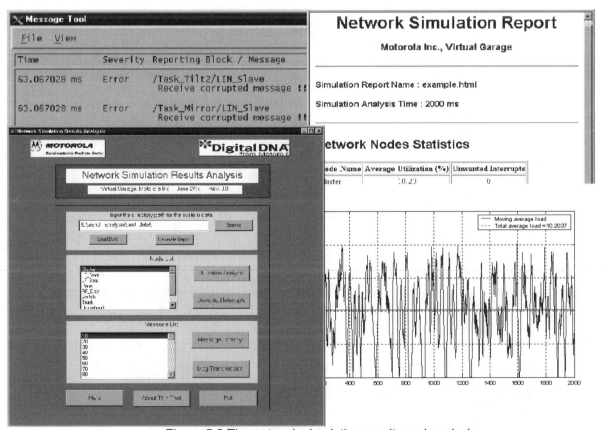

Figure 5.2 The network simulation results and analysis

2001-01-0065

Seamless Solutions for LIN

Hermann Kern, Heinrich Gschloessl and Wilhard von Wendorff
Infineon Technologies

ABSTRACT

Today's body and convenience applications in general directly control actuators and sensors from a single central electronic control unit (ECU). Future systems will be made of subsystem-clusters communicating via a local Class/A communication bus. This enables modular system design to reduce system complexity. For these types of new distributed applications the LIN bus is currently the most promising communication protocol.

To allow a seamless migration from existing centralized to these next generation clustered system developers require software and hardware products for a homogenous and transparent LIN bus communication.

Within this paper Infineons presents its new LIN solution package:

- details on Infineons newly developed software building blocks allowing seamless customer specific LIN implementations;
- Infineon Technologies new family of LIN line drivers supporting smart wake-up from low power mode and smart power supply control;
- details on the line driver circuitries, configurable SW building blocks, EMC results, and achievable system performance are disclosed.

INTRODUCTION

The electronic system complexity of a car is currently growing extraordinary fast. The system and vehicle suppliers are therefore going through a process the personal computer industry went through several years ago.

The main issue to solve is setting up a development process:

- controlling the rapidly growing complexity of hardware and software;
- achieving the required quality to allow mass production with a low number of faults;
- shortening the development cycle;
- competitor differentiation by increasing or enhancing functionality;
- building up a network hierarchy covering the entire car within the road environment.

Today's centralized system use dedicated communication lines to pass analog/digital data from a central ECU to actuator and sensor units or vice versa. Each relevant state or event is coded analog to a dedicated channel. Being an analog system with a limited transmission delay, the time delay for communication is neglectable and deterministic. But the complexity grows fast with the number of sensors and actuators to be controlled. This growing number of analog links is increasingly sensitive to noise pick-up and emits electromagnetic radiation.

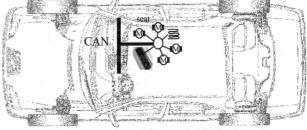

Figure 1: Central System Architecture

The automotive industry therefore follows the old rules of emperors: "divide and conquer".

To migrate a system from a centralized to a distributed system it has to be broken up into subsystems with common, standardized, and open interfaces.

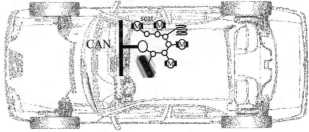

Figure 2: Distributed System Architecture

These newly introduced interfaces are a shared boundary between three or more subsystems. Rules (so-called communication protocol) of sharing this common resource have to be followed to allow proper communication of data across these interfaces. These protocols are implemented by dedicated hardware and communication software (hardware/software co-design)

The interfaces/protocols must support:

- transparent debugging, diagnosis, and maintenance during development and life cycle;
- flexible system distribution and assembly;
- synchronization of the asynchronous and de-tuned local clocks;
- easy to implement software and hardware interfaces;
- low cost interfaces;
- system composability (the ability to test a sub system and ensure the validity of these results within the overlaying total system);
- availability of devices supporting the interfaces;
- optimized wire harness (cost wise, space wise, and availability wise);

By breaking up such systems into smaller subsystems, the amount and length of the analog communication channels is decreased. Each of these subsystems communicates data to other sub-modules using the newly introduced multiplexed, commonly shared digital communication interface. Such an interface can be effectively implemented using an open standard field bus approach. The most promising field bus standard is currently the LIN bus, developed by an automotive industry consortium [2].

With the introduction of LIN bus as standard sub-bus the migration towards distributed systems will speed-up. LIN will open new system partitioning especially in the field of body and convenience applications. Door modules, Seat modules, and HVAC will be the applications using LIN first and the others will follow. But LIN is not going to replace CAN as communication path. LIN will be used as sub-bus below CAN.

The approach of distributed systems using sub-buses is not totally new but has not been realized so far because of cost reasons. Bus systems, which have been available, were high and low speed CAN, J1850, and K-Line. Now with LIN cost targets can be reached.

In contrast to CAN or J1850 the LIN requires no dedicated on-chip microcontroller communication module. LIN utilizes the standard serial communication interface (USART). That is one mayor point for the well-balanced cost/performance ratio of this recently introduced Class/A sub-bus.

LIN overcomes major issues when decentralizing a system:

- deterministic and real-time timeliness. CAN introduces a huge amount of jitter to the timeliness of

communication due to a complex mechanism of error concealment and multi-master capability not mandatory required for intra-system communication.

- cost and complexity of a node is high. CAN requires complex communication circuitry and precise resources as a high precision clock, power supply etc.
- LIN provides more extended composability enabling scaling/configuring a system into a variety of configurations.
- LIN especially targets the needs of intra-system communication (communication within a single system/application) where else CAN targets inter-system communication (communication between independent systems/applications).

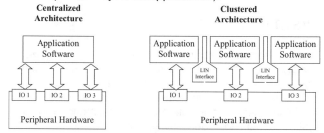

Figure 3: Structure of the LIN software driver

One of the challenges of the automotive industry is to be able to follow a smooth and clear road to migrate an existing centralized system to such a new decentralized field bus approach.

This migration requires:

- introducing new interfaces into the existing system (into hardware and software);
- distributing the single software to respective sub-systems;
- distributing the required "intelligence" (processing performance);

This touches all levels of design:

- the software design environment;
- the hardware building blocks (devices);
- the qualification and test environment;
- the debug and maintenance environment;
- the data for power and thermal budget calculations;
- the data for EMC requirement calculations.

Therefore exact knowledge of required communication resources (EMC, memory, bandwidth, power, and performance) is required. Infineon Technologies provides a toolbox to address the majority of these issues and to ease off the cumbersome path for system migration. This new toolbox simplifies the distribution of functions to a couple of electronic control units (ECU) hooked up by a system intra-communication bus (network within a single system/application) by providing flexible building blocks on different levels of system design:

- software modules supporting customer specific implementations;
- broad portfolio of integrated devices providing the respective resources of each sub-system;
- data on EMC, power, performance, memory requirements, etc.

So if a system developer has to migrate a conventional central system as sketched in Figure 1 into a distributed system, he has to identify and implement the new system internal interfaces. Infineon simplifies the task by providing software and hardware blocks ready to be used. These building blocks allow the decoupling and communication of the threads formerly running on a single ECU and after conversion running on different ECU. Designers no longer have to hassle around with the deep details of the LIN communication and partitioning, but can rely on the higher-level interfaces in software (application interface: API). The next step is to choose out of Infineon broad portfolio the appropriate hardware components. This hardware has to provide all the features used when designing centralized systems, e.g. synchronized power-up (smart power supply management), start-up synchronization etc. Therefore Infineon set up a family of different hardware to provide the respective features. Coming along with these new components are data as EMC behavior, power consumption etc. to fit the new distributed design to the given EMC, performance, and power budget.

Hopefully the last step of an engineer following this new seamless migration path is to head home early, because the system is running ahead of planned schedule. But wait, let's go through this flow and the respective components step for step.

LIN BASIC

LIN is a single master multiple slave bus system. According to that one master node and several slave build up the bus system. The master node contains a master task and a slave task, whereas slave nodes contain slave tasks only. Figure 4 illustrates the bus architecture.

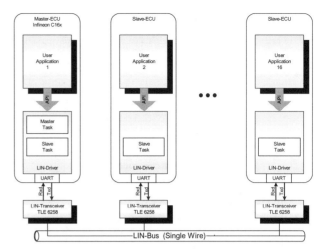

Figure 4: LIN bus architecture

A fixed message format is defined for the LIN protocol. Each LIN message starts with a header frame consisting of a synch break followed by a synch field and an identifier field. Only the master task may send this frame. Next 2, 4, or 8 bytes code the message data. A check sum field completes the LIN message (Figure 5). This response to the header is sent either by the slave task of the master node or by one of the several slaves nodes hooked up to the LIN bus.

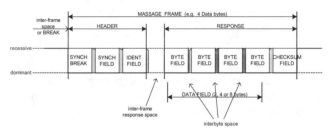

Figure 5: LIN message frame

Within a LIN bus system the master node initiates all message transfers.

Data exchange bases on a common hardware peripheral (serial communication interface) controlled by a dedicated LIN software driver. In opposite to CAN and J1850 the driver software handles the basic communication layers and takes care of message transfers, message filtering, and error detection (protocol handling). Two major modules, a master task and a slave task build up this driver software.

In general only the master task may send header frames and only a single slave tasks may reply a respective data field and check sum field. Certainly several slave tasks may receive one message. The master task controls the data exchange between master to slave, slave to master, and slave to slave. Figure 6 below illustrates the LIN communication paths.

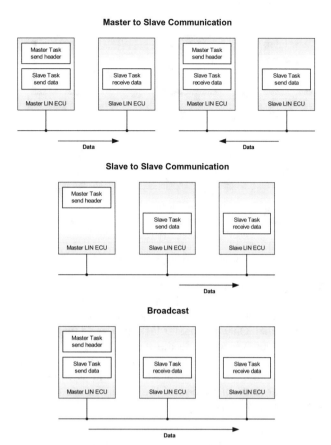

Figure 6: LIN communication paths

The user has to configure the system as such, avoiding several slave tasks concurrently responding to the same ID. Moreover the user is responsible for arranging an application specific network management.

LIN SOFTWARE DRIVER

The domain of LIN will be local sub-bus systems. Most systems using LIN bus will have a gateway to CAN. Infineon Technologies has a strong involvement to CAN providing microcontrollers with on-chip CAN controllers, standalone CAN controllers, and a family of CAN transceivers (high, low speed, and single wire CAN). Moreover Infineon Technologies will also support LIN based designs with LIN transceivers, 8, 16 and 32-bit microcontroller families and LIN driver software.

Following the implementation concept of Infineons new LIN software driver is presented and details are given for the microcontroller hardware, software, and performance requirements.

Configurable software building blocks handle the LIN protocol. LIN message frame handling is done autonomously by the LIN software driver. Operations on LIN are at disposal of the user and are initiated by API-function calls.

The LIN driver entirely encapsulates hardware modules and exclusively handles all on-chip peripherals of the Infineon C16x microcontrollers, which support LIN. Doing that makes life easy for the designer and he can focus on his application. Figure 7 illustrates this approach.

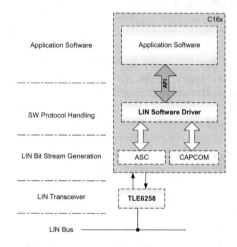

Figure 7: Structure of the LIN software driver

The LIN software driver builds on several state machines and functional blocks. Three main processes can be identified within the LIN software driver running as concurrent threads.

- a handler for message transfer that coordinates the correct generation of the message frame.
- a message filter that evaluates all received ID's within the slave task for their relevance and takes respective further actions if required.
- error detection process responsible for bus monitoring and detection of timing violations.

LIN SOFTWARE DRIVER PROPERTIES:

The message buffering will be discussed in detail hereafter to demonstrate the flexibility of the software driver architecture.

The LIN software driver provides a high amount of flexibility to the user in terms of LIN message buffering. The driver itself supports a single message buffer for all incoming and outgoing messages. It is up to the user to determine additional identifier specific message buffers, the respective location, and means of data integrity. Therefore the driver only requires a small amount of resources and provides a maximum of flexibility and applicability to the user. This concept of user definable message buffers allows e.g. receiving data objects directly from the LIN bus and transferring it to a CAN message buffer and vice versa.

LIN software driver properties are:

- LIN software driver and application code is cleanly divided into different libraries;
- strictly separation of specific application sections and LIN software driver data sections;
- enabling the user to handle interrupt service routine directly e.g. by a real-time operation system;
- allowing user definable message buffer;
- supporting LIN – CAN gateway functions.

A header file drives the configuration of the LIN driver. This allows a wide range of application specific configurations e.g. protocol and target specific settings. The workflow is illustrated in figure 8.

LIN protocol specific settings are:

- baud rate settings;
- configuration of Master or Slave LIN ECU;
- declaration of LIN Messages (ID allocation for send and receive messages).

Target specific settings are:

- allocation of Timer resources;
- allocation of serial interface resources;
- allocation of interrupt priorities.

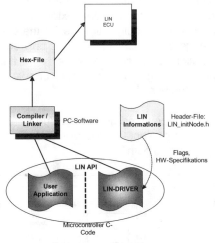

Figure 8: Configurable LIIN driver software

SOFTWARE DRIVER SERVICES

The Infineon Technologies software driver provides several services for a LIN implementation. All these services are under control of the user and invoked by service calls.

Two major categories of operations are provided:

1. Administrative services are used e.g. to configure or initialize the LIN software driver, respectively

- LIN driver initialization

2. Functional services making use of services the LIN software driver provide.

- Message transmission
- Message reception
- Message filtering
- Connect/disconnect the LIN node to the LIN bus
- Sending "*go to sleep mode*" command
- Sending "*wake up*" command
- Bus timeout detection
- Frame monitoring
- ID field calculation
- Data length extraction
- Check sum calculation
- LIN message scheduler

SOFTWARE DRIVER PERFORMANCE

The LIN protocol uses hardware and software resources of a microcontroller. Like in real life nothing is for free. In this chapter the required on-chip resources, memory consumption and CPU load will be discussed to provide the system designer a reliable data base. The following absolute numbers relate to a C16x 16-bit microcontrollers of Infineon Technologies clocked with 20 MHz system clock and executing code from on-chip memory.

LIN utilizes the serial interface for message frame generation and additionally occupies one timer channel for LIN bus timing monitoring e.g. for detection of not responding slaves or no bus activity. Both peripherals require an interrupt handler and most of the LIN software driver is executed within the interrupt handler of the serial communication channel peripheral.

The introduced software driver occupies less than 3 Kbyte of code memory and less than 100 bytes of data memory. These numbers include all the basic LIN software driver functions. Only application specific user LIN message buffers require additional memory space.

Figure 9 sketches the signals on the LIN bus (CH2) and CPU load on the C16x (CH1). This microcontroller is configured as LIN Master and executes a LIN master and slave task. The master task transmits the header and the slave task of the same ECU attaches four data bytes and the corresponding check sum byte. During each spike on CH1 the CPU is involved in the LIN software driver. States 1, 2, and 3 are handled by the master task. The slave task is involved during states 4 to 9. Message filtering is done within state 4. Error detection takes place at every state of the LIN frame generation.

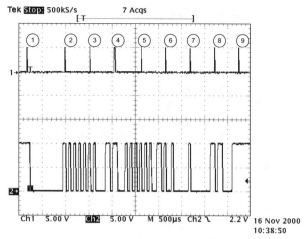

Figure 9: LIN buses signals

Master task:

1) Send-synch break
2) Receive-synch break and send synch field
3) Receive-synch-field and create / send ID-Field

Slave task:

4) Receive-ID-Field (message filtering), copy data to LIN transfer buffer and send first data-byte
5) Receive first data-byte and send second data byte
6) Receive second data-byte and send third data byte
7) Receive third data-byte and send fourth data byte
8) Receive fourth data-byte and calculate / send checksum-Field
9) Receive-Checksum

The LIN frame sketched in Figure 9 contains 4 data bytes and is running at baud rate of 19.2 kBaud. Its frame length is about 4.5 ms. The required CPU time for a LIN frame generation depends strongly on the number of declared ID's. In that example configuration 16 identifiers are defined. During the states 1 to 9 the CPU of the master LIN node is busy processing LIN driver operations. The overall processing time for the LIN software driver is 110 µs. While sending and receiving a LIN message frame the total CPU load of the LIN master node is below 3 %.

THE LIN TRANSCEIVERS

In general the physical layer of LIN is basing on an already well-known and approved standard: ISO9141 or also known as K-line or K/L-line. The ISO9141 network was mainly developed for transmission of diagnosis data within a garage. For that reason further improvement of the LIN physical layer performance respectively the transceiver is required in order to fulfil the requirements on a network that is operating while the engine is running but also during parking. Figure 10 sketches the main components of a typical LIN network.

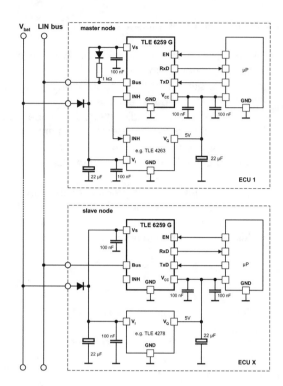

Figure 10: LIN network

To achieve a basic termination resistance for the recessive bus voltage level a 30kΩ pull up to the battery supply voltage is integrated in the transceiver ICs. Only the master node has an additional external termination of 1kΩ to achieve the required timings for the dominant to recessive edge of the bus signals. This only changes the total termination resistance negligible when the number of slave nodes changes due varying electronic equipment connected to the bus.

In case of an un-powered node (e.g. due to a blown out fuse) the bus is prevented from reverse currents by the diodes in the pull up path.

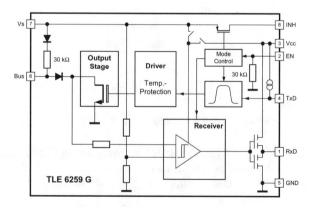

Figure 11: Block diagram of TLE6259

Infineon Technologies is going to offer two different stand-alone transceiver ICs for LIN applications: the TLE6258 and the TLE6259. Both are coming up in a P-DSO-8 package. The pin-out follows the proposal of the LIN consortium and represents the merger of the results of a pin FMEA as well as the requirements of EMC. In addition to the features of the TLE6258 the TLE6259 offers an inhibit output to control the voltage regulator that supplies entire ECU (Electronic Control Unit). When sending the TLE6259 to sleep mode via it's control input the inhibit output becomes high impedance by this switching off the voltage regulator. Now a wake-up of the ECU is only possible via a wake-up message on the LIN bus. During sleep mode the internal wake-up detection circuitry is supplied via the battery supply line. The current consumption is as low as 30µA in order to support the ever-decreasing quiescent current consumption requirements of today's and future cars. As soon a wake-up message is detected, the voltage regulator is enabled via the inhibit output pin and the wake-up is reported at the RxD output. The TLE6258 requires external 5V supply to detect bus wake-ups when switched to stand-by mode.

Compared to a differential mode as the CAN-system a LIN system is expected to have a much lower performance regarding EME (Electro Magnetic Emission). But the much lower maximum data transmission-rate in combination to limited slew rates of the bus signals is avoiding most of the problems arising in high-speed data transmission systems. Furthermore rounded corners of the signal edges also contribute to reduced emissions. To evaluate the EMC performance of the transceiver ICs on one hand the DPI method (Direct Power Injection) is used. RF interference is coupled into a LIN system to test the electromagnetic susceptibility. On the other hand the same test setup is used to investigate the electromagnetic emission performance. Figure 12 and 13 show test results for the first design step. Also the influence of additional capacitive bus-load is shown.

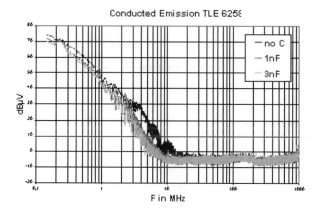

Figure 12: Electromagnetic emission TLE6258

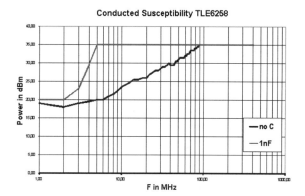

Figure 13: electromagnetic immunity TLE6258

CONCLUSION

Intersystem and intrasystem communication is evolving to one of the key technologies within future automotive systems. Infineon Technologies provides a broad portfolio of hardware, software, and application building blocks to reliably shorten the design cycle through robust network solutions, e.g. using CAN, J1850, ByteFlight, LIN, etc.

This paper presented the LIN family concept within Infineon Technologies portfolio. Software drivers allow a fast prototyping to investigate LIN architectures based on Infineon Technologies broad spectrum of integrated circuits. These modules allow power, EMC, and performance budget calculations to speed-up the migration from centralized systems to decentralized clustered system. So welcome aboard to Infineon's system world and make your sailing through the automotive communication highways a breeze.

REFERENCES

1. Murray, C. J.: "European carmakers get behind low-cost bus", EETimes, March 06, 2000.
2. von der Wense, H.-C.: "Introduction to Local Interconnect Network", SAE World Congress, Detroit, March 2000, Document No. 2000-01-0145, SAE Press.
3. von der Wense, H.-C.: "Introduction to LIN (Local Interconnect Network), http://www.lin-subbus.de/pdf/introduction_to_lin.pdf, March, 2000.
4. Web side of LIN: http://www.lin-subbus.org.
5. "LIN: Class-A Open Mux Standard Gains Momentum", Vol.13, No. 3, The Hansen Report on Automotive Electronics, Paul Hansen Associates, April 2000.
6. Web side of Infineon Technologies: http://www.infineon.com.

CONTACT

Hermann Kern
Infineon Technologies
Automotive System Engineering
St.-Martin.-Str. 76
81609 Munich, Germany
Tel.: (49) 89 234 83685
Fax: (49) 89 234 85150
Email: hermann.kern@infineon.com

Heinrich Gschloessl
Infineon Technologies
AI AP SYS PM
Balanstr. 73
81541 Munich, Germany
Tel.: (49) 89 234 22762
Fax: (49) 89 234 711898
Email: heinrich.gschloessl@infineon.com

Wilhard von Wendorff
Infineon Technologies
Automotive System Definition
St.-Martin.-Str. 76
81609 Munich, Germany
Tel.: (49) 89 234 83270
Fax: (49) 89 234 717830
Email: wilhard.wendorff@infineon.com

2000-01-0153

Introduction to Local Interconnect Network

Hans-Christian von der Wense
Motorola Munich

ABSTRACT

In June 1999, five major European car-manufacturers, one semiconductor supplier, and one tool vendor agreed on a specification for a class A multiplex protocol called LIN (Local Interconnect Network). This protocol has the following major features:

- Single Master / Multiple Slave Concept
- Low cost silicon implementation based on common UART/SCI interface hardware
- Self synchronization without crystal or ceramics resonator in the slave nodes
- Guaranteed latency times for signal transmission
- Low cost single-wire implementation (enhanced ISO 9141)
- Speed up to 20Kbit/s (limited by EMI)

This protocol offers for the first time the common base for class A multiplexing, supported by major car companies and being implemented by a large number of suppliers.

INTRODUCTION

In the last century the automotive industry has made great progress. Main factors in this progress have been electrical systems and electronics. In today's cars a high number of electrical motors, lamps and other actuators, sensors carry out important functions for functionality, safety and comfort. In an upper class car today there are more than 100 small electrical motors, which move seats, drive window lifts, flaps in the air conditioning, fans, windshield wipers, door locks and many other mechanical items (Figure 1: Automotive Body Electronic Applications).

In the early days of electrical comfort systems, the driver or passenger has controlled motors directly over switches and wires. With an increasing number of motors, this would mean to have a substantial amount of weight in the wiring because each of the wires has to be able to carry enough current for the particular motor.

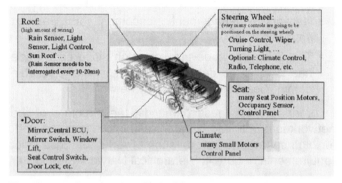

Figure 1. Automotive Body Electronic Applications

In recent years electrical control units (ECUs) have taken over this function. They are located remotely in the car and control a number of motors, which are in their area. A door, for example, could have one ECU that controls Window lift, panels, door locks, and the mirror. For the control of the passenger door from the drivers seat (and not only for that purpose), communication between the ECUs is necessary. This is done with a multiplexed network. A typical network is the Controller Area Network (CAN). A complete body electronic network system is displayed in Figure 2: Typical Automotive Body Electronic Network.

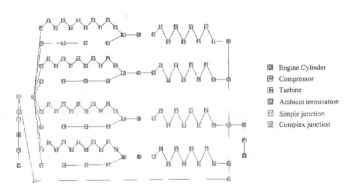

Figure 2. Typical Automotive Body Electronic Network

The latest trend is applying the ECU directly to the mechanical system: each motor get it's own ECU, which is directly mounted on the Motor. Or in other words: Actuators and sensors become intelligent. Such a mechatronic system is displayed in Figure 3: Mechatronic Stepper Motor.

Figure 3.   Mechatronic Stepper Motor

All these smart systems need to be connected to the main body network. This could be done directly, all nodes to one network. That has several disadvantages:

- The overall network traffic becomes very high
- Network management becomes an almost impossible task
- The costs per node are very high since each node has to fulfill the demand of high speed and reliable communication of a main body network.

In order to meet these demands, the structure of the network is partitioned hierarchically into a main network for high speed and several sub-networks with less strict requirements (Figure 4: Hierarchical Network Structure).

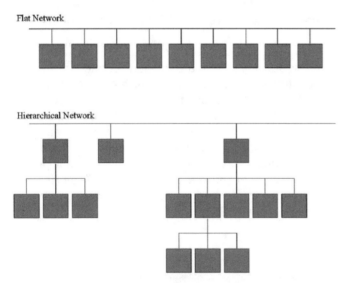

Figure 4.   Hierarchical Network Structure

Due to the increasing number of small nodes in the system the main demand of the automotive industry is to keep the costs of a subnet as low as possible. This allows adding functionality to the car at affordable costs.

There are already a high number of different solutions that have been incorporated in different cars. Yet, there has been no standard. In fact, in some cars several different sub-nets can be found - each department has it's own. This makes reuse of system resources impossible. Suppliers must also make their products compliant to their customers' product, which is difficult and expensive when no standard exists.

HISTORY

In fall 1998 the car manufacturers Audi, BMW, DaimlerChrysler, Volvo, and VW, the Semiconductor manufacturer Motorola, and the tool company Volcano Communication Technologies (VCT) met for the first time to tackle this problem. Volvo and tool supplier VCT already have had experience with a sub-bus system which has been implemented in the Volvo S80. This experience has been a valuable input to the discussion about LIN. The intention of this workgroup was finding a solution for a cost competitive implementation of a network that fulfills the requirements for the majority of applications and car manufacturers. The costs (and the performance) of this network should be clearly below the one of a main body network (e.g. CAN).

REQUIREMENTS

Electrical systems in cars have to fulfil many requirements. This is also true for a sub-bus system.

COSTS – One of the most important requirements is the cost competitiveness over traditional solutions, e.g. CAN or hardwired applications. Following aspects have to be taken into consideration:

Wiring – The classical CAN bus uses 2 wires, direct connection to the motors even more (e.g. stepper motors need 4 wires). A sub-bus solution should have less than 2 wires. Cost estimations regarding wiring talk about 0,50 - 1 $ per meter of single wire.

Semiconductor Die-Size – CAN communication needs specific peripheral devices that occupy silicon real estate. Due to their high complexity the size of CAN modules is considerably higher than the one of a simple serial interface. Also the voltage levels on the bus play an important role regarding the chip size: The CAN needs voltage regulators operating at 5V to drive the lines. A line based on battery voltage level (V_{BAT}) only needs I/O that is able to handle this „high voltage".

Clock Generation – A significant factor to reduce costs on a remote node is the number of required components. Typically an external clock generator is needed to support communication. Cost reduction can be acquired if the external crystal is replaced by an internal RC-oscillator. The communication protocol has to provide the means to recover the clock from a master node.

Simple Implementation – The communication solution should be implementable in a simple way. This reduces costs during development. There has to be a simple interface to the application program and the footprint of the communication software should be very small.

Simple Protocol – Simple actuators or sensors do not need a microcontroller. The protocol should be capable to be implemented in state machines.

Independence from Hardware Platform – The protocol should be microcontroller independent. This way also microcontrollers from different suppliers can communicate with each other. Car manufacturers and their suppliers have the full freedom to implement their sub-bus nodes.

SAFETY – The sub-bus solution must be conforming to automotive safety requirements. Erroneous messages must be detected with high probability so that the nodes can react properly. A safety level similar to CAN would be desirable but can only be achieved at relatively high costs. A sub-bus is mainly used for local communication. Therefore the safety requirements are clearly below the ones of CAN.

SPEED – Messages have to arrive at their destination within a specific amount of time. For example a window has to react within 100ms on a push of a button. Else the user might think the system does not work at all. Following aspects have to be taken into consideration:

Transmission Speed on the Bus – A high transmission speed on the bus is the basis for a high throughput. The speed may be limited by cost factors and by EMI aspects.

Length of Messages – An adaptive size of the message frame ensures that messages are transported more effectively. Frames with "unused data" are mainly avoided.

Message Overhead – All messages need a certain overhead. This overhead consists for example of identifier, synchronization efforts, arbitration efforts, check bytes etc. A reduced overhead increases the amount of data that can be put on the bus.

Message Frequency – The shorter the period in which a specific message can be transported, the shorter is the time between "pushing the button and reaction of the receiving system. If a specific message has to wait for a long time to receive a "slot" in the message flow the timing behavior of the complete system might be unacceptable. Each signal that has to be transferred on the bus should have it's own specific transmission frequency appropriate to it's application.

Predictability – In order to be able to specify a worst case timing, the system must be predictable. With non-predictable systems only a statement based on statistics can be made. Predictable systems offer (in the case that no error happens) an absolute worst case timing statement.

Extension of Main Bus – Sub-busses are used as an extension of the main-bus. Data should be transported transparently from the main-bus to the sub-bus and vice versa.

EXTENDABILITY – Car manufacturers offer their cars with different variants of equipment. There could be a door system that is equipped with a window lift, the next car is delivered without a window lift, and the third car has additionally a panel that controls the seats. All of these variants should be managed by the same bus systems. Additional nodes should be easily attachable without a complete redesign of the complete network structure.

There should be also the possibility to "hotplug" nodes e.g. for diagnostic purposes. The system should be capable to detect which nodes are participating in the communication. There are operating modes possible in which only some nodes of the network actually work. Others only work when the ignition is turned on.

THE SOLUTION WITH LIN

Almost any modern microcontroller features a serial interface (SCI -- Serial Communication Interface / UART -- Universal Asynchronous Receiver Transmitter). Therefore it makes sense to utilize this interface for a sub-bus concept. The serial interface is cheap and manufacturer independent.

The following concept has been developed:

- Standard low-cost serial interface
- Master-slave concept, to avoid complex arbitration.
- Clock recovery from the serial data line, in order to use R/C-Oscillators at the satellite nodes.
- 12V single wire interface, typically using the ISO 9141 standard
- 1/3 communication cost relative to CAN

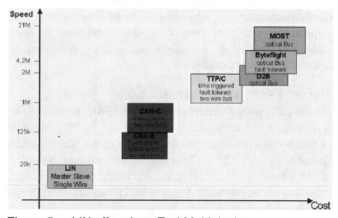

Figure 5.   LIN offers Low End Multiplexing

The system communicates via UART or SCI, which exists on nearly every microcontroller. The most common format for a data frame is Startbit, 8 Data bits, 1 Stopbit, no Parity. The number of data bits per frame is fixed to 8.

A complete message frame consists of synchronization sequence, identifier, 2 to 8 data bytes and a checksum.

The communication protocol works as follows (Figure 6: LIN Message Frame): The master sends a header, which consists of a frame synchronization sequence (basically a defined number of dominant bits), a synchronization byte for clock recovery and a message identifier. With the information of the identifier each slave knows whether it should transmit, receive or ignore data. Exactly one slave is allowed to send the data bytes and the check byte. Depending on the identifier the message is received by one or more slaves (multicasting is possible).

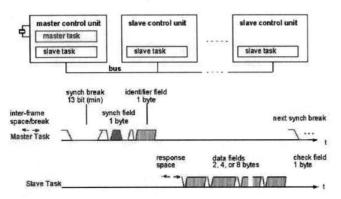

Figure 6.   LIN Message Frame

This message structure has the following advantage: Only the master node determines which message is to be placed on the bus. Arbitration does not take place. The master is scheduling the messages according to a table and controls the timing of the messages. This way the latency and transmission time of messages is well known. The complete system is predictable.

Disadvantage of this concept: when the master does not work, the complete sub-bus cannot operate. On a comfort bus this can be tolerated. In this case the comfort application (e.g. a window-lift) does not work. The application program has to make sure that a non-operating bus (e.g. an open wire) does not lead to unpreditctable operation.

FRAME SYNCHRONIZATION – Frame Synchronization is achieved with a synchronization BREAK. It consists of a minimum of 13 consecutive dominant bits. This number of bits is dependent on the inaccuracy of the slaves' clock generators (+/- 15%). The slaves can clearly detect a break condition if the detected dominant period is longer than any other possible dominant bit stream on the bus.

SYNCHRONIZATION BYTE – In order to retrieve the master clock from the data stream, the first byte of a message frame is a synchronization byte. This byte consists of a known pattern with a number of falling edges. The time between two falling edges is well known. The slave determines the time between two falling edges according to it's own time base and adjusts the baudrate. This way it can readjust it's serial clock base on every frame and compensate clock inaccuracies that might derive e.g. from temperature drifts. Therefore a crystal is not necessary.

IDENTIFIER FIELD – The identifier field (ID-field) is used as message identifier. With it the slave nodes know whether they have to send, receive or just ignore the data. The identifier contains also length information, which is coded into two bits. This divides the total number of 64 into 3 length categories: 32 2-byte identifier, 16 4-byte identifier and 16 8-byte identifier. This approach allows an easy implementation without a microcontroller. It also allows checksum control without the knowing any details about the message. Messages that contain information that occupy e.g. 1 byte, 3 bytes, or 5 bytes have to use the next possible message length category.

The remaining 2 bits of the ID-field are parity information to protect this highly sensitive information.

The sub-bus is intended to be used only locally. Therefore it is recommended not to use more than 16 nodes. Theoretically more than 16 nodes can be attached, this results in a higher electrical load. The message identifier consists of 6 identifier and 2 parity bits. This way up to 64 different identifiers are possible.

DATA FIELDS AND CHECKSUM – The sending slave puts the data bytes on the bus. The identifier determines the number of data bytes. The calculated checksum is sent afterwards. The receiving slaves compare the checksum and disregard the message on checksum error.

MASTER NODE AS SLAVE – In the very common case that the master node has to transmit or receive data to/from a slave node, the master node must execute the appropriate slave task ("The master is slave!"). Master in the nomenclature of LIN has to be understood as a master-task within the master node. The master node usually has also a slave task.

PHYSICAL LAYER

The use of a physical layer that is compliant to ISO 9141 bears several advantages: The behavior of the line is well known, existing line drivers can be reused, there are also some microcontrollers with integrated ISO 9141 interface on the market (e.g. Motorola's MC68HC05PV or MC33393TM). The wired-AND concept offers capability to clearly detect errors, especially bus collisions. The use of V_{BAT} eliminates the necessity of a high power voltage regulator.

However the ISO 9141 specification in its pure form is in some aspects problematic. A few enhancements ensure that data is transmitted properly:

In order to handle ground shifts the reception levels of the devices should offer a significant distance to the transmit levels. Therefore the LIN physical layer requires a threshold of a maximum of 40% V_{BAT} for low level and a minimum of 60% V_{BAT} for high level (Figure 7: LIN Voltage Levels).

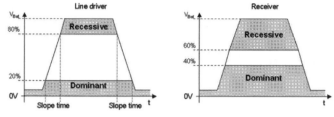

Figure 7. LIN Voltage Levels

In order to cope with EMI and on the other hand operate at sufficient speed, the slew rates on the bus must be between 1 and 2 V/µs.

The bus is specified to operate at speeds up to 20 kBit/s. The maximum wire length is 40m (for use in busses or trucks).

TOOLS

The acceptance of a bus system can be increased, if tool support is available. A set of tools is available including network configuration tools, node configuration tools and a LINalyzer, a tool to diagnose the traffic on the LIN Bus using the output of the network configuration tool. The LINalyzer is also capable to simulate not yet existing nodes on the bus.

CONCLUSION

This universal approach for a sub-bus fulfills all given requirements. As a multiplex network it offers more flexibility than traditional wiring. It is superior to CAN in the aspect of costs, offers safety and interoperability. LIN is already going to be used by several car manufacturers and suppliers. Originally targeted for body electronics, LIN will also be used in various other application domains inside and outside the car, such as alternater control in the engine compartment, sensor communication in industry, or lighting control inside buildings.

With the hierarchical approach it is possible to control the complexity of a network with more than 100 nodes. High-speed networks such as CAN will control the backbone of this system; the remote nodes will communicate via LIN.

REFERENCES

1. "LIN Protocol Specification", Revision 1.0, Audi AG, BMW AG, DaimlerChrysler AG, Motorola Inc., Volcano Communication Technologies AB, Volkswagen AG, Volvo Car Cooperation, 1999.

2. Hans-Chr. v. d. Wense, Dr. Will Specks, Dr. Andreas Krüger: "Anforderungen und Implementierung von Sub-Bussen in Hierarchischen Automobilnetzwerken", Embedded Intelligence , Nürnberg 1999.

CONTACT

Hans-Christian von der Wense works for the Strategy and Advanced Systems Lab at Motorola Munich. He is responsible for the LIN Program within Motorola. He serves also as the speaker of the LIN Consortium.

Information about LIN and LIN specifications can be obtained via http://www.lin-subbus.org.

Other Protocols

2005-01-1541

In Vehicle Communication With Solid Oxide Fuel Cell Using MODBUS Protocol

Bapiraju Surampudi and Pierre Gutierrez
Southwest Research Institute®

Rhys Foster
Acumentrics Corporation

Stanley Andrews
GD C4 Systems

ABSTRACT

Integration of fuel cell Auxiliary Power Units (APUs) into vehicles requires that the vehicle Electronic Control Unit (ECU) communicate with the fuel cell controller using appropriate protocol. The common communication standard used in automotive systems is the CAN protocol. Some fuel cell controllers originate from stationary power scenarios and communicate using process industry communication protocols such as MODBUS. This paper reviews the MODBUS TCP protocol and CAN protocol. It then describes the MODBUS TCP software driver development for QNX Real Time Operating System (RTOS) used in the SwRI® Rapid Prototyping Electronic Control System (RPECS). The architecture of the communication system on the vehicle and details of RPECS needed to understand the MODBUS drivers are described. Tools such as RPECS™ and EtherPeek™ (TCP/IP packet sniffing software) used during debugging are also briefly described. A new mapping method for conversion between MODBUS and CAN protocols is proposed.

This work has been developed at Southwest Research Institute on the SunLine Project, funded by US Army National Automotive Center.

INTRODUCTION

BACKGROUND

Truck accessories are critical to operation but reduce the over all efficiency of the vehicle. De-coupling the accessories from the engine operation facilitates utilization of more effective control strategies based on demand and performance. To this end, electrification of accessories is seen as one way of de-coupling accessories from the engine. A 5 kW fuel cell APU is used to power a 42 V electrical system that drives the accessories on this vehicle. The integration process of the fuel cell into the truck electrical system requires robust communications between the fuel cell controller and vehicle supervisory controller for coordination and seamless operation.

The US Army National Automotive Center has provided overall program direction and technical focus. The NAC focus is on technologies that have dual-use capabilities for the US Army and commercial industry. SunLine Transit Agency provides the commercial service demonstration location, with a special interest in developing clean fuel-efficient technologies for the transit and trucking industry. SwRI is the technical integrator, providing the engineering and fabrication responsibilities for the overall program.

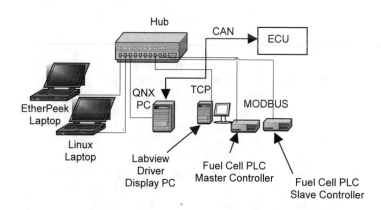

Figure 1 Communication Network on Sunline Truck

COMMUNICATION NETWORK ON TRUCK

The network on the Sunline truck is shown in Figure 1. A central hub located under the passenger seat networks all TCP messages. A QNX PC operating the SwRI RPECS system is the supervisory controller. It enables/disables fuel cell controllers to start/stop the fuel cell, communicates with the engine ECU, sends messages to driver display and controls water pump and air conditioning accessories on the truck. The protocol of communication between RPECS and fuel cell controllers is MODBUS TCP. The protocol of communication between RPECS and ECU is CAN (Controller Area Network). The communication between driver display running a labview™ interface on Windows™ XP and RPECS™ is accomplished using basic client server communication on TCP/IP. The RPECS developmental GUI (Graphical User Interface) runs on a laptop with Linux operating system. The etherpeek™ software operating on windows™ laptop was invaluable in debugging all communication protocols involved in the network by sniffing all packets crossing the hub.

ORGANIZATION

The paper is organized as follows:
- Section 2 introduces the tools used (RPECS™ and EtherPeek™ software)
- Section 3 introduces and describes MODBUS protocol and the software driver structure and development
- Section 4 compares CAN and MODBUS protocol and offers a structure for a conversion from MODBUS to CAN
- Section 5 gives summary and future work.

TOOLS

RPECS™

The RPECS™ (Figure 2) has a convenient PC/104 form factor, with compact packaging for shorter wiring harness lengths, while retaining full desktop processing power. It uses QNX™ 4.25: a hard, real-time, POSIX™ compliant operating system capable of supporting preemptive multitasking. RPECS™ is RS232/RS485 compatible and can communicate using TCP/IP, CAN J1939 protocol, or MODBUS protocol. It has a built in UPS, to handle voltage drop during cranking of the engine. It can tolerate supply voltage changes from 8 to 32 V DC and operate between -25 to 60 deg C ambient temperatures. A single FPGA provides flexible digital I/O, logic and capability to measure/send PWM signals. The Graphical User Interface (GUI) is installed on a laptop running Redhat™ Linux capable of communicating with QNX™ RTOS using TCP/IP. The hardware is easily extendible, to accommodate required analog inputs/outputs and digital inputs/outputs.

ETHERPEEK™

EtherPeek™ monitors and captures data packets exchanged between single or multiple sources and destinations on an Ethernet, Gigabit and Virtual Local Area Networks (LAN). The software captures data packets exchanged between a source and destination, apart from the type of network hardware and software installed on the LAN. EtherPeek™ can be configured to capture and interpret the various protocol layers of a captured frame, and to depict core information within the frame. EtherPeek™ has multiple view capabilities to display results; the views are: the packet capture list, single packet decode, and hex view of the raw data. Standard filters can be used to isolate communications between a source(s) or destination(s) on a network by node, protocol, packet content or error type. Advanced complex filters can be created using logic statements and additional filtering options. The graphing and trending features collect information, which are essential to identifying communication errors between nodes.

EtherPeek™ v5.0 was used to monitor, capture and decode data packets between RPECS and the MODBUS Master and Slave controllers. The computer systems were connected to an independent ethernet network configured for Network Address Translation (NAT) on the Sunline vehicle. The components of the network consisted of a Switched Workgroup Hub and a Windows client running Labview for the viewer display. It also consisted of SwRI RPECS, which contains a Linux client, a QNX client, and the MODBUS Master and Slave Controllers (Figure 1).

For a case in the point, when a command string was issued to the MODBUS Master and Slave controllers from the RPECS™ system, the reply packets could not be interpreted by the RPECS™ system. For the purpose of debugging communication packets between the MODBUS Master controller and the RPECS™ system, a Windows client running EtherPeek™ was connected to the Sunline vehicle network. A filter was created within EtherPeek™ to only permit IP, ARP, RARP and TCP packets to be monitored and captured. Several decoders were used for specific packets as an enhancement to the filter, which facilitated the debugging of packets. The communication packets between the MODBUS Master and RPECS™ system were saved to a file for analysis. The reply packets were also displayed on the RPECS™ display for comparison with the EtherPeek™ data. During the comparison of the data packets, it was discovered that the one data packet being sent to the RPECS display was out of sequence. The packet was the second to the last packet, when it should have been the initial packet received by RPECS™. This software is useful in establishing source that is out of specification when there are several supplier items in the network.

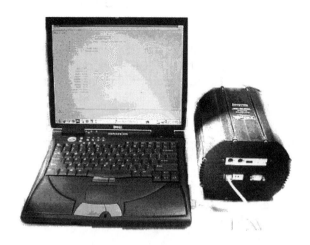

Figure 2 SwRI™ Rapid Prototyping Electronic Control System (RPECS™) uses the MODBUS Software Driver

MODBUS PROTOCOL AND SOFTWARE DRIVER

MODBUS PROTOCOL

MODBUS is a communication protocol with messaging structure developed by Modicon™ in 1979. It is typically used to establish client–server communication between different digital devices. MODBUS is used in stationary fuel cell controllers (eg. Advanced Energy Systems), process control simulation software, AC and DC drives in power/automation industry, electric utility substation measurements and welding applications.

MODBUS is an application layer messaging protocol, positioned at level 7 of the OSI™ model, which provides client/server communication between devices connected on different types of buses or networks The OSI™ model layers for MODBUS are shown below in Figure 3.

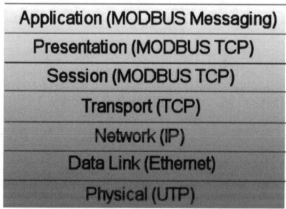

Figure 3 OSI Model for MODBUS

Typically MODBUS is a serial communication. With serial to Ethernet converters, it has also been used on TCP/IP networks. Conversion simply wraps up the serial packet. A gateway allows communication between several types of buses or network using the MODBUS protocol. It is a request/reply protocol and offers services

specified by function codes. MODBUS function codes are elements of MODBUS request/reply PDUs (protocol data units). MODBUS uses a big-endian representation for addresses and data packets. This means that the most significant byte (MSB) is sent first when the field length is more than one byte.

PACKET FORMAT AND BASICS

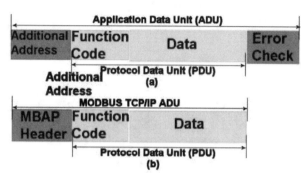

Figure 4 (a) General MODBUS packet (b) MODBUS TCP/IP Packet[1]

A general packet and TCP/IP packet format used in MODBUS protocol are shown in Figure 4. The MODBUS 'Protocol Data Unit' (PDU) is independent of the underlying communication layers. The mapping of MODBUS protocol on specific buses or network can introduce some additional fields on the application data unit (ADU), such as serial to Ethernet. The maximum size of the ADU is 256 bytes. There are three kinds of PDUs, namely MODBUS Request PDU, used for data/operation request; MODBUS Response PDU, used for acknowledgement and successful execution of request; and MODBUS exception Response PDU, used to report error or failure.

The client builds the MODBUS ADU and initiates a MODBUS transaction. The function indicates to the server what kind of action to perform. The MODBUS application protocol establishes the format of a request initiated by a client. The function code field of a MODBUS data unit is coded in one byte. Valid codes are in the range of 1 ... 255 decimal (128 - 255 reserved for exception responses). The function code is based on the type and location of information being exchanged. When a message is sent from a client to a server device, the function code field tells the server what kind of action to perform. Sub-function codes are added to some function codes to define multiple actions. A list of function codes can be found in references 1 or 2.

---

[1] Courtesy Modbus.Org

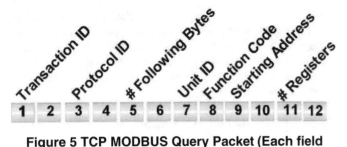

Figure 5 TCP MODBUS Query Packet (Each field represents 1 byte)

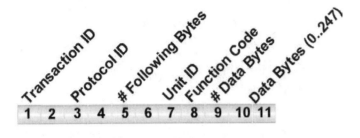

Figure 6 TCP MODBUS Response Packet (Each field represents 1 byte)

The MODBUS TCP query and response packet fields are shown in Figure 5 and Figure 6. Transaction identifier is useful to tag individual request if multiple requests from client are pending. Protocol identifier is set to zero for MODBUS. The number of following bytes includes unit identifier, and is useful for the receiver to read the message. The unit identifier is set to 0x00 for master fuel cell controller and 0x01 for slave fuel cell controller. Starting address is the where the required data is stored in a contiguous fashion. The number register represents the number of contiguous memory bytes that can be read. Booleans and 'reals' can be mixed if they are contiguous. The response packet is similar query packet until the function code. After the function code, the number of data bytes and actual data bytes are appended.

For example, a battery voltage query and response between RPECS and the slave fuel cell controller is shown below:

Query: 00 00 00 00 00 06 01 03 01 04 00 02
Response: 00 00 00 00 00 07 01 03 04 6F BC 42 18

The bytes 6F BC 42 18 must first be swapped to 42 18 6F BC, and then using IEEE754 format, are converted into a real number. In this case, it computes to a value of 38.109116, which was verified by actual measurement of voltage. If there has been error, the system will generate this packet response:

Response: 00 00 00 00 00 03 01 83 0B
0x80 is added to the function code field, and is followed by type of error or exception, which occurred. In this case, OB is defined as the 'gateway target device failed to respond'. A list of these exception codes can be found in reference 1.

SOFTWARE DRIVER FOR MODBUS TCP PROTOCOL

The software driver was programmed using C. Layers 5 and 6 (MODBUS TCP) of the OSI model (Figure 3) were programmed using basic TCP/IP socket programming methods. The RPECS was programmed with a client, and the fuel cell controllers acted as servers. Standard port 502 was used for socket binding, and static IP addresses were used for the servers and client. A gateway was used between fuel cell controllers and TCP/IP network, to wrap the serial MODBUS packets with TCP/IP headers. Etherpeek sniffer was used to monitor packet traffic and debug the code. A dedicated real time child process was spawned on the QNX RTOS, to isolate MODBUS delay effects from the rest of the vehicle control system.

Layer 7 programming primarily consisted of assembling of query packet and parsing the response packet. Arrays of appropriate sizes were used to represent the packets. Header files with structure constants, mapping all the bits and register memory, were created to represent the information in the two fuel cell controllers. Functions for MODBUS byte swapping and IEEE 754 real number format conversions were developed to decipher information in engineering units and report to the top level application (vehicle supervisory controller). A representative structure, which allows a linked list of contiguous variables to be easily accessed, is shown below:

```
typedef struct modbusvar
{
        char id;
        char address_hi;
        char address_lo;
        char type[8];
        float minValid;
        float maxValid;
        char units[10];
        int memOffset;
        struct modbusvar *next;
} modbusvar_t;
```

Each 'char' represents a byte of information, in which the memory offset, valid variable range, type of MODBUS register and the addresses are all registered.

CONVERSION BETWEEN MODBUS AND CAN

CAN PROTOCOL

CAN or Controller Area Network provides a low-cost, reliable and fast communication in automotive and industrial applications. The development of CAN began when electronic devices such as: engine management systems, active suspension, ABS, gear control, lighting control, air bags, stability control, steering control, central locking, and GPS, began to become prevalent in motor vehicles. The CAN is an advanced serial bus system that was developed by Robert Bosch GmBH™

Germany in late 1980s, and the International Standardization Organization (ISO) and Society of Automotive Engineers (SAE) have adopted it. CAN is an advanced serial bus system like the MODBUS protocol, although it primarily uses the Data Link Layer and physical layer in the OSI model.

CAN is a multi-master bus with open linear structure, one logic bus line and equal nodes. The number of nodes is not limited by the protocol. The bus nodes do not have a specific address like TCP/IP. The address information is embedded in the identifiers of the transmitted messages indicating the message content and priority. Nodes can be added and removed dynamically without disturbing communication of the other nodes. CAN also supports multicasting and broadcasting. CAN provides error detection and handling mechanisms like CRC check. It has good immunity from EMI. Erroneous messages are automatically retransmitted. Bit stuffing is used for synchronization process. It uses Non-Return-to-Zero or NRZ bit coding.

Twisted wire pair is the commonly used medium, with a maximum bus length of 40 meters. Data transfer rates can be as high as 1000 kbits/sec. To ensure short latency between request and reply, the maximum message length is limited to 8 data bytes. Collision of messages is avoided by bitwise arbitration, without loss of time. '0' is the dominant bit and '1' is the recessive bit. The dominant bit overrides the recessive bit when there is arbitration between messages.

The CAN packet format is shown in Figure 7. A CAN node generates the 'data frame' when transmitting data. When a CAN node is requesting data it uses a 'remote frame'. The start of frame is always a dominant bit for hard synchronization of all nodes on the CAN bus. The Identifier field and Remote Transmission Request bit (RTR) bit are called arbitration field. The identifier bit is 11 bits for Standard (V2.0 A) CAN and 29 bits for Extended (V2.0B) CAN. In general, the lower the identifier number then the higher the priority it will have. The RTR bit is dominant in data frames and recessive in remote frame; this lets the data frame have higher priority in arbitration for transmission on the bus. The IDE (Identifier Extension) bit, Reserved bit and Data Length Code (DLC) are called control fields. The IDE is dominant for standard frames and recessive for extended frames. The reserved bit is always dominant. The DLC specifies the number of bytes of data contained in the message. The data field can vary from 0 to 8 bytes. In remote frames, there is no data field. The CRC (Cycle Redundancy Field) is used for transmission errors. The CRC delimiter bit is always recessive. A transmitting node keeps the ACK (Acknowledge) bit recessive, while the receiving node sends out dominant ACK message, whether the others nodes are watching for it or not. The ACK delimiter is always recessive and cannot be overwritten. All 'End of Frame' bits are recessive, to mark the end of data frame.

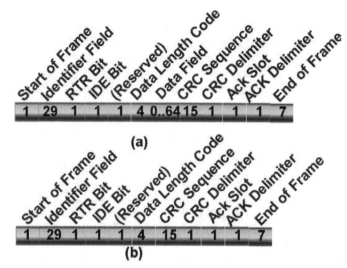

Figure 7 CAN Packet Format (Each field represents number of bits) (a) Data Frame (b) Remote Frame

A CONVERSION MAPPING BETWEEN MODBUS AND CAN

The MODBUS communication, being serial, can be slow (9600 bits/sec) compared to the 1 Mbit/sec peak CAN speed. A converter from MODBUS to CAN, and vice versa, will need to match the communication rate difference. Due to the small size of the CAN packets compared to the MODBUS ones, some decision making is required within the converter. A visualization of this black box converter is shown in Figure 8. This converter will require memory and a low-level processor for making some logic decisions. The entire logic of this converter can be tested on a Linux/QNX PC, before implementing it as an embedded controller.

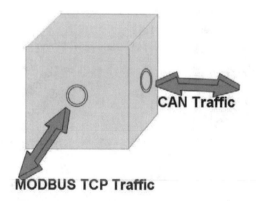

Figure 8 A Protocol Converter between MODBUS and CAN will allow for plug and play of fuel cells into Vehicle Applications

The mapping from CAN to MODBUS is shown in Figure 9. The transaction ID is always 0x00 since no queuing of messages will be allowed at the converter. The protocol ID will be 0x00 since it is MODBUS. The unit ID is restricted to 3 bits on the CAN identifier; 7 units can be connected to the converter at a maximum. The function codes from 0 to 127 will be encoded in 7 bits, since exception from 128 to 256 will be only present in the response packet. The starting address is transferred in its entirety of 2 bytes. The maximum number of registers is restricted to 8 (3 bits). If more than 8 bytes are seen, the converter will saturate it to 8. With the converter, it can request maximum of 7 contiguous bytes in the holding registers. These constraints allow us to define a unique identifier for every request, within a permissible range, and embed all-important information in the extended CAN packet; the identifier is stored in memory.

The mapping from MODBUS to CAN response is shown in Figure 10. Since the number of registers is restricted to 8 bytes, all the data from the MODBUS packet can be stored on the data field in the CAN packet. The identifier will be restored from memory after verifying that there is no exception or error in response. If there is an exception, the code is written on to the data field instead of the data.

The conversion of CAN to MODBUS and MODBUS to CAN are also shown in flowchart form in Figure 11 and Figure 12 at the end of the paper.

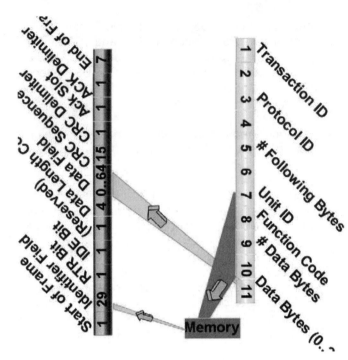

Figure 10 Converting MODBUS Response to CAN Response

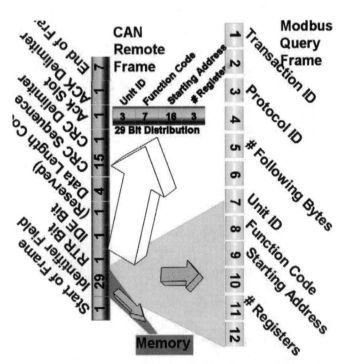

Figure 9 Converting CAN Query to MODBUS Query

SUMMARY

A brief review of MODBUS and CAN protocols is presented. A software driver for communicating with MODBUS protocol has been developed using 'C' for in vehicle communication between supervisory controller and fuel cell controllers. The software driver structure and development is described. A new method for conversion between MODBUS and CAN is delineated. This solution can be implemented in a converter device that allows plug and play of fuel cell controllers based on MODBUS in automotive CAN networks.

ACKNOWLEDGMENTS

The authors would like to thank the SunLine™ Services Group, The United States Army National Automotive Center (NAC™), General Dynamics C4 Systems™, Acumentrics Corporation™ for their contributions to the Sunline integration project.

REFERENCES

1. MODBUS Messaging on TCP/IP Implementation Guide, Rev 1.0 8 May 2002, Schneider Automation, MODBUS.ORG
2. MODBUS Application Protocol Specification, V1.1 12/06/02, MODBUS.ORG
3. MODBUS Application Protocol Specification, Rev 1.0 8 May 2002, MODBUS.ORG
4. Embedded CAN Tutorial, Embedded Systems Conference 2000

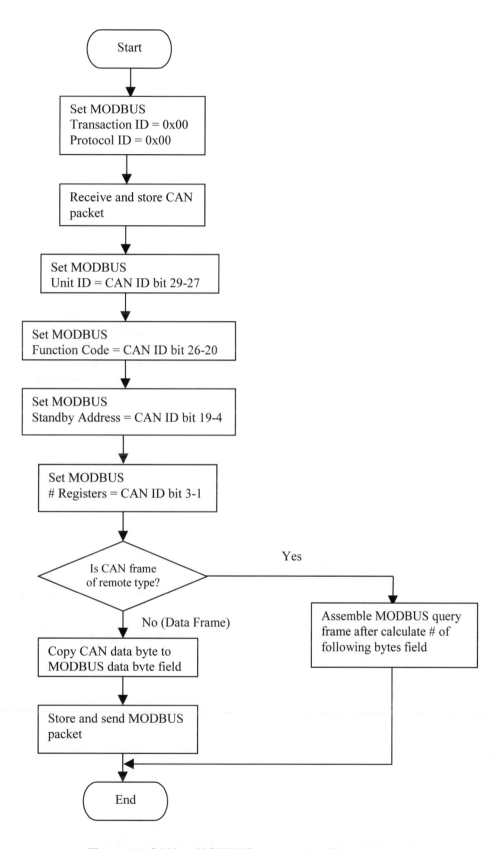

Figure 11 CAN to MODBUS conversion flow chart

309

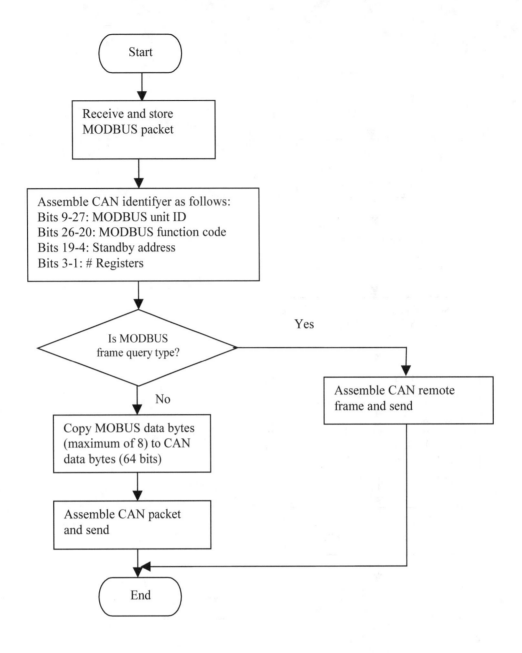

Figure 12 MODBUS to CAN conversion flowchart

Distributed Control System Development for FlexRay-based Systems

Pradyumna K. Mishra and Sanjeev M. Naik
General Motors Corporation

ABSTRACT

FlexRay is a new communication subsystem for future in-vehicle controls. There is a lack of mature model-based development methodologies to build complex FlexRay-based systems. In this paper we describe an end-to-end model-based development process for building a complex FlexRay-based distributed control system. We describe this in the context of safety critical x-by-wire systems for a realistic automotive application. This involves: control system modeling, functional simulation, and distributed software development. We first describe the process of functional and physical architecture design. Next we discuss the software development process dealing with software to hardware allocation, as well as scheduling of software and communication tasks on a time-triggered communication bus under stringent practical restrictions. We conclude by considering the integration issues relating to joint OEM/supplier development of distributed control systems. The above experience has been obtained during the model-based development of a FlexRay-based control system for a vehicle, which features by-wire steering, braking and propulsion.

INTRODUCTION

FlexRay is a new communication standard for future in-vehicle controls. This protocol provides flexibility and determinism by providing scalabale static and dynamic message[i] transmission. Due to FlexRay's time-triggered features and its possible use in safety critical applications, considerable software and developmental planning is required. Proper and optimal use of FlexRay in safety-critical applications requires construction of reliable embedded software. Model-based software development [2, 5, 9] can significantly increase the reliability and quality of software, as well as provide a proper development process. Some of the key benefits of model-based software development are "composability" and managing "complexity".

To develop reliable and efficient model-based software, one has to follow a rigorous software design process, given that the functional mathematical models are accurate. The functional models should capture the essential dependencies of the underlying physical architecture yet provide the abstraction to enable control system engineers to develop mathematical models without being overly concerned about the software and hardware implications. We believe that this is possible if the control system engineer assisting software development follows proper architectural and design guidelines. In this paper we intend to share our experience gained in the software development of a FlexRay-based control system for a complex X-by-wire vehicle[1] (Brake-by-wire, Steer-by-wire and Propulsion-by-wire). We also address issues faced by OEM and suppliers in the joint development of such systems.

In the development of our X-by-wire concept vehicle, we relied on a tool chain combining DECOMSYS[2], SIMULINK/STATEFLOW[3] and dSPACE[4] software tools for the basic controls software development. This tool chain can be used in a mechanized model-based developmental process for FlexRay-based rapid prototyping hardware. These tools automate part of the development process, yet significant development work has not been automated. We highlight the intricate and complex details of the automotive domain not yet captured by the above tool chain.

---

[1] This X-by-wire concept vehicle has a FlexRay network as a backbone and multiple CAN networks for communication.

[2] DECOMSYS provides a comprehensive set of tools for FlexRay-based software development.

[3] SIMULINK is a platform for multi-domain simulation and model-based design of dynamic systems. STATEFLOW is a design and simulation tool for event-based systems.

[4] dSPACE provides software and hardware products for rapid prototyping ECU development.

FLEXRAY-BASED SYSTEMS

INTRODUCTION TO FLEXRAY

FlexRay [1] is a new time-triggered communication system for high-performance in-vehicle applications. It supports both synchronous and asynchronous data transmission. Application-level tasks[ii] are executed and the related signals[iii] are communicated over FlexRay according to a predefined schedule. It combines the benefits of the familiar synchronous and asynchronous protocols. It supports data rates of up to 10 Mbits/sec per channel. FlexRay also supports fault-tolerant clock synchronization via a global time base. A fault-tolerant clock synchronization algorithm ensures the global time base. A modified version of the Welch and Lynch FT midpoint algorithm [10] is used to achieve this. This algorithm ensures that the global times of the different FlexRay controllers on each node do not differ by a significant margin, even in the presence of faults.

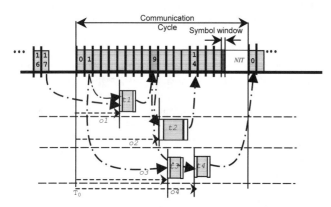

Figure 1: Example FlexRay task and communication schedule

In the above example (Figure 1), the FlexRay bus is configured for 18 static slots[iv] per communication cycle [v] (CC). In this example, we have the FlexRay bus configured for 10 Mbits/sec, a communication cycle of 1ms, a static slot size of 256bit (46µs), a dynamic segment of 0µs, a network idle time (NIT[vi]) of 163µs, and a symbol window of 9µs. Hence we have:

CC [1000µs] = Static slots [18] x Static slot length [46µs] + Symbol Window [9µs] + NIT [163µs]

This example consists of three nodes 1,2, and 3. Task *t1* executes on Node 1, task *t2* executes on Node 2 and *t3, t4* both execute on Node 4. Task *t1* reads data from Static Slot 17 of the *previous* communication cycle and Static Slot 1 of the *current* communication cycle. It performs computation based on that input data and provides the output to Task *t2* (on Node 2) and Task *t4* (on Node 3) via static slot 9.

All three nodes are synchronous at time T_0, which is the start of the communication cycle. Tasks in each node are triggered based on an offset relative to T_0. The offsets are o1, o2, and o3 for the tasks t1, t2 and t3 respectively. All tasks on different nodes agree on the time (are synchronized) and are able to communicate synchronously using FlexRay communication slots. Task *t2* collects data from Slot 9, hence Task *t2* cannot be allowed to start execution before Slot 9 is transmitted. If it starts executing before Slot 9 is communicated, then Task *t2* would use data communicated by Slot 9 of the *previous* communication cycle.

SOFTWARE AND PHYSICAL ARCHITECTURE

In a FlexRay-based system we define "host" as the part of an ECU where the application software executes. In this publication we will often use the term "node". We define a node as an entity that contains one host, one FlexRay communication controller[vii], two FlexRay bus drivers[viii], and other communication media like CAN transceivers, ADIO, PWM drivers etc.. A FlexRay node could also contain optional bus guardians. Each FlexRay physical channel is connected to one bus driver. The host interacts with the FlexRay controller via a communication handler interface (CHI) as defined in the FlexRay protocol specification [1].

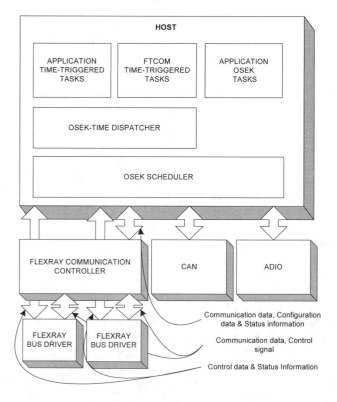

Figure 2: FlexRay based Software Architecture: Configurable by user

We will now describe the software architecture from an application developer's standpoint. At the highest abstraction of software design, we have time-triggered (TT) application tasks executing on the host. The host also contains a Fault Tolerant communication middleware layer that is compliant with the OSEKtime FTCom standard. This middleware layer is built on top of the FlexRay communication protocol, which provides standard API services like message exchange, integration and clock synchronization. Generally, the application TT tasks all have the same priority. The operating system uses a proprietary dSPACE kernel. This kernel is a pre-emptive "last come first served" scheduler. We permit pre-emptions in our system as long as they are all statically analyzed during design time. In this model, the OSEKtime dispatcher dispatches TT tasks based on a static schedule. However when two tasks are in contention for processing time, a "last come first served" scheduling strategy is used. A "first come first served" scheduling strategy is not used for the following reason: If one task runs unreliably and does not complete execution within its allocated time, this scheduling strategy will not allow any other tasks to pre-empt the unreliable one.

FlexRay distributed architecture consists of multiple nodes connected by dual FlexRay channels as shown in **Figure 3.** This FlexRay network is what we will refer to as a FlexRay distributed system. We use a linear passive bus topology, as shown here.

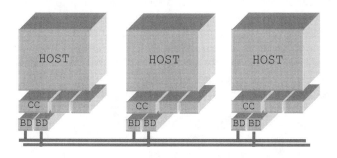

Figure 3: Three FlexRay nodes connected by a linear bus topology

We shall now describe the different functional architectures from the control system engineer's standpoint. **Figure 4** shows some of the basic different physical architectures for control systems. The centralized architecture is not well suited for fault-tolerant applications. The distributed architecture is very modular, but needs careful coordination between the different local controllers. In comparison, the hierarchical architecture offers benefits of both architectures. One of its shortcomings is that it involves more hardware. For our X-by-wire vehicle control system, a modified hierarchical physical and functional architecture was selected. The modification involves collocating different functionalities on each physical controller with the intent of reducing controller hardware. Even though the hardware architecture is distributed, the functional architecture is more hierarchical in nature.

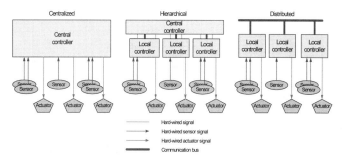

Figure 4: Different Functional and Physical Architectures

CHALLENGES IN MODEL-BASED SOFTWARE DEVELOPMENT FOR FLEXRAY-BASED SYSTEMS

Building reliable software for FlexRay-based systems involves building reliable controller models and then generating software from the controller models. In this section we list some challenges in building software from models:

- Providing end-to-end temporal guarantees under uncertain knowledge of execution times: While building a complex embedded system, it is difficult to predict the execution times of the SW components and the communication latencies. One has to make some educated guesses supported by analysis to ensure that end-to-end temporal guarantees are met. The models should be designed to be easily refinable once measurements are taken from the platform.

- Allocating software to hardware, global task scheduling and global message scheduling: Most previous work has approached this topic as multiple, isolated issues. While this leads to ease of handling, it results in sub-optimal solutions. Poor assignment of tasks to nodes causes unnecessary communication overhead. Inefficient communication is difficult to schedule, and middleware tasks have to perform extra packing and unpacking of signals. This consumes processor time, which directly impacts task schedules. Finding a good solution requires a unified approach where these "isolated" issues are targeted as one larger problem.

- Building an upfront global task and communication schedule with evolving functionality: FlexRay-based architectures might have software from different suppliers and OEM residing on the same host controller. Furthermore, different suppliers and the OEM may share the same Flexray communication network. Due to the nature of FlexRay, proper integration can only be guaranteed if computational resources (processing time) and communication resources (static slot identifiers) are distributed upfront between the suppliers. This distribution is possible only if application functionality is well-developed (e.g. known execution times, exact number and size of signals, task interdependencies etc.). This depends on the integrated architecture and how systems share resources, presenting a chicken and egg problem that requires iterative refinement.

MODEL-BASED FRAMEWORK FOR FLEXRAY-BASED SYSTEMS

In this section we will describe our design methodologies in terms of a model-based developmental process. It should be noted that it is not only a software development process, but also a process that touches control system functional design.

Various model-based processes have been proposed. The V process, which is quite popular, encompasses the entire development cycle. We shall limit the scope of our design to the left arm of the "V". DECOMSYS has proposed a model-based process called the "A" process in the publication "Model-Based System Development, An approach to Building X-by-Wire applications" [5]. This process has also been implemented as a tool chain within the V process, which has been used in our design.

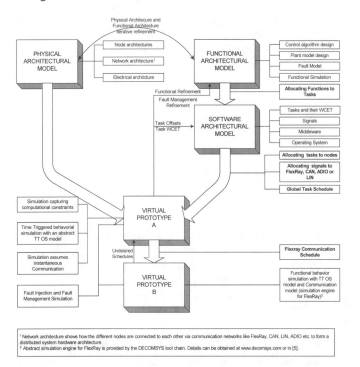

Figure 5: Distributed Systems Development Process

Figure 5 describes the steps involved in our design. We had to develop the functional architecture, software architecture and the physical architecture in modules. The fidelity of our design was verified and refined by simulations at various levels. The steps are highlighted

in bold letters. Each 3-D box stands for an important design stage, while the associated boxes denote the properties. It should be noted that design was done in sequential steps from top to bottom. In subsequent sections, we describe the challenges faced in each of the design stages in detail and how they were resolved. Feedback arrows from the virtual prototype to the other design stages capture iterative design refinement.

Our design intent has been to use software tool based synthesis and analysis, which reduces the amount of testing performed during implementation.

FUNCTIONAL ARCHITECTURE

A functional architecture is an arrangment of functions and their breakdown along with interfaces (internal and external) that either defines the execution sequencing, conditions for control and data flow, and the relative performance levels of achievment for a desired outcome, or that provides a desired capability.

There are numerous well-known tools for modeling and simulation for the automotive domain like Simulink/Stateflow, Ascet-SD[5] and SCADE Drive for automotive[6]. We relied on Simulink/Stateflow for our development. In a modeling environment like Simulink/Stateflow, a functional architecture is used primarily for modeling and design of control algorithms and verification of functionality using simulation. This architecture does not model or capture the low-level details of the software executing on limited hardware resources, e.g. finite but non-zero execution time, pre-emption, and communication delays etc. Figure 6 shows a screen shot of an X-by-wire vehicle control system. The model consists of two parts: Controllers (X-by-wire vehicle control system) and Plant (Vehicle plant model). The vehicle control system contains controller models that will ultimately be software executing on computing nodes. The plant models are required for functional simulation.

Prior to simulation, SIMULINK orders all the blocks based on dependency constraints (i.e. topological dependencies) and execution rates. This involves flattening all hierarchical blocks into a single flat list grouped by execution rates. The execution order is extracted from this list. Any block where the output is directly dependent on its input (i.e., any block with direct feedthrough) cannot be executed until the block driving its input has executed.

[5] Ascet-SD is a rapid developmental tool for modeling and software development. Further details can be obtained at http://en.etasgroup.com/products/ascet_sd/
[6] SCADE Drive for Automotive systems has been developed by ESTEREL Technologies. Further details can be obtained at http://www.esterel-technologies.com/

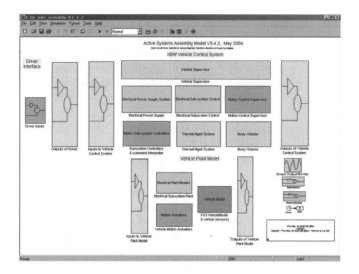

Figure 6: X-By-wire functional architecture in SIMULINK

The fidelity of this simulation in relation to the real execution behavior needs to be properly understood based on these important points:

- SIMULINK simulation block execution ordering might not reflect the true block ordering of tasks in a real-time distributed system.
- In SIMULINK all computations occur instantaneously and are not interrupted during their computation. This does not occur in a real system, where a task needs to be allocated a finite non-zero execution time which could be pre-empted during computation.
- All inter-block communication occurs instantaneously unless explicitly modelled otherwise.

The effect of the above three points needs to be accounted for with suitable analysis before generating code from the model. In order to verify the behavior of the control system accurately, we needed to capture some of the above constraints at the SIMULINK model level. In this section we briefly describe how we obtained a high fidelity simulation model in SIMULINK that captured software and hardware implications indirectly. This simulation model also provides the groundwork for proper iterative refinement.

The task schedule used in our development is a static schedule specified during design time. A static schedule is required for deterministic behavior. Preemption is sometimes disallowed in order to preserve determinism. We permit preemption as long as all preemptions are statically analyzed and fixed during design time. Hence, all tasks that get preempted are modeled as tasks that take a finite time longer than usual execution times.

If the task execution is unspecified, SIMULINK will determine the execution order of the subsystems (as explained above). In order to control the execution order, we built a separate scheduler using the SIMULINK "Functional Call Generator" . These blocks produce "task

triggers" at prespecified rates, and are connected to "GOTO" blocks. A "Demux" block routes the trigger signals to the appropriate subsystems when it is connected to the output of the "Functional call generator". The subsystem extracts its trigger signal via a "FROM" block. The topmost output signal from the "Demux" block triggers the first execution. The execution sequence then continues from top to bottom.

The details on how we obtain a static task schedule is described later.

Using this method, we simulate the functional architecture with subsystems having an execution order which is very similar to the final execution order. As mentioned earlier, this also lays the framework for iterative refinement, where the impact of changing task schedules on behavior can be examined in SIMULINK.

As mentioned in the introduction to this section (Points 2&3), SIMULINK assumes all computations to occur instantaneously without interruption and without any computational resource constraint. This is an abstraction from the real system, where a finite execution time needs to be allocated to a task which could be pre-empted during computation.

To model the finite computation time and communication delay, we used the Decomsys tool chain, which has an interface to SIMULINK. The computation tasks were annotated with "Worst case execution times" (WCET). Even though the tool does not model preemptions, we modelled pre-emptions as tasks with extended WCET. To elaborate further, each task that was preempted had a total execution time of its own added to the WCET of the task that preempted it. The Decomsys tools also provided simulation of communication synchronous time-triggered tasks and generic communication with or without delays. Further details on the process can be found in [5, 7].

In order to model real FlexRay communication delays, one has to obtain the FlexRay schedule. We shall address this problem in a subsequent section. However, once the static schedule has been determined, one can also simulate the functional architecture with real communication delays.

ALLOCATING FUNCTIONS TO TASKS

A functional architecture consists of functional blocks (which are hierarchical in nature) modeling the controller as well as the plant model (e.g., **Figure 6**). However, this architecture is used primarily for control system design and simulation. These functional blocks do not represent application software tasks[1]. We need to regroup "functional" blocks into "application task" blocks that symbolize "pieces" of software code. Essentially, we can now view each block at the model level as ultimately

315

a "piece" of software with clearly defined inputs and outputs.

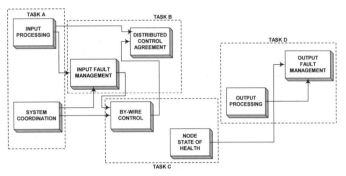

Figure 7: Allocating functions to tasks

The most effort-intensive part of collecting functions into tasks is to flatten the hierarchy. Hierarchy in SIMULINK is a visual organization to manage complexity, but the final execution is non-hierarchical. In our time-triggered system all software tasks that communicate over FlexRay have the same priority. The scheduler is a preemptive "last come first served" scheduler.

The functional architecture as shown in **Figure 7** consists of seven functional blocks. The functions are tagged to four tasks: A, B, C and D. The four tasks could be implemented on the same node or distributed on different nodes. Tagging two functions into one software task essentially implies one "piece" of software code where the two functions could be implemented as sequential blocks of software code. However, marking two functions as separate tasks essentially forms two "concurrent" entities.

The way in which functions are allocated into tasks greatly impacts the architecture and the performance of the system. In the above example it is also possible to have three different tasks, one for each subsystem. Having too much granularity increases the task and signal scheduling complexity. Composing functions into tasks was primarily based on experience and knowledge of the hardware architecture, however we can generalize some rules for allocating tasks to functions:

Rule 1: Any two functions which reside on separate computation nodes cannot be in the same task, i.e. only functions on the same computational node can be allocated to the same task.
Rule 2: Any two functions that act on the same inputs and have the same deadline for producing outputs can be combined into a task.

TASK ALLOCATION TO HARDWARE NODES

Tasks need to be allocated to nodes such that the resulting distributed embedded system satisfies all the temporal constraints of the design without exceeding the available resources. In addition to the temporal constraints, the allocation should be done in a manner

that optimally utilizes the resources (computing, memory, I/O, and communication bandwidth). For example, optimality criteria could be "obtaining a solution", "obtaining a solution under bounded uncertain design constraints", "even utilization of nodes", or "ease of future integration of tasks". However, "optimal" assignment early on during the design stage might not be possible since the complexity of the design becomes apparent only during the final stages of development. It is this "complexity" that cannot be modeled accurately during the early stages of design. In our application, the criteria for allocating tasks to nodes were I/O and communication constraints as well as fault tolerance requirements. It is also important to ensure that the allocations do not exceed the computational resources and do not overburden any particular node.

Tasks that do not require direct I/O, e.g. *Subsystem coordination* as well as *Supervisor* and *Diagnostic tasks,* are allocated to nodes based on their proximity to other tasks (already allocated). Additional requirements for allocation could be to evenly spreads out the computation load based or based on fault tolerance requirements.

GLOBAL STATIC TASK SCHEDULE

The global task scheduling problem consists of determining the execution order of all the tasks (over all nodes) based on knowledge of the periodicity, WCETs, task dependencies, deadlines, and processing power of the nodes. The nodal task schedule is the outcome of the global task schedule. Different scheduling methods for embedded real-time systems have been adequately addressed in [4]. This publication reviews alternate approaches to scheduling methods

Two approaches exist for scheduling tasks for distributed systems: *Static* scheduling and *Dynamic* scheduling. With static schedules, the execution order of tasks is strictly determinstic and does not change during execution (if the design assumptions are not violated). Within dynamic schedules, the execution order is determined during run time.

For our design, we used a static schedule to achieve determinism. All time-triggered tasks (SBW, BBW, PBW, Vehicle dynamics supervisor) are of the same priority. Certain fault tolerant background application tasks that check the health of the processors are of higher priority than the rest of the application tasks. The real-time kernel implements a preemptive "last come first served" scheduling policy. However, there is an OSEKtime dispatcher that triggers tasks based on a static schedule. As mentioned earlier, in our design we permit pre-emption of tasks as long as we make sure that we statically analyze all possible preemptions during the design stage. Dynamically scheduled systems are extremely difficult to tune and to analyze for safety.

Once tasks are allocated to the nodes, the task execution times need to be scheduled on each node. Two tasks in a global schedule need to execute sequentially if they are on the same node. On the other hand, they can execute concurrently if they are executing on different nodes. Since we assume all nodes to be connected to each other via FlexRay, there is a global concept of time for all the nodes. The existence of the distributed clock synchronization algorithm in FlexRay permits all nodes to have a single concept of time. We relied on manual scheduling, because of the lack of mature scheduling tools targeting FlexRay-based applications. In this section we also highlight certain practical constraints in the realm of fault tolerant by-wire control systems which have not been captured in earlier work. These have been extremely critical in the manual schedules.

We have different tasks executing at different rates. As required in a time-triggered system, the rates must be harmonics of each other. In order to better understand the intricacies involved in scheduling tasks, one has to understand the events that take place when two tasks communicate over FlexRay. **Figure 8** shows tasks executing on three nodes. Each horizontal row denotes tasks executing on a single node and shows the starting and ending times of tasks and communications. **Table 1** describe the activities associated with each time event. We have Task A1 executing on Node 1 sending data to Task C2 on Node 3 via a FlexRay slot. We have **S** and **R** as the Send and Receive FTCom middleware tasks.

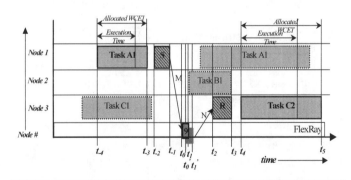

Figure 8: Time flow of tasks and messages in a FlexRay system

Timeline	Action
t_{-4}	Task A1 is triggered to start execution by the OSEKtime dispatcher.
t_{-3}	Task A1 is scheduled to complete execution.
t_{-2}	FTCom (middleware) *Send* task starts execution[7].

---

[7] Context switching time is modeled explicitly as an added WCET to the task that immediately follows the context switch.

t_{-1}	FTCom (middleware) *Send* task completes execution.
M	Data (to be sent) transfer between the main memory and the FlexRay CC buffer.
t_0	Start time for the scheduled FlexRay Slot 9 to send the message across the bus.
t_1	End time for the scheduled FlexRay Slot 9 to send the message across the bus.
t_0'	Time after propagation delay where Node 3 receives the first bit of the frame. [ix]
t_1'	Time after propagation delay where Node 3 receives the last bit of the frame.
N	Data (received) transfer between the FlexRay CC buffer and the main memory.
t_2	FTCom (middleware) *Receive* task starts execution.
t_3	FTCom (middleware) *Receive* task completes execution.
t_4	Task C2 is triggered to start execution consuming the data sent by Task A1
t_5	Task C2 is scheduled to complete execution.

Table 1: Timelines when two tasks communicate via FlexRay

The static task scheduling problem consists of determining the times t_{-4} and t_4 which are the triggering times of Task A1 and Task C2. However, we also have to determine the triggering times of the middleware tasks. Hence, the problem involves determining the time stamp offsets t_{-4}, t_{-2}, t_2, t_4 which are the starting times of Task A1, FTCom Send middleware task on Node 1, FTCom Receive task on Node 3, and the start time for Task C2 on Node 3. Sometimes, one triggers the Send (or Receive) FTCom task immediately after (or before) the task that sends (or receives) the data. This is typically not the case in our application, since the FTCom Send task multiplexes signals from a number of tasks (on the same node) and sends them over FlexRay in the same slot. A similar situation could occur on the receiving side. The timelines are actually offsets with respect to the start of the communication cycle, i.e. the start of the communication cycle in FlexRay is assumed as zero time. These offsets are tabulated in the OSEKtime dispatch table and the scheduler triggers tasks based on that.

In order to obtain the schedule we had to specify the WCETs of the tasks. We had to also determine the WCETs of the middleware tasks. There is significant difficulty in predicting the WCET of the middleware tasks. These WCETs depend on how many signals are packed into a slot and the sizes of the constituent signals. This, however, is not known before *we actually allocate signals into slots and schedule the communication slots.* Hence, we had to count on intiutive analysis, experience and iteratively refine our task schedules.

In the above example, we demonstrated the detail involved in scheduling two tasks involving two nodes communicating with each other over FlexRay. In our X-by-wire distributed vehicle we had over one hundred application and middleware tasks communicating with each other through numerous signals.

Fault tolerance can significantly limit the task-scheduling problem. As shown in **Figure 8**, replicated components need to be scheduled concurrently. It requires messages to be sent very close to each other for reasons of temporal message consistency. There are scenarios where voting occurs between nodes over multiple communication cycles.

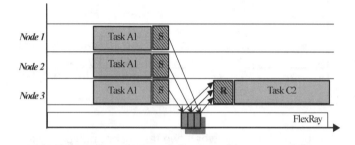

Figure 9: Effect of replicated fault tolerant tasks on scheduling

ALLOCATING SIGNALS TO FRAMES

In this section we shall discuss how we allocate signals into slots. Thus far, we have described the method of how signals need to be scheduled over FlexRay. We now determine how we actually allocate signals into the *same frame*[8]. In the subsequent section we shall schedule the frames and allocate FlexRay identifier numbers to the different frames.

FlexRay is a time-triggered fault tolerant communication bus. One complete instance of a communication structure that is periodically repeated in FlexRay is called the *communication cycle.* A communication cycle consists of a static segment, an optional dynamic segment, an optional symbol window and a network idle time segment. For the purpose of our analysis, the communication cycle shall consist of only a static segment and a network idle time segment.

We first had to choose the period of the communication cycle, which was a sub-multiple of the periods of all the tasks. The static segment of a communication cycle consists of *slots.* The size of each slot (**SS**= Slot size) can range from 0 to 254 bytes (0-127 two-byte words). In one communication cycle, only one node can own a communication slot. A slot consists of one message having only one *sender* and possibly many *receivers* (not the sender). A message contains multiple signals of varying lengths and types packed together.

Before packing signals into slots, one must decide on the optimum static slot size. The effect of the choice of static slot size is also dependent on the speed of the FlexRay bus. Even though the protocol specifies the maximum slot size to be 254 bytes, it is also limited by the available on-chip FlexRay buffer size. Our development buffers were restricted to 32 in number and 256 bits of memory each. **Table 2** summarizes the tradeoffs involved in selecting between different slot sizes.

	SMALLER SLOT SIZES	LARGER SLOT SIZES
1	Larger number of reads and writes required between host memory and FlexRay controller buffer ties up CPU time.	Fewer reads and writes required.
2.	More customized packing of signals into frames targeted to receivers[9].	Receiver might get many signals that it does not intend to use.
3.	Requires smaller buffer sizes	Requires larger buffer sizes
4.	Increased fragmentation: Signals need to be sent over multiple slots	Less fragmentation
5.	Possibly more economical packing of signals into frames.	Could possibly lead to wasted buffer space.

Table 2: Comparison of Smaller Slot Sizes with Larger Slot Sizes

FlexRay provides a great amount of flexibility in defining the size of the communication slot. Once the size of the communication slot is decided, it applies to all communication cycles and to all slots within a communication cycle. The granularity of the slot length depends on the Flexray communication configuration

---

[8] This is different from *actually scheduling a frame at a particular time,* i.e it is one problem to have 7 signals allocated to a frame and another problem to actually allocate the frame to FlexRay static slot number 9. It is the former that we are discussing.

[9] A node that receives a *static slot* must unpack all the signals and execute the Replica Determinate algorithm on them. If the slot contains signals that it does not need, it wastes resources trying to unpack them before discarding them.

parameters (e.g. bus speed, microtick[x] length, macrotick[xi] length), the signal list and properties. Various subsystems that share the FlexRay communication system may have a very different number and size of signals that need to be sent. Hence, agreeing on a unique slot size has an important impact on the design.

There are two key points in allocating signals to frames. All signals sent by one task to another at the same time may be allocated to the same frame. Signals that need to be sent at roughly the same offset (which might contain signals from different tasks on the same node) may be allocated to the same frame. Two signals that are meant for different tasks on different nodes should not be allocated to the same frame.

GLOBAL COMMUNICATION SCHEDULE

Once the global tasks are scheduled and signals are packed into frames, we need to schedule the frames into communication slots. This involves assigning a unique FlexRay identifier to each frame. The Decomsys tool chain has an automated scheduling wizard, but we found it unsuitable within the scope of our development (because of the scale of magnitude involved). One key problem is that the automated scheduling technique does not take into account task interdependencies when it comes up with its automated communication schedule.

Figure 10 shows Task A1 (periodicity p) sending a frame via slot number 19 to Task C2 (periodicity p). Ideally, we would want to send the frame in a slot between the end time of the middleware task S and start time of receiving middleware task R. If the frame is indeed sent in the time within this time frame, the reaction time will be simply the difference between the end time of Task C2 and the start time of Task A1. However, if we are unable to schedule the message within the desired time frame and we schedule the frame at a later slot, then the reaction time increases by the periodicity p. Let us define the term *"Communication Miss"*, denoting a situation where a frame in not able to be scheduled in the desired time frame. Communication misses exist in our design (due to stringent temporal constraints) and so the impact of each miss on the control system has to be analyzed. They lead to increased phase lag in the control system. For cases where these are part of a feedback loop, they impact stability margins.

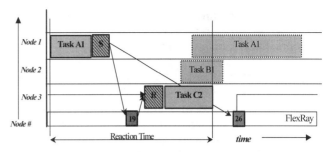

Figure 10: Effect of communication miss

The communication schedule needs to satisfy all the dependency constraints. Dependency violations could occur due to "Communication misses". In such cases, the effect of all such misses needs to be statically analyzed.

OEM/SUPPLIER SOFTWARE AND COMMUNICATION INTEGRATION

When it is used as a vehicle communications backbone, FlexRay has to be shared between electronic control units (ECU) from different suppliers. The situation becomes more complex when software from different suppliers shares the same ECU in addition to sharing the communication network. In this section, we shall highlight some of the integration issues related to FlexRay-based systems from the OEM-supplier co-development perspective. We will also outline how we dealt with them. **Figure 11** shows software from two suppliers, the OEM sharing three nodes and the FlexRay communication bus.

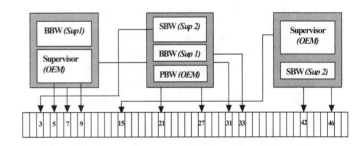

Figure 11: Supplier and OEM Software sharing nodes and the Communication subsystem

FlexRay being a time-triggered subsystem presents its unique set of integration constraints. In order to facilitate proper integration, one has to develop accurate models of the software and communication architecture during the earlier stages of the design. To be more precise, one has to determine the location of the supplier tasks on the nodes, the WCETs of the supplier tasks, the signal lists, and the task dependencies. The location of the software and WCETs is required to be able to divide the

319

computational resources of each node. The signal lists and task dependencies are required to distribute the communication requirements.

The integration problem of the entire vehicle subsystem was solved hierarchically. First, a high level architectural model was developed and the constraints arising from this model were provided to each subsystem supplier. Each subsystem supplier developed their subsystem independently without violating the specified constraints. The constraints were developed by us (OEM) based on some front-end architectural analysis. During the different stages of development, the constraints were refined iteratively.

Next, we took each of the individual supplier subsystems as well as our own system and merged the entire architecture together. During this merge, we captured the inter-subsystem and OEM-subsystem interactions. We also tuned the architecture whenever conflicts arose. Proper development of constraints during the earlier stages of the design leads to proper integration and proper verification of the system during the final stages of development. Significant architectural analysis needed to be done while developing these constraints.

We now highlight the steps involved

Step 1: The OEM performs front-end analysis and provides constraints or models to the supplier. Some of the constraints provided to each subsystem supplier are
1. Communication constraints:
   - FlexRay bus speed.
   - Communication cycle time.
   - Slot frame size in bits.
   - Number of FlexRay slots available.
   - Network topology and other low-level communication-related parameters.
2. Processor constraints
   - Number of processors allocated.
   - Percentage of processing power of each node allowed to be utilized by each supplier.
3. Allowable loop rates (harmonics).
4. Other input/ouput interface (ADIO/CAN) constraints.

Step 2: Each subsystem supplier develops its subsystem independently adhering to the constraints. It then gives the integrator:
1. The number of nodes on which software tasks are going to reside .
2. The intertask dependency.
3. The WCETs for each task on a node.
4. The local offsets of each of its tasks within its subsystem.
5. The signal list and signals packed into frames.
6. The frame identifiers for each frame used.
7. Other software architectural information.
8. The software models/object code which has the functionality hidden (which can also be used for functional simulation).

Step 3: The integrator merges the different supplier task and communication schedules with the OEM task and communication schedule:
1. Every task gets its share of computation time (modeled as WCET).
2. Every supplier message slot gets scheduled without violating any dependency.
3. Integration captures the intersupplier as well as supplier-OEM task dependencies.
4. Integration captures the intersupplier as well as supplier-OEM communications.
5. Integrator tunes the task and communication schedule of the entire system to resolve conflicts.

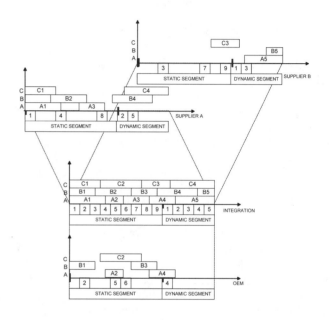

Figure 12: Supplier subsystems being merged into a global system

CONCLUSION

This publication describes a model-based software development approach for a distributed X-by-wire automotive control system utilizing FlexRay. Some of the challenges in this process are highlighted, including upfront integration, OEM/supplier collaboration, iterative hardware/software development, iterative communication and application software development. While several tools exist to enable different aspects of the overall process, the need remains for an end-to-end tool chain that addresses some of these real-world challenges in the automotive domain.

ACKNOWLEDGMENTS

The authors would like to acknowledge Thomas E. Fuhrman, Lawrence Peruski for their advice and the support of the management of GM R&D.

The authors would also like to acknowledge Mohan Sundar, Doug Cesiel, Arnold W. Millsap, Priya Narashiman and Tom. M. Forest for their assistance.

The authors would also like to thank our suppliers for joint discussions.

REFERENCES

1. FlexRay Consortium. FlexRay Communication System: Protocol Specification, 2004

2. Rajeev Alur, Thao Dang, Joel Esposito, Rafael Fierro, Yerang Hur, Franjo Ivancic, Vijay Kumar, Insup Lee, P. K. Mishra, George Pappas, Oleg Sokolsky. Hierarchical Modeling and Analysis of Embedded Systems, Proceedings of the IEEE, January 2003, Volume 91, Number 1.

3. Rajeev Alur, Thao Dang, Joel Esposito, Rafael Fierro, Yerang Hur, Franjo Ivancic, Vijay Kumar, Insup Lee, P. K. Mishra, George Pappas, Oleg Sokolsky. Hierarchical Hybrid Modeling of Embedded Systems. 1st International Workshop on embedded Software, October 2001 (LNCS 2211).

4. F. Balarin, L Lavagno, P. Murthy, A. Sangiovanni-Vincentelli. Scheduling for Embedded Real Time Systems. IEEE Design & Test, January 1998, pp. 71-82.

5. R. Nossal, R. Lang. Model-Based System Development- An Approach to building X-by-wire Applications. IEEE Micro, July/August 2002, pp. 56-63.

6. T. A. Henzinger, B. Horowitz, C. M. Kirsch. Giotto: A Time-Triggered Language for Embedded Programming, EMSOFT 2001, pp 166-184.

7. Roman Nossal. Meeting the Challenges in a collaborative OEM-supplier development of distributed embedded systems. SAE 2004 World Congress, 2004, SAE-2004-01-0278.

8. John Rushby. Partitioning in Avionics Architectures: Requirements, Mechanisms, and Assurances. NASA/CR-1999-209347, June 1999.

9. Edward A. Lee. Embedded Software- An agenda for Research, ERL Technical Report UCB/ERL No. M99/63, December 1999.

10. J. L. Welch and N. A. Lynch. A New Fault-tolerant Algorithm for Clock Synchronization, Information and Computation, vol. 77, April 1998, no. 1, pp 1-36

CONTACT

Please direct all questions to Pradyumna K. Mishra (Pradyumna.Mishra@gm.com) or Sanjeev M. Naik (Sanjeev.M.Naik@gm.com).

---

[i] *Message*: a collection of signals that is intended to fit the data payload segment of the FlexRay communication frame.

[ii] *Task:* piece of application software that can be executed as a concurrent unit.

[iii] *Signal*: application data produced and/or used by application tasks.

[iv] *Communication Slot*: an interval of time during which access to a communication channel is granted exclusively to a specific node for the transmission of a frame with a frame ID corresponding to the slot.

[v] *Communication Cycle* is one complete instance of the communication structure that is periodically repeated to comprise the media access method of the Flexray system. The CC consists of a static segment, an optional dynamic segment, an optional symbol window, and a network idle time segment.

[vi] *Network idle time* serves as a phase during which the node calculates and applies clock correction terms. It also performs implementation-specific communication cycle tasks.

[vii] A *Communication Controller (CC)* is an electronic component in a node that is responsible for implementing the protocol aspects of the FlexRay communication system.

[viii] A *Bus driver* is an electronic component, which consists of a transmitter and a receiver that connects a communication controller to one communication channel.

[ix] *Frame:* a structure used by the communication system to exchange information within the system. A frame consists of a header segment, a payload segment and a trailer segment. Application data is transmitted through the payload section.

[x] *Microtick:* an interval of time derived directly from the CC's oscillator. The Microtick is not affected by the clock synchronization mechanisms, and is a node-local concept. Different nodes can have microticks of different duration.

[xi] *Macrotick*: an interval of time derived from the cluster-wide clock synchronization algorithm. A Macrotick consists of an integral number of Microticks. The clock synchronization algorithm adjusts the actual number of microticks in a given Macrotick. The Macrotick represents the smallest granularity unit of the global time.

2003-01-1287

Improving Availability of Time-Triggered Networks: The TTA StarCoupler

Georg Stoeger, Alexandra Mueller and Sharon Kindleysides
TTTech Computertechnik AG

Leonard Gagea
TTChip Entwicklungsgesellschaft mbH

ABSTRACT

The Time-Triggered Architecture (TTA) provides many state-of-the-art mechanisms to guarantee fault tolerance and highest system availability, in part due to the use of a fault-tolerant communication protocol. However, some failure modes are known that cannot be tolerated by a fault-tolerant communication protocol alone and that can threaten the availability of distributed systems. The possibility of these failure modes occurring in safety critical applications like steer-by-wire or brake-by-wire without mechanical backup is not acceptable.

A dedicated device can be used to transform arbitrary node failures to failure modes tolerated by the Time-Triggered Protocol (TTP), eliminating failures that can lead to a loss of communication and thus to a loss of availability of the distributed system. In the star architecture, this functionality is concentrated in two redundant nodes (called TTA StarCouplers) placed in the center of a star configuration, providing highest system availability at lowest cost.

The paper describes the functional blocks of the TTA StarCoupler. A description of the specific faults that are addressed by this architecture, i.e., "SOS faults", "spatial proximity faults" and "babbling idiot faults", is given, and the impossibility to protect a distributed system against these faults without such dedicated countermeasures is shown.

INTRODUCTION

The TTP protocol has been designed with a clear focus on safety-relevant and safety-critical applications. It provides mechanisms that ensure consistent communication or, when not possible, prompt error detection and fast recovery from inconsistent states. The protocol and the formal correctness arguments brought forward for its core mechanisms are based on a simple design strategy: a static amount of the overall bandwidth (a "slot" for transmission) is allocated to each communication node, and an independent device called "bus guardian" ensures that each node only transmits during its assigned slot, and never outside of it.

BUS SYSTEMS CANNOT HANDLE ALL FAULT CLASSES

With error detection in the communication controllers (self-tests, synchronization tests etc.), in the sender (acknowledgment) and all receivers (transmission CRC, clique detection), TTP ensures that faults in the communication layer are reliably detected. TTP also provides a clear-cut set of fault tolerance capabilities: since all TTP mechanisms comply with the "single fault" hypothesis, any single faulty unit – node or channel – cannot interfere with the correct operation of all non-faulty units.

In a distributed system, however, there is a special (usually rare) class of faults that evade this fault hypothesis: asymmetric or "Byzantine" faults manifest themselves in a system so that several components at the same time become inconsistent. While such a fault on a single channel is transparently (i.e., the application does not necessarily notice it) tolerated by TTP, an asymmetric fault occurring on both TTP channels at the same time introduces an inconsistency or "clique" error in the network. The claim for "fully independent channels" is usually not fulfilled by realistic system implementations, so such dual-channel faults cannot simply be ignored. TTP has the capability to reliably detect such a dual-channel Byzantine fault but not tolerate it.

For fail-safe systems, the detection and subsequent fail-safe handling of such a fault is sufficient to meet safety requirements. For highly critical fail-operational applications (like fly-by-wire), it cannot be guaranteed that a single TTP bus provides sufficient availability because the occurrence of dual-channel Byzantine faults cannot be calculated or experimentally determined with sufficient confidence. The aerospace industry thus uses several independent buses (typically three or even four in parallel); this can be done with TTP, but the associated costs seem too high even for the most critical automotive applications like steer-by-wire.

Therefore, research was done to find an alternative solution to the (rare) problem of "dual-channel asymmetric communication faults". Two very different approaches are currently discussed in the automotive industry:

The "high-level error detection and recovery" strategy, pursued by some research groups, detects and resolves communication level inconsistencies by an application-specific software layer which performs retransmission of received redundant message copies, Byzantine voting, etc. The idea behind this approach is that any "lower layer" (i.e., communication) faults are automatically covered by a "higher layer" (i.e., application) level error detection.

The error detection approach taken by the current TTP protocol specification and the design philosophy for fail-operational systems according to the TTA principles specifies that error detection (or prevention) must be done as close to the source as possible, and as fast as possible to allow for most efficient response and recovery. Therefore, the TTP protocol includes error detection even for faults it cannot tolerate (often called "level II faults" in the related literature) and leaves error handling to the system level safety architecture, e.g., switching to a fail-safe state or performing system level recovery (similar to when a transient power failure or millisecond noise bursts have occurred) for fail-operational functions.

Some failure modes observed during fault injection experiments showed that even a complete permanent failure of an otherwise fault-tolerant TTP network can be "achieved" by Byzantine communication faults. The faults leading to such grave failure modes were called "slightly off specification (SOS) faults" due to their source, e.g., slightly off specification bit timing or slightly off specification differential voltage. In the TTP protocol, where correctness checks are also made on the transmission timing, additional sources for SOS faults exist, but the bit timing and voltage (i.e., the time and value domain of the physical bit encoding) alone offer ample possibilities for a potential permanent disruption of the communication. In a CAN network, repeating acknowledgment and error frames would be observed; in a TTP network, the clique detection would be activated continuously. In a network without error detection mechanisms for such faults, inconsistency would only be observed as "strange behavior" without a clear indication of the problem source on a higher layer, i.e., unpredictable but frequent loss of messages from one node on several (but not all) other nodes; the values based on these messages would be inconsistent between different nodes, leading to unpredictable and unforeseeable instabilities in the computations. True tolerance of such a fault is impossible unless several fully independent communication channels (i.e., using independent power supply, independent clocks, independent communication controllers) are available.

THE TTA STARCOUPLER

The TTA StarCoupler introduces an alternative approach to the challenge of asymmetric communication faults. Instead of detecting them after introducing a clique scenario, it prevents SOS transmissions from reaching any of the receivers. To do so, the TTA StarCoupler has to perform operations not unlike the bus guardian, but in addition to protecting the transmission timing from babbling idiot faults, it also has to protect the bus from potentially Byzantine bits (i.e., SOS faults on the bit coding layer) by re-shaping the timing and voltage of all transmissions, and from any higher-layer Byzantine fault. In case of the TTP protocol, this property is provided by the so-called "bus guardian window timing". With a "single fault" tolerant setup, which requires two redundant TTA star units, SOS faults are tolerated by removing them from the bus before they reach any receiver and inflict a clique scenario.

Figure 1: TTP cluster in star topology

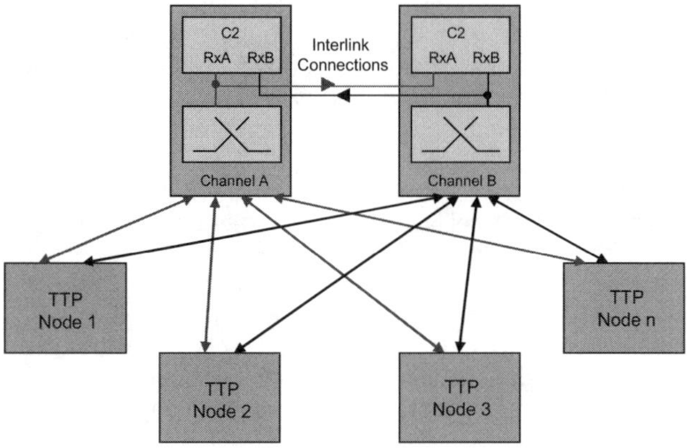

Connecting all nodes with a separate TTP controller forms the basis for a star architecture (figure 1). Each node is connected to one arm of the star. Since each node has two independent channels, two independent StarCouplers are required. This also provides safeguard against StarCoupler failure. If one of the StarCoupler's channels is faulty, correct transmission remains guaranteed via the second channel.

RESHAPE UNIT

A star architecture is also well suited for the avoidance of clique scenarios due to slightly-off-specification (SOS) faults. Distributed systems using a linear bus topology are inherently vulnerable to clique formation. If a certain physical value of the transmitted frame of one node is slightly outside the specified area, some nodes may still understand the frame correctly, whereas other nodes detect an erroneous frame.

There are

☐ Voltage level SOS faults, which can be due to a faulty line driver transmitting a voltage level near the threshold, and
☐ Time domain SOS faults, when the transmission timing is incorrect.

An SOS fault will lead to the formation of two groups (cliques), differing in their view of the node states (healthy or faulty). TTP has a clique avoidance strategy causing the TTP controllers of the smaller clique to enter the freeze state. Usually these nodes will reboot afterwards and reintegrate into the cluster. Nevertheless, for the sake of overall system safety, SOS faults should be avoided. To support this, the StarCoupler has a special block: the reshape unit.

Incoming signals are reshaped: decoded and encoded again. If the TTP controller in the StarCoupler detects an invalid frame, it will not retransmit to the other nodes. Thus they will get a consistent view of the membership of the sender. Even if one StarCoupler is faulty, consistent messages will be transmitted on the other channel.

SPATIAL PROXIMITY FAULTS

Imagine a car accident: easily a complete node could be destroyed which could cause a short circuit on the bus. In bus architecture, both, channel A and channel B will be affected and the cluster will be separated into two parts, which can no longer communicate with each other. In most cases, the application will not tolerate missing messages and might crash.

In star architecture, destruction of one node will only affect this arm of the star, and the rest of the nodes remain operable. The faulty arm is simply shut off. Even if one StarCoupler is destroyed, operation can still be guaranteed through the second one (therefore their location should be physically separated).

MONITORING FUNCTIONS

The StarCoupler implementation of austriamicrosystems does not require a CPU. The TTP controller in the AS8205 does all protocol jobs. Nevertheless, for bus monitoring purposes, a CPU can be externally connected as an option.

CONNECTING STARS – CASCADING

AS8205 also supports star cascading. This feature might be useful for a couple of reasons:

☐ The number of connected nodes exceed the number of ports of one StarCoupler (AS8205)
☐ The nodes are spatially distributed so that a cluster of nodes is in one location and the other nodes are located quite far from the first group. In that case, cascaded stars application can save cabling costs.

A TTP cluster with cascaded StarCouplers is shown in figure 2. Due to the frame processing time in the StarCouplers, propagation delay gets notably higher with each step of cascading. This is taken into account when designing the system with TTP development tools from TTTech.

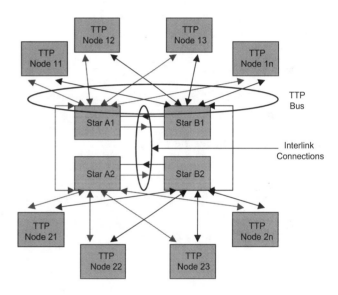

Figure 2: Cascaded StarCouplers

STARCOUPLER IMPLEMENTATION – THE TTP-C2X CONTROLLER AS8205

The TTP-C2X controller AS8205 forms the heart of a StarCoupler. The functional blocks are shown in figure 3. The implementation is based upon a design, licensed from TTChip, a TTTech subsidiary developing and marketing TTP controllers as technology-independent IPs.

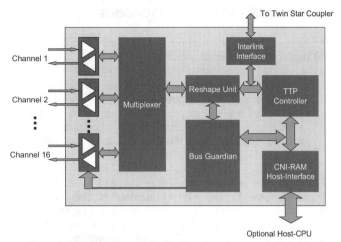

Figure 3: Block Diagram of TTA StarCoupler

AS8205 consists of a conventional TTP controller, switching, decoding and encoding logic. The TTP controller operates TTP protocol. In StarCoupler implementation it only analyzes the data, but does not actively participate (i.e. transmit) in the TTP network.

With the CNI (Communication Network Interface, implemented as RAM), the controller also provides a standard TTP interface to a possible host. Furthermore, the controller analyzes the frames whether or not they are syntactically correct to allow valid retransmission.

The switching logic mainly comprises of the internal MUX, the bus guardian and the reshape unit. The

enabled signal for the node that is supposed to send will allow the sender to write on an internal "shared channel" via the MUX. After the frame passes the reshape unit it will be broadcasted to all nodes.

CONCLUSION

Thus, the design intention (that has recently been validated in a first set of fault injection experiments) of the TTA StarCoupler is not to detect or tolerate, but to avoid the occurrence of asymmetric communication faults due to SOS-faulty communication controllers or other components on the communication layer. There is no formal guarantee for a 100% coverage of all possible sources of such asymmetric faults, and the TTP protocol will therefore continue to provide the formally verified clique detection mechanism even in TTA star systems. However, the chance for such a fault manifesting itself in a clique error is greatly reduced – to a level that is believed to offer sufficient availability for safety-critical applications in commercial aircraft.

CONTACT

Georg Stoeger
TTTech Computertechnik AG
Schoenbrunner Strasse 7
A-1040 Vienna, Austria
Tel.: +43 1 585 34 34-0
Fax: +43 1 585 34 34-90
E-mail: georg.stoeger@tttech.com

2002-01-0440

The Development of Communication Protocol for Real Time Control

Shumei Satoh
Toyo Microsystems Corp.

Fujio Matsui
Subaru

ABSTRACT

The new network protocol, Easy Control Network (EC-NET) and its communication control LSI have been developed, which realize the real time control over network. By solving the problems that CAN cannot solve, the EC-NET protocol has opened the practical path to the distributed network control system inside vehicles. The EC-NET protocol includes various features such as Non-prioritized token passing protocol, Improved communication quality with CMI code system, Network standard time, Memory mirror, etc.

INTRODUCTION

Various in-vehicle LAN protocols have been put forward since the 1980's. CAN has become the most common of all as control LAN. For effective control LANs, there have also been the new Byteflight and TTP protocols, and the TTCAN (improved CAN) protocol [1]. These protocols point to a powerful control LAN that combines a high-performance CPU and real-time OS communication module. We have aimed for a simpler yet faster high-function control LAN, and have developed the Easy Control Network (EC-NET) -- an improved Token Passing Bus Protocol – along with its communication controller LSI.

LAN FOR DISTRIBUTED CONTROL

The control technology trend keeps changing from concentrate to distribute, and distribute to concentrate. The last quarter of the 20th century was the time when microcomputers were adopted and accepted as stand alone systems in all the industry area. However, advanced computer technology has introduced new issues such as the inflation of system size and the increased system complexity. On the other hand, spreading communication technology enables to connect computers mutually, that is going to bring infinite possibility and revolutionary reform into the control technology.

The 21st century can be the computer communication age. The trend for control technology is moving from the closed system with stand alone control to an open system with distributed control, wherein controllers share the same data over a network to realize more efficient control [2](Fig. 1).

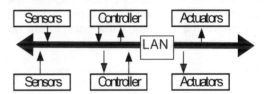

Fig. 1 Sharing on LAN

You may not want to increase the network traffic to share the data. Network for the distributed control should have the functions to share the data over network without increasing the traffic. You may not want to control the system by using data without time information, since you cannot tell how old the data is. Network for the distributed control should have the function to include precise time information in its data.

Reliable network protocol is mandatory for makers to provide real time distributed control systems. There are some problems that the popular network CAN confronts (Chart 1).

Weak Point	Possible Troubles
Priority based protocol	Unclear responsibility on each node
Inadequate error check	Malfunction
No time information	Complex to synchronize
Heavy software overhead	Need high end CPU and OS
Complicate usage	Need deep knowledge of network

Chart 1 Problem on CAN

RISK MANAGEMENT - It is very important to get rid of any risk from the actual product of the real time control system over network. Equality is the principle during the nodes on the network to simplify the system, but each node has to be responsible to its own behavior. Each node is controlled as a closed system and there will be no influence to the whole network system, that is the key point of the risk management. The intentional priority is not only against the principle of equality but also causes to be burdened with the risk. The control itself may have priority, but its result; the function of the product may not have such an idea of priority. All functions have the equal value to the users. Token passing protocol works well to eliminate the risk of priority.

RESPONSE TIME - Guaranteed maximum waiting time is one of the required elements in the control system. In the control system, the system design has to guarantee its function with maximum (not average) waiting time. The LAN transmission speed should be increased in order to shorten the maximum waiting time, i.e. to shorten the maximum waiting time. However, due to the collision detect mechanism, the prioritized CSMA/CD such as CAN has the limitation to improve its transmission speed. The Token Passing protocol is the choice to improve the response time since it has no limitation toward the transmission speed. [3]

SCHEDULING – The most of LAN's have priority, and that is one of the reasons why the importance of scheduling is stressed in the Control LAN.

Master-Slave - In the Master-Slave procedure, slaves carry out transmission in accordance with instructions from the Master. However in this procedure, data is concentrated on the Master, so the data processing load is heavy on the Master. It is furthermore heavier owing to the fact that the Master handles communication control.

Scheduling - In the Scheduling procedure, transmission request is not carried out above the frequency determined by scheduling management. The premise of the Scheduling procedure is to ensure that the length of transmission period is sufficient. If frequently transmitting updated data, priority must be lessened to prevent from being a general hindrance. However in low priority, updating data cannot be carried out frequently enough if other high priority transmission is frequently carried out. Advanced scheduling in the prioritized CSMA/CD is ultimately necessary. A shortcut to doing this is to introduce the OS and put matters in the hands of the Communication Manager. However, the overhead due to OS introduction increases, and it is furthermore necessary to introduce a high-performance CPU in order to provide the response speed required by the system. Even with the Token Passing Protocol, if the concept of priority is implemented in it, the waiting time is not precise and the anticipated responsiveness is not obtainable [3].

EASY CONTROL NETWORK (EC-NET)

We have stressed the importance of guaranteed maximum waiting time, and have developed an in-vehicle control LAN protocol; Easy Control Network (EC-NET)[4] using non-priority Token Passing as well as a Communication controller LSI (Fig.2 & 3).

The guaranteed waiting time eases the necessity of high-performance controller, and high quality control network can be built very easily.

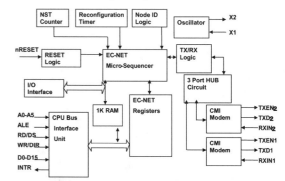

Fig. 2 Block Diagram of EC-NET Controller LSI

Fig. 3 EC-NET Controller LSI

Furthermore, if the amount of data transmittable at once is too small, waiting queue must be formed. It is necessary to control the transmission order [5], and the priority must be resolved inside the node before proceeding with transmission. (Fig. 4)

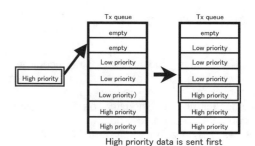

High priority data is sent first

Fig. 4 Common waiting queue management

Accordingly, the minimum buffer size in the EC-NET is set to 32 bytes, and the problem of waiting queue is resolved by transmitting the necessary data at once.

EC-NET CHARACTERISTICS

EC-NET is a distributed processing control LAN with scalable speed protocol. You can select transmission speed up to 5 Mbps in line with the system requirements.

HIGH RELIABILITY - The IC uses a non-priority Token Passing protocol and equal transmission of nodes is guaranteed without schedule management. Even if excessive transmission requests occur due to CPU error or program bug, because transmission cannot be carried out unless holding a token, bus occupation by a specific node can be prevented.

A simple equation provides of the system's maximum waiting time to transmit a packet.

$$(141.0 + 4.4 \times B) \times N \times 2.5/R \ [\mu S]$$

B: Maximum Tx data size (byte)
N: Number of nodes
R: Data rate (Mbps)

For example, Tx within 1mS is guaranteed in the system of 8 nodes with max.16bytes data @5Mbps as follows.

$$(141.0 + 4.4 \times 16) \times 8 \times 2.5/5 = 845.6 \ [\mu S]$$

Also, as the EC-NET does not need to detect collision nor to arbitrate bus access, there is a wide choice of communication media. Almost any communication medium most appropriate for the operating environment can be selected, thus making possible to connect to the same IC one or a mixture of the following: low cost RS485 or CAN non-isolated transceiver, isolation by highly reliable pulse transceiver isolated from GND voltage, or noise proof fiber optics. Also in addition to a CRC data checking, bit error detection and automatic readjustment is available using CMI Coding (Chart 2).

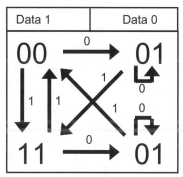

Chart 2 CMI Coding

Since CMI code prevents transformer saturation by toggling the polarity periodically, the transformer isolation can be realized with simple circuit.(Fig. 5) Fig. 6 is the newly developed transformer for automotive applications. It is a surface mount type pulse transformer with special magnetism as its toroidal core to make it small, high impedance, and low loss (Chart 3). It also has a choke coil for additional noise reduction (Fig. 7).

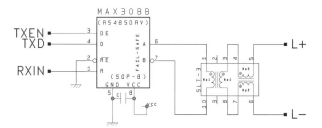

Fig. 5 Transformer Isolation

· Size (mm)
  1.7×10.0×6.0
· Weight(g)
  2.0

Fig. 6 Pulse Transformer

Operating Temp	-40^{0}C to +125^{0}C
Data rate	0.8Mbps to 2.5Mbps
Turns	1:1
Inductance	4mH min
Leakage	300nH max
Withstand Volt	AC1000Vrms

Chart 3 Transformer Spec

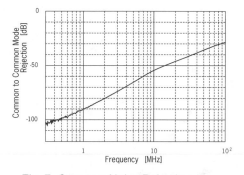

Fig. 7 Common Noise Rejection

HIGH FUNCTIONALITY - Transceiver functions are all automatically controlled by the communication IC itself, so communication control from the CPU is not necessary. When each packet is transmitted, the status of the reception node is automatically verified, so repeated transmission verification is not necessary. Packets not addressed to nodes are also automatically received. This data can be stored to a memory page assigned for each ID, so data is shared across the whole network. Also, the clocks of each node are automatically synchronized across the whole network, and it is guaranteed that the time information contained in all shared data is treated in the same way on any node.

USER FRIENDLINESS - A CPU is normally required to read and write the packet data to the communication controller. In stand-alone mode, the EC-NET transmits the input pin's status to the LAN, whereas it outputs from the output pins according to the received data from the LAN. When data processing by the CPU is not necessary, LAN nodes can be made of the EC-NET communication LSI alone.

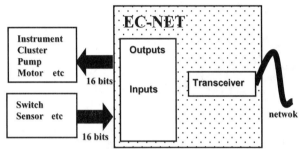

Fig. 8 Standalone Mode

COMMUNICATION PROTOCOL IMPROVEMENT

The EC-NET protocol is based on the Token Passing Bus Protocol [6] with some additional improvements.

BASIC FUNCTION - Nodes that have received self-addressed tokens can only transmit 1 packet of data before passing the token to the next node. (Fig. 9) Confirmation of passing the token is carried out by a network response after the token has been transmitted. When there is no response, it is inferred that there is no partner. The partner ID is increased by +1 and the token is re-transmitted. This function is repeated until the partner device is found.

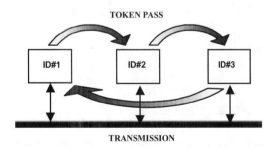

Fig. 9 Token Passing

ID DATA WIDTH REDUCTION - In general LAN protocols a large number of nodes are connected, so there is the tendency to have large ID data width. However, as the number of nodes increase in Token Passing Protocol, the increasing token lap time causes

the guaranteed response time to be deteriorated. Since the number of connected nodes can be fixed in the in-vehicle LAN, it is possible in the EC-NET to set up the system to limit the available ID to a necessary minimum number of bits.

ID ALLOCATION RESTRICTION - In a Token Passing Bus, all nodes must transmit the token to the node with the next largest ID. It takes time to find the next largest node when starting up the network, and it affects the responsiveness. In the EC-NET protocol, nodes are generally assigned an ID in a sequence from 1 upwards and the largest node number is designated. Together with these simple restrictions, responsiveness is maintained without any special function.

TOKEN PROTECTION - In Token Passing, the transmission right is controlled by a token. If the token is lost or damaged, the network function is temporarily suspended or a specific node is unchained from the network. There are two measures to prevent this.

1. Retry passing the token to the same ID again to prevent a node to be unchained by a single error.
2. Start sending the token always to node_ID+1. The nodes unchained can return by this.

FRAME FORMAT

The EC-NET uses the following five types of messages (Chart 4).

ITT | ALERT | EOT | NID | NID | ENDING |

FBE | ALERT | ENQ | NID | NID | ENDING |

PAC | ALERT | SOH | SID | DID | DID | C. P | DATA | CRC | CRC | ENDING |

ACK | ALERT | ACK | ENDING |

NAK | ALERT | NAK | ENDING |

ALERT   111111 (1 x 6bits)
EOT   04H
ENQ   85H
SOC   01H
ACK   15H
NAK   86H
NID   Next ID
DID   Distination ID
SID   Source ID
C.P   Continuation Pointer
DATA   Data
CRC   Cyclic Redundancy Check $(X^{16} + X^{15} + X^2 + 1)$
ENDING 000000000 (0 x 9bits)

Chart 4 Frame Format

The only one message; PACKET is user programmable. The other four types are automatically generated by the controller.

DATA TRANSMISSION PROTOCOL

The most of transmission procedure such as receiver verification before sending a packet, packet transmission, reception verification, and passing a token etc. are automatically carried out by the network controller without any CPU intervention (Chart 5).

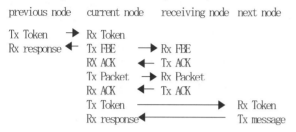

Chart 5 Automatic Data Transmit Procedure

At each packet transmission, the partner is verified before sending a packet. As replies from the receivers inside the same message frame are not requested in this procedure, synchronization is not necessary and long delay is permitted [7]. Due to this, accompanying the delay such as automatic error detection/correction by CMI coding, and the electro-opt conversion are possible in the EC-NET. High data rate up to 5 Mbps as well as long-distance communication is also possible.

NETWORK STANDARD TIME (NST)

In constantly amended data, it is important to know the data acquisition time. Because data in the network is received after a delay due to packet creation, transmission and reception processes, the process must be carried out to ascertain the point of data in time. All nodes in the EC-NET have clocks and can add the acquisition time into a transmission packet as data. These clocks are called "Network Standard Time" (NST). By automatically adjusting the speed in comparison with the time information transmitted by the Clock Master node, the whole network is always in synchronization (Fig. 10).

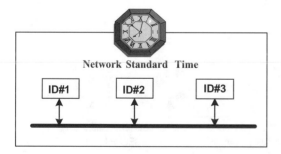

Fig. 10 Network Standard Time

MEMORY MIRROR

With the EC-NET, nodes other than the reception partners can also receive the packets. Most information handled by the in-vehicle control LAN is numerical data that is constantly amended. In processing this data, the newest data must be always prepared for when it is referred. In order to do this, it is preferred that data transmitted from nodes be always constantly monitored and that node data be constantly updated, as opposed to gathering data as necessary. In storing by ID the transmission each node receives from other nodes, a data mirror function is installed in the EC-NET that prepares the same data on any node (Fig. 11). By doing this, node data can be obtained with virtually no waiting time.

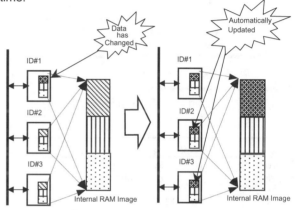

Fig. 11 Memory Mirror

DEVELOPMENT ENVIRONMENT

Operating environments including various standard bus boards, communication media boards and monitor tools (Fig. 12) are provided.

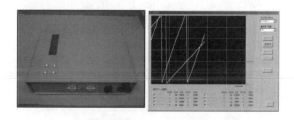

Fig. 12 Monitor Tool

The monitor tool has various functions for developing system, such as displaying communication data, operation information, and error status in real time (data capture function), generating emulation traffic data (debugger function), programming micro-controller's internal EEPROM over network (program function).

APPLICATIONS OF DISTRIBUTED CONTROL OVER NETWORK

GENERAL PURPOSE CONTROL SYSTEM FOR BUS BODY - The conventional bus body control system consists of wires to connect switches directly. The increasing demands of control have been introducing the problems; increasing cost of wire harness and its wiring labor. Connecting the general purpose standard control units by the EC-NET is the possible solution of the problems. The developed system is capable to control 240 inputs and 240 outputs over network. (Fig. 13) Each ECU is standardized as Chart 6.

EC-NET + standard ECU (15 nodes), LAN length 100m

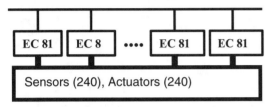

Fig. 13 Distributed ECU System for Bus Body

Item	Specifications
Program	Relay symbol method
Control	Cyclic calculation
Prog. Memory	FEEP ROM
ROM Size	2720 steps
Inputs	16 channels
Outputs	16 channels
Measure Cycle	1KHz
Analog IN	2 points
Diagnostics	Output, Network, etc.

Chart 6 Standard ECU for Bus Control

The control must be customized for every bus body, so each ECU's functions can be programmed by its remote programming function. After the programming, the programming monitor tool sends the program as commands. Each command has a node ID information built-in, so only the specified node takes the command and replies to it. This protocol is developed by MAiS, and the programming monitor tool generates the commands automatically from user's program. Train control system has the similar system built-in.

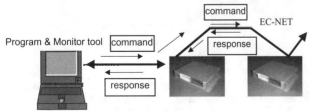

Fig. 14 Remote Programming for Bus

GENERAL PURPOSE CONTROL SYSTEM FOR AUTOMOTIVE - Automotive control has the same problems as bus body control. To simplify the control system and to standardize the control ECU, new technologies such as smart sensor and intelligent actuator are necessary.

One example of the distributed ECU system for automotive is capable to control up to 120 inputs and 108 outputs.(Fig. 15 and Chart 7) This specification should satisfy the actual control system's requirement, e.g. GM's Precept[8] uses 100 outputs, 70 digital inputs and 50 analog inputs. By assuming the data length to transmit is 12 bytes each, the whole system can update all the data in all ECU's in every 581.4μS without breaking down the system into several sub-systems.

EC-NET + standard ECU (6 nodes), LAN length 10m

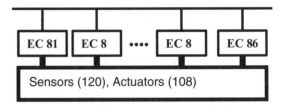

Fig. 15 Distributed ECU System for Automotive

Item	Specifications
Program	Subroutine method
Control	Micro calculation
Prog. Memory	Flash ROM
ROM Size	448 Kbytes
Inputs	12 channels
Outputs	4 to 18 channels
Measure Cycle	1KHz
Analog IN	8 points
Diagnostics	Output, Network, etc.

Chart 7 Standard ECU for automotive

MCM technology integrates bare chips as a standard ECU, that requires to be programmed over network. The external control unit can control software on all the ECU's. (Fig. 16)

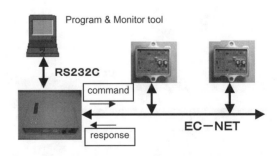

Fig. 16 Remote Programming for Automotive

CONCLUSION

CAN has been used as de-facto standard for low speed network. However, because of its high potential, the EC-NET cans be adopted as the high speed network for real time distributed control. The EC-NET has just started to be adopted as distributed control module, and it will be widely adopted in various applications. Chart 8 shows the comparison between the CAN and the EC-NET.

	CAN	EC-NET	Advantage of EC-NET
Protocol	CSMA/CD (Priority)	Token Passing	Low priority node may not communicate with CAN. All node are guaranteed to communicate with EC-NET.
CRC check	CRC15	CRC16	
Coding	NRZ	CMI	CMI is capable to detect and correct a bit error
MAX Data rate	1Mbps	5Mbps	EC-NET can communicate fast enough to support real time distributed control.
MAX number of nodes	30 (ISO11898 High Speed)	31	
MAX Frame length	8Byte	253Byte	Send large data at a time
Real time clock	None	1micro second precision	All the clocks are synchoronized over the network
Software load	Need CPU on every node	No CPU necessary in Standalone mode	Simple I/O without CPU. No CPU load for sharing data.

Chart 8 CAN vs EC-NET

CONTACT

Shumei SATOH
Toyo Microsystems Corporation (TMC)

TEL: +81-3-5487-0484
FAX: +81-3-5487-0490
E-mail: shumei@tmc.co.jp
URL: http://www.tmc.co.jp/ec-net/

Fujio MATSUI
Fuji Heavy Industries LTD. (SUBARU)

TEL: +81-422-33-7513
FAX: +81-422-33-7741
E-mail: matuif@tky.Subaru-fhi.co.jp

REFERENCES

1. Karl-Thomas Neumann et al. "Architecture Leadership in the Automotive Industry", SAE 2000-01-C067
2. Michel F.Sultan et al. "Smart Sensors for Future Robust Systems", SAE 2000-01-C055
3. Mitunori Kato et al. "Evaluation of Latency Time for High-Speed In-Vehicle LAN Protocols", JSAE paper 9432723
4. http://www.tmc.co.jp/ec-net/
5. Kiyoshi Inoue et al. "An Evaluation of Latency by Simulation for Distributed Multiplex System", SAE paper 910717
6. ANSI/ATA878.1-1999: "Local Area Network: Token Bus", http://www.arcnet.com/lit.htm
7. Seiji Nakamura et al. "The High-Speed In-Vehicle Network of Integrated Control System for Vehicle Dynamics", SAE paper 910463
8. Joe LoGrasso et al. "Low Power Flexible Controls Architecture for General Motors Partnership for a New Generation (PNGV) Precept Vehicle", SAE 2001-01-C060

2000-01-3503

Internet-based Vehicle Communication Network

John Skibinski
Eaton Corporation

Jim Trainor and Chad Reed
Fastek International

ABSTRACT

A number of different data networks have been implemented for electronic control unit communication in vehicles to date. Each network serves a particular need, such as low-cost networking of cab components or high-speed networking of powertrain components. Although each communication network performs its original purpose, the different communication networks, especially those using hardware-based messaging protocols, are expensive to integrate for information sharing and are not readily upgradeable with new messages. This is complicated by the growing number of different communication networks for vehicles, often driven by OEM and supplier technology consortiums rather than by end-user requirements. The result is added vehicle-support costs for the OEM, dealership and customer to maintain multiple networks.

The application of a world-wide internet-based messaging protocol standard, eXtensible Markup Language (XML), will be explored as an easily-integrated and field-upgradeable messaging protocol usable for a wide variety of vehicle networking applications. The advantages are unified rather than disparate access to all vehicle component information, software rather than hardware extension of vehicle system performance, and lower support costs by using globally-available internet-based technology rather than proprietary custom support systems. This translates into opportunities such as:

- Quicker drop-off of shipments having RF tags that transmit their ID to a yard computer system when delivered, causing an internet-based invoicing process to be initiated without the need for a clerical staff at either the shipper or the receiver.

- An automobile computer system that accesses a service center chain's schedule while in transit, then suggests to the driver the best times and nearest locations to repair the automobile component.

- A sensor or a switch that instead of sending an analog signal or closing a circuit, broadcasts an

Internet message routable across many different types of hardware networks.

EXtensible Markup Language has been implemented in many different industries. For the purposes of this paper and for demonstration at this conference, we will examine the use of XML in vehicle communication networking.

INTRODUCTION

Our vehicle communication networking proposal is based upon trends developing at two ends of the network communications spectrum. At the high end of the spectrum scale is the vast communications network known as the Internet. At the low end of the spectrum scale is the smallest communication node known as a single-chip micro-controller. Once the two ends of the spectrum converge, the possibility for a unified control network exists, independent of the hardware used.

The advent of Standard Graphics Markup Language (SGML) for the publishing industry led to the simpler Hyper-Text Markup Language (HTML) widely-adopted as an Internet page format standard. Out of SGML & HTML use grew the need for an Internet data exchange standard that would be widely-accepted, resulting in the eXtensible Markup Language (XML). Nearly every industry is engaged in defining one or more XML-based standards, because XML enables universal data exchange independent of the hardware used. Hence, unified data and format networks become possible.

The advent of the microprocessor led to its wide-spread adoption in programmable hardware controllers. Out of microprocessor use grew the need for real-time micro-controllers able to be networked within a system, resulting in high-speed eight-pin surface-mounted micro-controllers able to host within themselves a complete TCP/IP communications stack. As a result, many companies are using these tiny micro-controllers. One reason is because these IC's enable their product to directly communicate to the Internet. Hence, the horizon of networked real-time control has been vastly expanded.

By comparison, most vehicle control networks use a hardware-based address and data scheme – that is, all participants in the network communicate using a pre-defined multi-byte hardware address and data protocol. Over time, such a control network has a limited lifespan due to the pre-defined set of hardware messages, and invariably a new control network is defined – once again using a pre-defined multi-byte hardware address and data protocol. A vehicle control network using an XML-based software address and data scheme allows each participant to communicate using an XML-based software message. An XML message set can be extended indefinitely independent of the hardware used – and now the hardware technology is just emerging with the power to provide unified real-time control using XML.

The global Internet, when used at a local level, thus can become a real-time control network. Yet, the local level network is fully integrated with the global Internet - again independent of the hardware used. Hence, a unified control network emerges. We will demonstrate one application of an XML-based vehicle communication network for the multiplexing of vehicle switches.

XML-BASED CONTROL NETWORK PRIMER

XML is defined by the World Wide Web Consortium document REC-xml-19980210, "Extensible Markup Language (XML) 1.0". The following information is extracted from that document and made relevant to this topic for the purposes of developing our control network example.

Extensible Markup Language describes a class of data objects called XML documents. XML documents are made up of entities, which contain either parsed or unparsed data. Parsed data is made up of characters, some of which form character data, and some of which form markup. Markup encodes a description of the document's storage layout and logical structure, and in our control network example defines the set of valid messages within our control network language. Each participant in the network uses an XML processor software module to receive and send XML documents in accordance with their content and structure. An XML processor does work on behalf of the embedded application, and in essence is used to receive and send messages in our control network.

Typically, XML documents begin with a declaration that specifies the version of XML being used:

<?xml version="1.0"?>

The XML version number "1.0" refers to XML Specification Version 1.

The XML document type declaration is specified next, and it contains or points to markup declarations that provide a grammar for a class of documents. This grammar is known as a document type definition or DTD, but for the purposes of this paper is referred to as our control network language. (XML does allow the definition of language as English, French, etc., but we will not use that concept here.) The DTD schema defining our control network language is contained in:

<!DOCTYPE XMLNET "Version-1-0.dtd">

The DOCTYPE "Version-1-0.dtd" refers to our XML Network Version 1.0 schema, which defines the data storage layout and logical structure allowable for our control network language. In it simplest form, a DTD defines a network address or message data as a string of characters. In its most complex form, a DTD defines not only this information but the logic behind how the information is to be used by the application. This XML concept allows for the creation of an extensible control network language, a powerful concept, provided that the end application is able to parse the DTD and implement the additional new language elements. Even if the end application cannot use some of the DTD information, it can choose to ignore it without impacting its ability to communicate with legacy messages. Compare this to what happens when a hardware-based address and message data format are changed in an existing control network – network communications becomes completely disrupted.

The beginning of an XML element is marked by a start-tag, always starting with "<" followed by a label followed by ">". The end of an element that begins with a start-tag must be marked by an end-tag containing a name that echoes the element's type as given in the start-tag, always starting with "<\" followed by the same label followed by ">". The text between the start-tag and end-tag is called the element's content. A start and end tag combination for our control network language is:

<LED>1<\LED>

The "LED" element is defined to have a content of "1". In our network control language, the LED is turned on. The element LED data and its format must be defined in the DTD or else it is an invalid element. The content "1" data and format also must be defined in the DTD or else it is invalid. Once again, the use of a DTD schema is a powerful concept for defining and extending the control network data in a very structured manner.

Elements can be nested in XML. This is very useful for creating addressing / data schemas for our control network:

<Node1>

    <LED>1-0<\LED>

    <LED>2-1<\LED>

<\Node1>

Here, the communications node "Node 1" in our control network is addressed, and the message for LED "1-0" implies LED 1 Turn Off and LED "2-1" implies LED 2 Turn On, if the DTD schema defines that as the control network language meaning. Alternatively, another DTD schema could imply the same meaning but be written as:

```
<Node1>
    <LED>1-0, 2-1<\LED>
<\Node1>
```

Obviously, this is a more efficient transmission of information, but it requires more XML parsing overhead. That is a trade-off to make with an XML control network - message size versus parsing overhead - but these are the same issues with most real-time control network designs. The difference with XML is that the control network language is defined at the software level - not the hardware level - and so this optimization can be flexibly achieved over time via software upgrades.

We will now examine how to use XML in the simplest of control elements - a networked switch. This concept may seem farfetched - but then remember that continuing increases in computing power have been matched by corresponding decreases in price per MIPS - so the dawn of networked control elements on the Internet may just be upon us.

AN XML SWITCH LANGUAGE

A switch by definition has two states: open or closed. Our example of a "switch" is shown below, having two momentary pushbuttons with three lights that each indicates a switch state. One pushbutton switch is used to scroll through the three states, the other to select the state currently displayed. In essence, this is a three-position switch with three programmable lights. The lights in our switch are programmed to indicate the previous state while the state desired is scrolled to and selected. In addition, as a master computer on the network is implementing the state selected, the light blinks until the implementation is completed, at which time the light stays on. The XML for our switch operating in a polled network environment is described below.

Three Position Switch

USER INTERFACE – As the user presses the scroll pushbutton, the current state remains lit while the next state displayed in the sequence blinks. If the select pushbutton is pressed, the corresponding state is entered and continues blinking. The micro-controller performing these functions now remembers the previous state and the state desired, awaiting a poll from the master computer to first send the message, and then upon the next poll, receive a message that the master

computer has executed the request. If it has, the switch turns off the previous state indicator and turns the desired indicator on full with no blinking. The XML messaging for our switch is defined as follows.

MASTER COMPUTER REQUEST – The master computer on the network requests the status of the switch by sending the XML message below. This tells the switch to respond to the poll of the master computer. The master computer knows the switch by the software label "SWITCH1" as opposed to a pre-defined hardware address. Hence, there is a virtual "plug and play" addressing of the switches as well as an infinitely addressable number of switches available on the control network. The master computer is issuing a command with no data, so the message is sent with a null element as shown below:

```
<?xml version="1.0"?>
<!DOCTYPE XMLNET "Version-1-0.dtd">
<SWITCH1>
    <REQUEST><\REQUEST>
<\SWITCH1>
```

Note the use of the DOCTYPE. Once the switch has parsed the "Version-1-0.dtd" once, it remembers the XML language for this network. If the switch sees a DOCTYPE other than "Version-1-0.dtd", it knows to re-parse the document to learn the new XML language for this network. Hence, the control network language is software-extensible over time, regardless of the network hardware.

SWITCH RESPONSE – The switch responds to the master computer request by sending the XML message below. This example illustrates that the user has selected State1:

```
<?xml version="1.0"?>
<!DOCTYPE XMLNET "Version-1-0.dtd">
<SWITCH1>
    <RESPONSE>State1<\RESPONSE>
<\SWITCH1>
```

Note that the switch must respond with data and format allowed by the DOCTYPE. If the switch responds with anything else, the response is invalid. Also note that the switch is responding with DTD "Version-1-0.dtd". If the control network is updated to a new version of DTD with correspondingly new messages, and this switch was not upgraded for whatever reason, this switch still is able to communicate to the master computer because it responds within the context of its DTD. Thus, software extensibility of the control network language does not necessarily imply obsolescence of legacy switches.

MASTER COMPUTER SEND – The master computer can not only determine the switch status but control the switch lights as well. In this example, our micro-controller is programmed for off/blink/on lighting conditions. Adjusting the lights lets the user know that the desired state selected has been acknowledged, is being implemented and then has been implemented. Here, the first character 1 through 3 means which light to address, the "-" is a data separator, and the X represents a character between 0 and 9, where 0 means the corresponding light is off and 9 means the corresponding light is on. The values 1 through 8 are correspondingly increasing light blink rate for use as a progress indicator. The XML message for our switch is then as follows:

```
<?xml version="1.0"?>

<!DOCTYPE XMLNET "Version-1-0.dtd">

<SWITCH1>

    <LIGHT>1-X<\LIGHT>

    <LIGHT>2-X<\LIGHT>

    <LIGHT>3-X<\LIGHT>

<\SWITCH1>
```

Thus, the user scrolls and selects the desired switch state, the master computer polls the switch to determine the desired state, the master computer sends light blinking data about the desired state selection progress, and then once the master computer has executed the selection, again sends light blinking data to turn off the indicator for the previous state and turn on full the indicator for the desired state.

XML MESSAGING TRADE-OFFS – Note throughout this example the combined power of control network language simplicity, virtual "plug and play" addressing and message extensibility. It does come at a cost, requiring an embedded micro-controller to parse the XML messages, but that cost is decreasing with time. In addition, a machine-readable version of XML, Binary XML, can be used to dramatically increase communication and parsing efficiency.

AN XML SWITCH IN A J1708 NETWORK

XML by itself is a data andmat messaging language. XML messages can be asynchronously communicated via any wired or wireless network, and are completely routable across different networks without conversion of the messages. However, as a result of being completey isolated from any communications media, it still requires a hardware network to deliver its messages. Thus, XML messages require the foundation of at least Layer 1 through Layer 4 of the OSI Network Model to be sent from one point to another. For our example, we will utilize an SAE J1708 hardware network to deliver the XML messages defined above.

We shall define four modes for our vehicle communication network, Configure, Query, Send and Receive, to demonstrate the transmission of XML

Messages over a J1708-type hardware network. Here, we retained the physical Layer 1 and datalink Layer 2, but we have re-designed the network Layer 3 and transport Layer 4 to accommodate XML message communications.

CONFIGURE – The Configuration Mode allows for the self-identification of switches on the network after system power-up. This allows for the removal of old and the insertion of new switches onto the network. Essentially, this is "plug and play" capability. An OEM can install a switch in a vehicle without needing to configure its data, a dealer can replace a switch without needing to configure its address, and an end-user can install new switches to obtain new functionality.

The mode operates as follows. Upon power-up, all switches remain silent. The master computer places a preamble of eight escape characters on the network, causing all switches to enter the Configuration Mode. All switches then attempt to place a Configuration Mode identification frame on the network. The master computer then responds with a Configuration Mode acknowledgement frame for each identification frame it sees on the network.

Once a switch has received a Configuration Mode acknowledgement frame from the master computer, it is not allowed access to the network until the Configuration Mode has ended. If a Configuration Mode acknowledgement frame from the master computer is not received within several milliseconds after the switch has placed a Configuration Mode identification frame on the network, the switch will place another Configuration Mode identification frame on the network, in the attempt to get a Configuration Mode acknowledgement frame from the master computer. This attempt will continue a specific number of times.

The master computer will stay in Configuration Mode listening for Configuration Mode identification frames from the switches until the network is idle a specific number of milliseconds. All switches on the network stay in Configuration Mode until the following conditions are met:

- The switch receives a Configuration Mode acknowledgement frame from the master computer. In this instance, the master computer has identified the switch and assigned it a virtual address. This is the "plug and play" self-configuration of the switch's address.

- The switch has placed a number of Configuration Mode identification frames on the network and has not once received a Configuration Mode acknowledgement frame within a specific number of milliseconds from the master computer. In this instance, the master computer or the network or the switch could be experiencing a fault.

- The switch has seen an idle network. In this instance, the Configuration Mode has ended.

In the Configuration Mode, as the switches are powering up and attempting to self-configure, a contention-based protocol must be used to establish the virtual addressing needed for a "plug and play" network. As a result, the switches must broadcast their Configuration Mode identification frames according to the following contention-handling protocol.

- Upon power-up, the switch transmits its Configuration Mode identification frame only after its own quiet time has passed. One method to compute the switch's quiet time is use its serial number divided by a number for multiplication by the network frame time to get the quiet time. This scheme is similar to Ethernet's contention-handing scheme.

- Once a switch transmits its Configuration Mode identification frame, it must immediately check to see that the first byte placed on the network is correctly read back. If it is, transmission of the rest of the message continues. If not, there was contention between two or more nodes on the network. The switch must stop transmitting and wait until its quiet time again. After its quiet time has passed, it may transmit its Configuration Mode identification frame again.

- If a switch has attempted to transmit its Configuration Mode identification frame and lost contention several times, it must calculate a new quiet time, based upon the serial number divided by a number times two (the number of contention sets plus one), for multiplication by the network frame time, then added to the original quiet time.

- If the new quiet time continues to result in contentions, the switch continues to calculate new quiet times, based upon the serial number divided by a number times the number of the contention sets plus one, for multiplication by the network frame time, then added to the previous quiet time.

Eventually, all switches have broadcasted their Configuration Mode identification frame and have received their Configuration Mode acknowledgement frame, resulting in a virtually-assigned network of switches. This ends the use of the contention-handling protocol typical of an Ethernet network, which is not optimal for a real-time control network. As a result, all other modes operate using a polled network protocol.

QUERY MODE – The Query Mode allows for the rapid determination of switch state change. If the switch has not changed state, as indicated by a one status-byte response, the master computer does not spend any time on the switch regarding the Send and Receive Mode. Thus, most of the time, the master computer is rapidly polling every virtual switch address as fast as it can, stopping at a switch only if it has changed the status byte that signals a data exchange is necessary.

SEND MODE – The master computer uses the Send Mode to transfer XML messages to one of the switches

on the network. The sequence starts with a Connection Management Request to Send frame from the master computer. This is the start of a dialog between the master computer and a switch that ensures the XML message is delivered, similar to the packet data delivery schemes used for Layer 1 through Layer 4 in the OSI Network Model.

Once the Connection Management Request to Send frame is sent from the master computer, the addressed switch must respond with the Connection Management Clear to Send frame. Then, the master computer will send the XML message frame. Once the XML message has been sent, the switch must send one of two frames to the master computer. If the message was received by the switch without error, the switch will send a Connection Management End of Message frame. If not, the switch sends a Connection Management Clear to Send frame again to cause a resend of the message.

RECEIVE MODE – The master computer uses the Receive Mode to receive XML messages from one of the switches on the network. The sequence starts with the Connection Management Request to Send frame from the switch (which previously informed the master computer via the Query Mode that it wishes to transmit). Once the Connection Management Request to Send frame is sent from the switch, the master computer must respond with the Connection Management Clear to Send frame. Then, the switch will send the XML message frame to the master computer. Once the XML message has been sent, the master computer must send one of two frames to the switch. If the message was received by the master computer without error, the master computer will send a Connection Management End of Message frame. If not, the master computer sends a Connection Management Clear to Send frame again to cause a resend of the message.

CONCLUSION

In summary, the following has been learned and demonstrated:

- XML is an Internet data exchange standard that has been applied and demonstrated as an almost-real-time control network language. Thus, the goal of a unified control network is becoming attainable. This is supported by the implementation of TCP/IP communication stacks on $5 in quantity eight-pin micro-controllers, once again demonstrating Moore's Law corollary that increases in processing power results in decreases in price per MIPS.

- XML, by nature of the DTD Schema, allows for the extensibility of the network control language by software, thereby isolating the addressing and messaging from the hardware. In addition, as long as the embedded micro-controller has the parsing processor power, the control network lifespan is extensible beyond that of a pre-defined multi-byte hardware address and data protocol network. Considering that a $5 eight pin micro-controller today has an onboard 20MHz RISC processor, one can assume it has the processing power for several extensions to the network control language.

- XML is asynchronous and routable, which means that its messaging is transportable on the Internet (Ethernet), or on J1708 (RS-485), J1939 (CAN), or any other hardware network. As a result, XML messages pass across network boundaries without conversion. Consider the potential cost savings due to unification of the vehicle control networks, which today includes a variety of powertrain, dashboard, entertainment and wireless communications networks. Then include the potential cost savings due to supply chain integration, factory assembly test equipment and dealer servicing equipment. Then include the potential cost savings for the customer using the vehicle in an order fulfillment process - shipping, tracking, receiving and invoicing – for which XML is well-suited.

We have successfully implemented eXtensible Markup Language in our prototype vehicle communications network. It has performed more responsively than expected, especially the micro-controller communications nodes. The trend for XML, and especially Binary XML, is to become even more responsive, just as the Internet only five years ago hosted just HTML web pages and today is upgrading its bandwidth to host streaming video in real-time. XML-based streaming audio applications are already available in the form of speech recognition systems and Internet radio broadcasts, certainly more demanding applications than our switch communications network demonstrated here. As a result, we recommend a joint SAE/TMC Committee explore the potential applications and benefits of XML, in preparation for developing XML DTD Schema functional requirements for its participating industries.

REFERENCES

1. "Extensible Markup Language (XML) 1.0:", World Wide Web Consortium, REC-xml-19980210
2. "Serial Data Communications between Microcomputer Systems in Heavy-Duty Vehicle Applications", SAE J1708 Revision Oct. 93
3. "Joint SAE/TMC Electronic Data Interchange between Microcomputer Systems in Heavy-Duty Vehicle Applications", SAE J1587 Revision Mar. 96

CONTACT

John Skibinski

Eaton Corporation

4201 North 27th Street

Milwaukee, WI 53216

414/449-7753

jskibin@execpc.com

Jim Trainor

Fastek International

1350 Boyson Road

Suite B

Hiawatha, Iowa 52233

319/294-6664

jtrainor@fastekintl.com

Chad Reed

Fastek International

1350 Boyson Road

Suite B

Hiawatha, Iowa 52233

319/294-6664

creed@fastekintl.com

GATEWAYS AND MIDDLEWARE

2005-01-1696

Automotive Gateway Design Using Evolutionary Algorithms

Wolfgang Hauer
DaimlerChrysler AG

Hans Peter Großmann
University of Ulm

Günter Stöhr
University of Siegen

ABSTRACT

Because of the rapidly increasing amount of electronic components and busses in a vehicle, the use of gateways in Electronic Control Units (ECUs) becomes more important. The upcoming question is how to design an optimal gateway.

This paper describes a method for designing an optimal automotive gateway in an FPGA by using Evolutionary Algorithms (EAs). The complete gateway functionality is diagrammed in a specification graph which consists of a function graph and an architecture graph. The function graph describes the complete functionality of the gateway. The architecture graph shows the variety of the different implementation options of the mapped function graph. Each gateway task in the function graph can be realized either in a parallel way (different kinds of hardware implementations) or in a sequential way (software on a microprocessor core). Depending on the different realization methods, the tasks of the function graph are mapped on the individual implementation node of the architecture graph. The edges of the specification graph are weighted with the values which give input for the multi-objective optimization of the system. The optimization of the specification graph can only be solved in polynomial time with a nondeterministic algorithm (NP-hard problem). Therefore heuristics like EAs are used. The procedure provides a set of optimal implementations based on the multi-objective optimization which satisfy user defined constraints like for instance cost and performance. In this paper the method of designing a specification graph for an FPGA based automotive gateway is described. The assigned literature will be referred to for the optimization of the specification graph.

INTRODUCTION

The increasing amount of ECUs requires the split of the vehicle network into different bus systems based on the functionality and the location of several ECUs. The segmentation into a Powertrain and Chassis network is a common method in the automotive industry. The Powertrain network is mostly a high speed Controller Area Network (CAN) Bus with a transfer rate of about 500KBaud up to 1 MBaud. The Chassis CAN has a lower data rate of about 83,3KBaud up to 125KBaud. Figure 1 shows a part of a typical vehicle network topology.

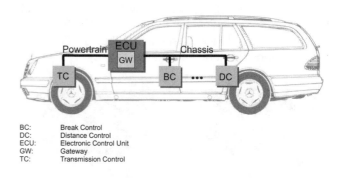

BC:	Break Control
DC:	Distance Control
ECU:	Electronic Control Unit
GW:	Gateway
TC:	Transmission Control

Figure 1: Vehicle Network

To connect different bus systems which communicate mutually, a gateway is required. Here it's shown how to design an optimal gateway in an FPGA using graph theory and evolutionary algorithms. This gives an example how to design an optimal automotive gateway which builds the communication node between the Chassis and the Powertrain CAN, based on different constraints like costs and performance. It is often realized in the Engine Control Unit, because the functionality of the engine controller requires information from both, the Powertrain and the Chassis CAN. The functionality of a gateway comprises of several tasks like message routing, signal routing, signal extraction, protocol conversion, diagnosis and network management, as shown in figure 2.

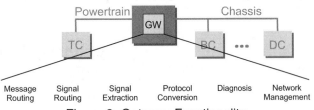

Figure 2: Gateway Functionality

The message routing routes a message from one bus to another without changing the message. Signal routing extracts information from one or more messages of a bus to build a new message and to send it to the same or to another bus. Signal extraction extracts message information for the own use in the ECU (e.g. the rpm signal of a specific CAN message on the Powertrain CAN is used for a special algorithm in the Engine Control Unit). The protocol conversion is necessary to route messages between bus systems with different transmission methods. For instance one bus protocol is a message oriented protocol and the other is a time division multiplex (TDM) based protocol. Diagnosis is responsible for user defined diagnosis functions. For network management see [3].

ISO/OSI-MODEL

The ISO/OSI-Model (International Standardization Organization/Open Systems Interconnect-Model) shows an internationally accepted method to describe the functionality of complex communication systems. The model is based on a hierarchical layer approach [4]. To describe an automotive gateway by means of the ISO/OSI-Model, the layers one, two and seven are used [5].

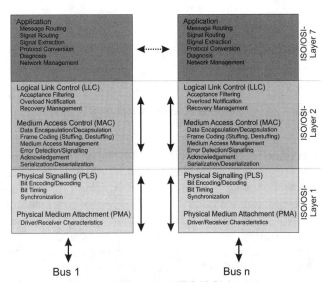

Figure 3: Gateway ISO/OSI-Layers

FPGA GATEWAY DESIGN

There are different implementation possibilities of an FPGA based gateway design. One is to design the gateway in terms of a pure hardware solution (VHDL) which could be optimized in respect of performance or area. The other is to implement a microprocessor IP core in the FPGA with several Bus IP Cores and program the gateway functionality in software (software solution). The third alternative solution is an optimized gateway design based on different constraints. This could be, based on the constraints, a pure hardware, a pure software or mixed solution. The optimal solution is reached by designing a specification graph for the gateway and optimizing the graph with evolutionary algorithms.

HARDWARE SOLUTION

In an FPGA based gateway design, the user can choose the granularity of several function blocks, which are needed to handle a message from a bus. The definition granularity is divided into rough granular and fine granular. Rough granular means that many functions are centralized in one function block. The amount of function blocks in the design decreases whereas in a fine granular constellation the amount increases. Here the functionality is split into four main function blocks plus one physical layer function block outside the FPGA. The four blocks are the basic constellation for the functionality of one bus in the FPGA, which is to read a message from the bus, extract the information of the message and decide the further handling of the message. To design a gateway, at least two parts of the basic constellation in figure 4 are needed. The Bus Basic Constellation is rough granular, because the system is easier to describe.

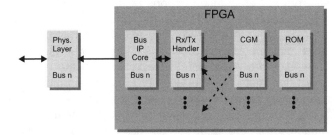

Figure 4: Bus Basic Constellation

In reference to layer one of the ISO/OSI-Model, the Physical Layer Driver In figure 4 represents the Physical Medium Attachment (PMA). The Physical Signalling (PLS) in layer one and the complete layer two are covered by the BUS IP Core while layer seven contains the whole gateway functionality. The Physical Layer Driver is responsible for the translation of a bus message on the physical side into a serial or parallel digital signal. The Bus IP Core is responsible for the handling of the bus protocol. It receives and transmits the bus messages from and to the Physical Layer Driver, buffers them, extracts the message information and signalizes the Rx/Tx Handler that a new message was received. The error handling like buffer overflow, bus timeouts and the handling of status information like receive and transmit interrupts is also part of the Bus IP Core task. If the Bus IP Core signalizes that a new message was received, the Receive and Transmit Handler gets the message from the IP core to forward it to the Configurable Gateway Manager (CGM). It also transmits messages from own or another CGM to the Bus IP Core of the same Bus Basic Constellation. The CGM function block contains the main gateway functionality for the connected bus system. Based on the received message identifier the CGM executes the functionality message routing, signal routing, signal extraction, protocol conversion, diagnosis or network management. The CGM gets the message handling information through encoded data in the ROM. The design of the CGM depends on the gateway constellation, the bus functionality and the user. For the purpose of better understanding I choose rough granular. In reality the CGM consists of several blocks, based on the implemented gateway- and message handling functionality of the respective bus system. This keeps the reusability of the individual function blocks. The ROM contains encoded information about the handling of the received message. The coding includes information such as destination bus and kind of the message type. That's how the CGM knows exactly, whether the received message is a message for message routing, signal routing, signal extraction, protocol conversion, diagnosis or network management. The encoding of the ROM is user defined.

Design Hint:

For a design with maximum performance, it is essential to think about mechanisms for accessing the ROM and its encoded information as fast as possible. In respect of access time, a minimum is reached by implementing the address bus of the ROM in terms of the received message ID (parallel). Like this the encoded information can be read out in three FPGA clock cycles (set ROM address, one wait state, read encoded information). The ROM in an FPGA can easily be created through a core generator and configured by a user defined ROM configuration file. The core generator makes use of the FPGA internal Block RAMs (BRAM) and thus avoiding the usage of additional FPGA area. Figure 5 is an example of a possible ROM coding for a gateway with up to 32 bus systems and 11 bit message IDs.

Assuming, figure 5 is the configuration ROM for bus system one and the received message has the ID 000h. The encoded information at ROM Adr. 000h says that the received message is a diagnosis message and has to be routed to the destination bus system three. If the received message has the ID 7FF, the CGM knows that it is a network management message for its own bus.

The width and depth of the ROM is defined by the developer of the system and can vary, based on the functionality and the design of the CGM. Each bus system has its own Config ROM. Thus, each gateway functionality is fully configurable.

Performance Optimized Design

In an FPGA the user always has to decide between area and performance optimization. A performance optimized design is necessary, if the connected bus systems have a relatively high baud rate and/or there are many bus systems connected to the FPGA. For a high performance design, the Bus Basic Constellations (Figure 4) have to be implemented in parallel. The amount of the constellations is proportional to the amount of connected bus systems. For a gateway functionality, the CGM of a Bus Basic Constellation will send the message to the defined target bus system as described before. A really important point in designing FPGA gateways is the calculation of the internal FPGA clock rate. For its calculation all gateway function blocks have to be taken into account. The minimal internal FPGA clock rate f_{FPGA} can be calculated as follows:

$$f_{FPGA} \geq f_{FPGA_{io}} \cdot N_{FPGA} \tag{1}$$

Where $f_{FPGA_{io}}$ is the FPGA input/output frequency at the Physical Layer Driver and N_{FPGA} the number of the FPGA internal clock cycles for the functional conversion. For a serial connection of the Physical Layer Driver to the FPGA, the FPGA interface speed is the maximum of the connected bus baud rate:

$$f_{FPGA_{io}} = max(\ f_{bus}\ ) \tag{2}$$

and for a parallel connection:

$$f_{FPGA_{io}} = \frac{max(\ f_{bus}\ )}{min(\ N_{bit/Msg}\ )} \tag{3}$$

Where $N_{bit/Msg}$ is the amount of bits per message. The internal FPGA message processing time in clock cycles N_{FPGA} depends on the maximum amount of clock cycles of the slowest function block:

$$N_{FPGA} = max(\ 2N_{busIP}, 2N_{RTH}, N_{MR}, N_{SR},$$
$$N_{PC}, N_D, N_{NM}, N_{ROM}\ ) \tag{4}$$

The receive and transmit procedure of the Bus IP needs equal clock cycles, as well as the receive and transmit procedure of the RX/TX Handler. This explains the factor two in formula 4. For the CGM only the functionality with the highest amount of clock cycles is relevant for calculation. Because of this the clock cycles for message routing

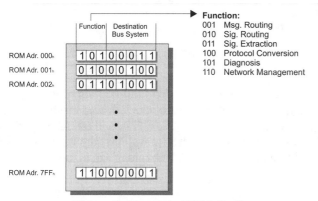

Figure 5: Example ROM Coding

(MR), signal routing (SR), protocol conversion (PC), diagnosis (D) and network management (NM) are also present in formula 4. To allow functional gateway modifications during runtime, which use additional clock cycles and to avoid failure caused by calculation inaccuracy of the internal design cycle time, an offset of twenty percent is beneficial. So the final FPGA design frequency $f_{FPGA_{Design}}$ is:

$$f_{FPGA_{Design}} = f_{FPGA} \cdot 1.2 \qquad (5)$$

Area/Cost Optimized Design

An area/cost optimized design is necessary for a real series product with a small FPGA. The area reduction is only possible, by replacing similar parallel working function blocks through sequential working units. Figure 6 shows such an implementation.

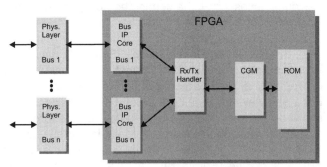

Figure 6: Area Optimized Gateway

This method is often used for designs with low baud rate busses and/or only a few bus systems are connected to the FPGA. Similar to the performance optimized design, the calculation of the internal FPGA design frequency is very important. Formula 1, 2, 3 and 5 are also valid for the area/cost optimized design. Formula 4 differs as follows. First, the gateway function is related to a message, which is received and transmitted on different bus systems. Therefore the amount of internal FPGA clock cycles is:

$$\begin{aligned}
N_{FPGA} = \quad & max(\ N_{busIP} \ ) + 2N_{RTH} + \\
& max(\ N_{MR}, N_{SR}, N_{PC}, \\
& \quad N_{D}, N_{NM} \ ) + N_{ROM} \qquad (6)
\end{aligned}$$

Second, the gateway function is related to a message, which is received and transmitted on the same bus systems. Therefore:

$$\begin{aligned}
N_{FPGA} = \quad & 2 \cdot max(\ N_{busIP} \ ) + 2N_{RTH} + \\
& max(\ N_{MR}, N_{SR}, N_{PC}, \\
& \quad N_{D}, N_{NM} \ ) + N_{ROM} \qquad (7)
\end{aligned}$$

As a result of the comparison of formula 4 with 6 and 7 it can generally be defined, that for parallel implementations of blocks only the block with the highest amount of clock cycles has to be taken into account. For serial block constellations, where a message is handled sequentially/serially (see figure 6), the clock cycles of each serial block have to be added up.

$$\begin{aligned}
N_{SerialDesign} &= N_{SB_1} + N_{SB_2} + ... + N_{SB_n} \qquad (8) \\
N_{ParallelDesign} &= max(N_{PB_1}, N_{PB_2}, ..., N_{PB_n}) \qquad (9) \\
N_{MixedDesign} &= N_{SB_1} + ... + N_{SB_n} + \\
& \quad max(N_{PB_1}, ..., N_{PB_n}) \qquad (10)
\end{aligned}$$

With "SB" for Serial Block and "PB" for Parallel Block.

SOFTWARE SOLUTION

There are several microprocessor IP Cores for FPGAs on the market. A software solution means that the gateway consists of a microprocessor IP core as its main unit plus the Bus IP Cores. The whole functionality is programmed based on the microprocessor in assembler or in a high level language like C. The example shown in figure 7 uses a 32bit RISC processor from Xilinx (MicroBlaze), which can be programmed in ANSI C.

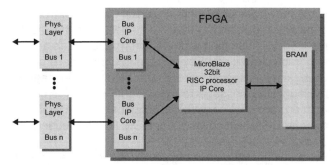

Figure 7: Software Solution

The advantage of the software solution is the simple realization of the functionality in C, especially for users who are not familiar with VHDL. Further the area in the FPGA is independent from the size of the source code, because the instruction and data code of the MicroBlaze is stored in internal block RAMs. The disadvantage is the determination of the internal FPGA clock cycles of the system, which has to be done via software synthesis. Therefore the developer has to be very familiar with the used microprocessor IP core. The source code has to be analyzed as well as the generated assembler code. Based on the generated assembler code and the clock cycles per assembler instruction (AI), the amount of clock cycles for the gateway N_{FPGA} can be calculated as:

$$N_{FPGA} = max(\ N_{busIP} \ ) + \sum_{i=1}^{n_a} N_{AI_i} \qquad (11)$$

n_a is the number of assembler instructions of the source code and N_{AI_i} the number of clock cycles for the individual assembler instruction. N_{FPGA} has to be calculated for the critical path of the software with the highest amount of clock cycles. The design clock rate is like in formula 1 and 5.

OPTIMAL FPGA GATEWAY DESIGN

The most frequently asked question is, how to design an optimal gateway and which functionality should be realized in VHDL and which through a microprocessor IP core? This question will be answered by the system synthesis theory from [1]. First a model to describe the tasks of the system synthesis is introduced. It consists of a function graph, which is comparable to a message data flow graph of the gateway, an architecture graph and a specification of mapping possibilities from the nodes of the function graph to the nodes of the architecture graph via edges.

Function Graph

The function graph is a variant of the problem graph which was defined in [1]. It is a directed graph and reflects the tasks of the gateway. In the described FPGA gateway application, the problem graph G_P defined in [1] can be reduced by its communication nodes. The communication nodes define the type of communication between the different nodes in the problem graph. In an FPGA the function blocks (tasks/nodes) are directly wired without using an internal bus. Because of this, the communication nodes can be removed.

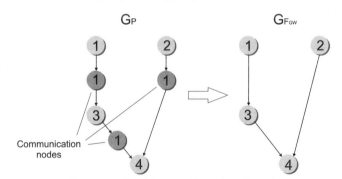

Figure 8: Problem to Function Graph

The resulting gateway function graph $G_{F_{GW}}$ is identical with a common control flow graph (CFG).

Architecture Graph

The nodes of the architecture graph are the functional resources of the design. They are possible realizations of the tasks in the function graph. The architecture graph $G_{A_{GW}}$ is also a directed graph and its nodes are either a microprocessor IP core or a VHDL function block (Hardware Modul). In special cases, the nodes could also be FPGA peripherals like ROM, multipliers or other parts, based on the tasks in $G_{F_{GW}}$ and the granularity of the described system. The reduction of communication nodes in the function graph leads to a reduction of communication resources in the architecture graph.

Gateway Specification Graph

The gateway specification graph $G_{S_{GW}}$ is a bipartite graph, which consists of a function graph $G_{F_{GW}}$, a gateway architecture graph $G_{A_{GW}}$ and a set of mapping edges $E_{M_{GW}}$. The mapping edges between the two graphs describe all possible mappings of the tasks from the function graph to the resources of the architecture graph. In other words: The edges represent user-defined mapping constraints in the form of a relation: "can be implemented by " [2]. To show the procedure of designing a specification graph, an area/cost optimized design with two bus systems was chosen. The granularity of the function graph is an important point during the design step. The developer has to verify which functions

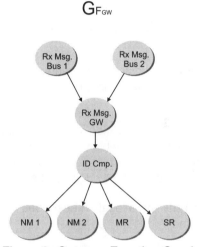

Figure 9: Gateway Function Graph

can be combined to a function graph task node. The granularity has to be chosen in a way, that offers different implementation possibilities in the architecture graph (hardware or software). $G_{A_{GW}}$ consists of different types of potentially allocatable components like processor IPs, hardware modules (HWM) or memory components. Each node of the function graph will be mapped to one or more components in the architecture graph. The mapping is realized via edges. Each edge has a weight which represents the effort of implementation of a function graph node to the mapped resource in the architecture graph. The effort could be area, power, performance or other characteristics. Figure 10 is an example of a gateway specification graph, with edge weighting of performance in clock cycles and area in slices [Delay/Slices] for an area/cost optimized design. Here the CGM contains the functionality of network management for two busses, message routing and signal routing. The values of the edge weighting [delay/slices] are randomly chosen. If you choose CAN for the bus systems, the specification graph will be a common gateway for an engine control unit. The gateway specification graph suits perfectly to find proper architectures and mappings of an FPGA based automotive gateway solution. In 10, all possible mappings of $G_{F_{GW}}$ are shown. The system synthesis aims to optimize the specification graph and to show one or a set of optimal gateway realizations based on user

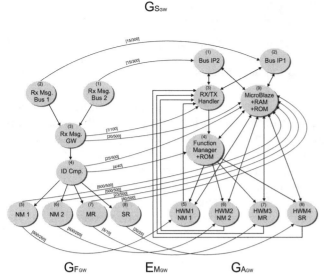

$G_{S_{GW}}$

$G_{F_{GW}}$ $E_{M_{GW}}$ $G_{A_{GW}}$

Figure 10: Gateway Specification Graph

constraints. Further design steps require the introduction of some terms and definitions [2].

- activation: The activation of a specification graph $G_{S_{GW}}$ is a function which activates or deactivates edges and nodes of $G_{S_{GW}}$. An activation describes their use. In figure 10 all edges and nodes are active.

- allocation α: The allocation describes the subset of all activated nodes and edges of $G_{A_{GW}}$.

- binding β: The binding is a subset of all activated mapping edges.

- feasible binding β: A feasible binding is a binding that fulfills the following points:

  - each activated edge starts and ends at an activated node.

  - for each activated note, exactly one outgoing edge is activated.

  - for each activated edge either both operations are mapped onto the same node or an additional activated edge to handle the communication exists. The determination of a feasible binding is NP-complete.

- feasible allocation α: A feasible allocation is an allocation that allows at least one feasible binding.

- schedule τ: The schedule is responsible for the calculation of the execution delay of the mapped function graph nodes to a specific resource.

The most important task of the system synthesis is the partitioning. A specification graph consists of functional objects (function graph nodes) and system components (architecture graph nodes). The allocation α and the binding β of a gateway specification graph can be defined as a

common partitioning problem, which is an NP-hard problem. A partitioning algorithm maps each functional object onto exactly one system component and searches for the optimal solution based on the user constraints (edge weighting). There are some standard algorithms to solve such partitioning problems, which are split into the following groups:

- heuristic methods

  - constructive methods like random mapping, hierarchical clustering, ...

  - iterative methods like Kernighan-Lin algorithm, simulated annealing, ...

  - evolutionary algorithms

- exact methods

  - Integer Linear Program (ILP)

For the automotive gateway, the problem of system synthesis is solved with evolutionary algorithms. The procedure how to do this is specified in detail in [1] and [2] and will not be described here. To apply this procedure it is important to know how to design a gateway and its specification graph which is described in this paper. The solution of the synthesis is displayed in picture 11 as a two dimensional pareto front.

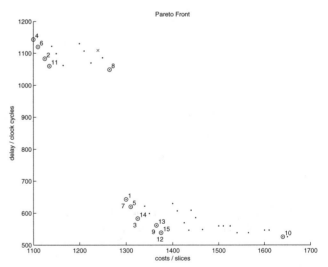

Figure 11: Two dimensional Pareto Front

The dimension depends on the amount of the parameters to be optimized. Here the parameters are costs and delay. Costs are proportional to the amount of area in the FPGA and delay is the reciprocal of the performance. With the pareto front the developer of a system is able to design an optimal gateway based on his constraints. The example gateway specification graph $G_{S_{GW}}$ is a very small one with a rough granularity. This is done because of easier visualization. This graph could also be solved with a deterministic algorithm. For real implementation the CGM functionality like for instance network management will be refined into an extra graph, so the granularity

348

of the specification graph will increase, which tends to use heuristics like evolutionary algorithms to solve the optimization task.

For the system synthesis we designed a tool with Matlab, which calculates the pareto front (limited to two dimensions) with evolutionary algorithms. The specification graph is the input to the program and consists of two adjacency matrices. The output is shown in figure 11. To determine an optimal solution, the pareto implementations are numbered in figure 11. To know what kind of implementation a pareto optimal implementation is, the program displays additional information about the allocation of each pareto optimal individual. The output is in ASCII:

Individual 4:
Allocation : 1 1 0 0 0 0 0 0 1
Binding   : 1 2 9 9 9 9 9 9 9
Delay   : 1143
Costs   : 1100

Individual 7:
Allocation : 1 1 0 0 1 0 0 0 1
Binding   : 1 2 9 9 5 9 9 9
Delay   : 643
Costs   : 1300

Individual 15:
Allocation : 1 1 0 1 0 1 1 1 1
Binding   : 1 2 9 4 9 6 7 8
Delay   : 539
Costs   : 1375

The individual number describes the pareto point in figure 11, the allocation is the binary coded information of the used components in the architecture graph (see numbering in figure 10). "1" stands for allocation/activation. The binding visualizes the mapping from the function graph nodes to the architecture graph resources. Binding 1 2 9 9 9 9 9 9 means, that node 1 from the function graph is mapped to node 1 of the architecture graph, node 2 to node 2 and all other nodes are mapped to node 9 (MicroBlaze). The pareto optimal implementation Individual 7 for instance is given figure 12.

CONCLUSION

Heuristics like evolutionary algorithms are becoming more and more important for NP-hard problems. The use of these heuristics for an FPGA based automotive gateway design is a new method. The design of a perfect specification graph is one of the most important tasks during the system synthesis. To find the right granularity of the graphs for an FPGA-design requires profound knowledge in HW/SW Codesign and embedded system design.

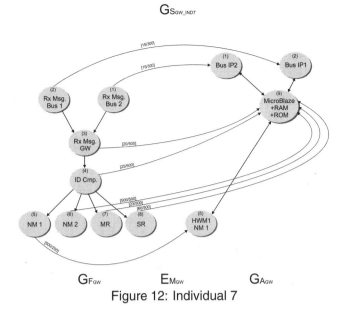

Figure 12: Individual 7

The introduced examples of different hardware and software designs together with the constraints area and performance, are the basics for an efficient automotive gateway design.

Additional steps can be taken to refine the specification graph to increase the granularity. This will result in more possible pareto optimal solutions. The evaluation of the weighting of the edges between the function graph and the architecture graph via hardware and software synthesis offers considerable potential for research of an FPGA based gateway solution.

REFERENCES

[1] J.Teich. Digitale Hardware/Software-Systeme. Springer, Berlin Heidelberg 1997

[2] J.Teich and T.Blickle. System-level synthesis using Evolutionary Algorithms. In Journal Design Automation for Embedded Systems, 1997

[3] OSEK/VDX. Network Management. Concept and Apllication Programming Interface. Version 2.5.3, 26th. July 2004

[4] ISO/OSI Open System Interconnect Basic Reference Model. ISO Reference number ISO 7498

[5] Wolfhard Lawrenz. CAN Controller Area Network Grundlagen und Praxis, 2000

349

CONTACT

DaimlerChrysler AG
Wolfgang Hauer
Engineer
Research and Technology Powertrain Control (REI/EP)
HPC: T723
D - 70546 Stuttgart

Phone: +49 711 17 47051
E-Fax: +49 711 3052136699
E-Mail: wolfgang.hauer@daimlerchrysler.com

University of Ulm
Prof. Dr. Hans Peter Großmann
Department of Information Resource Management
Albert-Einstein-Allee 43
D - 89069 Ulm

Phone: +49 731 50 22502
Email: hans-peter.grossmann@uni-ulm.de

University of Siegen
Prof. Dr. Günter Stöhr
Department of Mechanics and Control Theory
Paul-Bonatz-Str. 9-11
D - 57068 Siegen

Phone: +49 271 740 3313
Email: guenter.stoehr@uni-siegen.de

2002-01-0264

The Application of Middleware to In-Vehicle Applications

Hasina M. Abdu and David H. Yoon
University of Michigan-Dearborn

ABSTRACT

With the increasing presence of the Internet in today's applications, ranging from legacy enterprise systems to handheld devices and home appliances, there is an increasing need for a generic middleware that can enable the interoperability between heterogeneous systems. In the specific case of in-vehicle networks, a generic middleware would enable the vehicle/driver to interact with a diversity of applications, including legacy enterprise systems, other embedded systems and wireless ad-hoc and ubiquitous applications. This paper is the result of an initial investigation into the requirements for such a middleware, and possible directions to be taken for its implementation.

INTRODUCTION

Recent advances in technology and the emergence of Internet-based applications have resulted in a distributed application environment that differs from the typical enterprise application scenario. Formerly independent applications must interact to access and share functions and data stored in heterogeneous database [13]. Collaborations with other enterprises, mergers, acquisitions, and the Internet further diversify this scenario. A variety of performance issues must also be considered: a fluctuating and unpredictable number of users, the need to satisfy increasingly demanding Quality of Service (QoS) requirements, and the need for interoperability between existing legacy applications and new web-based applications. To make the situation more complicated recent trends such as nomadic mobility, ubiquitous and ad-hoc computing, telematics and real-time applications impose an additional set of constraints on existing systems.

Middleware solutions such as CORBA [28], DCOM [31] and Java RMI, despite providing a framework to design and implement distributed applications, do not support the ever increasing set of requirements imposed by existing technology. In addition, these solutions are not interoperable, i.e., an application developed in a CORBA framework cannot communicate with an application under DCOM or Java RMI, unless a specialized bridge or gateway is used. What is needed is a generic middleware approach that is robust, flexible

and scalable enough to support the rapidly changing trends in technology, without losing the interoperability with existing legacy applications. This *new-generation* middleware should enable the design, implementation and integration of heterogeneous applications, ranging from legacy enterprise systems to embedded in-vehicle applications. In addition, a varying set of QoS requirements specified by users and applications should also be satisfied.

This paper discusses the role of the *new-generation* middleware in enabling open and distributed in-vehicle applications. The following issues, along with possible solutions, are discussed:

- Need to satisfy QoS (fault-tolerance, real-time, etc.)
- Support for wireless communication with remote systems.
- Support for a programming model that enables the design and implementation of in-vehicle applications and integration with external applications.
- Need for efficient management of in-vehicle systems and applications

The rest of this paper is organized as follows. The next section justifies the application of middleware to in-vehicle applications, focusing on their main requirements. The *Related Work* section describes related work in applying existing middleware to in-vehicle applications, as well as approaches that focus on achieving interoperability between heterogeneous application frameworks. The *Proposed Approach* outlines the main issues that are being investigated by the authors and possible approaches to be taken. A summary of this initial investigation and future directions for this research are presented in the *Summary* section.

IN-VEHICLE APPLICATIONS: MOTIVATION AND REQUIREMENTS

One of the motivations for a framework for open in-vehicle applications is to efficiently communicate with external applications via the Internet. This would enable applications such as remote traffic control or remote fault detection and monitoring of the sensors in a car. Figure 1 illustrates an example in which the car driver can obtain information on traffic ahead of him/her through a wireless connection to the Internet. In Figure 2, sensors in the back of the car can detect objects when parking or backing-up. This information can be transmitted to the engine or break, slowing down or stopping the car if necessary. Remote management of in-vehicle networks is another application, as illustrated in Figure 3: management stations located at remote dealers can monitor and locate faults in different parts of the network, without being restricted to a given location, e.g., mechanic or dealer. The challenge does not lie on

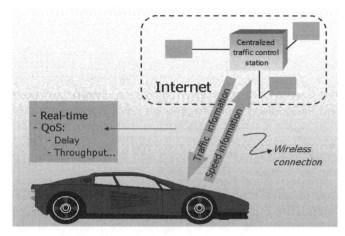

Figure 1 - Traffic control application

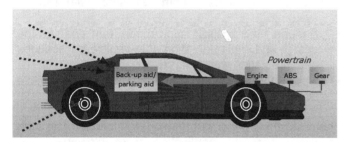

Figure 2 - Parking aid application

the applications themselves (in fact, some car manufacturers have implemented part of these applications), but in creating a framework that supports the modeling, implementation and deployment of a variety of applications. This framework will be the *middle tier* between the in-vehicle network and external applications, thus referred to as middleware. A generic middleware serves as a software bus connecting different application components, as illustrated in Figure 4. The main advantage of middleware approach over custom designed applications is that the former can free software designers from developing custom communication layers between processes. In addition, a middleware can provide a generic and structured approach to:

- Intranet/Internet connectivity
- Fault detection, isolation and recovery

- Satisfy Quality of Service (QoS) requirements, such as fault-tolerance and real-time constraints.
- Enable access to services under different framework/architecture/implementation

A properly designed solution offers additional benefits [24]: *composability*: large systems can be built in an incremental manner by integrating a set of well-specified and tested subsystems; *extensibility*: existing systems are open to incremental modification and extensible without some predefined upper limit; and *maintainability*: it is possible to implement well-defined error-containment regions and to achieve fault tolerance by replicating nodes.

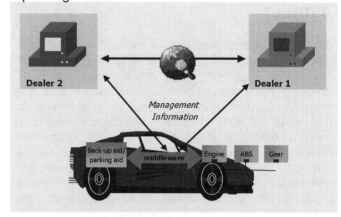

Figure 3 - Sample management scenario

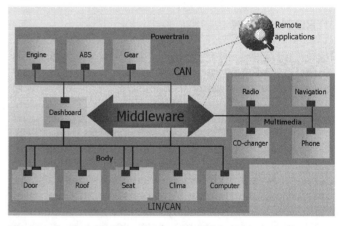

Figure 4 - A middleware for in-vehicle networks

The described functionality of a generic middleware cannot be achieved without software support. Due to the stringent conditions of real-time embedded systems, what is known as the embedded software crisis, open distributed frameworks cannot be employed as easily as in enterprise systems. In other words, middleware solutions such as OMG's CORBA, Microsoft's DCOM and Sun's Java RMI, cannot be directly integrated into embedded systems due the following:

- Most of these technologies assume a point-to-point, connection-oriented, transport protocol, whereas embedded systems are based on group communication.

- As opposed to enterprise systems, embedded systems have reduced resource footprint.
- Real-time embedded systems require predictable behavior, which is not supported by existing middleware solutions.

Another drawback of utilizing existing middleware solutions is that they are not interoperable, i.e., an application under a CORBA framework cannot communicate with an application under Java RMI or DCOM framework, unless specific bridges/gateways are deployed.

There have been attempts to tailor CORBA to CAN (Controller Area Network) systems used in cars. In addition, OMG has recently released a new version of CORBA for real-time systems [30], *Real-time CORBA*, which also supports the specification of QoS requirements. However, the suitability of CORBA to the restricted world of in-vehicle applications remains questionable, especially with the advent of more recent XML-based approaches that focus on interoperability and connectivity, and that consume fewer resources. Related work that address these issues are discussed in detail in the next section

RELATED WORK

We divide related work into two main groups: those that focus on tailoring existing middleware such as CORBA to embedded systems, and those that focus on creating new solutions that enable interoperability with heterogeneous applications.

TAILORING CORBA - Distributed object computing has been a promising approach to support the implementation of complex distributed applications. At the heart of contemporary distributed object computing models are *Object Request Brokers* (ORBs), which facilitate communication between local and remote objects. ORBs eliminate many tedious, error-prone, and non-portable aspects of creating and managing distributed applications. One of the widely used ORB models is the Common Object Request Broker Architecture (CORBA) [28], which is standardized by the Object Management Group (OMG).

Since 1991, CORBA has been the *de-facto* middleware used in building distributed enterprise applications. Language, platform and location independence enables the communication between objects implemented in different languages, running on different platforms and on different hosts in a transparent manner. To access a service offered by a remote server, a client only has to make a local method invocation and the ORB will take care of locating the server, transferring parameters and returning results. If client and server are located in different platforms, the ORB will also take care of any required conversions.

The interoperability between ORBs from different vendors is possible through the *General Inter-ORB Protocol* (GIOP). A specific example of GIOP is the *Internet Inter-ORB Protocol* (IIOP), that includes a mapping to TCP/IP.

CORBA's language independence is achieved through the *Interface Definition Language* (IDL). An IDL file contains the specification of attributes and operations that can be invoked by a client on a given server. The specification also includes exception values, type definitions, constants and operation signatures. The IDL compiler generates *client stubs* and *server skeletons* based on the specified information. Stubs and skeletons serve as the "glue" connecting remote application processes, and can be generated in various languages, including C, C++, Ada, Smalltalk and Java. Figure 5 illustrates the main components of a CORBA framework.

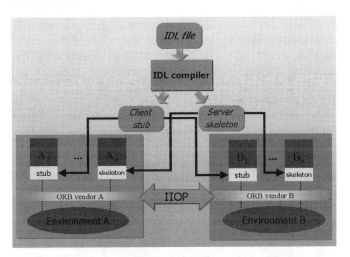

Figure 5 - Stubs, Skeletons, ORB, IDL and IIOP

CORBA was developed for enterprise systems that do not have the resource-constrained environment of embedded systems. In addition, group communication and real-time requirements need to be addressed. Some of the work that addresses these points are briefly described below.

Group Communication - CORBA is based on the connection-oriented transport model and an object reference denotes only a single CORBA object, thus resulting in a point-to-point communication. Embedded systems such as CAN offer a consistent broadcast mechanism in a straightforward manner via a serial broadcast medium and non-destructive priority-based bus arbitration. It also supports the producer/consumer model of data transmission, which is often referred to as the publisher/subscriber communication model [25]. In this context, a producer of a message is totally unaware of its consumers and simply broadcasts messages over the bus without specifying their destinations. Consider, for example, stopping a set of motors. When issuing a stop command, it is not of interest to address a specific motor, rather it must be ensured that all relevant motors receive the command. Similarly, when reacting to a stop command, it is not of interest which controller has issued that command. On a more abstract level, a sensor object triggered by the progression of time or the occurrence of an event spontaneously generates the respective information and distributed it to the system. It can thus be considered as a producer. There have been different approaches towards including group communication and publisher/subscriber support to CORBA. In [24], a CAN-based transport protocol is designed to support group communication. This protocol makes use of the CAN identifier structure to implement a subject-based addressing scheme, which supports the anonymous publisher/subscriber communication model. [24] also proposed an abstraction scheme called *invocation channel*, which denotes a virtual communication channel connecting a group of communication ports and a group of receivers. A *conjoiner* object, responsible for group management, dynamic channel binding and address translation, is also proposed. An invocation channel is uniquely identified as a channel tag in an IDL program. One of the main drawbacks of this approach is that interoperability between different ORB implementations is lost due to the elimination of the connection-oriented point-to-point communication services.

Another approach, combining broadcast and filtering approaches, is proposed in [22]. In [7] an *Object Group Service* (OGS) aimed at facilitating the parallel processing of CORBA operation calls to a set of consumers is proposed.

Real-time support. The goal of real-time is achieving predictability in the behavior of some external attributes of a system, such as response time to inputs [10]. Obtaining this predictability (or ``end-to-end predictability'' of the system) requires that all the components in the system behave predictably. Thus, if CORBA is to be used in real-time systems, its behavior must be predictable. Making the CORBA component of a real-time system predictable means making the time of operations invoked on CORBA objects predictable.

There have been different approaches towards achieving predictability in CORBA and other frameworks [16,20]. In [21], a framework for invoking real-time objects on a CAN bus is proposed. The first byte of the arbitration field in a CAN message is used to define three levels of priority that can be assigned to any message: *hard*, *soft* and *no* real-time constraint. The deadline or scheduling of a message is done based on this priority value.

In [22] the use of asynchronous operations, as opposed to the synchronous request-response nature of CORBA and JAVA RMI is proposed. The motivation being the fact that synchronous operations result in blocking the caller until a response is obtained from the receiver, thus not resulting in a ``predictable'' behavior. An asynchronous or event-driven paradigm would not block

the sender and would contribute in maintaining an end-to-end predictability.

In [23] an API for real-time Distributed Object Programming is proposed, which enables deadline imposition for arrival of results from an invoked object method, time-triggered actions and non-blocking invocation of object methods.

An extended IDL enabling the specification of timing constraints in a CORBA program, as well as ``fast-track messages'' used for time critical real-time traffic, are proposed in [19]. The ``fast-track messages'' can bypass layers of software and be sent to guarantee predictability.

Finally, we have OMG/s *Real-time* CORBA [30], with the goal of synchronizing the ORB operations with those of the underlying Real-time Operating system's environment in order to make operations predictable.

Reduced Resource Footprint - With the goal of reducing the resource footprint of CORBA, OMG has launched *Minimum* CORBA [29], primarily intended for embedded systems. The approach was to exclude dynamic aspects of standard CORBA, that are not required for embedded systems, e.g., the Dynamic Invocation Interface and the Interface Repository that supports it.

In [24], a resource-conscious customization of CORBA is proposed for CAN-based embedded systems, by reducing the size of data representation and by customizing GIOP with simplified message types and reduced headers.

FOCUS ON INTEROPERABILITY - The main drawback of existing middleware solutions is the lack of interoperability. For example, an application under a CORBA framework cannot communicate with an application under COM/DCOM, unless a specialized bridge or gateway is used. The lack of interoperability between existing middleware solutions is a challenge in achieving truly open distributed frameworks. This section outlines some of the work that has been done to address this problem.

One of the most recent approaches towards integrating existing applications over the Internet is the Simple Object Access Protocol or SOAP [11]. SOAP is a relatively new development born to meet the needs of developers who found DCOM and CORBA difficult to deploy in a world of the Internet and firewall requirements. SOAP is intended for ease of use and greater platform independence. Although DCOM and IIOP offer a broad range of services (albeit a bit complex) some developers consider that in 80 to 90 percent of potential applications they would gladly trade some features for a simpler technology that "guarantees the greatest interoperability in conjunction with simplicity" [11]. SOAP was developed to meet those needs.

In simplistic terms, SOAP is an open protocol specification that defines a method for using Remote Procedure Calls (RPC) with HTTP as the underlying protocol and XML as the data serialization format. The simplicity is introduced in that XML is the wire protocol, using its text-based format for data encoding. Every method call is translated into XML at run-time. To guarantee uniformity, the SOAP protocol requires adherence to a generic XML vocabulary. The data is parsed at the server side SOAP gateway. Information is encapsulated in an HTTP request in the form of a POST packet and then sent to a server. The calling browser utilizes a small number of HTTP headers that signal the target that a SOAP service is being transported by the HTTP packet. The text-based information makes developer comprehension and troubleshooting much easier. SOAP focuses on the familiarity of HTTP and XML systems to programmers and developers. Some of the main concerns/drawbacks with this approach are:

- Overhead - there is a high overhead associated with parsing ASCII requests. In addition, exchanging
- XML data can be a great toll on the limited bandwidth available in applications running in embedded/real-time systems.
- Simplicity - SOAP cannot perform the same variety of services DCOM or IIOP can perform. Instead, this component technology is evolving in response to immediate development needs in the current marketplace.
- Quality of Service and other requirements- it is not clear whether SOAP can support the specification of QoS or real-time constraints.

Others approaches towards interoperability include:

- XOIP (XML Object Interface Protocol)[26] – describes a way in which heterogeneous networked embedded systems can interface to a variety of distributed object architectures using XML.
- Generalized dispatch mechanism [17] – a flexible notion of dispatching is used to integrate objects belonging to different models, systems, and paradigms.
- OMG's Interworking Architecture [27] – specifies support for two-way communication between CORBA objects and COM objects. The goal is that objects from one object model should be able to be viewed as if they existed in the other object model. For example, a client working in a CORBA model should be able to view a COM object as if it were a CORBA object. Likewise, a client working in a COM object model should be able to view a CORBA object as if it were a COM object.
- Wireless applications for automobile applications [35] – the possibility of using Bluetooth with CAN systems for automobile applications is explored. However, the question is whether Bluetooth can offer the end-to-end predictability required for CAN systems.

SUMMARY – This section summarized two types of work: tailoring CORBA to support embedded real-time systems; and achieving interoperability between heterogeneous systems. The CORBA based solutions, despite attempting to address the needs of embedded systems such as in-vehicle networks, have the following drawbacks:

- They are not interoperable with other systems. For example, the management scenario illustrated in Figure 3 cannot be implemented if the management applications at the dealers are not running under a CORBA framework.
- The *Real-time* and *Minimum* CORBA versions, the only standard solutions, are still in early stages of deployment. Unless thoroughly tested with in-vehicle applications, it remains unclear if the complexity and overhead of CORBA is adequate for the resource-constrained environment of in-vehicle networks such as, for example, CAN applications.

The XML-based solutions and other approaches to interoperability, despite addressing the interoperability issue, present the following issues:

- QoS – The current XML-based approaches do not support the specification and enforcement of QoS requirements. In the case of in-vehicle applications, requirements include, for example, the specification of real-time constraints.
- Network bandwidth - The XML approach may be an overhead for the limited bandwidth available in in-vehicle networks as well as in wireless connections to external applications, such as the scenario illustrated in Figure 1.
- Programming model – as opposed to CORBA, SOAP and other XML based approaches do not provide a programming model to create distributed applications; being restricted to existing applications.

PROPOSED APPROACH

Unlike related work that propose a framework for in-vehicle applications based on tailoring CORBA to CAN systems, we investigate the use of an XML-based middleware to achieve interoperability between in-vehicle applications, legacy enterprise systems, other embedded systems and wireless applications such as ad-hoc and ubiquitous systems. We also investigate methods to include the specification and enforcement of QoS requirements, a programming model that will enable the design and implementation of new distributed applications and a framework for efficient management of in-vehicle networks and applications (Figure 6). The next sections will outline the main points being investigated in this work.

XML AND QOS – Future in-vehicle applications will consist of a combination of embedded systems, wireless and multimedia applications. For example, the future automobile may enable the passengers to view videos or join video conferencing through a wireless connection from an embedded browser. Despite the shortage of the underlying resources, such as network bandwidth, automobile drivers and passengers hope to receive consistent, predictable, timely and reliable services. Their hope can only be realized with the efficient specification and provision of Quality of Service (QoS). The term "QoS" involves [15] the specification and provision of *predictability*, *continuity* and *accessibility*. [15] defines *predictability* as the ability to maintain the contracted QoS and have minimum probability of QoS violations during the resource fluctuation period. *Continuity* is defined as the ability to degrade gracefully and adjust resource allocation distributions dynamically to tolerate transient resource scarcity. *Accessibility* is the ability to access the server from a wide range of devices, including PCs, workstations, cellphones, and PDAs.

A wealth of research work has been done to support the QoS specification for different types of applications. Recent approaches include solutions for setup and enforcement of QoS in middleware systems [8], [14], [15], [34]. An XML based QoS specification for multimedia applications is proposed in [15]. Figure 7 illustrates two examples of specifying QoS parameters for two in-vehicle applications: remote traffic control and remote video transmission. Delay (or response time), throughput and jitter are examples of QoS parameters.

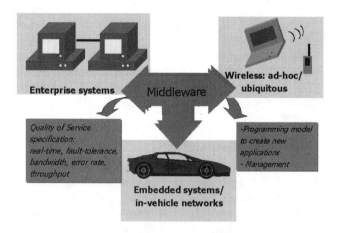

Figure 6 - Proposed framework

Our approach should include the following:

- A QoS enforcer that can detect illegal QoS specifications made by users.

- A QoS specification tool that facilitates the specification task, by relieving user from dealing with low-level system and resource information. This may involve the specification of *levels* of QoS,

which can later be mapped to numbers, based on specific system parameters and low-level resource information.

```
<App name = "Remote traffic control">
    <QoS>
        <Delay> 10 </Delay>
        <Throughput> 230 </Throughput>
    </QoS>
</App>
```

```
<App name = "Remote Video TX">
    <QoS>
        <Delay> 10 </Delay>
        <Jitter> 30 </Jitter>
    </QoS>
</App>
```

Figure 7 - Using XML to specify QoS

DEALING WITH LIMITED NETWORK BANDWIDTH – One of the main drawbacks of using XML for in-vehicle networks is the overhead that it can have on network bandwidth. In-vehicle networks and embedded systems in general have serious bandwidth constraints. These constraints are also present when communicating with remote applications through a wireless connection.

We propose the use of a compressed format of XML, such as the Wireless Markup Language (WML) used by the Wireless Application Protocol (WAP). WML is a tag-based browsing language with screen management, data input and hyperlink capabilities. It is based on XML and inherits technology from HTML. It was designed to satisfy the resource constraints of wireless devices: small memory, limited CPU, small screen and limited input mode. These constraints are similar to the constraints found in embedded systems. WML's compressed and reduced format results in reduced network traffic and better caching. The use of WML can also be linked to the use of embedded browsers for in-vehicle applications.

The use of WML presents the same obstacle as XML: the need to specify and enforce QoS. We are currently looking into extending WML to support QoS specification and enforcement, based on the same requirements described in the previous section.

PROGRAMMING MODEL – Related work in interoperability with XML focus on communication between existing applications, but do not address the need of creating new applications.

Existing middleware frameworks are *process-oriented* methods based on a client server programming model, e.g., CORBA, COM, DCE [32], resulting in synchronous communication and tightly coupled systems. They are also based on point-to-point communication, instead of the publisher/subscriber interaction found in embedded systems.

XML, on the other hand, is a *data-centric* method for getting disparate systems to interoperate. Unlike, a client-server model, XML is message oriented, and promotes building loosely coupled systems.

The challenge in creating a programming model for a generic middleware is to give the user the choice of a programming model based on the application in question. For example, embedded systems are based on publisher/subscriber message oriented interaction, whereas many legacy enterprise applications are based on a client-server model. The proposed middleware should, therefore, support both communication paradigms. In this context, the authors are looking into the following question: how can XML be used to design and implement new applications under either communication paradigms?

MANAGEMENT OF IN-VEHICLE APPLICATIONS – Management consists of monitoring, analyzing and controlling the behavior of network components, to ensure that they behave as expected or specified by users. In this context, the proposed middleware should support two levels of management: internal management and remote management.

Remote management was illustrated in Figure 3: fault/failure of sensors and/or controllers in an in-vehicle network is detected at remotely located sites, without restricting the user to a given location, e.g., mechanic or dealer. This involves a wireless communication to a remote system, as well as satisfying QoS requirements such as the maximum delay time in obtaining a diagnostic or in locating a fault. Another issue to be considered is the choice of which remote location to connect to. This will depend on where the car is located, the type of diagnostic, the resource to be managed, and the requested QoS. The middleware should be equipped to locate the dealers or remote sites based on these requirements. We are currently examining previous work in management optimization ([1], [2], [3], [4], [5], [6]) to model this problem and include such specification in the XML/WML schema. For example, in addition to the desired QoS, an XML specification can include the type of diagnostics or resource being managed.

Internal management, on the other hand, involves management from within the in-vehicle network. In the particular case of CAN in cars, network management is implemented by CAN application layers ([9],[12], [18],[33]) following a master-slave configuration. The master node polls the slaves for a given set of information. If the information indicates a problem, the corresponding alert message is given to the user. Complex analysis or control functions that change the

state of a component as a result of analyzing the monitored information are not implemented. In addition, most of the existing approaches are based on a master-slave format, being a centralized approach. We propose a distributed management approach, where one or more managers monitor, analyze and execute control functions on the different parts of the network. The number of managers and how their functions are distributed will depend on the network in question, the desired QoS and other requirements. Figure 8 illustrates a distributed management scenario.

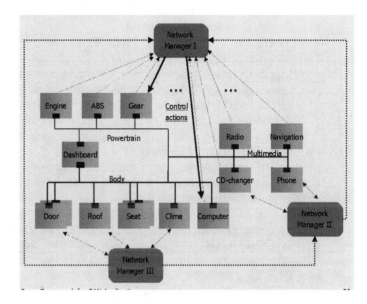

Figure 8 - Distributed management example

As in the case of remote management, the data to be collected, the analysis to be done and actions to be taken can be included in the XML/WML specification.

SUMMARY

This paper introduced the need for a generic middleware and its use in in-vehicle applications, enabling the seamless communication within the vehicle and with remotely located heterogeneous applications. Due to the stringent nature of embedded systems, there are many requirements to be satisfied. In addition, existing middleware frameworks cannot be easily adapted to the constrained world of in-vehicle networks.

We examined existing work on tailoring CORBA to CAN systems, as well as XML based approaches. We proposed the implementation of an XML based middleware that supports QoS specification and enforcement, satisfies resource constraints, supports multiple programming models and enables efficient management of in-vehicle network and applications. To achieve this goal, the following are the steps to be taken:

• Investigate the use of a compressed form of XML, such as WML, to address the problem of limited resources in embedded systems as well as wireless connections.

- Investigate existing work in QoS specification and enforcement, such as [15], to implement a framework for efficient specification, enforcement and execution of generic QoS specifications.

- Examine the author's previous work in modeling efficient management configurations based on a set of requirements. This will enable the user to:

  - Chose which resource to be managed
  - Specify QoS for management
  - Specify the management actions to be taken in case a problem is detected
  - Specify constraints on where management agents can be started or which dealers should be contacted based on location, or available bandwidth.

- Programming model – investigate extensions to be made to XML/WML to support different communication paradigms, such as client/server, message passing, multicast, etc.

- Compare the efficiency of an XML-based approach with a CORBA-based approach. The authors are currently working with *Real-time* CORBA, to evaluate its suitability for in-vehicle applications. The following will be examined:

  - How easily can QoS requirements be specified?
  - How is group communication implemented?
  - Is the overhead, if any, justifiable?

Both approaches will be compared in terms of:

- Resource consumption
- Scalability
- Interoperability
- Flexibility
- Complexity and ease of use

REFERENCES

[1] Abdu H., Lutfiyya H., Bauer M., *"Investigating Monitoring Configurations"*, Proceedings of ACM Symposium on Applied Computing, Pennsylvania, pp 366-373, 1995

[2] Abdu H., Lutfiyya H., Bauer M., *"A Model for Adaptive Monitoring Configurations",* Proceedings of the 6th IFIP/IEEE International Symposium on Integrated Network Management, pp 371-384

[3] Abdu H., Lutfiyya H., Bauer M., *"Modeling Management Requirements in Distributed Systems",* Submitted to the Journal of Parallel and Distributed Computing, 2000

[4] Abdu H., Lutfiyya H., Bauer M., *"Towards Efficient Management of Distributed Systems,"* Submitted to the Journal of Computer Networks

[5] Abdu H., Lutfiyya H., Bauer M., *"A Model for Efficient Configuration of Management Agents in Distributed Systems,"* Submitted to the Elsevier Journal of Performance Evaluation

[6] Abdu H., Lutfiyya H., Bauer M*., "Towards Efficient Resource Management in Distributed Systems",* Proceedings of the IEEE International Parallel and Distributed Processing Symposium (IPDPS) 2001-April, 2001 - San Francisco

[7] Axel M. A., *"Design and Implementation of a CORBA-based Object Group Service Supporting Different Data Dispatching Strategies"*, http://citeseer.nj.nec.com/335099.html

[8] Becker C., Geihs K., *"Generic QoS Support for CORBA"*, Proceedings of the International Symposium of Computers and Communication (ISCC 2000), IEEE CS Press, Los Alamitos, Calif., pp. 60-65, 2000

[9] CAN in Automation (CiA), *"CAN Application Layer for Industrial Applications"*, DS 201-201 Version 1.1, 1996

[10] Currey J., *"Real-time CORBA Theory and Practice*: *A Standards-based Approach of Distributed Real-time Systems"*, Embedded Systems Conference – San Jose, 2000

[11] DevelopMentor, IBM, Lotus Development Corporation, Microsoft, UserLand Software, *"SOAP: Simple Object Access Protocol"*, http://www.w3.org/TR/SOAP

[12] Etschberger I. K., *"CAN-based Higher Layer protocols and profiles"*, http://www.ixxat.de/english/knowhow/artikel/hlp.shtml

[13] Geihs K., *"Middleware Challenges Ahead"*, IEEE Computer, Vol 34:6, pp 24-31, 2001

[14] Gill C. D., Levine D., Schmidt D. C., *"Towards Real-time Adaptive QoS Management in Middleware for Embedded Computing Systems"*, Fourth Annual Workshop on High Performance Embedded Computing, MIT Lincoln Laboratory, 2000, http://www.ll.mit.edu/HPEC/

[15] Gu X., Nahrtedt K., Yuan W. Wichadakul D. Xu D., *"An XML-based Quality of Service Enabling Language for the Web"*, Journal of Visual Language and Computing, special issue on multimedia languages for the Web. Dec. 2001

[16] Hong S., *"Coping with Embedded Software Crisis using Real-time Operating Systems and Embedded Middleware"*, Invited for presentation at IEEE Asian Pacific ASIC (AP-ASIC) Conference, Korea, 2000

[17] Hurwitz B. *et al.*, *"Generalizing Dispatching in a Distributed Object System"*, ECOOP'94, pp. 450 ff, http://www.ifs.uni-linz.ac.at/~ecoop/cd/papers/0821/08210450.pdf

[18] J1939 Committee Draft, *"SAE: Recommended Practice for a Serial Control and Communication Vehicle Network"*, 1996

[19] Jeon G., Kim T. H., Hong S., *"Seamless Integration of Real-time Communications into CAN-CORBA with Extended IDL and Fast-Track Messages"*, Proceedings of IFAC Workshop on Distributed Computer Control Systems (DCCS), Australia, 2000

[20] Jeon G., Kim T. H., Hong S., Kim S., *"A Fault Tolerance Extension to the Embedded CORBA for the CAN Bus Systems"*, Lecture Notes in Computer Sciences, 2000

[21] Kaiser J., Livani M., *"Invocation of real-time objects in a CAN bus-system"*, IEEE International Symposium on Object-oriented Real-time Distributed computing, 1998, http://citeseer.nj.nec.com/kaiser98invocation.html

[22] Kaiser J., Mock M., *"Implementing the Real-time Publisher/Subscriber Model on the Controller Area Network (CAN)"*, Proceedings of the 2nd Int. Symp. On Object-Oriented Real-time distributed Computing (ISORC99), France, 1999, http://www.infomatik.uni-ulm.de/rs/core/isorc99.ps

[23] Kim K. H., *"APIs for Real-time Distributed Object Programming"*, IEEE Computer, pp 72-80, 2000

[24] Kim K., Jeon G., Hong S., Kim S., Kim T., *"Resource-conscious Customization of CORBA for CAN-based Distributed Embedded Systems"*, In the Proceedings of 2000 IEEE International Symposium on Object-Oriented Real-time Distributed Computing , Newport Beach, pp 34-41, 2000

[25] Kim H., Jeon G., Hong S., Kim T. H., Kim S., *"Integrating Subscription-based and Connection-oriented Communications into the Embedded CORBA for CAN Bus"*, Proceedings of 2000 IEEE Real-time Technology and Applications Symposium Washington DC, 2000

[26] Nielsen M. K., Jorgensen A. B., *"XOIP – XML Object Interface Protocol"*, Center for Object Technology, COT/3-34, 2000

[27] Object Management Group (OMG), *"Interworking Architecture"*, CORBA V2.3, 1999

[28] Object Management Group (OMG), *"The Common Object Request Broker: Architecture and Specification Revision 2.4"*, OMG Technical Document formal/00-11-07, 2000

[29] Object Management Group (OMG), *"Minimum CORBA"*, OMG Document formal/00-10-59, 2000

[30] Object Management Group (OMG), *"Real-time CORBA"*, OMG Document formal/00-10-60 edition, 2000

[31] Platt D.S., *"Understanding COM+"*, Microsoft Press, Redmond, Wash., 1999

[32] Rosenberry W., Kenney D., Fisher G., *"Understanding DCE"*, O'Reilly & Associates, Sebastopol., Calif., 1992

[33] Schofield M., *"An Introduction to CAN – The serial data communication bus"*, http://www-ife.tu-graz.ac.at/Local/Tech/Can-Bus/schofield.htm

[34] Vanegas R. *et al.*, *"QuO's Runtime Support for Quality of Service in Distributed Objects"*, Proceedings of IFIP International Conference in Distributed System Platforms and Open Distributed Processing, Springer-Verlag, New York, pp. 207-223, 1998

[35] Wunderlich H., Schwab M., *"The Potential of Bluetooth in Automotive Applications"*, Embedded Systems Conference – San Jose, 2000

Gateway between CANopen and ISO 11992 or SAE J1939

Holger Zeltwanger
CAN in Automation

ABSTRACT

J1939-based networks are used in most of the in-vehicle networks in trucks and buses and in some off-highway vehicles. In Europe, many of the super-construction manufacturers like to use off-the-shelf, price-competitive CANopen devices originally developed for other applications fields. CANopen is a standardized CAN-based application layer and profile specification. In order to standardize the gateway functionality, a CANopen standardized truck gateway profile is introduced.

INTRODUCTION

In Europe, CANopen is already used in several truck-based control systems such as cranes and snowplows, in off-road vehicles such as mining machines as well as in some military vehicles. The German Bundeswehr is going to use CANopen as standard network for superconstructions on trucks as well as tanks. In North America, the non-profit Industrial Truck Association (ITA) has decided to recommend CANopen for forklifts powered by electrical motors.

The CANopen application layer provides standardized communication services for transmission of process data and service data as well as some emergency, synchronization, and time-stamp data. In addition, CANopen provides standardized network management information including error control services and messages. The CANopen truck gateway profile defines a standardized translation of ISO 11992 / SAE J1939 parameters to CANopen application objects and vice versa.

The gateway profile is based on the parameter specification of ISO 11992 part 2 (braking and running gear equipment), part 3 (equipment other than braking and running gear), and part 4 (diagnostics). Parameters required by superconstruciton manufacturers and not yet spüecfied in ISO 11992 or J1939 will be submitted to ISO for standardization. The ISO 11992 standard specifies equivalent parameters to J1939. However, ISO 11992 and J1939 are not application layers by means of the Open systems interconnection (OSI) reference model. Therefore these standards are not at all compliant with the ISO reference model. They are describing parameters that can be regarded as an application profile.

CANOPEN COMMUNICATION SERVICES

In each decentralized control application there are different communication objects required. In CANopen all these communication objects are standardized and well described in the Object Dictionary. The CANopen Object Dictionary is accessible by a 16-bit index and in the case of Arrays and Structures additionally by an 8-bit sub-index. This dictionary also describes all application objects of the device. Application objects may be specified by a CANopen Device Profile or by an Application Profile. In addition, a device manufacturer may define non-standardized application objects, but then this device will not be interchangeable with one of the same class from another company.

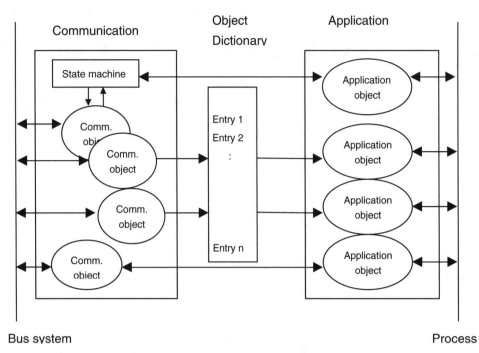

Figure 1: CANopen device model

Index (hex)	Object
0000	not used
0001-001F	Static Data Types
0020-003F	Complex Data Types
0040-005F	Manufacturer Specific Complex Data Types
0060-007F	Device Profile Specific Static Data Types
0080-009F	Device Profile Specific Complex Data Types
00A0-0FFF	Reserved for further use
1000-1FFF	Communication Profile Area
2000-5FFF	Manufacturer Specific Profile Area
6000-9FFF	Standardised Device Profile Area
A000-FFFF	Reserved for further use

Table 1: CANopen object dictionary structure

There are four classes of communication objects standardized in CANopen. Process Data Objects (PDO) are mapped to a single CAN frame using all 8 bytes of the data field to transmit application objects. Each PDO has a unique identifier and may be transmitted by only one node, but it can be received by more than one (producer/consumer communication). PDOs may be transmitted in different modes that is driven by an internal event, by an internal timer, by remote requests and by a sync message received from a specific node. The default mapping of application objects as well as the supported transmission mode is also described for each PDO in the Object Dictionary. CAN identifiers for PDOs

have by default high priorities to guarantee good real-time performance. The system designer can configure an inhibit-time for each PDO. The inhibit-time forbids this object to be transmitted within a specific time.

Which application objects are transmitted within a PDO is defined in the PDO Mapping Object. It describes the sequence and length of the mapped application objects. A device that supports dynamic mapping of PDOs must support this during the Pre-Operational State. If dynamic mapping during Operational State is supported, the SDO Client is responsible for data consistency.

The second class of communication objects are Service Data Objects (SDO) transmitting configuration data, which is sometimes longer than 8 bytes. The SDO transport protocol allows one to transmit objects of any size. The first byte of the first segment contains the necessary flow control information including a toggle bit to overcome the well-known problem of double received CAN frames. The next three bytes of the first segment contain index and sub-index of the Object Dictionary entry to read or write. The last four bytes of the first segment are available for configuration data. The second and the following segments using the same CAN identifier contain the control byte and up to seven bytes of configuration data. The receiver confirms each segment, so that there is a peer-to-peer communication (client/server). In the future CANopen will allow also a fast SDO transfer confirming not only each segment but also the complete object.

The third class of communication objects is the network management objects: the Nodeguarding Object and the NMT Object. The Nodeguarding Object is CAN frame with 1 byte remotely requested by the NMT master node. The data byte contains a toggle bit as well as 7 bits indicating the state of a node. The Nodeguarding Objects are transmitted periodically. The guarding time is also specified in the Object Dictionary and can be configured by SDOs. In addition, there is the Life Guarding Time specified, after which the NMT master should remotely request the NMT slave node. This guarantees that even in the case that the master is gone the nodes can react in an user-specified way.

The NMT Object mapped to a single CAN frame with a data length of 2 bytes. It has the identifier 0. The first byte contains the command specifier and the second contains the node ID of the device that must perform the command (in the case of node ID 0 all nodes have to perform the command). The NMT object transmitted by the NMT master forces the nodes to transit to another state. The CANopen state machine specifies the Initialization state, the Pre-Operational State, the Operational State and the Stopped State. After power on each CANopen node is in the Initialization State and transits automatically to the pre-operational state. In this state Sync objects and node guarding is provided, also the transmission of SDOs is allowed. If the NMT master has set one or more nodes to the operational state they are allowed to transmit and receive PDOs. In the stopped state no communication is allowed except NMT objects.

The Initialization state is divided in three sub-states in order to enable a complete or partial reset of a node. In the Reset_Application sub-state the parameters of the manufacturer-specific profile area and the standardized device profile area are set to their default values. In the Reset_Communication sub-state the parameters of the communication profile area are set to their power-on values. The third sub-state is the Initialization State, which a node enters automatically after power on or after reset communication or reset application. Power-on values are the last stored parameters.

CANopen defines also three specific objects for synchronization, emergency indication, and time stamp transmission. The Sync Object is broadcast periodically by the Sync Producer. This object provides the basic network clock. The time period between Sync messages is defined by the Communication Cycle Period Object, which may be written by a configuration tool to the application devices during the boot-up process. There can be a time jitter in transmission by the Sync Producer due to some other objects with higher priority identifiers or by one frame being transmitted just before the Sync Object. The Sync Object is mapped to a single CAN frame with the identifier 128. By default, the Sync Object does not carry any data, but it can have up to 8 bytes of user-specific data.

Emergency Objects are triggered by the occurrence of a device internal fatal error situation and are transmitted from an emergency client on the concerned application device. This makes them suitable for interrupt type error alerts. An Emergency Object may be transmitted only once per 'error event'. As long as no new errors occur on a device no further Emergency Object must be transmitted. Zero or more Emergency Object consumers may receive these objects. The reaction on the emergency consumers is not specified. CANopen defines several emergency error codes to be transmitted in the Emergency Object, which is a single CAN frame with 8 data bytes.

By means of the Time Stamp Object a common time frame reference is provided to application devices. It contains a value of the type Time-of-Day. This object transmission follows the producer/consumer push model. The associated CAN frame has the identifier 256 and a data field of 6 bytes length.

GATEWAY SPECIFICATION

The truck gateway profile to CANopen has to define the PDO transmission types and the default PDO mapping. The most important part of the gateway specification is the CANopen-compliant description of all parameters. In SAE J1939 and ISO 11992 all parameters have a fixed physical value. In CANopen each application object has the very same format by default. However, the format is optionally configurable, so that in the CANopen network there may be used another physical unit. In this case, the gateway has to transform the physical units.

All PDOs are transmitted and received asynchronously. The default mapping is not yet specified. The J1939 parameter mapping to the CANopen Object Dictionary is done by using the ISO 11992-parameter structure. There are only such parameters passed by the gateway to CANopen networks that might be of interest of super-construction control systems. In the other direction, only that information may be passed to the J1939-based network, which is allowed by the vehicle manufacturer. In addition, also information request services to the J1939 network may be restricted regarding the available network resource. If the data is requested by SDO communication, there will be an abort message indicating the reason. Remotely requested by PDO communciation will not indicate why a specific parameter is not available.

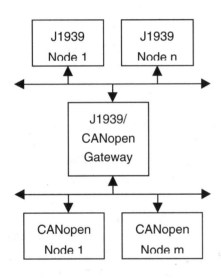

Figure 2: If both networks using the very same application object profiles, the system designer can decide, which functionality is implemented on which network

The application objects are made in CANopen by two different objects: The first one represents the variable value and the second describes the scaling and offset of this value. If the scaling and offset objects is not implemented the default values shall apply. The default resolution and default offset on the CANopen network is the very like the ISO 11992 / SAE J1939 parameter resolution. If the object is implemented, the CANopen system designer can re-define scaling and offset values accordingly to the application requirements.

Example: Engine coolant temperature

ISO 11992 description

Data length	1 byte
Resolution	1 °C/bit (-40 °C offset)
Data range	-40 °C to 210 °C
Type	Measured

CANopen description

OBJECT DESCRIPTION

Index	60XXh
Name	Engine_temperature
Object code	Variable
Data type	Unsigned8
Category	Optional

ENTRY DESCRIPTION

Sub-Index	0h
Access	Read only
PDO Mapping	Optional
Default Value	No

OBJECT DESCRIPTION

Index	60YYh
Name	Scaling_and_Offset
Object code	ARRAY
Data type	Signed16
Category	Optional

ENTRY DESCRIPTION

Sub-index	0h
Description	Number_of_entries
Entry Category	Mandatory
Access	Read only
PDO Mapping	No
Value range	1 to 5
Default value	No

Sub-Index	1h
Name	Resolution_numerator
Entry category	Optional
Access	Read/Write
PDO mapping	No
Value range	Signed16
Default value	+1

Sub-Index	2h
Name	Resolution_denominator
Entry category	Optional
Access	Read/Write
PDO mapping	No
Value range	Signed16
Default value	+1

Sub-Index	3h
Name	Maximum_value_range
Entry category	Optional
Access	Read/Write
PDO mapping	No
Value range	Signed16
Default value	+210

Sub-Index	4h
Name	Minimum_value_range
Entry category	Optional
Access	Read/Write
PDO mapping	No
Value range	Signed16
Default value	-40

Sub-Index	5h
Name	Offset
Entry category	Optional
Access	Read/Write
PDO mapping	No
Value range	Signed16
Default value	-40

CONCLUSION AND FUTURE ENHANCEMENTS

The CANopen truck gateway profile specifies a standardized interface between ISO 11992-based networks and CANopen networks. Because both are based on Controller Area Network (CAN) the hardware requires only two transceiver and controller chips compliant to ISO 11898 and a microcontroller running both protocol stacks and the gateway software. Of course, the gateway software may be different regarding the chosen J1939-based network. In case of ISO 11992, also the transceiver is different. The gateway profile will also specify diagnostic data.

The non-profit CAN in Automation (CiA) international users and manufacturers group is going to specify additional device profiles for diesel engines, electronic gear and other equipment used in vehicles. All these profiles will be compliant with the J1939 application profiles. Beside this default behavior, the system designer can configure another behavior in order to tailor the system according to the application requirements. With the gateway profile there is a smooth migration strategy to CANopen possible.

CONTACT

Holger Zeltwanger, CAN in Automation (CiA),
Am Weichselgarten 26, D-91058 Erlangen (Germany),
Phone +49-9131-69086-0, Fax +49-9131-69086-79,
Email: headquarters@can-cia.de, URL: http://www.can-cia.de

REFERENCES

1. ISO CD 11992: Road vehicles – Interchange of digital information on electrical connections between towing and towed vehicles – Part 2, Part 3, and Part 4
2. SAE J1939/71: Vehicle Application Layer
3. CiA DS-301, Version 4.01: CANopen application layer and communication profile
4. CiA DR-303-2, Version 1.1: CANopen representation of SI units and prefixes.

ADDITIONAL SOURCES

1. 1st to 6th international CAN Conference (iCC) Proceeding published by CiA, Erlangen (Germany) 1994-1999
2. CAN Newsletter, volume 1 to 9, published by pz marketing, Nuremberg (Germany) 1992-2000

2000-01-1312

CPJazz – A Software Framework for Vehicle Systems Integration and Wireless Connectivity

M. Tim Jones
CellPort Labs Inc.

ABSTRACT

Integration of new technology into vehicles continues at a rapid pace. New technology includes not only deeply embedded devices for vehicle systems management but also operator interfaces such as navigation systems, voice-recognition/text-to-speech interfaces and integration of consumer electronic appliances such as the Personal Digital Assistant (PDA). Standardization efforts have attempted to bring these technologies together through common buses, but because these devices have different capabilities and requirements of their own this is most likely impossible. The CPJazz framework was designed to provide integration of disparate devices and protocols through a dynamic and scalable component-like architecture. In this paper the capabilities and services provided by CPJazz for vehicle-based software integrators will be discussed as well as the identification of key scenarios that demonstrate its contributions to the vehicle and wireless environment.

INTRODUCTION

A number of trends are occuring in the evolution of consumer and commercial vehicles. These include the integration of:

- Numerous Vehicular Networks,
- Consumer Electronic Subsystems, and
- Wireless Network (IP) Connectivity.

These trends create new challenges such as the need for resource sharing, interconnectity and security. CPJazz, an architecture for in-vehicle computing and integration, was designed to address these challenges.

CPJazz is a middleware layer designed to simplify the development of complex multi-protocol, multi-application communication systems. CPJazz builds services on a POSIX compliant operating system to provide a component-like model for application development. This paper discusses CPJazz and its applicability in development of networked vehicle systems.

CPJAZZ ARCHITECTURE

The CPJazz architecture is a message-passing framework that provides communication capabilities between software components in addition to a set of components that provide a specific set of services such as bus interfaces or device interfaces.

CPJazz provides a set of communication APIs that provide component connectivity. The APIs are layered providing two methods of communication. The first is known as the "Bus" API that provides message-based communication. The second is the "Stream" API that provides communication between two dedicated end-points.

Broadcast and multicast (group) communication is provided through the Bus API as well as the ability to identify services provided by components. Synchronization on services or the presence of components is also provided at the Bus layer.

As discussed, CPJazz provides the basic mechanisms for communication between components. Standard platform components provide the necessary functionality for vehicle and communications systems development.

ANATOMY OF A CPJAZZ COMPONENT – CPJazz components share a common structure: a native interface (the components native protocol), the CPJazz interface (implemented through the Bus or Stream API) and the internal processing element (performs the internal processing of information). Each component is an independent program, whose linkage to other components is provided through the CPJazz API.

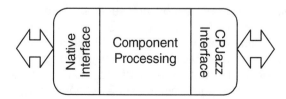

Figure 1. Basic Component Structure

Examples of CPJazz components include an IDB Protocol Manager (arbitrates between the ITS data bus and the CPJazz communication medium), a Text-to-Speech Device Manager (arbitrates between the CPJazz bus and the Text-to-Speech Codec) and an Internet Protocol Manager (arbitrates between IP and CPJazz). CPJazz treats all interfaces the same (such as IDB or IP links) so that communicating between them is simplified.

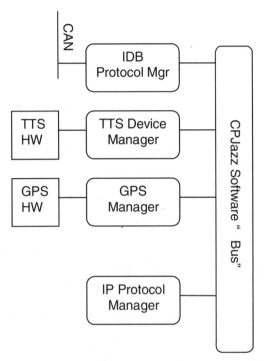

Figure 2.　CPJazz architecture showing the software "bus" which provides the basic communication capabilities and a set of components that provide specific interface functionality.

CPJazz provides a number of communication capabilities while minimizing overhead. These capabilities include:

- Name-space management (static and dynamic names)
- Service identification and lookup
- Component synchronization

Every component in the system must have a name in order to communicate. Names can be statically defined (well-known services such as GPS) or dynamically assigned by the bus (such as client components that use services but do not export any services). Services can be identified to CPJazz using a simple text name that allows other components to find it using the service API. Finally, synchronization services are provided to permit components to wait for a particular component to be available before performing bus-related operations.

As shown in the diagrams, this structure is very conducive to protocol translation. Packets of information arrive through the native-side of the component and are translated into the CPJazz-based protocols on the　software

bus side of the component. Alternatively, information flowing from the CPJazz side of the component is translated into the native protocol of the particular component. This general structure simplifies the bridging and routing of information between disparate protocols and devices.

Using the facilities of the underlying POSIX operating system, the architecture is very dynamic supporting the registration and deregistration of components. This permits dynamic scaling and reconfiguration of CPJazz.

BRIDGING OVER CPJAZZ – The CPJazz architecture simplifies the routing of data from disparate protocols. For example, the GPS Manager component could be configured by an external device on the CAN bus to emit location data at some frequency. The GPS Manager component creates a thread that communicates with an associated CAN protocol thread. Location data flows from the GPS thread to the CAN thread through the bus API. The CAN thread, having knowledge of its private interface, emits the collected GPS data to the appropriate destination. Additionally, an external application could connect via wireless IP to the CPJazz host and request a specific IDB message. This results in the creation of an IP Protocol Manager thread that communicates to an IDB thread. The IDB driver is registered with the particular message of interest and data automatically routed from the IDB thread to the IP Protocol Manager thread when available through the bus API.

KEY ARCHITECTURAL ELEMENTS

A platform providing communication between individual components is in itself not novel. The key to the CPJazz architecture is a set of components that provide key functionality within the vehicle domain. This section discusses some of these components.

LINK MANAGEMENT AND DYNAMIC SELECTION – The wireless market is currently segmented and exists as different offerings based upon bandwidth, reliability and geographic availability. For these reasons, connectivity to a vehicle may utilize different links based upon its geographic location as well as the desired application constraints. These include cost, reliability and availability of the link.

Dynamic Link Selection solves this problem by intelligently selecting the most optimal link (given the available options) for the current desired constraints in a policy-based fashion. The CPJazz model utilizes a market-based model whereby Link Broker components (attached to each available link) bid on the right to enable their link based upon their constraints compared to the application desired constraints. The Dynamic Link Selection component broadcasts the bid defined by particular application(s) and then accepts bids back from the Link Broker components. The closest bid (based upon constraints represented by a bid vector) is chosen and the particular Link Broker notified to enable the link.

This model matches a market-based model for the sale of goods in which data represent goods, consumers represent applications and producers represent the available links. This approach has been used successfully in scheduling and resource allocation systems for many years with roots in Contract Net Systems. [1, 2]

The logistics of managing modems, overlaying network stacks and overall link maintenance is supplied by the Link Manager. The Link Manager provides the ability to script a modem to a particular mode (such as CDPD or CSC) and then overlay the appropriate link-layer protocol (such as SLIP or PPP). The Link Manager may also be driven by applications to force a particular link outside of the control of dynamic link selection.

MANAGED FLOWS – A common design pattern for data movement through a network is for the data to be altered in some desired way. In many cases, the modification of the data can be invisible to the producing component. Take for example the migration of data from an onboard computer to an external host over the Internet. It may be desirable for the data to be first encrypted (for security purposes) and then compressed (for bandwidth savings).

The Flow Manager was designed for this purpose to provide a transparent construction of a data flow between a producing component and an anticipated destination component such as a procotol manager.

Using the transparent and dynamic features of the CPJazz architecture, the flow manager can automatically create intermediary components that provide the necessary services on the data flow.

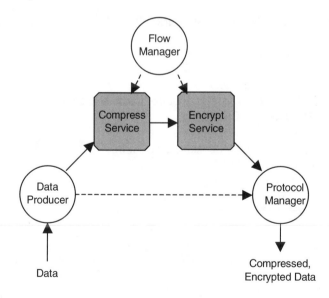

Figure 3.   Flow Management Example illustrating the transparent generation of services to build into a flow between a data producer and consumer. The ability to install user-defined intermediary components is also permitted to further tailor the managed flow process.

SECURITY – Building a networked vehicle computer implies accessibility that in turn mandates validation for access. CPJazz provides features for both Internet access as well as disparate bus access and therefore the ability to grant or deny access is a strict requirement.

The CPJazz paradigm for external access is that any access, which is not expressly permitted, is prohibited. This paradigm requires that access be defined per interface in very specific terms. This can represent a large administration task for nodes but is necessary for both node and data protection.

CPJAZZ SOFTWARE CONSTRUCTION

CPJazz, being a component-like model, borrows heavily from the Component Based Design methodology. In this model, system construction leverages heavily from existing software components to provide desired functionality within the system. For example, a GPS Manager component provides GPS data in a standard bus API form that other components may interrogate. The requesting component therefore does not need to understand various GPS protocols, but instead the abstraction of the protocol provided by the GPS Manager component.

This model is desirable for many reasons. One advantage is containment of complexity. A system is built from a set of components (either pre-built or custom designed for the system's purpose). The system is therefore built from a number of simpler components that integrate to provide more complex behavior. The CBD model has been called "the industrialization of software development". [3] This lesson of moving from custom development to development based upon the assembly of prefabricated components was learned from the manufacturing industries long ago. The benefits and leverage of time-to-market and cost reduction are clear.

Fast prototyping is also simplified with the model. The ability to collect a number of pre-fabricated components that are used by a custom component for a data management or control task means that systems may be more quickly developed and verified.

Another advantage is the natural firewall to changes. A component does not need to know that another component was redesigned or that it now speaks to different hardware as long as its basic interfaces (via CPJazz) remain the same.

This component-like model provides another step towards large-scale reuse for software development which can lead to higher quality, higher reliability and lower development costs.

SAMPLE SCENARIOS

WIRELESS VEHICLE MONITORING – The ability to monitor vehicles for performance, tracking and/or operational characterization helps fleet-based businesses increase efficiency. The CPJazz architecture and supporting components provide the vehicular interfaces (such as OBD, GPS and others) as well as common interfaces to wireless links. With dynamic link selection, the ability to monitor remote vehicles also becomes more cost-efficient with the concept of policy-based wireless link selection. Providing vehicles with multiple links provides for distribution of data based upon criticality or cost. Therefore, non-critical data can be stored for distribution when a low-cost link (such as spread-spectrum or Bluetooth) is available or critical data can utilize a transient link such as a satellite phone or CSC modem.

VEHICULAR RESOURCE SHARING – As vehicles incorporate more technology (non-critical, such as navigation systems), the ability to share some of the common resources becomes attractive. For example, a GPS receiver is useful not only for an inboard navigation system but also as a time source. A Voice Recognition device is useful to any device that can benefit from voice commands such as a cellphone or entertainment radio. The CPJazz architecture provides the bridging glue in this application to tie the non-critical disparate subsystems together (regardless of physical bus used) so that cooperative use is possible. The bottom-line is subsystem reuse in vehicles that can represent significant cost reductions.

RESOURCE-CONSTRAINED EMBEDDED VEHICLE SYSTEMS – Embedded system developers commonly attempt to size the resources needed in a system in order to reduce the overall cost of the hardware components. These components include not only FLASH, EEPROM or RAM but also the performance rating of the processor. CPJazz supports these efforts through its dynamic abilities. Systems that include transient components (components that are enabled to perform a task, and are then removed from active execution) can benefit from CPJazz. For example, consider a vehicle that includes a docked cellphone (connected to a vehicle bus) that provides data services. A component operating on CPJazz could request, through the link manager, that the cellphone be enabled for use through a PPP link. The link manager would build the PPP-based network stack over the particular bus and make it available for the requesting component. Once the link was no longer needed, the link manager would tear the network stack down and remove it from active memory making the memory and processing capacity available for other component's use.

CONCLUSION

This paper discussed the CPJazz architecture and its applicability to embedded vehicle systems development. CPJazz is based upon a component model for software development where each component is an individual executable that provides a common API interface over the CPJazz communication framework. Key features of the architecture such as link selection and managed flows simplify the job of applications developers through transparent services. The dynamic nature of the architecture also supports systems with minimal resources.

REFERENCES

1. A. Goscinski, 'Distributed Operating Systems: The Logical Design', Addison-Wesley, 1992
2. R. G. Smith, 'The contract net protocol: High-level communication and control in a distributed problem solver', IEEE Transactions on Computers, 29(1980).
3. K. Short, 'Component Based Development and Object Modeling', Sterling Software, February 1997.

CONTACT

CellPort Labs Inc. currently employs Mr. Jones in Boulder, Co as the principal software engineer for the Mobile Networking Technologies group. He can be reached at mtj@cellport.com. CellPort Lab's URL is http://www.cellport.com

DEFINITIONS, ACRONYMS, ABBREVIATIONS

API: Application Programmers Interface
CDPD: Cellular Digital Packet Data
CSC: Circuit Switched Cellular
GPS: Global Positioning System
IP: Internet Protocol
NMEA: National Marine Electronics Association
OBD: OnBoard Diagnostic
PDA: Personal Digital Assistant
POSIX: Portable Operating System Interface
PPP: Point to Point Protocol
SLIP: Serial Link Interface Protocol
TTS: Text-To-Speech
VR: Voice Recognition

NETWORK/PROTOCOL TESTING

2005-01-1535

LIN Protocol Conformance Test

Wolfhard Lawrenz and Thomas Roskam
C&S group at the University of Applied Sciences

ABSTRACT

Networking is an important technology in today's vehicles: Up to 100 electronic control units are interlinked through various communication protocols such as CAN, LIN, MOST, etc. As such communication protocols represent a "standard" component in vehicles which is assumed to provide a specified functionality regardless who the manufacturer is. Unfortunately 10 years of practical experience show that implementations of communication protocol specifications done by different manufacturers in most cases are not compliant a priori. Luckily experience of automotive OEMs together with their suppliers and silicon manufacturers have proven that conformance testing is a very effective means to avoid interoperability problems in mixed suppliers automotive systems.

The following paper describes for LIN protocol the process of conformance test specification, conformance test suite implementation and the experiences gained while performing such tests.

INTRODUCTION

Networking in vehicles has become the essential backbone for interlinking distributed electronic control units - ECUs - into a "system". LIN protocol is a highly cost effective means to network ECUs in body electronics applications such as doors control with power windows, power lock and various controls for further applications such as seat positioning, mirror positioning, etc. Typically, any of these networked systems is a multi vendor system consisting of ECUs from different 1st tier suppliers, while the various ECUs typically make use of standard LIN communication modules being supplied by different 2nd tier suppliers, which are the semiconductor and/or software manufacturers.

Although LIN is considered a "standard" communication protocol, the interpretation of the specification by each of the various implementers may be "slightly" different therefore inducing an interoperability risk in the network between the communicating ECUs; see fig. 1. As this may cause serious problems in distributed control

adequate countermeasures are required. One of the countermeasures is to provide upfront a specification which itself does not bare the risk of being interpreted differently. Unfortunately this is difficult to achieve. Current specifications for various reasons are typically written in natural languages, therefore being ambiguous, un-precise, not comprehensive, etc. Going towards formal and executable specifications, which per se are not ambiguous, "more" precise, "more" comprehensive is a must for future actions, but does not solve the problem either, because further questions arise, e.g. which is the appropriate abstraction degree offering enough freedom for the implementers but being precise enough to avoid interoperability risks, etc.

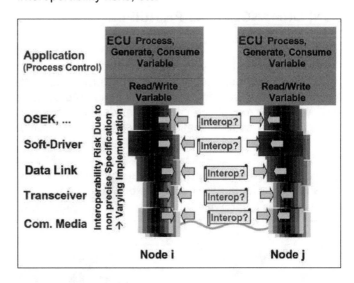

Fig. 1: Interoperability Risk in Networked Systems

Currently a very effective means to reduce the interpretation and thus interoperability risks of standard components is testing. Conformance testing is mostly applied for that purpose to check whether the component complies to the given specification. Conformance test experience in automotive for almost 10 years checking various communication protocols on various OSI communication layers – see fig. 2 – has shown that the vast majority of first test runs of protocols failed. After the detected problems were corrected by the 2nd tier suppliers, typically no problems were reported when the components were applied in the 1st tier suppliers modules and then integrated by the automotive OEMs. This positive experience resulted in the meantime that many OEMs worldwide require conformance tests to be passed before new communication components are accepted. Thus conformance tests are a proven practice to assure mixed suppliers networked systems interoperability and therefore contribute significantly to enhance the quality of automotive electronic systems.

OSI	CAN		TTCAN	FlexRay	LIN	MOST	Other
7	C&S: OSEK/VDX NM, ...				C&S	C&S: Functional	C&S: Various
...				...	...	...	...
2	C&S soft-com driver		C&S[4+]	C&S[5+]	C&S[***]	C&S	C&S ..
	C&S robustness						
	C&S Processor Interface	C&S Gateway	C&S[4+]				
	C&S ISO 16 845 enhanced		C&S[4+]				
	ISO 16 845		– ISO 16 845				
1	Transceiver		C&S	BusGuard + Transceiv C&S[4+]	C&S[***]	C&S	C&S ..
	GIFT – C&S[*]	Ford_US – GIFT – C&S[**]					
	Low-Speed	High-Speed					
"0"	C&S Net-Topology Signal Behaviour Simulation						

[*] sponsored by Audi, BMW, DaimlerChrysler, PSA, VW
[**] sponsored by Ford_US
[***] sponsored by LIN Consortium
[4+] sponsored by Semiconductor Manufacturer
[5+] in cooperation with FlexRay Consortium

Fig. 2: C&S Protocol Conformance Test Chart

CONFORMANCE TEST PRINCIPLE: METHODOLOGY

Conformance Testing of communication protocols is based on the corresponding ISO 9646 Coordinated Test Method standard; see fig. 3.: Communication protocols – and even other modules – can be considered as a set of layers. For LIN communication protocol two layers have been specified so far:

- Physical layer, comprising the transceiver, the media and the corresponding interface between the two
- Data link layer which is often referred to as the protocol layer

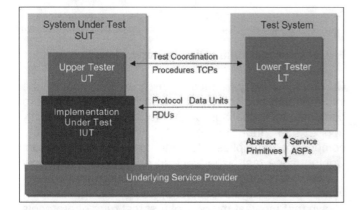

Fig. 3: ISO 9646 Coordinated Test Method

Testing typically follows the layered architecture; see fig. 3. The considered layer under test is referred to as "Device or Implementation under Test – DUT or IUT". The surrounding upper and lower layers represent the interface to the application environment. In the given ISO test architecture a typical problem is depicted: One of the DUT interfaces is directly accessible – in this case it is the interface to the lower layer – while the interface to the upper layer is only indirectly accessible. The intention of the ISO 9646 method is to install on top of the layer, which is indirectly accessible, a representation of the upper layer functionality – later on referred to as the upper tester – and perform the access to the upper layer tester through the DUT itself and the finally directly accessible other layer.

Testing based on direct and indirect upper and lower layer tester is very typical for communication protocols and other standard modules. For example LIN data link layer tests are performed that way: In a micro controller implementation LIN data link layer is typically based on the existing UART (Universal Asynchronous Receiver and Transmitter) in conjunction with a dedicated software module. As the upper interface of the software module obviously cannot be accessed directly without any performance problems, on top of that LIN data link layer software a correspondingly adapted "Upper Tester Software" is loaded. The test execution finally stimulates and monitors the upper tester through the DUT itself and finally through the lower layer interface, which is directly accessible through the UART.

LIN DATA LINK LAYER TEST METHOD

The described LIN data link layer test process already points out the problems of the final test implementation and the corresponding high responsibility of the performing test house. As these indirect tests must comply with the functional and performance requirements of the specified tests and as the upper layer tests are executed – stimulated and monitored – through the DUT itself, there is obviously a risk that the test execution may be influenced, even falsified. As such the implementation of the tests, the adaptation of the tests to the individual DUT-environment is a very sensitive and highly responsible task. As such C&S's experience has helped to identify problems which are due to non-experienced test implementation and low-experience test houses.

LIN FULLY INTEGRATED SLAVES TEST METHOD

The indirect test technique even becomes more indirect, when the so called "LIN Fully Integrated Slaves" are to be tested. In this case only the lower interface of the transceiver is directly accessible. No upper tester can be implemented as these modules are so-called "Closed Modules". In this case the upper tester must be implemented externally. Its operation is maximum indirect, as the upper interface of the DUT can only be accessed from the lower side indirectly through transceiver and DUT and from the higher side indirectly through the "filtering" Inputs/Outputs and the application itself.

This makes it obvious how sensible the test adaptation and test execution processes are; see the corresponding paragraph below: Closed module testing.

LIN TRANSCEIVER LAYER TEST METHOD

Testing of LIN transceiver layer as an isolated component is rather simple and low risk, as both ends of this interface per definition are directly accessible. If LIN transceiver testing is to be performed in a Closed Module environment, correspondingly the above mentioned technique must be applied.

CONFORMANCE TEST PRINCIPLE: TEST EXECUTION AND TEST CASE DERIVATION

ISO 9646 Coordinated Test Method standard – see fig. 3 – can be considered as a static data flow architecture of a conformance tester. This architecture is the basis for test execution principle and related test case derivation.

TEST EXECUTION: DYNAMIC DATA FLOW

Based on that architecture fig. 4 depicts a corresponding dynamic data flow architecture.

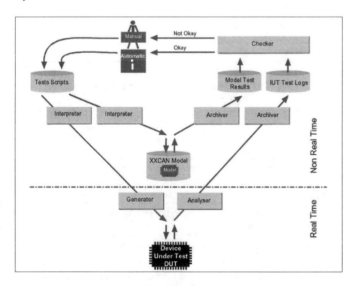

Fig. 4: Tester Dynamic Data Flow

Each test case is translated to the real time capable lower/upper tester interfacing the DUT. The related tester generator stimulates the DUT, while the corresponding tester-analyzer monitors the responses of the DUT and stores the results in a data base. In a second step the same stimulus is applied to a model, representing the "ideal" behavior of the device under test. Correspondingly the responses of the model are stored in a data base, from where they are compared to the responses of the actually measured results. If the comparison is positive the applied test case gets a "Pass", otherwise a "Fail".

In case of "Fail" further checks are performed to identify the potential source of the "Fail". On the one hand it may be the tester itself, which was detected to be defect. On the other hand a DUT defect may have caused the problem. In this case the test person typically analyzes the observed effect in more details and typically extracts from the test scenario a minimum stimulus sequence, which provokes the observed potential malfunction of the DUT. This stimulus then is handed over to the implementer of the DUT, who applies the sequence to his implementation model. While simulating the model at low abstraction degree and high time resolution the implementer should be able to observe the critical behavior, to identify the wrong part in his implementation, and to take the correspondingly action to correct the malfunction.

TEST CASE DERIVATION

LIN specification is written in English language. Therefore no formal methods can be applied to derive tests automatically. Empiric but systematic, so-called "Black Box Testing" is the adequate process to derive a conformance test specification. Based on this experience C&S developed a so-called SOVS-process (System Operational Vector Space) which was applied when contributing to the LIN test specification processes at the European LIN Consortium and the US SAE J2602 committee. The SOVS process consists of the following steps:

(1) Cut the system into subsystems – layers as mentioned before – for the purpose of easier handling, better exchange of parts of the system, smooth upgradeability, etc.; see fig. 1

(2) Identify test cases for the DUT layer. The basic idea behind this process is that tests must represent a realistic worst case application scenario of the layers surrounding the DUT. The multitude of tests at the end represent a reasonable variation of the scenarios to the limits of the DUT specification.
→ This step is a quasi arbitrary process driven by experience in testing and component application.

(3) Describe each test case by a repetitive identical format, such as e.g.:
"Purpose of the test case"
- Initialise DUT
- Apply a test step
- Analyze the response of the DUT with the expected result of the test step
→ This enforces a systematic easy-to-read and implement structure.

(4) Group the identified test cases into classes given by the so-called frame work of the so-called "System Operational Vector Space – SOVS"; see fig. 5.
→ The SOVS groups all tests into subclasses subdivided by a set of (hopefully orthogonal) vectors representing each an independent application constraint – in accordance with the given DUT

specification. Once the vectors have been identified the overall test space SOVS results from the vectors product. Each of the empirically found test cases represents one spot in the SOVS.

As such the SOVS then can be used as a systematic "tool" to identify the good, over- and under-specified areas; see step (5)

- Derive the parameters characterizing each vector
- Describe each of the empirically found test cases – see step (2) – as a unique set of values assigned to the parameters, grouped into the vector classification

(5) Identify over-specification or lacking tests by parameter value variation.

→ The goal is to derive a sufficiently exhaustive but minimal set of test cases covering the application scenarios under the conditions and limits given by the functional specification

- add lacking but necessary test cases
- delete redundant test cases

(6) Approve the "completeness" of the selected test cases by practical application; see fig. 5.

→ Typically practical systems will show anomalies in practical customer applications. Unfortunately this may occur as the tests cases had not been formally derived as there was no formal description of the component a priori.

- Careful analysis of these behaviours may disclose that some test cases may be lacking to assure interoperability under these circumstances
- Refine test specification as shown on steps (3) and (4)

System Operational Vector Space = SOVS =
{Node_Mode} x {Node_Timing} x {Frame_Type} x {Failure}

with e.g.

{Node_Mode} =
{Master/Slave, Sleep/Wake-Up/Awake, ... }
:
:
{Frame_Type} =
{Normal/Extended/Command, Reception/Transmission, ..}
:

Fig. 5: System Operational Vector Space - SOVS

CONFORMANCE TEST IMPLEMENTATION

A typical result of an SOVS driven test specification is shown in fig. 6. The LIN consortium appointed C&S to chair the corresponding LIN Test Specification Task Force. The Task Force was constituted from experts providing different technical backgrounds such as vehicle manufacturer, semiconductor manufacturer, test tool manufacturers. In parallel C&S supported the test

standardization process of the SAE J2602 Task Force, which has recently finalized the test specification.

DATA LINK LAYER TESTER

The standardization of the European LIN Test Task Force took more than 2 years, providing tests for the evolving LIN protocol with its versions 1.3 and 2.0. Fig. 6 depicts the main headlines for LIN version 2.0. Fig. 7 correspondingly depicts the typical format of an individual test case.

Fig. 6: LIN Consortium Layer 2 Test Spec (SOVS)

State	Description	
Set Up	IUT as	Slave
	Con-figura-tion	See table 1-1, TST_FRAME_2
	Baud Rates	According to the IUT specification
System Init	IUT initialisation required before test.	
Test	The Test System as master sends TST_FRAME_2 with the baud rate specified in the IUT specification and re-served identifier (2-15).	
Verification	The IUT must answer to the master request with negative response in RSID. $(\text{Test System baud rate} - F_{TOL\_RES\_SLAVE}) \leq$ measured slave baud rate in the first byte of the slave answer $\leq$ $(\text{Test System baud rate} + F_{TOL\_RES\_SLAVE})$ $(\text{Test System baud rate} - F_{TOL\_RES\_SLAVE}) \leq$ measured slave baud rate in the last byte (checksum) of the slave answer $\leq (\text{Test System baud rate} + F_{TOL\_RES\_SLAVE})$	

Fig. 7: LIN Consortium Layer 2 Test Case 2.8

The architecture and the final implementation of the tester – see fig. 8 and fig. 9 – follow the basic static and dynamic architectures as discussed above and shown in fig. 3 and fig. 4. The basic concept of fig. 8 is driven by the idea to apply standard test components, while providing a clear separation of the time critical and non time critical parts of the tester and the DUT itself with its indirect upper tester access.

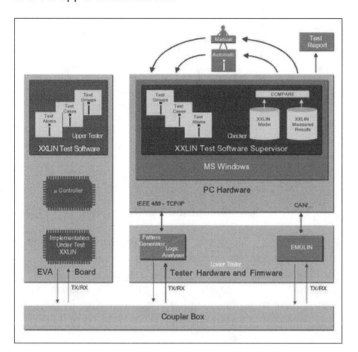

Fig. 8: LIN Layer 2 Tester Architecture

The PC – see fig. 8 – is responsible for the offline preparation of test stimuli, the offline DUT-model checking and the offline evaluation of the test results coming form the model test and the real DUT test. Furthermore the PC offers a platform to the test expert to analyze a problem in more details, if a fail resulted from a test. The execution of a test is performed correspondingly to the process mentioned above:

1. Test expert first analyzes a new DUT and provides the necessary test adaptations. The adaptations are typically required, because each DUT typically has some individually personalized features, such as register sets, etc. As such the test expert decides which subset of tests is useful and which upper tester software is required. Accordingly the UT software is written, compiled and checked. All this is provided to the PC, which then is started to perform the tests.
2. PC identifies the actual test to be performed
3. PC produces upper tester software and upload the UT SW through the DUT; note: This method assumes that the DUT provides a minimum functionality upfront.
4. PC produces and downloads a lower tester - LT - real time stimulus and translates it to the format required by the corresponding lower tester generator.
5. PC starts LT to execute the test.
6. LT generator executes the stimulus in real time while the LT analyzer grabs all data, which represent the response of the DUT in real time.
7. PC stores the analyzer data in a data base

8. PC perform the non real time execution of the same test case on the DUT ideal model and store the result in the data base.
9. PC checks both results offline and decides on test "Pass", "Fail" and puts the result to the test record.
10. If a Fail had been assigned the test expert decides, what to do. If a Pass had been assigned the PC selects the next test and continues step 4.
11. PC produces after all tests have been executed a report on the results. As a summary a so-called "Authentication Document" is issued, stating for the component xyz the LIN conformance tests version abc have been executed, giving the test groups results as PASS or FAIL. The latter document is typically used by the company, which had ordered the test in order to hand that over to the 1st tier supplier or the OEM. The detailed test report is used for trouble shooting.

The final implementation of a conformance tester – see fig. 9 – follows the architecture given above fig. 8. The very critical issue to be recognized is the fact that all the required performance constraints while executing tests must be assured. Furthermore it is highly recommended to apply remote controllable, standard calibrated measurement tools only for the real time lower tester. This offloads the test house from proving the technical quality and assure the test customer to achieve a good quality. As shown in fig. 9 the tester consists of a PC, specifically designed minimum HW/SW coupling box and a standard Agilent remotely programmable generator/analyzer tool.

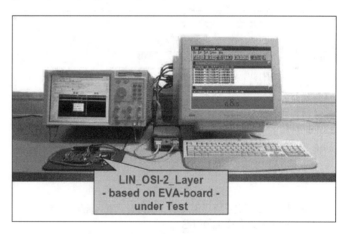

Fig. 9: LIN Layer 2 Tester Implementation

TRANSCEIVER TESTER

Specification of the tests had been carried out as mentioned above. C&S contributed as chairman of the European LIN Test Task Forces and brought in its experience to the SAE J2602 Task Force. Both Test specifications are finalized in the mean time. Fig. 10 and 11. give examples of the European LIN Tests.

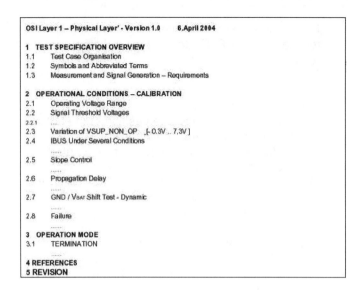

Fig. 10: LIN Consortium Layer 1 Test Spec (SOVS)

IUT node as	Master and slave ECU Transceiver	Test case 2.1.x (2 cases)
Parameter	V_SUP	See Table 2.1
	Bus Load	1nF (1%), 1kOhm (0.1%)
Test Steps	A voltage ramp is set on the VSUP as defined on Table 2.1. The LIN signal is driven with a 10kHz rectangular signal with a duty cycle of 50% and a voltage swing of 18V. The IUT must be in operational / active mode	
Response	The RX pin of the IUT has to show the 10kHz signal.	
Reference	LIN Physical Layer Specification, Rev. 2.0, Ref. 3.1.1 and 3.1.2	

# test	V_{SUP} Range	Signal Ramp
2.1.1	[7,3V...18V]	0,1V/s
2.1.2	[18V...7,3V]	0,1V/s

Fig. 11: LIN Consortium Layer 1 Test Case 2.1

The architecture and implementation of the transceiver tester is (almost) identical to the Closed Module Tester; see below.

CLOSED MODULE TESTER FOR INTEGRATED SLAVES/MASTER TESTS

As mentioned above Closed Module Testing requires a very indirect testing method; see figure 12. As the closed module per definition does not allow the implantation of an upper tester SW the indirect test methodology must be applied. The device under test can only be interfaced through the external available pins: LIN data, Input/Outputs, Power Supply. The actual tests to be applied to the upper interface of the buried LIN data link layer can only be accessed through the pins mentioned above in conjunction with the application itself, which represents a "falsifying" supplementary layer causing the indirect test technique. Obviously the adaptation of the tests to these additional constraints requires high attention of the test expert. As typically the application

implementation itself is confidential, a questionnaire was developed for the implementer helping to perform the adaptation process.

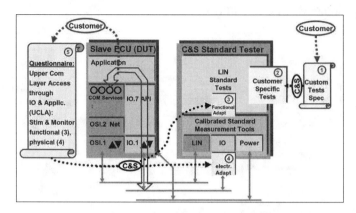

Fig. 12: LIN Slave Adaptation to Test Suite Principle

Fig. 13 and 14 show the architecture and related implementation of the tester. The same principles as mentioned above should be recognized: Use of standard, calibrated, remotely programmable measurement tools is highly recommended in order to safeguard a high quality tester.

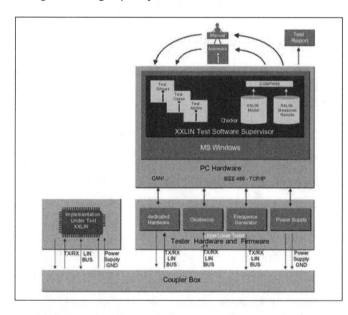

Fig. 13: LIN Closed Module (Slave/Master) Tester Architecture

The tester implementation as shown in fig. 14 is applicable for Closed Module tests as well as for physical layer tests. In the latter case, only a subset of tools is required. Analog and digital data and signal generators and analyzers are required to stimulate and monitor in real time the DUT's LIN data line, the DUT's IO lines and the DUT's power supply.

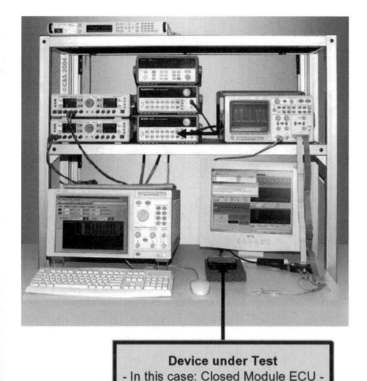

Device under Test
- In this case: Closed Module ECU -

Fig. 14: LIN Closed Module (Slave/Master) Tester Implementation

CONCLUSION

In the meantime more than 40 LIN devices have been tested including isolated transceivers, isolated data link layer and closed modules slaves and masters. So far, the above-described methodologies as well as the related test specifications and the corresponding tester implementations have turned out to be satisfactory.

Up to now all tests only had been performed for the European LIN specification. In the mean time the SAE J2602 tests have been implemented on the same test suites' architecture. As such no principle problem to perform SAE LIN tests have been identified.

Conformance is typically targeted at checking the compliance with a given specification. But the ultimate goal in distributed systems is typically Interoperability Tests, which is finally based on conformance tests, but goes further: Interoperability tests should check that corresponding "standard" components – such as network components, etc. – change states simultaneously (in a reasonable time frame), when initiated by state change of one of them. Unfortunately the corresponding "Interoperability Characteristics" up to date are still not thought about and are not stated in the components system behavior specification so far. This is a very important, not to neglected subject for future standardization actions.

REFERENCES

1. W. Lawrenz: Networked Systems High Level Design & Test Philosophy and Tools; SAE Conference and Show Detroit, paper 950296, 27.03. - 03.03.95
2. M. Scheurer: Entwicklung einer Methode zur systematischen Ableitung von Testfällen für Konformitätsprüfungen in der Informationstechnologie ; Diploma Thesis at Fachhochschule Wolfenbuettel, c&s group, Germany, July 2000
3. Abramovici, M.A. Breuer, A.D. Friedman: Digital testing and testable design; Computer Science Press, 1990.
4. A.L.Courbis, J.F. Santucci, N.Giambiasi : Automatic Behavioral Test Pattern Generation for Digital Circuits; 1st IEEE Asian Test Symposium, Hiroshima, Japan, November 1992, pp. 112-117
5. S.J. Chandra, J.H. Patel: Experimental evaluation of testability measures for test generation; IEEE Trans. On CAD, Vol. 8, N°1, pp.93-97, Jan. 1989.
6. K.T. Cheng, A.S. Krishnakumar: Automatic functional test generation using the Extended Finite State Machine Model; 30th ACM/IEEE Design Automation Conference, USA, 1993
7. J. Magnier: Représentation Symbolique et Vérification Formelle de machines séquentielles; State Thesis University of Montpellier II, France, 1990
8. K. Hoffmeister: Applikations-bedingte Kommunikationsanforderungen im verteilten Kraftfahrzeug-Echtzeitsystem und deren Testbarkeit; Diploma Thesis at Fachhochschule Wolfenbuettel, c&s group, Germany, July 2000
9. W. Lawrenz, Anne-Lise Courbis, Janine Magnier: Car System Conformance Testing, International Conference Systems Engineering and Information&Communication Technology, Nimes TIC 2000, September 2000
10. W. Lawrenz: LIN Conformance Testing, 1st International LIN Conference at Böblingen Germany, September 2002
11. Grzemba, editor: LIN-Technologie; Paragraph 11, W. Lawrenz et al: Testing, Vogel Verlag 2004
12. LIN Protocol Standard, LIN consortium www.lin-subbus.org
13. SAE J2602 Recommended Practice, LIN Network for Vehicle Applications; Draft of March 31, 2004,

CONTACT

Wolfhard Lawrenz is head of the Wolfenbuettel University C&S group research entity, while Thomas Roskam is LIN Project Mgr. at C&S. C&S is specialized in automotive networked systems with special emphasis on conformance testing of communication protocols and standard software/hardware modules. For more information please contact:
www.cs-group.de
Lawrenz@cs-group.de   Th.Roskam@cs-group.de

2005-01-1283

Reducing Costs Associated with Validating ECUs and Systems in the Increasingly Networked Vehicle

Mike Staszel
Vector CANtech, Inc.

ABSTRACT

Testing is an intrinsic, vital and on-going part of the product development process, particularly in the development of automobiles and automotive systems. Validation testing of vehicle electrical systems and their components is especially critical, and is expanding rapidly with the growth of electrical and electronic features offered in the automotive marketplace. In the face of this expansion traditional approaches to ECU and Systems validation are being seriously challenged by cost pressures in two areas: The cost of basic test equipment and test services labor costs, and the costs associated with the expansion of ECU I/O count and functionality in support of these features.

The evolution of multi-featured in-vehicle networking from its basis as multiplex wiring has also expanded systems complexity dramatically, which has compounded the problem. Paradoxically, however, this new form of complexity can also reduce test requirements, hence test and validation costs, to a significant degree. The key to reducing test costs in increasingly complex systems is to exploit the ability of the distribution process to make each of the testable components in the system simpler.

INTRODUCTION

Why test?

The development of any product requires verification of conformity to specifications and robustness in design. Testing allows the design engineer, test engineer or test technician to confirm that an ECU and/or system performs as intended, i.e. the product conforms to specifications. More specifically, it provides confirmation that it can execute the functionality it was created to provide, and that it will successfully accomplish its task over its entire lifetime and through all conditions for which it was designed.

Manufacture of a high-volume product demands uniformity in order to ensure first-run quality (the closest match to design specifications) over the entire production run, from start to finish. Generating first-run quality out of a production line eliminates the costly inefficiencies arising from reworking products that aren't quite correct as they come off the line, and scrapping products that cannot be reworked economically. Testing is included within the manufacturing process for use in establishing and maintaining uniformity control in production.

All development and production strategies rely on testing for the feedback required to develop, produce and refine their products. Efficient development methodologies match testing scope and depth with the desired complexity and required robustness of the product in order to test the product optimally yet complete the test regime in a timely fashion. This process maximizes test efficiency while minimizing the cost of the resources required.

Efficient production strategies seek to employ as little testing as possible, as a result of their mandate for efficiency, in order to support the uniformity that is their target. Historically, the most effective of these rely on statistical process control (SPC) as the means for managing uniformity in production. It is methodologies like SPC that dictate the scope and depth of testing to be used.

As products ECUs and Systems are no different in principle. In practice however, most are far more sophisticated than the typical kinds of products managed by most efficient development and production strategies.

TRADITIONAL TEST SCENARIOS

VALIDATING ECUs

The process of validating automotive ECUs generally involves exercising their functional capabilities while attempting to place them under controlled conditions that accurately represent those they will encounter in the target production vehicle. In this way the device under test can be scrutinized scientifically in its "natural" environment. How close to perfect this representation must be is the subject of much conjecture, and is highly dependent on the complexity of the target system.

For complex inputs and outputs like Exhaust Gas Oxygen (EGO) sensor inputs and fuel injector drivers, elaborate simulations of the corresponding production components are often used. However, to save money test engineers have frequently attempted to use actual production components instead wherever possible. Invariably this sort of simplification results in a test that does not mimic the real world well, if for only one reason: it cannot accurately represent the spectrum of variability encountered over the entire production run covering every part of the same design.

Take the case of the very simplest of I/O, the digital input. Responding to the state of a signal that has only two possible values, set at perhaps +12V and Ground, it would appear that the application of these discrete voltages by any power supply would be sufficient to represent the equivalent signals generated for the ECU's use by something elsewhere in the vehicle. However, an old axiom has it that every digital problem reduces to an analog one when problems begin to appear in the vehicle. In other words, even simple digital inputs have analog aspects that must be considered and accounted for.

Expanding this case to each input or output on each ECU in the vehicle highlights the degree to which test systems must be meticulously designed in order to avoid missing a failure. The only certain method for minimizing the need for this level of detail is the reduction of I/O counts themselves.

VALIDATING SYSTEMS

Historically ECU validation and Systems validation have been treated as one and the same in the automotive world. This is because each new system added to a vehicle has usually been built upon a single, and intimately related, ECU at its heart. Engineers have found that each new feature proposed is most easily developed as a separate "overlay" to the existing vehicle electrical design. Thus each is generally assigned a new ECU and added incrementally to the existing array of electronic features already fitted to its vehicle. This concept has worked well until recently because these new and independent systems have been developed, refined and put into production as self-contained, stand-alone solutions having a minimum of interaction with the remainder of the vehicle's systems.

Quite some time ago a more formal adoption of the concept of Systems began to be emphasized, focused primarily at ensuring thorough consideration of the effects of the external components making up each system, as well as interaction with the other systems in the vehicle, in addition to the ECU itself. Prior to this time detailed study and characterization of these external components and effects was frequently forgotten or unconsciously minimized by ECU-focused engineers. These oversights have frequently led to unfortunate results, primarily because no ECU operates in a vacuum in the increasingly complex vehicles that are being designed and built.

More recently, the Systems approach has returned to prominence because a migration away from traditional discrete wiring is occurring. The migration first introduced, then facilitated the expansion of, computer-style networking in vehicles. This type of networking, originally called multiplex wiring, was first introduced as a means of reducing I/O, more specifically those inputs and outputs used as interconnects between ECUs. In practice it has resulted in a corresponding reduction in wiring cost, and so its use has been expanded dramatically over the last several years. Successful implementation of in-vehicle networking requires a systems focus because it raises the level of interdependency between ECUs on the vehicle.

As a result of the emergence and recent reinvigoration of the Systems focus, and the clear need to test ECUs as part of the system to which they belong, more often than not, the validation of an ECU must go hand in hand with the validation of the system in which it resides.

TEST STRATEGIES

Of all design, development and manufacturing tasks, testing is perhaps most critical because of its ability to confirm the successful transfer of theory into practice. For this reason it is conducted periodically throughout the process of designing, releasing and manufacturing a vehicle from start to finish.

During the early part of the design effort testing usually involves the simple confirmation that desired outcomes result when designs are run through their operating regimes. These early tests are very frequently ad hoc, informal, and not usually conducted according to a detailed time line.

The first formal testing event in most development programs occurs when the entire design is completed. At this point Design Verification (DV) tests are created and conducted according to a detailed formal plan established prior to beginning the development process.

The second formal testing event occurs with the startup of production. Process Validation (PV) tests confirm the ability of the manufacturing process to meet its target production goals. This is essential to the establishment of a controlled production environment. Since many aspects of the design of ECUs and electrical/electronic components have an effect on manufacturability, this test set also provides feedback on the design process.

The PV test suite is also important for the maintenance of the controlled production process after startup. In today's quality/cost-conscious environment, some production processes employ Statistical Process Control (SPC) as the means for managing the production process, and ensuring controlled production, in a cost-conscious fashion. The initial PV suite is used to validate every ECU (also known as 100% Inspection) prior to and concurrent with startup. Afterward SPC allows it, or the relevant portions of it, to be applied to samples drawn at random from the production stream according to a pre-established plan, rather than testing every part. This reduces test expenses significantly, while simultaneously ensuring optimized quality.

Another formal testing event is known by the generic term End-of-Line (EOL). End-of-line tests are usually part of a 100% inspection program. By definition 100% inspection is at odds with the premise within SPC that only random samples of production output need to be tested to verify conformance to specification in a well-controlled production process. Thus, the existence of EOL testing is an admission that SPC is very difficult or even impossible to successfully carry out with some products.

Of the thousands of parts that make up a typical vehicle, experience has proven that ECUs exhibit this characteristic most often. As a result of the receipt of too many bad ECUs, i.e. those that made it past their respective production screening systems without being detected, OEMs frequently mandate EOL testing for most of the electronic components they buy. It is particularly true for complex ECUs that their inherent complexity makes it difficult for their manufacturing processes to hold all of their characteristics in control using SPC or by any other means.

COMMON TYPES OF TEST SYSTEMS

The ad hoc testing that is conducted during the earliest stages of a development program is a special case in which test equipment is generally not considered a system but more of a collection of independent items, brought together temporarily by the design engineer to serve his/her purpose. Historically, little automation has been used (although it is frequently usable paradoxically). This reinforces the idea that ad hoc testing is not formal. Unfortunately, it also can lead to the incorrect assumption that it is not important as well.

DV testing is generally the first testing conducted within a development program to benefit from the construction and use of a formal test system. The main advantage of the introduction of formality is repeatability, ensuring consistent results that the engineer can trust. In recent years DV systems have come to be increasingly automated, in order to improve throughput as well as to ensure repeatability.

Some development teams create a rudimentary DV test system concurrent with the start of product development,

essentially developing product and test in unison. Others argue instead that the test system should be developed independently so as to be considered "unbiased", hence more effective at uncovering design faults.

Process Validation (PV) testing is a manufacturing development process intended to examine the variability of the produced parts, and not necessarily the robustness of the basic design. Although most are developed independently, some PV test systems are built directly upon the DV testers that immediately precede them in the development chain. Similar in construction to DV testers, PV testers feature additional capabilities necessary for tracking and comparing key characteristics and unit-to-unit parametric measurements statistically.

Inasmuch as PV systems are constructed to conduct the initial PV exercise that occurs with the production launch of the product, they are also frequently called upon after launch. Most often this occurs when a change is made to the product design, either intentionally to accommodate new requirements or correct errors, or less intentionally when the characteristics of constituent parts change as a result of a change in their manufacturing process. In these cases, a full or partial PV is usually re-run to determine the new state of control of the ECU production process. Within a process controlled using SPC this most often involves temporarily terminating the sampling program and substituting a 100% inspection regime until the new state of control is established and stable.

EOL test systems have many of the characteristics of PV systems, and in many instances are actually the PV systems themselves. The difference is primarily one of perception. PV is the predominant term for production lines run by SPC, while EOL generally applies for lines run with 100% inspection of all parts produced.

All of the test systems discussed thus far are simultaneously complex and very much specific to the product. For these reasons they are almost always custom built only for their intended use. Bringing things full circle however, engineers who conduct the ad hoc development tests characteristic of the earliest parts of the development process do not usually benefit from the advanced features built into DV, PV and EOL test systems.

The construction of dedicated automated test systems for use in ad hoc testing has generally been considered too expensive for widespread application. As with DV, PV and EOL testers, ad hoc testers are generally very expensive because of the custom requirements of the test suite. However it is also true that these requirements are sometimes less custom than in the test systems found farther downstream. Consequently, over the years several attempts have been made at the design and marketing of generic automated test tools for ad hoc testing use, some better suited than others for this type of work.

One example is generically called the Plant Modeler. The term "Plant" here is borrowed from manufacturing engineering and refers to all aspects of the system surrounding the ECU under test excluding the actual ECU itself. Hence the ECU and its plant form the complete system under study. A Plant Modeler is a computer-based test platform that generates a simulation, based on the description of the plant it has been given, that is essentially identical to that seen in the real world. All Plant Modelers are programmable and modifiable, and in theory can present the ECU under test with an environment identical to that it will see in the actual vehicle, including the variability typically seen across the breadth of the entire production run.

COMPONENTS OF THE TYPICAL TEST SYSTEM

Testing in its most basic terms consists of exercising the functionality of the device under test, confirming the conditions that are applied, measuring the resulting state of the ECU (and the other components within and outside of the system for a deep system-level focus), comparing the results to a previously generated list of acceptable outcomes, and repeating this process for all functions.

Thus a key part of the process is the application of representative inputs, and the presentation of representative outputs, to the ECU in order to place it in a controlled state that accurately depicts the real-world conditions it will encounter in use. Test engineers usually refer to these inputs and outputs generically as "loads", and their consideration, design and construction is a key part of the process of developing any test system.

It is often tempting to use real sensors and actuators as loads. Following the line of thought that results in the acceptance of this logic, then it is logical to assume that test systems are simple to build and inherently accurate because they use real parts. However, in spite of being "real", such parts can only substitute for one of a range of possible acceptable parts that could be used for the purpose. This is because all parts exhibit some variability. To be cost effective, ECU-based systems are designed to accommodate this variability. In order to test most comprehensively this entire range must be traversed so as to ensure that all components in it, especially the ECU, can accommodate the variability.

A more accurate, and recent, method for presenting a representative environment to an ECU under test involves synthesizing, or simulating, the characteristics of all loads wherever possible. Simulated loads have more flexibility and help test the device under test more thoroughly because they can be programmed to simulate a part whose characteristics fall anywhere in the range of variability the system is designed to accommodate. The added flexibility that results is accompanied by substantial additional complexity in setup and control. For this reason these loads are almost always set up and controlled by a computer.

For this reason, along with several others, all modern test systems feature substantial computer-controlled operation. Computers in testing are used to command loads to a particular configuration and also to command test instruments to conduct measurements. They are also ideal, however, for recording and analyzing the results, and particularly for choreographing all of the above correctly with carefully controlled timing when called for.

Most importantly, while humans can generally accomplish all of these same tasks, computers have excelled at conducting them repeatably, time after time, test after test, part after part. This consistency has become critical to the process of finding problems quickly and easily when parts fail a test, hence saving valuable time and money.

The last key part of a typical test system is the mechanism used to handle the connection of loads, measurement devices and power to the device under test. When an engineer sets up and runs an ad hoc test, all electrical connections required to conduct the specific test are usually made by hand, set up and broken down function by function as the test suite is run through its sequence. When a computer runs an identical test it needs a method for handling the electrical interconnects automatically, since it doesn't have hands with which to make and break connections.

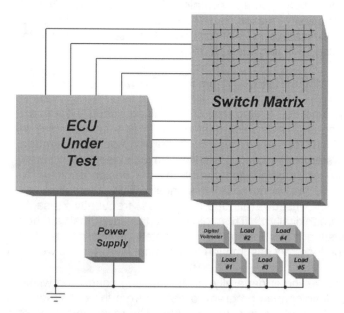

Figure 1 - Typical Switch Matrix and Connections to Device Under Test

In modern test systems this task is usually assigned to an electro-mechanical mechanism called a switch matrix (also known as a cross-point switch). Switch matrices allow fixed and costly loads and measurement assets to be selectively connected, broken down and reconnected to the I/O pins of the device under test by computer control. The term matrix has been applied because the mechanism consists of an array of switching relays, each of which can be activated independently, or with others in concert, to establish the electrical path between the test equipment and the device under test. The matrix, under command of the computer running the test, is responsible for setting up and breaking down every connection needed during the execution of the test suite.

When comparing computer-controlled testing to that conducted by a human, the fact that the computer requires a switch matrix is the only nontrivial difference. While computers themselves are generally inexpensive and ubiquitous, giving them "hands" in the form of switch matrices amounts to additional cost. Historically this cost has generally been accepted because the speed and consistency advantages provided by computer control usually outweigh the cost of the matrix.

TRENDS

Five basic trends are having, and will continue to have, a strong effect on the future of ECU and Systems testing:

The centralized focus lives on – In spite of the observed increase in the use of In-Vehicle Networking, most engineers continue to follow the centralized control model. This is mainly because organizational and technical inertia makes it hard to change. In some cases, however, the resistance to change is more tangible. With Powertrain control systems, for instance, engineers will likely remain focused on centralized control to a great degree because the process of certifying the operation of Powertrain control ECUs per federal and/or state requirements is simpler, hence lest costly.

The expansion of vehicle functionality and the corresponding proliferation of I/O will continue – Feature complexity in vehicles continues to rise at a significant rate. Almost all feature additions/upgrades require electrical circuits to be added above and beyond that already found in a typical vehicle. The addition of extra I/O increases the production cost of the typical ECU. It also has a significant impact on testability.

Use and complexity of simulated loads will grow – Simulated loads are far more powerful than production components conscripted into test service. Some art is necessary in the construction of suitable simulated loads, over and above the basic science, because mimicking the physics of real parts can be tricky, i.e. expensive. However, the ability to produce controlled variability in the components of the system surrounding the device under test produces tests capable of confirming that first-run quality is guaranteed. Avoidance of the cost of post-production scrap and rework is more than offset by the increased cost of the test system.

Cost pressure is increasing relentlessly – Cost sensitivity is rising in general throughout the automotive industry. Testing is not immune. Both fixed (equipment) and variable (personnel) costs will continue to rise over time. Variable costs can usually be addressed by building in more comprehensive automation, but increasing fixed costs can only be kept in check by the use of systems and concepts that simplify the test process.

Distributed Systems are becoming more commonplace – In-Vehicle Networking continues its expansion, as more and more ECUs are finding their way onto networks within vehicles. Specialized subnets are beginning to appear in production, and network complexity continues to grow. Smart sensors and actuators are novelties, as they have been since they were first proposed, but are closer to volume production than ever before as the cost of network interfaces continues to fall.

IMPACT OF IN-VEHICLE NETWORKING

CONTROL NETWORKS, MESSAGING AND SIGNAL TRANSPORT

Automotive networks generally are of the command-and-control type. Unlike home or business local area networking (LAN), or location-to-location wide area networking (WAN), they are not typically used to carry large data files, e-mail messages, or Internet content within the vehicle. Instead, they handle small data transfers called "messages", which are typically used to convey commands from one place in the vehicle to another or to retrieve measurements taken by ECUs or sensors. They have evolved out of a concept called multiplex wiring, which is the basic technique for using a single set of wires to carry multiple signals.

The initial goal of multiplex wiring was the practical elimination of wires, interconnects, and harnesses through their replacement, where possible, by a vehicle data bus. This action altered the process of exchanging data between ECUs, between sensors and ECUs, or between ECUs and actuators. As a result, data exchange no longer consists of activating output pins and sending out digital or analog signals, but has evolved into a process of constructing messages containing representations of those signals and sending them out over the new vehicle network or data bus instead. However, as with networking elsewhere, substitution of a network for discrete wiring brings much more than cost savings on wiring. The data channel provided by the network supports many more opportunities for feature expansion, functional efficiency improvement, and systems cost savings than immediately meet the eye.

Control networks are beginning to make a significant impact on the electrical systems of most vehicles. With the expansion of In-Vehicle Networking a major shift in the topology of the typical electrical system is underway, one that will have a profound impact on the test process, and ultimately shape the cost and complexity of testing.

DISTRIBUTED I/O AND DISTRIBUTED FUNCTIONALITY

The concept of In-Vehicle Networking has evolved substantially since multiplex wiring was first introduced. However, it has generally followed a path that has emphasized a controlled and deliberate rollout of practical applications rather than an unchecked explosion of theoretical capabilities. Three steps in this evolution process have been identified.

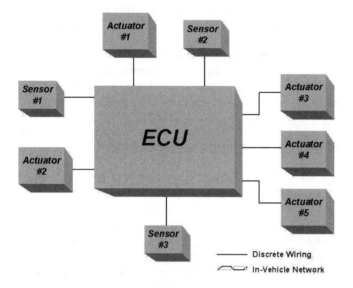

Figure 2 – Typical Centralized Control System

Sending signals over the bus eliminates dedicated I/O in a dramatic way as wires formerly dedicated to carrying them in traditional fashion are removed in favor of using the shared data bus. The bus has the effect of becoming a common pathway between devices, essentially allowing the small number of I/O dedicated to it to be reused over and over again.

The next major step was the introduction of the concept of distributed I/O. After implementing the first multiplex wiring systems engineers determined quickly that vehicle data buses could facilitate, at the system level, much more than simple communications between ECUs and components. They discovered that it was possible to use data buses to fundamentally alter the topology of the vehicle's electrical system, placing inputs physically near the sensors and input devices they connect to, and outputs adjacent to the actuators and output devices they drive.

The process of generating electrical systems with distributed I/O in this fashion is typically called geographical partitioning. I/O are moved from a centralized controller to a location in the vehicle much closer to the loads being queried or driven. This action has the effect of dramatically reducing wire harness length and makes the most sense currently in body and chassis control systems where functionality is exercised frequently across large areas of the vehicle.

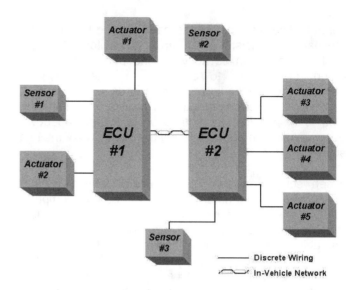

Figure 3 – Control System with Rudimentary Distributed I/O

At this point in time, given the introduction of more sophisticated data buses in vehicles, the concept is transitioning into the last major step in its evolution, Distributed Functionality. Distributed Functionality expands the concepts behind Distributed I/O by partitioning functionality as well as I/O geographically. This step involves widespread streamlining of existing vehicle functionality and the addition of new features and functions by splitting up the responsibility for these functions between ECUs.

Taken to its extreme, this process facilitates the long-sought concept of smart sensors and smart actuators. Both feature traditional sensors and/or actuators fitted with at least some measure of functional control. Rather than placing this control in a separate ECU, smart devices split it up among themselves, reducing system complexity through the elimination of centralized ECUs.

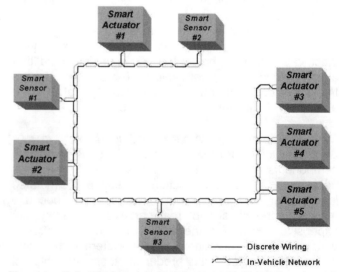

Figure 4 – Fully Distributed Control System with Smart Sensors and Actuators

Distributed I/O is critical to accommodating the increasing numbers of I/O required to support the newest features being added to vehicles, hence the concept of reducing test cost and complexity due to I/O expansion.

Distributed Functionality is critical to reducing ECU complexity, leading to the reduction of test complexity and the elimination of non-value added costs resulting from what is essentially overtesting (100% Inspection).

REDUCTION OF TEST COMPLEXITY AND COST

Since test costs are a significant percentage of the cost of developing and producing electrical systems, the reduction in test costs that accrues from the transition to distributed systems is worth considering in detail. The key to reducing test costs in increasingly complex systems is to exploit the ability of the distribution process to make each of the testable components in the system simpler.

Simpler components are easier to test because they have fewer characteristics to monitor. While the total number of characteristics for a given system will not change substantially in the transition from fully centralized to full distributed, there will be significant reassignment from ECUs to Smart Components, and from discrete I/O to signals embedded in messages sent across the network. These phenomena will result in a reduction of testing complexity and cost overall for two reasons:

1.) Cost of Testing to Support Quality -- The inherent complexity of most ECUs, especially those with high I/O counts, makes it difficult if not impossible to successfully implement SPC in their production processes. This leaves costly 100% inspection as the only means of guaranteeing quality in those processes. By eliminating complex ECUs the simpler components that remain become excellent candidates for SPC and its inherent cost savings, because there are fewer key characteristics to monitor for each one.

2.) Core Cost of Automated Testing -- Systems that use an In-Vehicle Network extensively have fewer traditional I/O, in many cases dramatically fewer. These systems are much less costly to test than those interfacing via many individual wires, because of the opportunity to use smaller, less costly switch matrices instead of larger, more expensive ones. The cost of switch matrices quadruples (approximately) for each input added, because each new input must be switched across all outputs, not just one corresponding new one.

Additional costs not related to testing are also saved, for instance the cost of the complex I/O drivers inside ECUs used to handle the interface to individual wires, but these are outside the realm of testing.

ISSUES

The exploitation of in-vehicle networking for test process improvement and reduction is promising, but faces several issues:

The dramatic expansion of in-vehicle networking that will accompany a shift from fewer, larger ECUs to more smaller ones will require the auto industry to accelerate its process and people improvement efforts. Although most test systems are currently configured to interface with and test network interfaces, most engineers and technicians that work with ECUs are not inherently trained in control networking techniques, hardware, systems or tools, since they typically regard them as simple inputs and/or outputs.

Additionally, in the real world the process of making components smart is not trivial. Suppliers of the existing versions of these components are usually not inherently knowledgeable about the concepts and methods behind the networking technologies necessary to elevate their components to the smart category.

Furthermore, advocates of large, centralized ECUs maintain that the costs associated with adding electronics to components that do not already have them are substantial. They often point to the results of simple thought experiments as evidence that cost savings are not always so clear-cut because they may simply be cost redistributions.

A good example of such an experiment is the observation that the simple addition of a regulated power supply to each component to support the addition of the circuitry that would make it 'smart' is actually a systems cost increase. This situation occurs with this line of thinking because the single power supply found in a large centralized ECU becomes a larger number of power supplies, one in each Smart Component. There are several more of these sorts of analyses that leave the final analysis unclear at the present time. As a result, the widespread introduction of smart components, and the cost savings in systems testing that can accrue from them, is far from certain.

CONCLUSION

Visionaries present during the infancy of multiplex wiring often predicted that smart sensors and actuators, and fully distributed systems would be the norm rather than the exception by the year 2000. In spite of the promising possibilities observed back then, the fully distributed system has not yet reached a level of ubiquity in 2005. Many factors have contributed to a slower pace of adoption than originally forecast, not the least of which is the auto industry's insistence on controlled and deliberate innovation.

This does not mean, however, that increased distribution has reached a peak that cannot be surmounted if a good reason is put forth for accelerating the transition. Tradeoffs are made constantly in the development of all products, including automobiles. Almost all are based on cost, or if not, at least with cost heavily considered. The cost of testing is a substantial portion of the cost of manufacturing a product, especially if 100% Inspection is employed. Reductions in the cost of testing through increased distribution have the potential to drive the change necessary for the sweeping adoption of distributed systems across the board, and the reaping of the test benefits that they bring.

ACKNOWLEDGMENTS

The author would like to thank and acknowledge Bruce Emaus, Vector CANtech's resident expert on the process of distributing functionality over In-Vehicle Networks.

REFERENCES

1. SAE Paper No. 2004-01-1725, "Hardware Synchronization Techniques of Analog, Digital, and CAN Signals for Device Validation"
2. SAE Paper No. 2004-01-1724, "Testing Networked ECUs in a Virtual Car Environment"
3. SAE Paper No. 2001-01-0061, "Automated Test System for Electronic Control Units and Prototypes of Them in Vehicle Networks"

CONTACT

For more information or to contact the authors, please email info@vector-cantech.com.

DEFINITIONS, ACRONYMS, ABBREVIATIONS

Characterization: The process of measuring, recording, and analyzing the characteristics of an ECU or System in a detailed fashion.

Distributed Functionality: Control functionality that is split between the distributed components of a system, for instance between Smart Sensor and ECU or between two ECUs.

Distributed I/O: I/O purposely disbursed from a central location, in a single ECU.

DV: Design Verification; A test regime used to characterize and evaluate an ECU's or system's conformance to specification. Ensures that a product performs as designed. Tests the design by validating the parts made from it.

ECU-Focused: Having the characteristic that it considers only the ECU and not the components to which it is connected or responsible for the control of.

ECU Validation: The process of confirming that an ECU's design conforms to specification.

EOL: End-of-Line test; A test regime employed after the manufacturing process to confirm that parts are good. Typically a stopgap measure to ensure that zero bad parts are shipped. Does not eliminate scrap and rework because it does not prevent bad parts from being manufactured.

Functional Partitioning: The process of splitting the logic or programming that controls a function, assigning portions of control responsibility to multiple ECUs or smart components.

Geographical Functional Partitioning: Functional partitioning executed specifically in order to place inputs and outputs as near as possible on the vehicle to the sensors or actuators they are to connect to. Typically inputs are assigned to an ECU physically near the sensors connected to them, and outputs to one physically near their respective actuators. Signals from/to these I/O are carried by the in-vehicle network between their corresponding ECUs. Control responsibility can be in either ECU, or both, or can reside in the sensor or actuator itself if it is 'smart'.

I/O: Input/Output; Each member of the collection of inputs and outputs assigned to an ECU or system.

I/O count: The number of I/O. Typically used in describing or analyzing the complexity of a function or ECU by indicating the number of I/O required to execute a function or within an ECU.

Loads: The inputs and outputs connected to a device under test. Loads represent their production counterparts during the test process. Can be 'real', meaning an actual production part selected as a best-choice stand in, and capable of representing only one of the entire set of production parts; or 'simulated' (synthesized), meaning created to represent any and all actual parts as closely as possible.

Multiplex Wiring: The early term for In-Vehicle Networking. The word 'Multiplex' comes from an early technique for sending multiple signals over a single wire or data channel, usually by requiring each to be sent in sequence during a dedicated and well-defined period of time assigned to it.

Parametrics: Detailed measurements that are taken and used in the process of characterizing the device under test. Employed for more detailed analysis than that conducted during simple pass/fail screening of parts.

Plant: A term sometimes used to define the entirety of a system less its ECU. Everything in a system, connected to operate as designed, except the ECU. Taken from the term 'Physical Plant' as originally applied in industrial engineering to machinery (plant and equipment) without its controls.

Plant Modeler: One type of generic tool/tester for creating and deploying the collection of all simulated loads required to represent the system that is connected to an ECU in its intended application. For use in testing/evaluating the ECU in its system context.

PV: Process Validation; A test regime used to characterize and evaluate a manufacturing process's conformance to specification. Ensures that the process is capable of producing identical parts consistently, with all conforming to specification. Tests the manufacturing process by measuring the consistency of the parts produced.

Smart Actuator: An actuator fitted with a connection to the In-Vehicle Network, and a level of logic or programming, in order to operate autonomously as an actuation and control unit. Control can be totally within but is most often shared over the network with an ECU, or with a corresponding smart sensor for optimum cost efficiency.

Smart Components: Sensors or actuators fitted with connections to the In-Vehicle Network, and a level of logic or programming that enables them to provide some measure of control autonomously.

Smart Sensor: A sensor fitted with a connection to the In-Vehicle Network, and a level of logic or programming, in order to operate autonomously as a measurement and control unit. Control can be totally within but is most often shared over the network with an ECU, or with a corresponding smart actuator for optimum cost efficiency.

SPC: Statistical Process Control; a manufacturing process control technique offering guaranteed first-run quality for all parts made by the process by monitoring and controlling the key characteristics of those parts. Eliminates scrap via the ability to detect that the process is deviating from the norm without allowing any bad parts to be made. Allows corrective action to be taken before the process goes out of control.

Switch Matrix: An organized collection of relays (electrically controlled switches) configured in an array with rows and columns, and connected so that inputs along one axis of the matrix can be independently and selectively connected to outputs along the other. It allows limited or expensive test equipment to be switched, usually under computer-control, from pin to pin of the device under test while it is being tested. Also known as a Cross-Point Switch.

System: Generically, an ECU and its sensors and actuators taken, considered and analyzed together.

Systems-Focused: Having the characteristic that it considers both the ECU and the components to which it is connected or for which it has control responsibility.

Systems Validation: The process of confirming that a system's design conforms to specification.

2005-01-1484

A Multi-hop Mobile Networking Test-bed for Telematics

Rahul Mangharam, Jacob J. Meyers, Ragunathan Rajkumar and Daniel D. Stancil
Department of Electrical & Computer Engineering
Carnegie Mellon University

Jayendra S. Parikh, Hariharan Krishnan and Christopher Kellum
General Motors Corporation

ABSTRACT

An onboard vehicle-to-vehicle multi-hop wireless networking system has been developed to test the real-world performance of telematics applications. The system targets emergency and safety messaging, traffic updates, audio/video streaming and commercial announcements. The test-bed includes a Differential GPS receiver, an IEEE 802.11a radio card modified to emulate the DSRC standard, a 1xRTT cellular-data connection, an onboard computer and audio-visual equipment. Vehicles exchange data directly or via intermediate vehicles using a multi-hop routing protocol. The focus of the test-bed is to (a) evaluate the feasibility of high-speed inter-vehicular networking, (b) characterize 5.8GHz signal propagation within a dynamic mobile ad hoc environment, and (c) develop routing protocols for highly mobile networks. The test-bed has been deployed across five vehicles and tested over 400 miles on the road.

INTRODUCTION

The commoditization of high-speed wireless interfaces and low-cost Global Positioning System (GPS) devices provides the opportunity to deploy a range of useful and practical inter-vehicular communication applications. A test-bed has been developed primarily to evaluate the feasibility of high-speed networking between vehicles for emergency and safety notification and to support multimedia telematics applications. The focus is to develop and deploy a multi-hop wireless routing platform to facilitate on-road testing. We observe significant divergence in the performance of multi-hop wireless routing protocols between simulation and implementation and therefore stress the importance of on-road testing. In addition, channel measurements and propagation analysis have been carried out to further understand the nature of the communication environment. This paper discusses the platform deployment process and shares our driving experiences. Based on the network performance analysis and insights from the test-bed, a vehicular networking architecture is being developed across multiple networking layers targeted specifically for inter-vehicular communication.

Consider the case when a vehicle accident on a freeway blocks two of the three available lanes. Most freeways rely on other drivers to notify the highway patrol, which in turn updates the freeway management system and messages are eventually broadcast on electronic signs along the highway. While the turnaround duration for accident notification may take considerable time, the traffic buildup is rapid and consequent blocking of emergency vehicles is still an unsolved and pressing problem.

Now consider the response when a subset of the vehicles has onboard inter-vehicular networking capability. From the accident site, alert messages are broadcasted and directed to all approaching vehicles. As the range of each wireless interface extends to less than a few hundred meters, it is essential to employ a multi-hop routing protocol to communicate 2-3 miles down the freeway. By using this mechanism to broadcast the event and its position information, approaching vehicles are notified of the hazard and possible congestion. As the broadcast message ripples through the traffic at high speed, approaching vehicles can use the next exit to plan a detour.

While traffic and alert notification systems such as OnStar [1] have been deployed, they are centrally managed entities. In the above example, we observe that the problem is local and requires action within the vicinity of the accident. Attempting a cellular call (used by OnStar) to each of the hundreds of vehicles in a particular region is not cost effective. Furthermore, the alert notifications are relevant only during the event lifetime and should be targeted specifically to approaching traffic.

In the following section, we discuss the range of vehicular applications and their unique networking challenges. We then describe the multi-hop mobile vehicular test-bed, our design decisions and driving experiences.

MULIT-HOP VEHICULAR NETWORKING

The primary goal of the multi-hop mobile test-bed is to provide insights towards the design of networking infrastructure necessary to support a range of inter-vehicular applications.

APPLICATION CATEGORIES

We focus on four application categories that encompass a broad range of possible services.

Emergency and Safety Messaging

Messages reporting vehicle accidents, sudden breaking, oil spills and other critical events need to be disseminated instantly and for the duration of the event's lifetime. As shown in Fig. 1, such messages are broadcast but within the scope of relevant vehicles approaching the event's position. Emergency messages are pertinent within the relative region of the event. This may be determined by restricting the message to be accepted and forwarded by only those nodes within a geographic region. In addition, navigation information such as vehicle direction and planned route may be leveraged. As the messages are critical, they must be delivered with the highest packet priority.

Traffic and Congestion Updates

These are semi-critical messages with updates on road conditions and congestion information. They may be delivered by fixed infrastructure-based transceivers or via mobile gateway vehicles equipped with both inter-vehicula wireless interfaces and cellular-data connections. Traffic and congestion updates are pertinent in the general scope of a region. A region may be described by a set of GPS coordinates. For example, a circular region may be described the GPS coordinate marking its center and the radius in meters. All vehicles within this region will accept and forward updates. Vehicles outside the designated region will drop any region-specific updates received. Updates will be delivered as broadcast packets.

Multimedia Telematics Applications

Feature-rich and interactive applications such as inter-vehicular voice communication, video streaming, file transfer and collaborative driving applications fall within the scope of telematics applications. Using such applications, users can form or join public and private groups of vehicles. For example, a group of friends driving on a day trip will be continuously informed of the position of their friends' cars and can maintain a voice and chat channel to communicate collectively to the group. Such applications may be scaled to the enterprise level for fleet management or connect all vehicles heading to a particular destination. Telematics applications require a robust network and transport layer where connections are reliable and may be suspended and resumed smoothly based on the connectivity between interested parties. Furthermore, as all vehicles are connected across one or more hops, it is essential that the routes selected satisfy minimum admission control requirements and are self-healing.

Commercial Announcements

Travelers may choose to subscribe to commercial announcements regarding parking lots with empty slots, regional boarding and lodging information and other travel-related advertisements. Such applications require a subscription service and also a lightweight acknowledgement scheme to estimate the size of the targeted population.

In order to analyze the performance of the above application categories, we evaluated an application from each category over our test-bed. The next section describes the test-bed design followed by the unique challenges vehicular networks pose over traditional topology-based ad hoc networks.

TEST-BED DESIGN AND DEPLOYMENT

TEST-BED HARDWARE

The multi-hop wireless vehicular networking test-bed developed at Carnegie Mellon University (Fig. 2) employs a Differential GPS (DGPS) receiver with a magnetically mounted antenna. An onboard computer with a modified mini-PCI IEEE 802.11a wireless [2] interface forms the main entity of the setup. The physical layer has been modified to emulate the Dedicated Short Range Communication (DSRC) [3] standard specifications with a

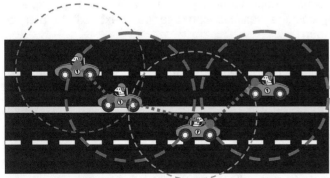

Figure 1. Vehicle-to-vehicle multi-hop networking showing each vehicle's wireless range

Figure 2. Portable multi-hop mobile test-bed with DGPS receiver and onboard computer

10MHz signal bandwidth and operates at a variable carrier frequency which includes the 5.85 – 5.925 GHz spectrum.

Vehicles communicate with each other via a magnetic-mounted wireless antenna. The range of the wireless link is approx. 300m for line-of-sight reception. To facilitate multimedia applications, the test-bed includes a voice headset and a camera. All devices are powered by the vehicle's DC power system via the cigarette lighter, utilizing DC-DC power converters as needed. The equipment fits neatly in a plastic molded case and is easy to carry and quick to set up.

TEST-BED SOFTWARE

All onboard computers run RedHat 9 Linux (kernel 2.4.18-3) as this provides a fertile platform for network protocol and application development. We primarily use three layers of software on the test-kit. The software is built from open source libraries and is available for free.

Mapping and Communication Software

In order to visualize the current position of the vehicle, we adapted the RoadMap tool [4] for our test-bed. We added runtime display capability for multiple vehicles so the movement of each vehicle can be tracked as we drive. We added communication capability so that each vehicle's

onboard computer acts as a server and accepts connections from other vehicles. Each vehicle runs a User Datagram Protocol (UDP) client thread to connect to all other machines in the test-bed. As the connections are at the socket level, the application manages the end-to-end data exchange between the client and server. The underlying kernel-based networking software handles multi-hop routing along the set of links between the client and server. Using this client-server setup, each vehicle exchanges its GPS information (position, speed, direction, etc.) and its network information (packet sequence number, fragmentation, frame length, etc). We implemented our own packet headers for efficient exchange of position and network information. GPS coordinates are computed five times every second and have an accuracy of ≤ 2m. Consequently packets are exchanged five times every second.

All GPS and network information exchanged between vehicles is logged by each machine. This enables us to playback the route driven and visualize the vehicles on a vector-based rendering of the map traversed. The maps use TIGER/Line 2002 data files available for free from the U.S. Census Bureau [5]. In Figure 3, we observe the playback of a trip with five vehicles (top right) illustrating the vehicles connected, a panel to send emergency messages, a playback control (bottom left) to speedup or

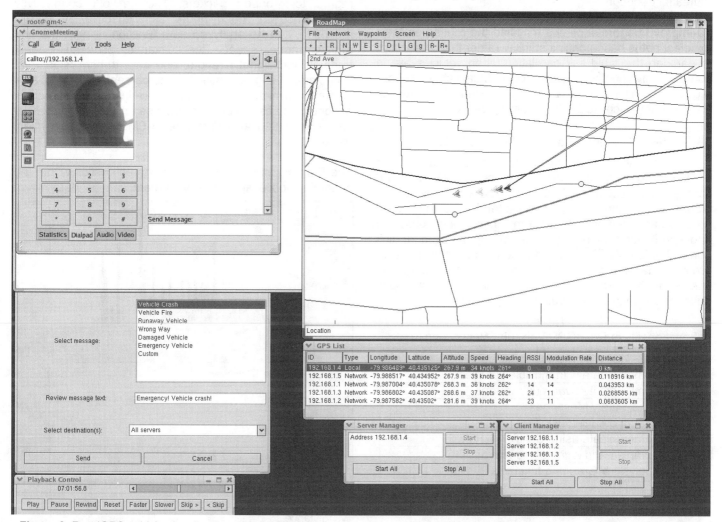

Figure 3. RoadGPS vehicle visualization with GPS tracking, network connectivity and audio/video/messaging communication tool. A playback control is provided for viewing logged trips.

slow down the playback, client and server connections and a multimedia application with voice and video (top left)

Table I lists the current functionality of our mapping and communication application – RoadGPS. All position and networking information within the multi-hop network on the road can also be channeled to the Internet via the 1xRTT cellular connection. This way, our team at the GM Technical Center in Warren, MI can monitor all of the vehicles and the data transmitted between them as they drive in Pittsburgh, PA or anywhere else with cellular network access. The ability to monitor the network in real-time assures that data is correctly being logged and allows real-time network performance and signal propagation analysis.

Multi-hop Networking Software

We tested several existing ad hoc routing protocols such as DSR [6] and AODV [7] kernel implementations. While the protocols provide connectivity across multiple hops, their performance in a highly mobile environment was unreliable. Most ad hoc routing protocols have been tested primarily through simulation or in small test-beds (< 10 nodes) with low mobility. In a highly mobile environment such as inter-vehicular communications where connectivity changes often, these protocols were unable to reconstruct routes fast enough.

To improve performance, we modified the routing protocol to select paths based on link stability rather than hop-count to dampen the rapid link oscillation when vehicles are in contact only briefly. Several conceptual changes are required to tailor multi-hop routing for vehicular networks and they are discussed in the following sections.

Multimedia Conferencing and Application Software

Audio and video streaming was carried out using conferencing software based on the H.323 standard libraries. The reception of the audio and video codec was clear but suffered from an extended delay when reconnecting temporarily broken links. New buffering implementations to address the frequent but brief link connection are necessary. In addition, a new session layer protocol that can suspend an open connection when the link is briefly disconnected and resume the flow upon subsequent connection would be very useful.

Table I
Functionality of RoadGPS Vehicular Networking Tool

1	Map & Display multiple vehicles
2	Communicate with multiple vehicles
3	Communicate over multiple hops
4	Send Safety Messages
5	Stream data & music files
6	Communicate using Cellular 1xRTT
7	Complete trip logging and playback
8	Analysis & Graphing functions

DRIVING EXPERIENCES & ANALYSIS

The multi-hop networking test-bed has been deployed across five vehicles provided by GM. The group of vehicles has been driven over 400 miles each. We chose four environments: urban (densely populated and crowded with several high-rise buildings such as downtown Pittsburgh), rural (flat with fairly open roads), highway (high speed driving along I-79) and city driving through Pittsburgh. In general we were able to maintain connectivity across all vehicles over 1.2km using multi-hop routing. This shows the basic usefulness of multi-hop networking as the range of each vehicle's wireless connection is limited to 300m under good conditions. The terrain of the city of Pittsburgh is quite hilly and offered us several non-line-of-sight opportunities where multi-hop routing proved to be very useful in maintaining connectivity.

We have developed a data analysis toolkit in MATLAB to analyze connectivity, signal-to-noise ratio, error rates, and data rates over different distances, speeds and link transmission rates. A comparison of logged packets transmitted from each local node with packets received by each of the other nodes in the network, combined with the GPS data from both nodes, allows a detailed statistical analysis of the network's connectivity performance with regard to dropped packets. The network connectivity performance for each node, as measured by packet reception, can be analyzed as a function of absolute and relative node speed, the distance between nodes, transmission signal strength, and data transmission rate. For example, Fig. 4 shows that the percentage of total sent packets which are dropped increases roughly exponentially as the distance between nodes increases for data transmitted at a data rate of 6MBPS, as one might expect.

The 5.8 GHz signal propagation characteristics within the dynamic mobile ad hoc environment can be determined

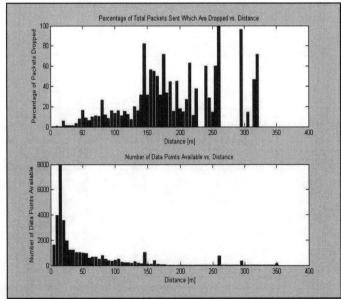

Figure 4. Percentage of total packets sent which are dropped vs. distance

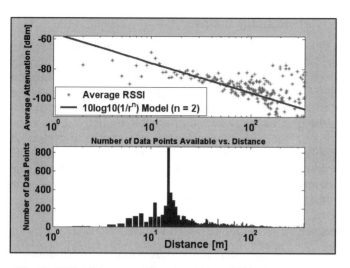

Figure 5. Total Average Measured Signal Attenuation and Free Space Propagation Model vs. Distance

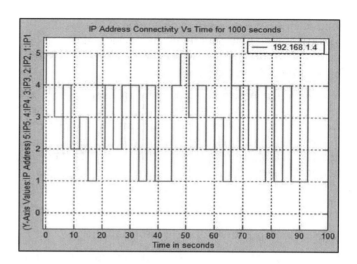

Figure 6. Graphical representation of a vehicle's routing table over time

using calibrated measurements of received and transmitted signal strengths in conjunction with the GPS data. The signal attenuation between each node can also be determined and modeled as a function of absolute and relative node speed, the distance between nodes, transmission signal strength, and data transmission rate. For example, the initial data shown in Fig. 5 suggests that signal attenuation can be roughly modeled using the free space propagation model

$$P_s \cong \frac{1}{r^n} + Offset$$

where r = distance between nodes, and n ≈ 2, as one might expect.

Fig. 6 provides a graphical representation of the routing table over time for source node with IP address 192.168.1.4 (referred to as IP4 here) attempting to communicate with destination node IP2 across one or more intermediate nodes. The y-axis plots the connectivity to nodes with IP addresses IP1, IP2, IP3 and IP5 with respect to IP4. A connection to IP4 is represented by a line passing through 4 on the y-axis. We observe that initially, IP4 is connected only to IP5 and sends the packet to IP5. IP5 forwards the packet to IP3 which in turn forwards it to IP2. This is the situation when the cars were parked in the order of their IP addresses and were in range of only their immediate neighbors. However, at a later time, we see considerable activity as cars overtake each other, or are separated by traffic lights or when direct connection is obstructed by passing large transportation carriers.

Our results are based on a modified implementation of AODV routing protocol. In the following section we outline several aspects where a new class of routing protocols needs to be developed to address the different aspects of vehicular networks.

UNIQUE CHALLENGES FOR VEHICULAR NETWORKS

Several designs for infrastructure-less "ad hoc" wireless networks have been proposed over the past decade [6].

They predominantly belong to the class of Topology-based protocols where a source node, s, requires a communication path with destination, d. As the range of the wireless interfaces is limited, the communication path will generally traverse multiple intermediate nodes, each of which relay messages. Such protocols assume every node is described by a logical (named) address (e.g IP address) and require a full search of the network to find the destination node. This search procedure (or link flooding) generally does not leverage any topological structure of the network and therefore floods the network with broadcast requests to find the destination node. Examples of such protocols are: Dynamic Source Routing (DSR), Ad hoc On-Demand Distance Vector Routing (AODV) and Optimized Link State Routing Protocol (OLSR) [6].

Are Topology-based Routing Protocols suitable for Vehicular Networks?

Several Topology-based ad hoc routing protocols have been shown to work for small and lightly mobile networks. We outline several factors in the routing protocol requirements for vehicular protocols that are not addressed by existing protocols as the challenges and applications addressed are largely different.

1. High Mobility

Ad hoc networking protocols have traditionally been simulated with average speeds less than 30kmph [6] and have shown to perform poorly under higher mobility conditions. In contrast, a vehicular network is by definition always mobile with automobiles moving in excess of 140kmph and reaching relative speeds over 300kmph. This **10x** difference of node speed requires rewriting the protocol to not be limited by (a) handshaking (e.g. route-request/route-reply), (b) formation of end-to-end connection oriented communication along fixed intermediate nodes (e.g. source routing), and (c) rebroadcasts of end-to-end route information (e.g. route-broken alerts). Furthermore, there is a need to test the correctness and performance of routing protocols through implementation rather than simulation. On-road implementation and testing bring about anomalies such as obstructions (buildings), highly dense networks (city driving), sparse networks (highway driving), etc.

2. Broadcast over Unicast

In a vehicular network, emergency messages and traffic updates are relevant to all nodes in the vicinity of the event. There is a need to communicate with nodes based on their **relative position**, destination, speed, direction, etc. The addressing scheme of the routing protocol must therefore take into account packet broadcast as the common way to communicate than using unicast messaging.

3. Geocasting with Topology Information:

As vehicle paths are constrained by the network of roads, connectivity and flooding will benefit from leveraging knowledge of the **traffic topology**. Most Topology-based protocols do not assume any known network structure and flood the entire network. This severely limits their scalability to less than 30 nodes. Furthermore, as messages have a geographic relevance, it is essential to use node position as a key network construct.

4. Intersection-Intersection Routing Vs. Vehicle-to-Vehicle Routing:

In order to exploit geographic information, several greedy location-aware routing protocols have been proposed. These however fall short when a road bends and the path to the destination is not as the crow flies. Greedy location-aware routing protocols attempt to reach an intermediate node only if it is geographically closer to the destination, and therefore the packet delivery rate is severely limited by their use of local information. On the other hand, using limited navigation information, packets can be routed from intersection-to-intersection rather than from vehicle-to-vehicle. This eliminates the need to keep track of the addresses of intermediate hops or to maintain a destination-based routing table. Topology-based protocols do not exploit any such information and are based solely on logical addressing. Using intersection-based routing, vehicles are used opportunistically if they lie along the region of interest. Furthermore, by fixing routing to static intermediate points, vehicles moving outside the area of interest will not adversely affect the path.

5. Message Lifetime:

Topology-based protocols aim to eliminate duplicate messaging and stop sending a message once an acknowledgement is received. In a vehicular network, if an incident occurs at a particular location, there is a constant need to update all vehicles approaching the incident site. It is therefore necessary to continue broadcasting the message to all new nodes until the incident lifetime has expired. Furthermore, as the network connectivity may be sparse, it is useful to hold the message until a new neighbor is detected. Topology-based protocols employ a "hot-potato" approach to routing where once a packet is forwarded it is forgotten. Message lifetime support is crucial to maintaining alert status and also increasing message delivery rate in sparsely connected networks.

6. Dirty Routing Tables:

With vehicles traveling over 140kmph, a node 1km away will be within 5m of standing traffic in less than 25 seconds. Such quick changes in the network neighborhood will result in drastic changes in routing table information. For example, a vehicle leaving downtown and entering a highway will corrupt the routing tables of all highway nodes in its vicinity with information of nodes just traveling in downtown. There is a strong need to maintain location-based routing information and ensure its freshness. Topology-based routing protocols assume routing updates occur at a much slower pace.

ADDRESSING FOR VEHICULAR NETWORKS

We now focus on the addressing requirements for vehicular networks. In addition to sending messages targeted to a particular vehicle, there are several instances where a message is targeted to a group of vehicles that meet a particular criterion such as current position within region of interest, speed, direction, etc. Three addressing schemes have been identified to support named, geographic and node property-based addressing. The following addressing schemes will be incrementally deployed in the test-bed.

A. Named/Assigned Addresses:
For unicast messaging and to identify vehicle-type groups, Assigned Addresses provide a single vehicle logical naming such as IP address, VIN, MAC address, etc. An example is: "Message for Buick FJF2323"

B. Geographic Addresses:
For messages focused on a particular geographic region, vehicles may choose to accept, forward or drop packets based on their geographic attribute.
1. **Absolute Geography**
- **Direction:** "Message for - All neighbors driving east on I-90"
- **Navigation:** "Message for - All cars headed for Exit 22"
- **Region:** "Message for Downtown Pittsburgh"

2. **Relative Geography**
- **Hop count:** "Message for - All cars within 6 hops radius"
- **Speed:** "Message for - All cars moving within +/- 8mph of my speed"

C. Property-based Addresses:
For messages focused on an attribute of a vehicle:
1. **Connectivity:** "Message for any mobile gateway"
2. **Vehicle Type:** "Message for Trucks only, or GM vehicles only, or 2-wheelers
3. **Vehicle Dynamics:** e.g. Absolute Speed: "Message for vehicles moving at 40kmph +/-5kmph"

We observe that the vehicular network routing protocol will need to support all three addressing schemes and ensure their coexistence. Therefore, based on the application, an appropriate addressing scheme will utilize broadcast, multicast groups and unicast for a well-defined subset of vehicles.

CONCLUSION

We describe the design and deployment of an experimental multi-hop mobile test-bed for vehicular networks. The goal of the test-bed is to get insights into the vagaries of the wireless channel and routing protocols to deliver both mission-critical emergency messages and interactive telematics applications. Our current deployment includes a DGPS receiver, an onboard computer, a modified 802.11a radio card, a 1xRTT cellular-data connection and multimedia peripherals. Our experiences from driving the test-bed over 400 miles indicate that traditional applications and network infrastructure are usable. However, to meet the demands for a scalable and stable vehicular network both routing protocols and transport protocols need to be developed to address the dynamism of a highly mobile and lightly connected network of vehicles. We provide a list of technical requirements for the design of routing protocols and highlight the unique addressing capabilities necessary. As future work, we aim to develop and deploy the ideas presented in this paper.

ACKNOWLEDGMENTS

The authors would like to thank Daniel Weller and Suchit Mishra for their valuable inputs.

REFERENCES

1. OnStar Safety System. http://www.onstar.com
2. IEEE 802.11a, Part 11: High-speed Physical Layer in the 5 GHz Band, Supplement to IEEE 802.11 Standard, September 1999.
3. Dedicated Short Range Communications. http://www.leearmstrong.com/DSRC
4. Roadmap. http://sourceforge.net/projects/roadmap/
5. Census 2000 TIGER/Line Data. http://www.esri.com/data/download/census2000_tigerline/
6. J. Broch, et al. A performance comparison of multi-hop wireless ad hoc network routing protocols. In Proc. Mobicom, pages 85--97, 1998
7. Charles E. Perkins, Elizabeth M. Belding-Royer, and Samir Das. "Ad Hoc On Demand Distance Vector (AODV) Routing." IETF RFC 3561

Integrated Test Platforms: Taking Advantage of Advances in Computer Hardware and Software

Mark D. Robison
Uson L.P.

ABSTRACT

Ongoing hardware, software, and networking advances in low-cost, general-purpose computing platforms have opened the door for powerful, highly usable, integrated test platforms for demanding industrial applications. With a focus on the automotive industry, this paper reviews the pros and cons of integrated test platforms versus single-purpose and stand-alone testers. Potential improvements in in-process testing are discussed along with techniques for effectively using such testing to improve daily production quality, to maintain high production rates, to avoid unplanned downtime, and to facilitate process and product improvements and refinements through the use of monitoring, data collection, and analysis tools.

INTRODUCTION

Manufacturing test platforms come in many shapes and sizes, as does the concept of integration. Integration can be in the form of combining different test technologies onto a single platform in order to improve production rates, or it can be in the form of sharing or multi-tasking advanced test controllers in order to minimize capital investment. Alternatively, effective integration can exist only at the conceptual level by integrating only the test data, resulting in a greater understanding of the total picture. This paper touches on each of these in turn.

IN-PROCESS TESTING

In-process testing is key to product quality and consistency. The standard focus of in-process testing is accuracy and repeatability, which is key to maximizing product quality while minimizing costs. Often overlooked though, is the opportunity for an organization to extract the maximum value from its test data for the benefit of both the manufacturing process and the product design. Cost effective improvements that result in more accurate test data, more efficient collection of test data, or better use of test data can be leveraged to create a competitive advantage and to improve profits. Several enabling technologies now exist to do just that.

ENABLING TECHNOLOGIES FOR IMPROVED TESTING

Technological advances in personal and office computing continue to create new opportunities for extending the capability and value of in-process testing. These advances include:

- Low cost, high speed computing platforms
- Low cost, high speed data networks
- Low cost, ultra high capacity storage devices
- Advanced, easy to use operating systems
- Software advances that facilitate the rapid creation of solid products that simplify even the most complex of tests.

The resulting improvement opportunities exist at four levels:

1. Individual tests
2. Individual test stations
3. The complete production line
4. Multiple production lines; remote or local

TEST LEVEL IMPROVEMENT OPPORTUNITIES

Individual tests can be improved by improving the test method, the test implementation, or by some combination of these. Improved test methods sometimes result from revolutionary scientific advances, but, more often than not, improvements are evolutionary in nature. Advanced leak testing[1] is one example of the later. By using high-resolution sensors, increased data collection, and incrementally smarter algorithms, test time can be reduced while simultaneously increasing test accuracy. For some applications, reduced test time can mean the elimination of an entire test station, for example, two stations to perform a given test instead of three stations. Table 1 summarizes test accuracy and test duration improvements achieved for several applications when advanced leak testing technology has been applied.

Table 1: Classical Leak Detection versus Advanced Leak Detection							
Part	Component	Leak Test Method	Trials	% Time Reduction	% Tests Within		
					± 2.5%	± 5%	± 10%
Engine Model 1	Cylinder Head	Classical	400	Baseline	N.A.	N.A.	89
		ALD	400	3%	N.A.	N.A.	99
Engine Model 2	Oil Cavity	Classical	100	Baseline	50	52	99
		ALD	100	40%	90	99	100
Non-engine	N.A.	Classical	60	Baseline	N.A.	N.A.	N.A.
		ALD	60	27%	60	90	100

Implementation improvements facilitated by the aforementioned enabling technologies include:

- Clear, intuitive user interfaces to minimize training costs and reduce operator error
- Test-to-test consistency using presentation standards, again to minimize training costs and reduce operator error
- Built-in diagnostics and troubleshooting tools for rapid detection and resolution of problems
- Advanced logging and data export options that enable process and product improvements through both real-time and after-the-fact data analysis.

STATION LEVEL IMPROVEMENT OPPORTUNITIES

Very significant, on-going savings can be achieved by doing more at a particular test station. Such station-level integration, where multiple tests are performed either concurrently or nearly concurrently by overlapping portions of different tests, have the potential for higher production rates through reduced overall test time and reduced inter-station transfer time. Direct savings also result from fewer test stations and reduced real estate requirements.

Disadvantages of station-level integration include increased test station complexity, which is primarily a development concern, and increased dependency upon a single station, which can be addressed by an adequate spares policy.

Example: Station-level integration

One example of station-level integration is to combine leak testing of the water jacket and oil cavity with green-engine compression testing.

During the machining of the engine components, cylinder block, and cylinder heads, they are leak tested to assure there are no leaks in the water cavities or in the oil cavities.

As the engine is assembled, the engine must be tested to verify the assembly of the seals, gaskets, and plugs.

The main areas to be tested are the water cavity, the oil cavity, and the compression of the power stroke of the engine. These tests can all be performed in one test station using state-of-the-art multi-channel testers, such as the Uson Vector.

The test process first rotates the engine to establish the initial conditions for the compression test. Transducers at each spark plug opening measure the pressure rise due to the compression of the cylinder while torque and position sensors monitor the crankshaft. Pre-programmed limits for each cylinder as well as a detailed master "signature" of all sensor signals are compared against results for the part under test as each cylinder experiences its compression cycle, yielding quick test results. A failed compression test can abort the test sequence or continue with the leak tests in order to gather more information about the nature of the defect.

The compression test is exited with the crankshaft properly positioned for the subsequent leak tests of the water cavity and the oil cavity. Portions of these tests can be overlapped to minimize test time.

Because the oil cavity is the larger of the two cavities, the tester fills the oil cavity first. During this step the tester monitors the water cavity for a pressure increase in order to detect cross-wall leakage between the two cavities. Following this step, normal leak tests are performed concurrently on each cavity. Figure 1 illustrates the timeline for the test steps.

LINE LEVEL IMPROVEMENT OPPORTUNITIES

Test controllers, running gigahertz-class processors, are capable of supporting multiple test stations concurrently. This approach diverges significantly from the current business model used by large manufacturers, but deserves consideration, as the potential exists for reducing initial deployment costs by tens of thousands of dollars. For example, a ten-channel test platform could support ten distinct tests. These tests could be distributed between one to ten different test stations.

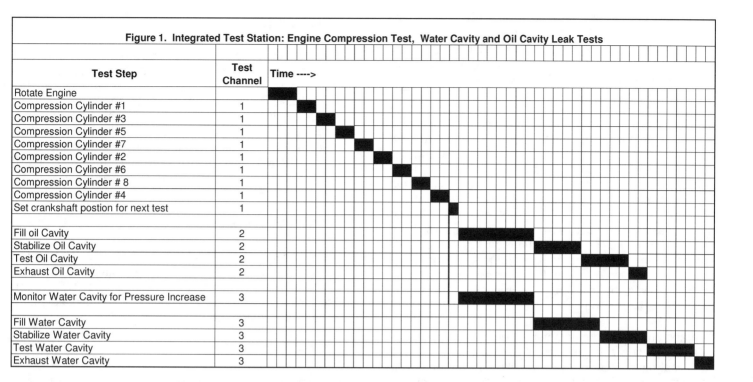

Figure 1. Integrated Test Station: Engine Compression Test, Water Cavity and Oil Cavity Leak Tests

Test Step	Test Channel	Time ---->
Rotate Engine		
Compression Cylinder #1	1	
Compression Cylinder #3	1	
Compression Cylinder #5	1	
Compression Cylinder #7	1	
Compression Cylinder #2	1	
Compression Cylinder #6	1	
Compression Cylinder # 8	1	
Compression Cylinder #4	1	
Set crankshaft postion for next test	1	
Fill oil Cavity	2	
Stabilize Oil Cavity	2	
Test Oil Cavity	2	
Exhaust Oil Cavity	2	
Monitor Water Cavity for Pressure Increase	3	
Fill Water Cavity	3	
Stabilize Water Cavity	3	
Test Water Cavity	3	
Exhaust Water Cavity	3	

The economic benefit of replacing ten $20,000 testers with a single $150,000 tester is obvious.

The introduction of a single point of failure for the affected test stations merits concern, but an adequate spares policy can mitigate this risk. Because of the fewer number of components, the actual mean time between failures for the system as a whole is reduced.

A second form of integration at the production line level is data integration: the creation of an integrated, global view of the test results and associated test data. From such a perspective, subtle trends can be detected that would otherwise go unnoticed, enabling emerging problems to be preemptively identified and corrected.

Data integration has only modest costs associated with it, yet it can result in tremendous savings both short term and long term by minimizing down time and facilitating product changes for improved manufacturability.

RESULTS MONITORING AND ANALYSIS

Regardless of the level of integration, monitoring and analyzing test results can yield significant gains. Benefits arise from monitoring test results real time as well as from after-the-fact analysis of data collected over weeks, months, or even years.

REAL-TIME MONITORING

A well-implemented and utilized information system can help improve production quality by enabling the timely detection of process level problems, and having a clear indication of the source of process problems can minimize the time needed to correct a problem. By proactively monitoring quality-impacting trends as they happen, higher production rates can be achieved by correcting emerging problems before they result in product rejects.

Example: Early Detection Of Trends

Assume that for a particular test, a part is completely acceptable if its test results are below 100. However, by design, and confirmed by historically collected data, good parts, on average, pass the test with a value of 80. Furthermore, the tests are repeatable to within ±10% of the average value. By continuously monitoring this average over a period of hours, days or weeks, a monitoring system can detect emerging problems, such as excessive machine wear. When the running average reaches a configured threshold, say 90 for this example, the monitoring system automatically notifies a supervisor or maintenance personnel via email or other mechanism, supplying sufficient details to enable investigation of the situation. With such early detection, corrective action can be taken before *any* parts are rejected. All of the parts may still be passing the test but by a much smaller margin. Because of the normal variability in the results, casual observation of the data by an operator would likely not detect the worsening condition.

HISTORICAL DATA ANALYSIS

An important trait of a test results monitoring and analysis system is the ability to archive detailed test data and interim test results. By collecting and analyzing interim test results, as well as final test results, valuable insight can be gained into the types of problems that occur and the relative frequency at which they occur. Armed with this information, product improvements aimed specifically at reducing or eliminating those problems can be implemented resulting in increased manufacturability and fewer product rejects.

The accumulation of historical data also creates a detailed audit trail that can be beneficial for in-depth studies of process or product defects. The data archive can also be used for management reports and for tracking various improvement initiatives. Zoom-like capability in the analysis tool can facilitate rapid identification of problems, such as a misaligned pallet, by highlighting deviations at a high level, and then allowing an increasingly granular view of the test results. For example, if a weekly report summarizing the percent rejects by day shows an anomaly on Friday, then a "drill-down" feature would allow Friday's test results to be readily viewed by operator, by batch, or by pallet, allowing the underlying cause to surface.

Perhaps one of the greatest yet underappreciated values to accumulating a large body of test results is that the statistical distribution of those results can be more fully understood. Understanding the true distribution of test results is key to optimally setting pass/fail criteria, which in turn affect the number of good parts falsely rejected and the number of defective parts erroneously accepted.

Example: Needless Part Reworking

To minimize falsely accepted parts, reject limits are typically set lower than the true design requirements dictate. If this limit is not set optimally, an excessive number of false rejects can occur causing needless rework and retesting. By reviewing historical data to fully understand the distribution of test results, an optimal reject limit can be determined to minimize wasted effort.

CONCLUSION

The right combination of test equipment and tools can provide insight into process problems and potential product improvements that would otherwise be unavailable. Having insight into such subtle trends gives you the ability to preemptively solve problems for greater profitability. While low cost and stand-alone testers have their place, combining powerful and flexible test equipment with integrated information management will be the distinguishing hallmark of companies that survive today's competitive climate.

ACKNOWLEDGMENTS

Special thanks go to Carl Aquilino, President of Uson, and Dan McCauley, Director of Automotive Programs at Uson, and Alan Campbell, Vector Product Manger at Uson, all of whom have contributed to the concepts and content presented here.

REFERENCES

1. System and Method for Leak Rate Testing During Adiabatic Cooling, U.S. Patent 6,741,955 B2 dated May 25, 2004.

CONTACT

Mark D. Robison, V.P. of Engineering
Uson L. P.
8640 North Eldridge Parkway
Houston, TX 77041
USA
Phone: 281-671-2000
Email: mrobison@uson.com

2004-01-1724

Testing Networked ECUs in a Virtual Car Environment

M. Plöger, J. Sauer, M. Büdenbender and J. Held
dSPACE GmbH, Paderborn, Germany

F. Costanzo, M. De Manes, G. Di Mare, F. Ferrara and A. Montieri
ELASIS, Naples, Italy

ABSTRACT

Modern vehicles use electronic control units (ECUs) to perform their functions. The ECUs are becoming more powerful and more complex in terms of their functionality. Moreover, many functions are distributed across several interconnected ECUs. It is this interconnection via data bus (CAN) that enables sensors, calculated information and actuators to be shared. ECU manufacturers eliminate many errors during the project and testing phases of a single ECU, but many others are very difficult to detect without performing testing at integration and system level.

This paper describes a concept and a powerful tool, VirtualCar, which allows a wide range of automatic tests to be performed on networked ECUs.

The VirtualCar tool presented is based on hardware-in-the-loop (HIL) technology from dSPACE. It was designed and realized in a joint project by ELASIS, a FIAT-owned engineering company; dSPACE GmbH, Germany; and TESIS GmbH, Germany. More precisely, it represents a complex system for connecting and testing all the networked ECUs in a modern car of the compact class. This involves more than twenty ECUs divided into two separate CAN networks.

INTRODUCTION

For many years, the development of new vehicles has been characterized by the ever increasing use of electronic control units (ECUs).

As legislation on environmental protection is repeatedly stiffened, e.g., CARB's OBD II standard, EOBD in Europe, mandatory reduction of fuel consumption, more and more complex engine controllers are required. Automatic gearboxes with new transmission concepts are also being increasingly used in medium-sized and compact cars. Electronic systems from the field of vehicle dynamics (ABS, ESP, ASR) are very often standard equipment in modern cars. Even for car body and convenience, ECUs have become indispensable.

Thus many functions, e.g., seat movement, side view mirror movement, interior/exterior illumination, parking assistant and dashboard, are realized by means of ECUs.

Implementing these complex functions is feasible only if the control units are interconnected via busses. This data bus networking of ECUs in the vehicle enables the sensor system, computed data, and the actuator system to be used jointly by a variety of functions. Typically, modern vehicle concepts consist of two or three different CAN networks. Particular ECUs, connected to more than one network, serve as gateways between the networks in these configurations.

Figure 1 shows the network system of a Fiat Stilo, a current representative of the compact car class.

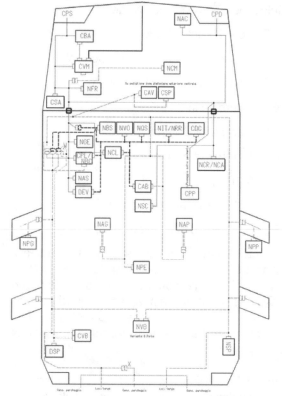

Figure 1: Electronic architecture of the Fiat Stilo

The Fiat Stilo vehicle contains the following ECUs:

- NBC: Body Computer Node
- NPG: Driver Door Node
- NPP: Passenger Door Node
- NSP: Parking Sensor Node
- NPE: Passive Entry Node
- NAG: Driver Seat Node
- NAP: Passenger Seat Node
- NBS: Steering Column Lock Node
- NVO: Steering Wheel Node
- NQS: Dashboard Node
- NIT/NRR: Infotainment Node/Receiver Node
- NAC: Adaptive Cruise Control Node
- NVB: Trunk Node
- NCL: Air Conditioning Node
- NAS: Steering Wheel Angle Sensor Node
- NGE: Electric Power Steering Node
- NFR: Brakes Node [VDC, ABS, ASR]
- NCM: Engine Control Node
- NCA/NCR: Automatic Transmission Node /Selespeed Transmission Node
- NSC: Gear Selector Node
- CAV: Alarm Sensor / Roof Unit Control
- CSP: Rain / Sun Sensor Unit Control
- DEV: External Light Commands
- CSA: Alarm Unit Control
- CPP: Tyre Pressure Unit Control
- CAB: Air-Bag Unit Control
- CPS: Left Headlight Unit Control
- CPD: Right Headlight Unit Control

The ECUs have CAN controllers (nodes) and are distributed on two CAN networks. The low-speed CAN network, the B-CAN, is connected to all body and comfort ECUs. The powertrain and vehicle dynamics ECUs are connected to the high-speed C-CAN bus. The body computer forms the gateway between the two CAN networks.

ECU manufacturers eliminate many errors during the project and development phases of the single ECUs. One of the standard tools, which has been widely used for years, is hardware-in-the-loop technology. This particularly applies to all powertrain and vehicle dynamics ECUs. However, there are many other errors which cannot be detected without performing tests at integration and system level. This means that the complete system of networked ECUs must be tested. This task is typically performed at the OEM or, as in this case, at an engineering company (ELASIS) working for the OEM (Fiat).

TESTING ECU NETWORKS

CONVENTIONAL TEST METHODS

Before the first vehicles prototypes are available, tests on ECUs (hardware and software) and other electrical components are performed on "static" benches that comprise the networked ECUs, the actual wiring harness and some of the sensors and actuators, i.e., the dashboard, the electrical motors of the seats, the control switches, etc.

These benches (see Figure 2) are normally used to test electrical actuators, simple sensors, and wiring harness, and to perform functional tests on the car body electronics and the self-diagnosis software. The network management, gateway functionality and CAN physical level are also tested.

In order to perform these tests, breakout boxes are added to introduce electrical faults, power supplies are used to generate ground shift presence, and some tools for CAN network and diagnostic lines analyses are used. During vehicle development, the individual components are gradually replaced by prototypes that have previously undergone thorough testing.

Tests are performed manually, so they are not fully reproducible and automatic test report generation is not possible.

Finally, on the static bench it is not possible to have complete test coverage because plausibility conditions are not simulated, especially for powertrain and chassis systems.

Figure 2: Conventional ECU test bench

The testing of networked ECUs addresses a specific set of questions which cannot be answered well enough by conventional test methods:

- Has CAN communication been implemented correctly by all the ECU manufacturers?
- Do the ECUs detect errors in CAN communication properly and does the resulting ECU behavior comply with the OEM's specification?
- Does the network management work correctly in all the ECUs? Do all the ECUs enter sleep modes as specified?
- Do the ECUs make their sensor signals available to other ECUs via the CAN bus fast enough?
- Does the gateway between different subnetworks work correctly?
- How fast do actuators, driven by their ECUs, react to demands from other ECUs via CAN?
- Do the on-board diagnostic functions work as required?

A number of requirements for the HIL test system can be derived from these questions.

The user's requirements for the VirtualCar test system are described below. They are fairly representative of other test systems for networked ECUs:

- All pertinent ECU power drivers and signal outputs must be read in by the test system. It must be possible to capture the signals and store them in files if required.
- The test system must be able to stimulate all the ECU inputs.
- Real electrical fault insertion capability is required on ECU outputs in order to verify how the VirtualCar system reacts to the insertion of known faults. For ECU inputs, electrical faults can often be stimulated by software.
- The test system must be able to log all CAN messages between the ECUs. To investigate the behavior of the CAN network, the test system must be able to perform the following tasks:
  - Manage standard and extended identifier messages
  - Trace and record on all of the CAN lines simultaneously with time stamps
  - Send predefined messages interactively
  - Generate triggers on start of frame for detailed analysis
  - Measure the time elapsed between a certain message with identifier "x" and a message with identifier "y"
  - Simulate the messages received and transmitted by nonexistent nodes and react to external triggers (events) or to events on CAN lines
  - Suppress all CAN messages sent by one or more ECU

- Modify specific signals inside CAN messages and if necessary calculate a new checksum
- Generate hardware errors on the CAN bus (e.g., by inserting additional capacitors or resistors between the CAN lines, generating error frames, destroying CAN messages at arbitrary bit positions)
- It must be possible to verify network management functionality: sleep mode, alive mode, wake up mode.
- A diagnostic serial line is available on many of the ECUs constituting the test system. During test execution, it is necessary to interface the ECUs through this line to request diagnostic services and get diagnostic responses from the ECUs. In this particular case, the ability to interface to the ISO9141 serial line is required. Diagnostic communication protocols must be implemented based on this layer.
- From the ECU's point of view, the test system must behave like a real car. This requires real-time capable models of all controlled systems, especially for the engine, transmission, vehicle dynamics and some of the body/comfort components.
- For manual interactive operation of the system, the experiment software must be powerful and flexible, but also easy to handle. The ability to automate the overall test system is crucial. For such a large system particularly, it is necessary to have powerful automation software with a well structured automation concept.

ELASIS was looking for an HIL simulator which was able to fulfill as many as possible of these and other requirements. After evaluating the HIL simulators available on the market, ELASIS opted for the dSPACE Simulator over other HIL vendors' products. One reason for this decision was good experience with using dSPACE Simulator Mid-Size for single ECUs (engine controllers) and for a small ECU network (engine and transmission controllers). In addition, ELASIS had had positive experience of the entire dSPACE tool chain for ECU development. The experiences of other parts of the FGP group also contributed to the decision.

VIRTUALCAR TEST SYSTEM FOR NETWORKED ECUS

HARDWARE SYSTEM SETUP

The main objective of the first phase directly after project start was to define the detailed system specification of VirtualCar. A team of experts from ELASIS and dSPACE worked closely together on this for several weeks. The results were summarized in the system specification document. This also included minor changes to the requirements contained in the original requirements specification document. This chapter provides the hardware and software details of the VirtualCar system.

The VirtualCar is based on dSPACE Simulator Full-Size. The main reasons for choosing the Full-Size technology were:

- High level of reusability and modifiability for all hardware, particularly with respect to later adaptation to new projects
- The large number of input and output signals required
- The need to contact a large number of ECUs

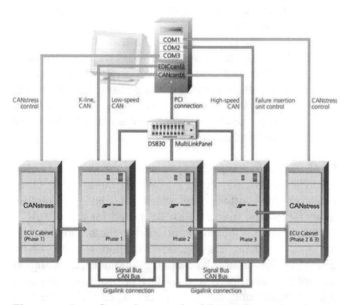

Figure 3: Overview of VirtualCar hardware components

Figure 3 gives an overview of all hardware components used in VirtualCar.

The heart of VirtualCar is formed by three 19" cabinets. One cabinet is dedicated to the B-CAN ECUs. The C-CAN network is divided into two cabinets. One is for the vehicle dynamics ECU, ACC and steering angle sensor. Since this sensor is a regular node on the C-CAN network of the Stilo, it is also integrated in the VirtualCar as a real part. During the concept phase it was decided not to integrate the electric power steering node as a real part but to simulate it by generating its CAN messages with the VirtualCar. The decision took into

account the work involved in integrating a real power steering node and the benefit to be expected of it. The engine controller and the ECU for controlling transmission are connected to the third cabinet.
Each of the Full-Size cabinets is constructed similarly and contains:

- The high-speed real-time processor for calculating the dynamic models, standard I/O and CAN traffic
- I/O cards for acquisition and generation of discrete signals, boards for receiving and transmitting CAN messages. A special HIL board (the DS2210 HIL I/O Board) was used for the engine control unit. This contains special functions for generating and reading crank-angle-based signals with high accuracy and convenience.
- The multiple modular dSPACE signal conditioning components
- FIU and load boards to make real or simulated loads available on ECU outputs and generate electrical faults.
- Integrated break-out boxes for accessing discrete signals. Since the number of signals for B-CAN ECUs is very large, and the signals can be accessed via terminal strips on the ECU cabinet, there are no break-out boxes in the HIL cabinet for the B-CAN ECUs.

In addition to these universal contents, there are further components integrated in the HIL cabinets:

- One common power supply (battery simulation) for the whole VirtualCar
- An adaptation system for the ESP controller. Since this ECU drives the hydraulic valves via integrated coils, the adaptation system is necessary for sensing the magnetic field of the coils by Hall sensors.
- A/W bus gateway:
  This component, developed by NSI, France, serves as a gateway between CAN and the Fiat A- and W-busses. Using this component makes it possible to generate messages on Fiat busses by sending specific CAN messages on a private CAN. The box translates these CAN messages into corresponding A-bus or W-line signals.
- CANstress system:
  CANstress, a combination of hardware and software from Vector Informatik, can introduce several types of CAN faults which cannot be generated by the dSPACE CAN boards. Its functions include generating fault frames and the inserting capacitors between CAN high and CAN low. A separate CANstress box is used for each CAN network (B-CAN and C-CAN).

All the ECUs for the B-CAN network and the ECUs for transmission and engine are located in separate ECU cabinets.

The host PC is used both to develop models and to configure and run the experiment and the test automation software. The PC contains a CANcardX for accessing the CAN busses (B-CAN or C-CAN) for particular purposes like providing bus statistics. Additionally, the PC contains an EDICcard2 from Softing for accessing the K-Line for diagnostic services.

Figure 4: VirtualCar in the laboratory

The following list of input and output channels used in the VirtualCar (Figure 4) clearly demonstrates the complexity of the system:

- 88 ADC channels
- 99 DAC channels
- 366 channels for digital I/O
- 6 resistor simulation channels
- 10 PWM input channels
- 10 PWM output channels
- Further special channels for ignition, injection, crank, and cam
- Four different CAN controllers, two for each CAN network, for providing a gateway and fault simulation in the Simulator

The total number of ECU pins connected to the test system is about 900.

SPECIAL HARDWARE SOLUTIONS
Fault Simulation Concept on CAN

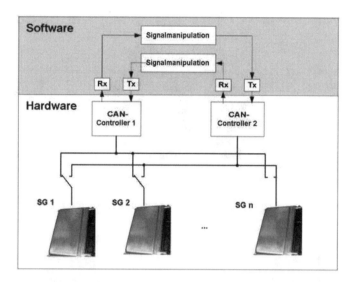

Figure 5: CAN fault gateway

Figure 5 shows how the requirements for testing fault behavior on the CAN network are met. The solution shown in Figure 5 is implemented both for the B-CAN network and for the C-CAN network.

Two CAN controllers for each network are implemented in the Simulator. Each ECU can be connected separately to either of them. All messages received on one controller are immediately passed to the other controller via the Simulink® model. This ensures that each ECU receives the CAN messages of the other ECUs even if they are connected to different CAN controllers. The delays that occur are so slight that the ECUs are not affected by them. Since transmission is fully under software control, it is easy to generate fault cases down to message or individual signal level.

Stimulation of Steering Wheel Sensor

The steering wheel sensor (LWS3 , Bosch) is a regular node on the C-CAN network. It was therefore a requirement to include it in the VirtualCar as a real component.

The LWS3 steering angle sensor contains two AMR (anisotropic magneto resistance) elements. Above each element, there is one gear wheel with a permanent magnet, whose rotation depends on the steering angle. The gears rotate at different speeds, resulting from their different numbers of teeth. Each AMR element consists of two independent Wheatstone bridges, which are used to generate an output voltage according to the angle of the associated gear wheel.

Since the rotation of the steering column is simulated by the vehicle dynamics model, it is necessary to stimulate the sensor in such a way that the microcontroller recognizes a correct rotation angle.

For this task, the gear wheels were removed and the inputs of the AD converters directly stimulated by analog outputs of the Simulator, in the form of sine and cosine signals with a specified phase relationship.

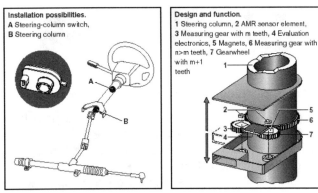

Installation possibilities.
A Steering-column switch,
B Steering column

Design and function.
1 Steering column, 2 AMR sensor element, 3 Measuring gear with m teeth, 4 Evaluation electronics, 5 Magnets, 6 Measuring gear with n>m teeth, 7 Gearwheel with m+1 teeth

Figure 6: LWS3, Bosch steering angle sensor.

High-Speed Data Acquisition

Another requirement of ELASIS was to have eight additional, analog inputs with much higher sampling frequencies than the standard I/O channels related to the real-time model. The inputs can be connected flexibly to any of the electrical signals to make a precise analysis. High-speed ADC measurement is realized by reading the ADC channels in a separate task, which is calculated with a sampling period of 50µs (20kHz). This task is triggered by a separate timer interrupt on the corresponding processor board and has the highest priority. The task also includes a separate capture service to trace the data with a sampling period of 20 kHz.

The sophistication of dSPACE software (Real-Time Interface, experiment software) makes it possible to implement this special feature easily within the Simulink block diagram and to access the high-speed signals together with all the other signals in the same ControlDesk experiment.

SOFTWARE COMPONENTS

Real-Time Software

Most parts of the controlled system and the I/O of the VirtualCar are specified in MATLAB®/Simulink®.

The multiprocessor blockset RTI-MP makes it possible to specify all models for the complete system within one Simulink model. Figure 7 shows the top level of this model.

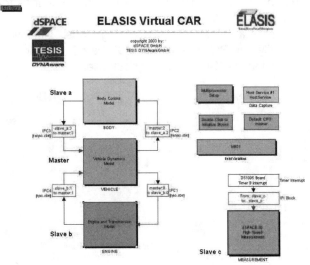

Figure 7: Simulink model, top level

It mainly consists of four subsystems, one for each processor:

- Body electronics (BODY)
- Vehicle dynamics (VEHICLE)
- Engine and transmission (ENGINE)
- High-speed analog measurement (MEASUREMENT)

The subsystems for body, vehicle and engine are each divided into one part for the complete I/O model and another part for the corresponding dynamic models. The body's I/O part benefits from a certain model structure inside the I/O model. The structure developed by dSPACE contains a very clear and logical hierarchy. This makes it easy to find I/O signals from among hundreds of other signals.

To simulate the dynamic behavior of the diesel/gasoline engines and the three-dimensional movement of the car, use is made of the well-tried Simulink models en-DYNA and ve-DYNA. (TESIS GmbH, Germany). Ve-DYNA was enlarged by a module for the simulation of ACC traffic. The model for automated manual transmission (Selespeed) was developed by ELASIS.

Body Models

Unlike applications for the powertrain or vehicle dynamics, car interior comfort frequently requires only a very simple model of the controlled system. Complex, nonlinear, high-order differential equations are replaced by small kinematic function models. For example, to control the windshield wipers, the ECU expects the moving wiper to pass through the park position regularly, depending on the wiper speed. The feedback comes from the "wiper park position" signal, which has to be activated cyclically.

Here are further examples of such comfort models:

- Movement of the seats:
  The driver/passenger seat nodes control the positions of the backrest, height and slide motors by means of the input signals provided by their Hall sensors. As these electrical motors are replaced by high-impedance simulated loads, a kinematic model which simulates the movement and generates Hall sensor signals according to the movement is needed.
- NCL (climatic controller):
  The NCL controls the positions of four motors by means of the input signals provided by their position potentiometers. As these electrical motors are also replaced by simulated loads, a kinematic model for simulating the movement of the four motors is necessary. The potentiometer signals are generated according to the simulated positions.
- Additionally, the NCL regulates the position of the air recirculation flap by observing the current flowing through the controller output pins. The end positions are detected by a high current after a specific period of low current. To simulate this behavior, an easy time-based model was implemented.
- The side view mirror movement is controlled by the door ECU, by observing the position potentiometers. The real electrical motors are replaced by kinematic models.
- The real mechanism for locking/unlocking and opening/closing is reproduced by a more complex functional model.

Experiment Software

The entire graphical user interface for manual operation is implemented with dSPACE ControlDesk. Many well-structured layouts, partly with photorealistic visualizations, enable the user to handle the system and manage the real-time experiments (see Figure 8).

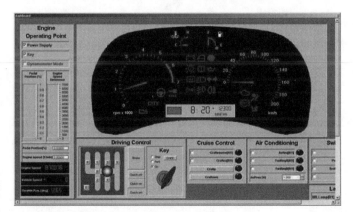

Figure 8: Main ControlDesk layout

Especially for testing vehicle dynamic functions or ACC functionality, it was decided to integrate a 3-D online animation tool. The tool, from the dSPACE tool chain, is MotionDesk. It enables the movement of the car to be visualized in a virtual world.

THE FULL BENEFIT WITH TEST AUTOMATION

Although it is possible to perform many tests manually, the full benefit is reaped only when automation is introduced. Since an automated test requires less time than the same test executed manually, more tests can be performed in a given period of time. This leads to broader test coverage and a greater test depth, meaning that more tests can be done and more details can be tested.

At ELASIS, two different approaches are used to run automated tests with the VirtualCar.

To test diagnostic functions, ELASIS uses TestPlatform, an object-oriented test system developed by ELASIS over the last two years for testing engine control units on dSPACE Simulator Mid-Size. The well-structured object-oriented design enables the software to be adapted to the VirtualCar, so many existing tests can be reused.

For functional tests of the body and comfort electronics, a new test project based on the new dSPACE software AutomationDesk is under development. Both approaches are described below.

ELASIS TESTPLATFORM

The first objective was to define a generic test procedure for the self-diagnosis software of the ECU with respect to project-specific requirements and international standards.

The test procedure was derived from the test specification and from the knowledge of the experts who perform these tests every day on the car.

Originally, the test procedure was then implemented on dSPACE Simulator Mid-Size by developing a software environment in Python code for test creation and management. An object-oriented approach allows a dynamic and hierarchical test structure and automatic test report generation. The main advantages of this approach are the maintainability, reusability and repeatability of tests.

Self Diagnosis Software

The ECUs are programmed with self-diagnosis software (on-board diagnosis), which allows management of the faults arising in the plant to be controlled (i.e. engine, transmission, etc.). In addition, the ECU passes the descriptive and standardized diagnosis trouble code (DTC) of the detected faults to diagnostic communication software via K-line or via CAN bus.

Fault detection and the DTC management are specified by the European On Board Diagnosis (EOBD) standards in Europe and the California Air Resources Board (CARB, OBD II) in the US. These rules have been included and extended in the self-diagnosis specification (SDS).

These are some of the functions covered by a generic engine SDS:

- *ReadDataByLocalIdentifier* (RLI)
  Check ECU variables accessibility and conversion
- *Inputoutputlocalidentifier* (IOLI)
  Check ECU actuators
- *StartRoutineByLocalId*
  Launch procedures (i.e., self-learning)

- *Other KeyWord2000 Services*
- *Scantool Services*
- *DTC*
  with the basic OBD structure:
    o Drive into specified operating point
    o Activate electrical\logical\model fault
    o Read out ECU diagnostic memory
    o Evaluate test by comparing the detected fault with the expected fault
    o Generate report automatically

has been improved to include the EOBD test concepts (see flowchart in the next section).

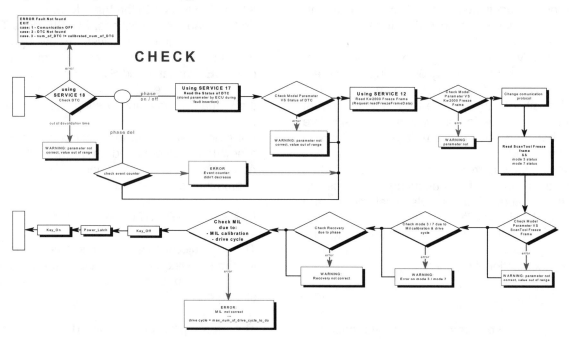

Figure 9: Flowchart for CHECK Item

Test Requirements

The previous items concerning the DTC have been formalized in a flowchart that describes the steps to be performed. The steps have to be performed for each DTC (i.e., P0201), for each fault symptom (e.g., short to GND, short to battery) and for each test condition (detection rules), e.g., power-on, cranking, engine run and vehicle run.

When a DTC, a fault symptom and a detection rule have been defined, a generic test could be performed as follows:

- Fault insertion with the correct fault symptom
- Check
- Fault off
- Check
- Fault insertion with different fault symptom
- Check
- Fault off
- Check

Figure 9 shows in detail how "check": is structured. The CHECK block has these items:

- Check DTC using Service 18
- Check model parameter VS status of DTC using Service 17
- Check model parameter VS KW2000 freeze frame using Service 12
- Check model parameter VS ScanTool freeze frame
- Check ScanTool mode 3/7 due to MI calibration & drive cycle
- Check recovery status
- Check MI status due to MI calibration & drive cycle

These steps are performed several times during test execution, depending on the test phases:
- Fault ON
- Fault OFF
- Change symptom ON
- Change symptom OFF

- Fault DEL (warm-up cycles to check the decrement of the event counter)

and on the MI Calibration:

Figure 10 shows in detail how the overall generic test is structured.

The Test Automation Architecture

To implement the presented test structure, object-oriented test software was designed and implemented. This has the following important advantages over a purely procedural solution:

- The test code for several OBD software versions of several ECUs (engine, transmission,...) from several suppliers can be managed, maintained and reused efficiently.
- The same test structure can be reused for functional tests on networked ECU.

The software named TestPlatform is based on the obect-oriented script language Python. To close the diagnosis loop, another third-party tool is used: Diagnosis Tool Set (DTS) from Softing. Use is made of the following dSPACE test automation libraries:

- DSTLib for accessing DTS
- IOCI-LIB for accessing the electrical failure insertion unit of dSPACE Simulator
- RTPLib for accessing the real-time model variables
- WordLib, ExcelLib for automatic report generation

TestPlatform has two main substructures:

- The **ECU tree**:
  A hierarchy of all the ECUs that can be tested.

- The **CAR tree**:
  A hierarchy of all the cars that use an ECU combination from the ECU tree.

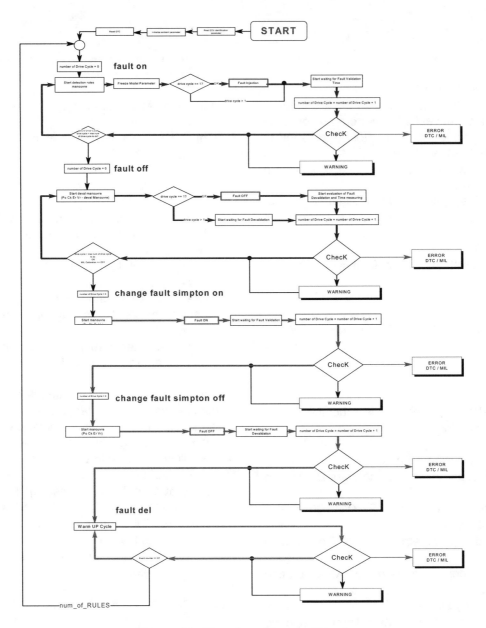

Figure 10: Structure of overall test

The main idea is that in the ECU tree, each ECU has its own tests and parameters but also inherits all the properties from the parent ECU. The same mechanism works for the CAR tree. All the maneuvers and distributed tests for the networked ECU can be defined at root level.

For example, the maneuver *VehicleChange_Gear*() is defined for a manual gear box at CAR root level, but can be overwritten for a car with an automatic gear, without changing the test implementation. In the same way, the

functional and distributed tests of networked ECUs in one car are used by all the cars derived from it.

TestPlatform starts by defining a root class named ECU, in which all the items of the SDS are present.

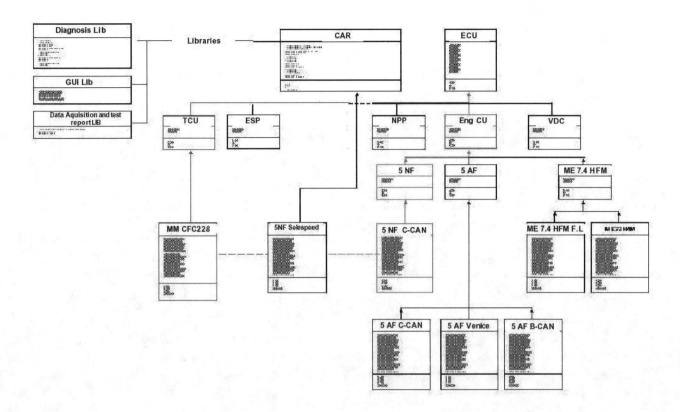

Figure 11: UML general class diagram for HIL TestPlatform

The UML diagram shown in Figure 11 is intended for general use and shows how the platform has been planned and how it can be extended.

A new car, including the desired ECU, is constructed dynamically. Once the necessary ECU modules are available, a new car consisting of a subset of the ECUs can be defined. For example, the 5NF Selespeed car consists of a Magneti Marelli (MM) 5NF engine management control unit and an MM CFC228 Selespeed transmisson control unit.

Some tests need very particular maneuvers for each detection rule. For example, some tests run only if they detect a specified engine speed. It was therefore necessary to create a structure that allows maneuvers (customized for each test) to be defined in the hierarchy. This was done by defining some subclasses. These use the inherited basic maneuvers defined at the car level.

DTC Parameterization

The lower levels of the hierarchy have all the functionality of the upper levels, so the parameters of a test can be changed at any level, but if a major change in test structure is needed, the same test can be overwritten.

Test SDS: Report Generation

Each test produces two files, a text file to trace the diagnosis communication flow and the simulation status MATLAB file in which the model variables are stored. In addition to these files, each test session produces a final Microsoft Word report. The completion of the tests is reported to a predefined list of addresses by e-mail.

Graphical User Interface (GUI)

Figure 12 shows the main layout of the TestPlatform environment. It is built automatically in ControlDesk according to the test. It provides information on:
- The running test (in the calibration frame)
- The previous test result (in the LED state on the DTC frame control)
- The current maneuver (in the customized instrumentation in the top part of the window)

The DTC frame control gives all the information about the test. It makes it possible to:

- Check the detection rule
- Enable/disable the fast modality (to do only 2 warm-up cycles in the Del phase)
- Read the LED states to know the result of the test during the session and
- Read the validation / devalidation time.

413

The DTC calibration display gives all the calibration information about the currently running test. It is refreshed for each test.

All the access functions to write the calibration parameters to this frame are provided by the GUI class.

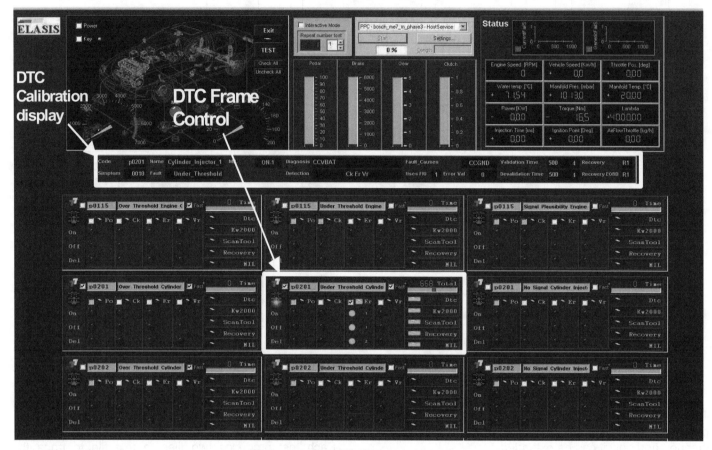

Figure 12: TestPlatform main layout

With the customized instrumentation, it is possible to read the state of the model, e.g., engine speed, water temperature, and manifold pressure.

TESTAUTOMATION CONCEPT BASED ON DSPACE AUTOMATIONDESK

A second approach to test automation is currently under development(*Figure 13*). This will be used mainly for functional tests. It is a graphical approach based on AutomationDesk, which is the new dSPACE test automation tool. It has the following advantages over the pure Python approaches:

The integrated test project management enables the user to store and manage tests, test data, test results and test reports in one place. This facilitates close integration into the development process.

The library concept consists of global built-in libraries for basic functionality, global custom libraries for extensions to the automation functionality, and project-specific libraries. Test components and tests can therefore be reused easily and conveniently.

Result management and report generation guarantee test reproducibility, data security, and consistency.

The deployment of AutomationDesk in the VirtualCar project is planned for 2004.

PROCESS INTEGRATION

Today the development cycle for a new car is typically 36 months. Throughout the development process, there are certain phases in which the combination of ECUs is tested with the HIL system. To ensure that the system adapted to the ECUs is available in time for these phases, close cooperation between the HIL system supplier and the OEM is necessary. The HIL system must be constantly available during the test phases. It must run 24 hours per day, up to 7 days per week.

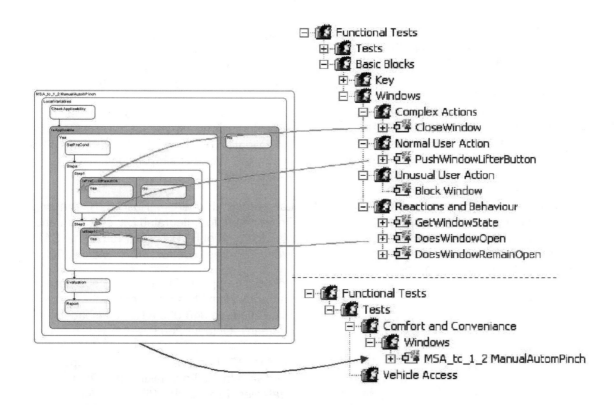

Figure 13 AutomationDesk: Libraries and graphical test sequence

ELASIS decided to gain experience with VirtualCar before using it in a tight project schedule. So at the beginning, the system was configured for the Stilo ECUs at a late phase in the development phase.

Now the system is in full operation, it will be updated for a complete new car .

In this new project, VirtualCar will be used from the very beginning. At the same time as the ECU suppliers develop the first version of their ECUs based on the ECU specifications, the virtual car will be updated according to the same specifications. It is essential to ensure that the HIL system update will be finished and new automatic tests implemented by the time the first versions of the ECUs are available. The new ECUs can then be tested intensively for a period of four months.

APPLICATION EXAMPLES ON VIRTUALCAR

One example of automated testing on the virtual car is testing indicator light activation. In the Stilo, this function is distributed across the body computer (NBC), the dashboard ECU (NQS) and the adaptive cruise control ECU (NAC).

The NBC receives the request for indicator light activation from the indicator switch and calls the relevant command. It sends the message `TurnLightFaultSts` on low-speed CAN and the message `TurnSignalSts` both on low-speed and on high-speed CAN.

The messages on low-speed CAN are used by the NQS to show the driver that the indicator light has been activated, both in the nominal case and in the fault case.

The message on high-speed CAN is used by the NAC if the Adaptive Cruise Control was inserted.

The TestPlatform structure was used for testing cases in which the driver activates the indicator light in nominal conditions (fault off phase) and in fault conditions (fault on phase).

The fault selected is DTC 9007, taken from the specification of the NBC, which describes the detection of an electrical fault in the indicator light.

The test consists of:

- Inserting an electric fault (connection to Vbat) for the indicator light by means of the FIU
- Investigating by means of the DTS whether the NBC detected the fault correctly and wrote it to the fault memory correctly
- Reading through the correct message flow on the CAN lines
- Establishing consistency between the indicator light indication on the dashboard and in the CAN messages.

CONCLUSION

The ever-increasing complexity of electronic systems in modern vehicles requires new ways of developing and testing ECUs and their functionality. After a discussion of the questions arising when networked ECUs are to be tested, the requirements for a corresponding test system were derived. VirtualCar, a complex and powerful tool for testing the entire ECU network of a Fiat Stilo based on dSPACE hardware-in-the-loop simulators, was presented as a solution.

Two complementary approaches to automated tests were presented. These approaches provide a reduction in test execution time, reliability of tests due to the repeatability of external and internal conditions, and the ability to perform more exhaustive tests by easily modifying the test conditions. Moreover, the risk of human error due to stressful, repetitive work is reduced.

REFERENCES

1. Schütte, H.; Plöger, M.; Diekstall, K.: Wältermann, P.; Michalsky, Th.: Testsysteme im Steuergeräte-Entwicklungsprozess. Automotive Electronics, pp. 16-21, March 2001

2. Gehring, J.; Schütte, H.: Automated Test of ECUs in a Hardware-in-the-Loop Test Bench for the validation of Complex ECU Networks. Proc. of the SAE World Congress, Detroit, USA, March 2002

3. Lamberg, K.; Richert, J.; Rasche, R.: A new Environment for Integrated Development and Management of ECU Tests. Proc. of the SAE World Congress, Detroit, USA, March 2003

4. Köhl, S.; Lemp, D. ;Plöger, M. : Steuergeräte-verbundtests mittels Hardware-in-the-Loop Simulation. pp.948-955,ATZ , Wiesbaden, Germany, October 2003

5. Di Mare, G.; Ferrara, F.; Scala, S.; Sepe, E.: Hardware In the Loop testing of EOBD strategies. Proc. of the 15th IFAC World Congress, Barcelona, Spain, July 2002

6. Caraceni, A.; De Cristofaro, F.; Di Lieto N., Ferrara, F.: Gasoline Rapid Control Prototyping System for Fiat Punto 1242cc 16v. Proc. of the AVEC '02 Congress, Hiroshima, Japan, September 2002

7. Caraceni, A.; De Cristofaro F.; Ferrara, F.; Philipp, O.; Scala, S.: Benefits of using a real-time engine model during engine ECU development. Proc. of the SAE World Congress, Detroit, USA, March 2003

8. Gruber, J; Steering Wheel Angle Sensor for Vehicle Dynamics Control Systems. Proc. f the SAE World Congress, Detroit, USA, February 1997

Interoperability of Networked Components Requires Conformance Test of Communication Interfaces

Wolfhard Lawrenz

c&s group at University of Applied Sciences

ABSTRACT

Actually automotive industries is in a dilemma: Further customer satisfaction requires further features to be built into cars and this requires more electronics in cars. All this currently happens at a rapidly increasing pace and thus increasing networked cars electronic control systems complexity correspondingly. Networking obviously requires interoperable communication modules consisting of communication hardware and software. And here is the problem: As network protocols are specified mostly in natural languages such as English, no precise and non-ambiguous specification exists. As such implementers may understand the same specification differently. And as a result realizations of the same protocol may behave differently under certain operational conditions. As a consequence mixed suppliers control modules may not be able to communicate properly.

Conformance testing is the solution to this problem. The following paper explains the general procedure how to derive conformance tests under the constraints of verbal non precise device specification. All this is explained with the example of deriving and implementing tests for CAN protocol transceivers. Finally a more general c&s group developed procedure is outlined to derive tests and how to implement them on the basis of a standard ISO tester architecture. Last but not least a road map is given on the conformance tests implemented for various vehicle communication protocols by c&s group.

INTRODUCTION

Networking nowadays is a standard technology applied in cars. There are various kinds of communication protocols applied, which each meet the different application requirements as for safety, body electronics, power train, x-by-wire, etc. The goal of networking is to provide a means to exchange variables between distributed control applications. As such a door controller may read the position of a switch and exchange the proper value with the seat positioning controller thus moving the seat properly. Obviously controllers are provided by various module manufacturers containing communication modules which are supplied by various software and hardware manufacturers. Given that final user expectation would be that all these applications can communicate with no problems requiring that the underlying communication components all are "interoperable".

But interoperability of communication components may be a problem: Although standards exists in SAE and ISO very often these standards are not precise enough. Standards are mostly written in natural languages and thus contain specifications which are ambiguous, incorrect or contradictory. Taking this into account there is a high risk that mixed components networked systems may end up in communication problems. This is even more true today, as networks tend to integrate plenty of nodes such as 30 and more in one network branch.

A very effective countermeasure to this dilemma are conformance tests, checking the properties of components against a "well defined" set of test cases and thus ensuring that different suppliers' devices work are properly interoperable under operational conditions as specified by the tests.

Over recent years various efforts have been undertaken to specify, implement such kind of interoperability tests. c&s group has been working on such tests for various protocols such as CAN, eBUS, TTx, LIN, etc. Some of the specifications in the mean time have become standards such as ISO 16845 CAN (layer 2) Conformance Test.

LAYERED COMMUNICATION STACK

There is an ISO-OSI model structuring any communication path into 7 different layers. As such the various layers may be implemented by different suppliers. A complete communication path in a control module may then consist from different suppliers at the various levels.

A communication implementation not always fills all the 7 communication layers, if there is no need for certain functionalities. Figure 1 depicts a typical automotive scenario.

Layer 1 specifies the so called "Physical Layer". This comprises e. g. the actual wiring of the network and the

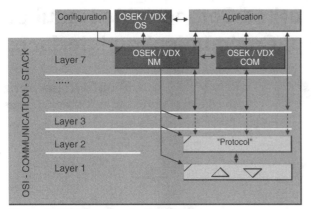

Figure 1: OSI Communication Stack

so called "Transceiver", serving as the interface between the wiring and the higher layer. Layer 2 very often is referred to as the protocol layer, which is CAN in this case. On top of CAN typically there are some further communication drivers in layer 2. These communication drivers typically adapt the various communication register structures of CAN to the next higher layers.

On top of layer 2 all the adjacent layers very often in in-vehicle communications are empty, no functionality like different data paths required. The final layer 7 – Application Layer – represents the interface to the user (software). In this case the OSEK/VDX implementation is depicted. As such OSEK/VDX specifies to blocks of interface. The NM – Network Management – continuously checks the health state of the whole communication path and reports it to the user application. The COM – Communication – module finally the services for the exchange of variables between the different user applications.

With regard to conformance tests all the implementations of the various layers must be checked, as discussed below.

CONFROMANCE TESTS SPECIFICATION – BASIC TECHNIQUE

Baring in mind the constraint of insufficiently specified components and the requirement that these components should be interoperable, one way out of this dilemma is to apply "sufficiently exhaustive" conformance tests to verify the desired interoperability of components. As no sufficiently detailed and formal description of the component exists a so called black box test technique is applicable. Test stimuli are applied from outside of the component – which is referred to as Implementation under Test IUT – to the accessible inputs, the responses are read from the outputs of the IUT and compared to what is understood from the specification.

In a first step in order to derive a first guess of a sufficiently exhaustive set of test cases an empiric driven selection of a set of test cases, representing practical conditions under which a component later on would work, is a recommendable practical approach. In a

second step the individual elements of the above set are organized in such a way that they each would result as the parameterized product of a set of – mostly desired – orthogonal vectors. These vectors may be defined application driven. The parameterized product of the vectors defines the so called System Operational Vector Space – SOVS, comprising all so far known possible operational conditions under which IUT is expected to be functional in the specified way. In a third step a subset of the SOVS is selected empirically, driven from application experience, or even formally. This subset is then expected to be Sufficiently Exhaustive Minimal Set of (conformance) Test Cases SEMSTC on one hand while comprising not too many tests cases on the other hand resulting in a still acceptable test time.

Obviously the approach mentioned above is a practically viable procedure to derive a sufficiently exhaustive set of test cases. But this procedure of course incorporates the risk that the selected subset of tests does not cover sufficiently the behavior of the component. There are several ways to optimize the selection of test cases – see below:

- The selection of the of SEMSTC should be supported by formal methods
- The application of the SEMSTC in conjunction with the practically gained experience that in real application cases will be detected that tested components will show deficiencies in their desired interoperability will lead to an iterative optimization process influencing the re-specification of the test cases and the component. The newly detected problem is analyzed leading to:
  - A redefinition of the SOVS, their parameters and SEMSTC in such a way that the new case systematically is a member of the sets mentioned above.
  - A redefinition of the specification of the component, leading to a more precise, more detailed, more formal result, without necessarily being too stringent to the implementers. Based on that "more" formal methods could be applied to derive tests which would then lead to a better SEMSTC. The black box test method would then turn more into a gray box test technique.

CONFORMANCE TESTS SPECIFICATION – EXAMPLE: CAN TRANSCEIVERS

The Transceiver is the interface between the analog signal transmission over the physical communication media and the digital "world" towards the higher OSI-layers. The latter could comprise – as it is the case for CAN – lines for digital signals such as Receive_Data, Transmit_Data, Error_Signal, etc. Towards the physical media there may be analog lines such as CAN_High, CAN_Low, etc. The redundancy of these lines is used for differential mode signal transmission which provides a good means for high common mode noise injection immunity. In case of a failure – short circuit or broken wire – the redundancy is used together with the failure detection mechanism of the fault tolerant transceiver, to switch off the defect line, communicate with a reduced signal to noise ratio and signal the error to higher layers. Finally there are power supply lines for supplying the component with energy. There are means provided to switch the component from normal mode into low power or power off mode. For more details refer to [4].

SYSTEM OPERATIONAL VECTOR SPACE - As an example for the purpose of doing interoperability tests on standard transceiver components a SOVS is given, derived from experience, see table 1. This experience has been gained from practical applications at the automotive companies sites' and the experience in testing and organizing tests within the research work at c&s group. [1] For further details refer to [5], [6].

Transceiver-SOVS =	{ System Configuration } x { Communication } x { Power Supply } x { GND Shift } x { Op. Modes } x { Failure }

Table 1: Transceiver System Operational Vector Space

The individual vectors can be broken down into parameters as shown in table 2.

System Configuration	
Baud rate	.... 5 kBd ...... 125 kBd
Termination	Calculated total termination = ... 100Ω ......
Topology	Bus, ring, star, mesh, ...
Composition	Homogeneous, heterogeneous (ratio), ...
Number of nodes	1, 2, ...40, ...
Environmental conditions	Temperature (...20°C...), moisture, shock, ...
......	

Communication	
Nodes' interaction	• Logical ring: • node x receives token • node x transmits token to node x+1 • after 1 cycle all nodes transmit 1 message leading to an arbitration conflict • ... • Arbitrary communication .... • ....
Identifier	Any, special, ...
Data	Any, nodes reference, ...

[1] Funded to c&s group as a research grant by Audi, BMW, DaimlerChrysler, PSA, Volkswagen

......	

Power Supply	....

Ground Shift	....

Operational Modes	....

Failure	

Single bus failure	• no failure
	 • short circuit:
	    • CL_Vx(up)@Rx with:
	      • Vx = [.. -3V .. +18V ..]
	      • Rx = [.. 0Ω .. 50kΩ ..]
	    • CH_Vx(up)@Rx
	    • ....
	 • open circuit:
	    • CL_OW@Rx(up)
	    • CL_OW@Rx(down)
	    • ......

1,5 bus failures	• apply CL_BAT + CL_CH then remove CL_BAT • apply CL_GND + CL_CH then remove CL_GND • ...
2 bus failures	• apply CL_BAT + CL_CH • apply CL_GND + CL_CH • ...
n bus failures	• ...
Location of failure	At node 1, 2, ...
....	

Table 2: System Operational Vectors and their Parameters

SUFFICIENTLY EXHAUSTIVE MINIMAL SET OF TEST CASES - The individual test cases then would result from – theoretically – the combinatorial product of any set of values assigned to the parameters defining the vectors. Apparently this would lead to a very large number of test cases which would be difficult to execute in a reasonable amount of time. Therefore a reduced subset of vector ranges must be carefully selected, This selection is empirically driven from experience, reducing the combinatorial product to a "realistic worst case" test scenario SEMSTC. In this case the group of automotive companies funding the c&s research work agreed on the following reductions:

- System Configuration = constant standard network with 100 kBd, .. , 40 nodes, ..
- Communication = constant standard with a fixed functionality logical ring communication ....
- ....
- Failure = reduced parameter sets such as: single and 1.5 failure types only, location of failure between node 39 and 40, ...

419

In this case the number of combinations shrinks down significantly because the are only 4 of the 6 vectors to be varied combinatorial as 2 of the vectors are kept constant. Furthermore the numbers and the variation range of the parameters is significantly shrunk. All this led to the final specification of transceiver conformance tests; for more information see [5].

Table 3 shows a typical example for a tests case specification "Test case 4.3.6.7: CH_Vx(down)@Rx". This case checks the behavior of the IUT in case of a short circuit failure. Therefore the IUT first is set into an initial state (repetitively) in which a specific condition is applied to the IUT: short circuit resistor and voltage against which the short circuit is applied. Then the actual test step is applied, which is a standard communication round, passing a token from one node to the next through out the standard test network. After each test step(s) the behavior of the network is checked, i.e. if an error is detected by the transceiver and into which state it is switched. The responses read are compared to the operational requirements and thus a decision is taken whether the test is passed or failed.

All the individual tests are specified to be carried out as so called homogeneous tests. This means that all tests as specified above are carried out on a homogenous standard network which consists of 40 nodes with all transceivers of the same brand. Then a second test run is performed with a first step inhomogeneous standard network is performed where the specific transceiver under tests is checked with regard to its interoperability against a so called "golden node transceiver", which currently is a Philips 1054 type. Finally a third test run is executed in an inhomogeneous standard network, where the device under test is mixed with 3 other types of transceivers. This gives an additional higher degree of confirmation of interoperability of mixed systems proper operation.

Typically set up and execution of these tests takes approximately 2 months. In order to shorten the time for first feed back to the semiconductor manufacturer in 1 week time frame a first rough run through a shortened set of homogeneous tests is performed. As such some (mayor) deficiencies or malfunctions may be detected in an early phase and the semiconductor manufacturer may already very early work on corrections of the device.

TESTER IMPLEMENTATION – The principle of the realization of the tester then is shown in figure 2: A tester supervisor – typically a PC – holds all the software for the test cases and the software to perform check off line the test results with the required operation. Furthermore the supervisor controls the so called upper and lower tester which serve as the interfaces towards the device under test. These interfaces implement all the physical signal adaptation as well as the time buffering. As such they contain e.g. the programmable devices to provide the varying short circuit resistors, the

correspondingly varying voltage generators, the memorizing

Constants	Power Supply = 12V / GND Shift = 0V / Op. Mode = Normal	
	Test procedure:	Short Circuit Failure (CL_Vx, CH_Vx)
	Operating area:	Short Circuit Operating Area
Initial State	System Configuration:	constant as specified in *Standard Net*
	Communication:	constant as specified in *Standard Net*
	Op. Mode :	Normal Mode
	Power Supply:	12V
	GND Shift:	0V
	resistor range for error generator R/U:	
	rx_start	: 0Ω
	rx_stop	: 50KΩ
	rx_next	: depending on steps

range		step
0Ω -	10Ω	1Ω
10Ω -	50Ω	5Ω
50Ω -	250Ω	10Ω
250Ω -	1.000Ω	50Ω
1.000Ω -	10.000Ω	1.000Ω
10.000Ω -	50.000Ω	10.000Ω

	voltage range for error generator R/U:	
	vx_start	: 16V
	vx_stop	: -3V
	vx_next	: depending on steps

range		step
16V -	-3V	0.1V

Test Steps	execution of communication as specified in *Standard Net*
Response	test results must match the operating areas as defined for short circuit failures

Table 3: Test Case "Short Circuit"

oscilloscopes, etc. The IUT finally is implemented through a standard network of 40 nodes. Each nodes contains one transceiver – IUT – which itself communicates with a CAN node and a micro controller. The micro controller runs part of the upper tester software which implements the token passing mechanism, the signal evaluation logics and the communication path to the supervisor.

All tests are executed fully automatic. This is a very important feature for any conformance tests as such tests do not depend on the actual condition of a tester person and his capability to observe and evaluate test measurement results. As such repeatability of tests is ensured any time any place. The results are all registered on the PC. If a problem is occurred the operator of the tester is alarmed to check the problem and communicate it to the supplier of the IUT. Along with that the actual test scenario is delivered which enables the manufacturer of the possibly faulty component to check for the problem and take countermeasures. At the end a final test report is produced by the tester which together with a summarizing Authentication Report is handed over to the manufacturer of the component.

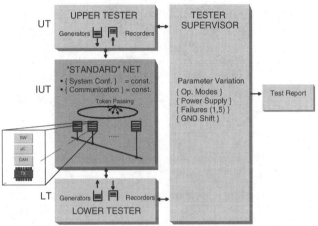

figure 2a: Transceiver Tester Architecture

figure 2b: Transceiver Tester Implementation

CONFORMANCE TESTS – FURTHER OSI-LAYERS AND PROTOCOLS

Specification of conformance tests and the design of the corresponding tester as it is done at c&s group follows a very straight forward well structured process.

CONFORMANCE TESTS SPECIFICATION – Given the constraints above that the specification of the device

under tests – DUT – typically is given in a verbal manner, the following procedure is basically the c&s standard applied principle and process:

1. Definition of System Operational Vector Space
   * Preface:
     * All the definitions and specifications following are of course derived from the specification of the behavior of the device under test – DUT.
     * If this specification shows deficiencies, ambiguities, incorrectness's, contradictions, etc. additional specifications through SOVS must clarify the problems.
     * As the latter refinement/enhancement step is very essential to the interoperability of a "standard" component this critical procedure must be always done in consensus of a group of involved and competent partners – may be standardizations organization o corresponding group.
   * Definition of vectors of SOVS for DUT
   * Definition of relevant parameters for each vector
   * Definition of value ranges for all parameters
   * Definition of responses of DUT
2. Definition of Sufficiently Exhaustive Minimal Set of Test Cases
   * Definition of a reduced, practically "sufficiently" worst case value range of parameter values
   * Specification of granularity of steps "practically" sufficient to check variables ranges of parameters
   * Derive from there individual test cases and specify them in a (semi-) formal repetitive form based on a standard procedure:
     * Specify test environment constants
     * Specify set up step, which initializes the device under test into a well defined internal state
     * Specify a (or a sequence of) test step(s)
     * Specify the required responses of DUT
3. Update of test specification
   * Field experience typically observes special cases where the observed behavior may deviate from expectation and requirements. If these cases can be reproduced and if they can be considered relevant to the operational area of the devices then correspondingly modified and/or further test cases must be specified. As this procedure is very sensitive towards the mixed systems interoperability standards, a group's consensus – as said above – is a must before modifications are introduced.
   * Field experience typically is reported to the c&s group for further process by:
     * automotive manufacturers and suppliers systems experience
     * semiconductor manufacturers experience
     * c&s test executions and networked system design.

CONFORMANCE TESTS IMPLMENTATION – The conformance tester itself consequently realizes in a structured way the above mentioned specification of the tests

1. Analysis of the functional and performance requirements of all individual tests
2. Group the test cases into classes with
   - High time resolution requirements
   - Medium time resolution requirements
   - Low time resolution requirements
   - Off line parts of the tests. This typically comprises test case generation and evaluation of the observed results of the tests
3. Assume a standard tester architecture corresponding to ISO 9694 Coordinated Test Method to be applied for implementation of the tests. This consists of the following 3 levels – see fig. 2 and 3:
   - Supervisor, which typically is a PC. The PC typically can be considered as an off-line device as compared to the real time requirement given by the test cases' specification. As such the PC:
     - stores all the test cases
     - controls and downloads the test stimuli to the upper and lower testers
     - stores all the test results delivered by the upper and lower tester
     - evaluates the results of the test measured with the results required and thus prepares the final test record with the detailed and global results "test failed/passed".
   - upper tester, which is the "upper OSI layer" interface between the DUT and the supervisor. This interface must provide the correspondingly required electrical and real time capabilities.
   - lower tester, which is the "upper OSI layer" interface between the DUT and the supervisor. This interface must provide the correspondingly required electrical and real time capabilities.
4. Map the classes onto the above mentioned blocks of the standard tester architecture consisting. This process shall be guided by the constraints in order to facilitate the design and achieve optimal flexibility, minimum development costs and best quality:
   - Shift as much functionality as possible into software, running on standard hardware for the supervisor, the upper and lower tester
   - Apply standard off the shelf hardware for the implementation of the required features and real time constraints. Standard hardware typically is PC and measurement tools such as programmable data generator and analyzers, voltage generators, oscilloscopes, etc. Standard Communication between the blocks should be implemented on standard protocols such as Ethernet, IEC 488, etc.
   - Minimize the hardware to be designed especially for this purpose
5. All tests are executed fully automatic. This is a very important feature for any conformance tests as then such tests do not depend on the actual condition of a tester person and his capability to observe and evaluate test measurement results. As such repeatability of tests is ensured any time any place and a good quality is guaranteed.

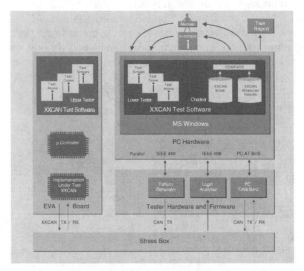

Figure 3a: c&s CAN Tester Architecture

Figure 3b: c&s CAN Tester Implementation

CONFORMANCE TESTS REALIZED AT C&S – In 1995 c&s group was founded at the University of Applied Sciences in Wolfenbuettel, Germany. This was due to Motorola, charging c&s with a first research project to specify and implement the first time conformance tests for CAN OSI layer 2. This project lead together with especially Rober Bosch GmbH and Dassault Electronique finally to the actual ISO CAN test standard ISO 16 845. In the mean time c&s has evolved world wide to the mostly required test site world wide. Some automotive manufacturers do not accept CAN devices delivered by their module supplier except they have been tested by c&s. Furthermore c&s has been asked to re-use its communication protocol conformance test knowledge to apply it other CAN OSI layers and to other types of protocols. All of these activities have been done in very close cooperation either with automotive manufacturer and or semiconductor manufacturers.

As such for instance the Low Speed Fault Tolerant CAN Transceiver Tests have been developed in the ICT consortium under the support of Audi, BMW, DaimlerChrysler, PSA, Volkswagen; the High Speed CAN Transceiver Test have been sponsored by Ford_US, a further High Speed Transceiver Test will be specified in the ICT group. The TTx protocol tests are done in cooperation with semiconductor manufacturers. LIN tests are specified under the umbrella of the LIN Consortium and in special cooperation with car and semiconductor manufacturers. Further protocol tests have been developed for instance for eBUS protocol together with corresponding industries; others on request.

C&S CONFORMANCE TEST IMPLEMENTATIONS					
OSI	CAN		TTx	LIN	Other
7	OSEK/VDX NM, ...				
....					
2	c&s soft-com driver		c&s ..	c&s 2 ..	c&s 3 ..
	c&s robustness				
	c&s Processor Interface	c&s Gateway	c&s TTx		
	c&s ISO 16 845 enhanced				
	~ ISO		16 845		
1	Transceiver		c&s 4 ..	c&s 5 ..	c&s 6 ..
	ICT-c&s 7	Ford_US-c&s			
	Low Speed / High 8 Speed	High Speed			

Table 4: Overview on Existing Communication Protocol Conformance Tests

Tests are all performed in real time. That is why a high number of tests can be executed in rather a short time frame, typically in less than 2 months. This is the reason why simulation phases, which typically have been carried out at the semiconductor manufacturer's site and which are supposed to be quite exhaustive, can not cover so many test cases as if testing is carried out in real time. The difference is in the range of several tenth of orders of magnitude. As such the devices under test – DUT – must be provided as such that they can be run in real time. Therefore typically device manufacturers supply test modules either in an early design phase in

[2] currently under design
[3] ask c&s for more information
[4] currently under design
[5] currently under design
[6] ask c&s for more information
[7] sponsored by Audi, BMW, DaimlerChrysler, PSA, VW
[8] currently under design

an emulated form for instance in an emulator box – such as Quick Turn Box – or as an FPGA-based implementation. As testing includes a customer coaching process possible malfunctions are discussed with the supplier who then can rather easily modify his implementation and deliver an updated solution for the continuation of the test procedure. The final test though typically is carried out on (first) silicon.

The actual c&s capability in protocols conformance testing is shown in table 4.

CONCLUSION

Conformance testing is a good way – and may be the only way – to safeguard actually interoperability in networked systems with mixed suppliers' communication modules, as mixed suppliers' modules system is a must from the OEMs' point of view, because of their second source policy.

Although conformance tests specification and the related tester implementation are living from the engineering genius and practical experience, the back bone of the process is based on a very structured approach, the System Operational Vector Space. As such the overall procedure is a mix of an empiric and systematic process.

The reason to all this is due to the dilemma that there is no "precise" specification of the device to be tested, yet. The solution could be to require formal, executable specification from which in an automatic, formal process test cases could be derived. Although nevertheless this is a very desirable situation because of many reasons, it bears several problems. Very often first inventers of a new solution do not want to be to clear outside of their company. They want to maintain their advantage against their competitors as long as possible; by the way is a push-pull situation as standardization of communication interfaces of course requires an "open" standard as soon as possible. Furthermore formal specifications do not guarantee that some "clever" implementers would realize some features in an optimized way, resulting un-willingly in possibly interoperability conflicts. This has been the case for instance with OSEK/VDX Network Management implementations. Although the underlying specification is written in a formal language, c&s group detected in their non formally derived tests severe problems.

Summing up the recommendation for future progress in coming closer to an easier way to interoperability is to go both ways in parallel: Try to achieve formal, executable specifications for the devices. Derive from there on one hand test cases automatically. On the other hand apply the SOVS process with reduced test vector findings while shrinking the vectors parameter ranges to the "practically" sufficient worst case. Super-impose both results. Enhance them continuously by further test cases derived from special observations from practical systems applications.

As a result test specifications must be published as standard. Further more obviously these standards are "living" standards, as they tend to be updated, modified, enhanced by experience.

Looking at the test cases for a device from another view angle, they could be taken as another type of specification for the component itself. Whenever device manufacturers enter the design process they typically ask front up for the test specification. Baring this in mind test cases' specification should conform to the device specification itself. As such both ways correspondences should be maintained anytime. Consequently this implies that device standards would be "living" standards, too. How can we cope with that in our "traditional" standardization process ?!

REFERENCES

[1]   W. Lawrenz: Networked Systems High Level Design & Test Philosophy and Tools; SAE Conference and Show Detroit, paper 950296, 27.03. - 03.03.95

[2]   W. Lawrenz, editor: CAN - Controller Area Network from Theories to Application; Springer Verlag, 1997, ISBN 0-387-94939-9

[3]   ISO 11898 Road Vehicle – Controller Area Network (CAN), part 1; International Standard ISO 11898, 1999

[4]   Specification of a Fault Tolerant CAN Transceiver; GIFT Working Group (Audi, BMW, DaimlerChrysler, PSA, Volkswagen, Infineon, Motorola, Philips, STMicroelectronics), chaired by c&s group at Fachhochschule Wolfenbuettel, Germany, 2000, www.cs-group.de

[5]   Test Specification for a Fault Tolerant CAN Transceiver; ICT Working Group (Audi, BMW, DaimlerChrysler, PSA, Volkswagen, Infineon, Motorola, Philips, STMicroelectronics), chaired by c&s group at Fachhochschule Wolfenbuettel, Germany, 2001, www.cs-group.de

[6]   M. Scheurer: Entwicklung einer Methode zur systematischen Ableitung von Testfällen für Konformitätsprüfungen in der Informationstechnologie ; Diploma Thesis at Fachhochschule Wolfenbuettel, c&s group, Germany, July 2000

[7]   Abramovici, M.A. Breuer, A.D. Friedman: Digital testing and testable design; Computer Science Press, 1990.

[8]   A.L.Courbis, J.F. Santucci, N.Giambiasi : Automatic Behavioral Test Pattern Generation for Digital Circuits; 1st IEEE Asian Test Symposium, Hiroshima, Japan, November 1992, pp. 112-117

[9]   S.J. Chandra, J.H. Patel: Experimental evaluation of testability measures for test generation; IEEE Trans. On CAD, Vol. 8, N°1, pp.93-97, January 1989.

[10]   Z. Manna, A. Pnueli: How to cook a temporal proof system for your pet language; Report n°STAN-CS-82-954, Department of Computer Sciences, Stanford University , USA, 1982

[11]   K.T. Cheng, A.S. Krishnakumar: Automatic functional test generation using the Extended Finite State Machine Model; 30th ACM/IEEE Design Automation Conference, USA, 1993

[12]   M. Larnac, V. Chapurlat, J. Magnier, B. Chenot: Formal Representation and Proof of the Interpreted Sequential Machine Model; EUROCAST'97, LNCS 1333, Springer, Las Palmas de Gran Canaria, Spain, 1997

[13]   J. Magnier: Représentation Symbolique et Vérification Formelle de machines séquentielles; State Thesis University of Montpellier II, France, 1990

[14]   K. Hoffmeister: Applikations- bedingte Kommunikationsanfor-derungen im verteilten Kraftfahrzeug-Echtzeitsystem und deren Testbarkeit; Diploma Thesis at Fachhochschule Wolfenbuettel, c&s group, Germany, July 2000

[15]   W. Lawrenz, Anne-Lise Courbis, Janine Magnier: Car System Conformance Testing, International Conference Systems Engineering and Information&Communication Technology, Nimes TIC 2000, September 2000

CONTACT

Prof. Dr.-Ing. Wolfhard Lawrenz
Director c&s group
at
University of Applied Sciences
Salzdahlumerstrasse 46/48
D-38302 Wolfenbuettel, Germany

Email: W.Lawrenz@FH-Wolfenbuettel.de
Web: www.cs-group.de

2001-01-2771

Test Strategy and Test Case Design for an Automated, 'To Order', Configuration System for an Electronic Controller

Eric Thomas Swenson, Joseph J. Kartje and William Sears
International Truck and Engine Corp.

ABSTRACT

The test strategy, test design and test case development for a 'to order' configuration system for a configurable electronic control system for a heavy-duty vehicle are discussed. The issues in test case design, to maximize test coverage for a highly combinatoric configuration capability, are reviewed. Then, the role played by the Electronics Integration Test Station [EITS] to evaluate test items as an element of test case design and conducting the tests is reviewed. EITS provides a hardware-in-the-loop test facility for evaluating vehicle electronic control systems. Finally, the coverage achieved is compared with the coverage required as determined by analysis of the order content of vehicles and the analysis of the configurable electronic control system design information.

1. INTRODUCTION

International Truck and Engine Corporation developed a multiplexed electrical system. The multiplexed electrical system was launched into production with the recent introduction of a new line of Medium Duty models, referred to as MD 107 here. The multiplexed electrical system was designed to be configured to match each vehicle order. The evaluation of the multiplexed electrical system required that a method and means be developed to assess the 'to order' configuration system.

This paper discusses the evaluation of the 'to order' configuration system. The evaluation seeks to assure that the system provides accurate configurations for each vehicle order. Section 2 reviews the major system elements, providing a context for the evaluation problem. Section 3 discusses the development of the test cases used to exercise the configuration means, providing design criteria for the test cases. Section 4 discuses the results of analyses performed to review the test cases. Section 5 summarizes the test case design problem.

2. PROBLEM STATEMENT

This section summarizes the major systems being integrated to develop and execute the tests of the 'to order' configuration system. First, the vehicle system supported by the 'to order' vehicle system is summarized. Second, the major functions of the 'to order' vehicle configuration system are reviewed. Third, the Electronics Integration Test Station, used to evaluate the 'to order' configuration system, is summarized. Finally, a problem statement is described. These elements provide the background for the test case design and test case coverage analysis provided in Sections 3 and 4.

MULTIPLEXED OPERATOR CONTROL SYSTEM SUMMARY - Multiplexed operator controls are being implemented for the MD 107 model line of medium duty trucks. The controls design is very flexible. To achieve this flexibility, the control system is provided the configuration of an individual vehicle, to assure its proper operation.

Figure 2-1 illustrates the multiplexed operator controls, and their interaction with the engine, instrument cluster, and antilock brake systems. Up to 24 configurable switches can be installed in a vehicle using one or two switch pack modules. The switches direct the Electrical System Controller or ESC to operate the equipment installed on the vehicle. The number of switches installed is minimized by a priority scheme that assigns functions to switch locations. A given switch function may be assigned at one of a number of switch locations.

Example functions controlled by the configurable switches include Fog Lights, Heated Mirrors, Two Speed Axle Shift, and Rear Axle Differential Lock. The ESC also supports functions such as Headlamps, Parking lights, Daytime Running Lights, Stop, Turn, Hazard, Cruise Control, Windshield Wiper and Washer, and other operator controlled features. Over 148 software features and configuration parameters have been defined for vehicle models to be supported by the ESC. 85 of these were determined to apply to the MD 107 model.

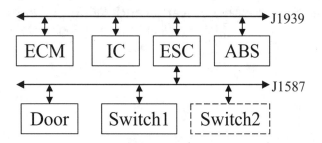

Figure 2-1 Multiplexed Electrical System Block Diagram

'TO ORDER' CONFIGURATON SYSTEM – The 'to order' configuration system for the ESC is called the Configuration File Compiler or CFC. The CFC the supports the ESC and the multiplexed electrical system, by providing the configuration required for each vehicle.

The Configuration File Compiler produces resolved object files to be loaded into the ESC. The CFC combines selected segments of object files, resolves symbolic references and storage allocations to produce the binary image segments for a particular vehicle. Individual segments have predetermined address ranges where they reside in the flash memory of the ESC. The segments are embedded, using S-record format, into an XML document for data transfer to assembly operations. Together, assembly processes and an internal load routine in the ESC accept this input and store it into the microprocessor's flash memory. Figure 2-2 summarizes the information flow into the ESC.

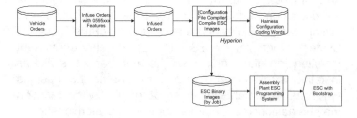

Figure 2-2 Configuration File Compiler Abstract Information Flow

The Configuration File Compiler provides vehicle configurations in a two-step process for the vehicle orders provided as input. In the first step, the required software content is identified for the vehicle. In the second step, a configuration file is produced for each vehicle ordered. These files are loaded into the ESC during vehicle assembly. The process produces log files and error messages that show the software features identified for the ESC and any configuration errors in the software feature content.

ELECTRONICS INTEGRATON TEST STATION (EITS) SUMMARY - The Electronic Integration Test Station provides a Hardware-In-the-Loop (HIL) test and evaluation platform for heavy-duty vehicle electronic control systems. HIL test techniques use mathematical performance models to predict the vehicle's (or system's) behavior. This predicted behavior is fed back into the inputs of the control system. The control systems outputs are then analyzed versus the design parameters of the control system to assess the control system's performance.

Some of the most important inputs to a control system are the vehicle operator's inputs. The initial design of the EITS focuses on presenting the multiplexed vehicle controls with individual operator actions and measuring the ESC's response. Figure 2-3 shows a block diagram the current system. The EITS provides methods to simulate the vehicle operator's input and assess the vehicle's response to these inputs, in an end-to-end manner. This end-to-end structure implements an automated stimulus response system.

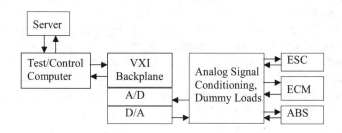

Figure 2-3 EITS Block Diagram

CONFIGURATION FILE COMPILER EVALUATION PROBLEM SUMMARY - There are three major concerns regarding the proper functioning of the CFC.

- First, does all the required functionality for the vehicle's ESC reside in the binary code segments?
- Second, were the binary code segments properly linked with the kernel?
- Third, were there any unintended side effects from a particular combination of features?

The theoretical limit of ESC feature combinations is literally in the millions. A more practical number of combinations was estimated to be approximately 356 powertrain combinations * 10 options (2^{10}) or 364,544 potential test configurations for the MD 107 models including V8 and I6 engine configurations. The reduction in complexity is due to the mutual exclusion among feature options.

Automated test sequences can provide a standard for comparing different methods of producing binary code segments for the ESC. The issue lies in designing and constructing test cases that leverage the automated test sequences. These issues include:

1. Defining methods to assess technical performance
2. Defining sufficient test cases for the Configuration File Compiler,
3. Converting test cases into input for the CFC,
4. Maintaining the intended content of a specific test case before and after it is compiled into binary data for the ESC,
5. Completing automated stimulus -- response test sequences for each individual features that is included in the test cases, and
6. Combining individual feature test sequences into a test sequence for each individual ESC configuration test case.

This paper focuses upon the first two issues, the test case design and its evaluation. Section 3 reviews the design and construction of the test cases used to evaluate the CFC. Section 4 reviews the results of the test case design.

3. TEST CASE DESIGN

This section discusses the design approach, design criteria, and construction of the test cases used to evaluate the Configuration File Compiler.

TEST CASE DESIGN APPROACH - A White Box test case design was not possible at the time the test design task was started. The information to generate white box test cases was incomplete. Black box test techniques were then adopted to generate the test cases for the CFC. The black box approach is well matched to the capabilities of the Electronics Integration Test Station, because of its capability to evaluate the end-to-end performance of ESC software features.

Automated stimulus and response techniques, like those provided by the Electronic Integration Test Station can clearly be applied to this problem.

- Measurement of responses is precise and repeatable.
- In general, the operation of individual software features is independent.
- Intended interactions among features are designed into the requirements of the feature.
- Automated test sequences for individual features are available to evaluate them, across different ESC configurations.

Error input for the evaluation of the CFC was not defined in this test design task. A separate test design task describing guidelines for the construction of error input for the CFC, based on missing software features and conflicting software features was prepared and

forwarded to the CFC development team for use in their unit and integration testing for the CFC.

TEST CASE DESIGN CRITERIA – This subsection describes general requirements for test cases. This is followed by assumptions and inferences comprising a design rationale for specific aggregate properties for the collection of test cases.

In general, the test cases must select features that comprise a complete vehicle order, so they may be routed as input into the CFC. An estimate of the expected configuration outcome must be prepared to permit its evaluation. This estimate is needed to compare with the software features selected for the vehicle by the CFC, and to plan the execution of the automated test sequences on the Electronics Integration Test Station.

There several goals that the CFC Test Cases seek to address. 1. Provide a wider variety of features than that experienced by the engineering development and manufacturing process validation builds. 2 Exercise all available ESC software features within CFC. 3 Cover all important combinations. These goals can be more formally stated as follows.

Let $\bar{X}_v$ and μ_v represent sample means and the population mean for the number of software feature codes in a vehicle. $\bar{X}_{4p}$ represents the sample mean for the number of software features within the early MD 107 builds. $\bar{X}_t$ represents the sample mean for the test cases developed to evaluate the CFC. The following assumptions provide a rationale for the test case design.

1. The number and combinations of ESC Electronic Features fairly provide a figure of merit as to the relative complexity of one ESC configuration to another.

2. A stress test approach is more appropriate than a nominal data approach when evaluating the performance of the ESC Kernel and CFC's preparation of ESC configuration files.

3. The larger the number of features in an ESC Configuration the greater the 'stress' on the kernel to execute all the features within the loop time and on CFC to initialize operating constants, assign memory, and organize sequences of logic into blocks.

4. All combinations of following features should be checked including: Brakes, Automatic or Manual Transmission, Axle Effects, Fog Lamps and Daytime Running Lights.

5. The most high value option after the items summarized in assumption 4 is HVAC Diagnostics.

6. Based on 1, 2, and 3, $\mu_v \leq \bar{X}_t$ is the desired relationship between the population and test sample mean number of electronic features.

7. Random selection methods minimize the likelihood of unintended patterns in test cases.

TEST CASE CONSTRUCTION - The cases were designed based on Product Content Variation, using cumulative probability distributions and samples from a uniform random distribution to generate feature content for synthetic orders. The test cases were generated using a spreadsheet.

To satisfy item 4 above, the following features were expanded into all possible combinations, yielding 352 test cases overall. The possible choices for each category appear in parentheses behind it.

- Engine (International 466)
- Transmission (Manual, Automatic1, Automatic2)
- ABS (Hydraulic, Air, Air with Automatic Traction Control)
- Rear Axle (Single, Single with locking differential, 2 speed ,Tandem with Power Divider lockout, Tandem with Power Divider Lockout and Locking Differential)
- Fog Lamps (Yes, No)
- Daytime Running Lights (Yes, No)

This expansion was recorded in a spreadsheet and used to guide the selection of sales feature codes. For sales feature categories that were fixed by the expansion, the selection of the sales feature code for that category followed a nested if pattern. The nested if pattern for the selection of the transmission sales code is provided as an example below.

If test case transmission = manual

 Then transmission sales code = manual_sales_code

Else If test case transmission = automatic1

 Then transmission sales code = automatic1_sales_code

Else transmission sales code = automatic2_sales_code;

For optional software features, nested if statements modeled a cumulative probability distribution. A Uniform random number sample, or URNS, was drawn for each unique selection decision. The cumulative probability distribution was evaluated against the sample, selecting a sales code among a set of mutually exclusive alternatives. This process is modeled by the nested if statements in the example below.

If URNS Feature1 < ThresholdA for Feature1

 Then Feature1 Sales Code = Sales Code 1A

Else if URNS Feature 1 < ThresholdB for Feature1

 Then Feature1 Sales Code = Sales Code 1B

Else Feature1 Sales Code = Null;

In the example above, the nested if statements implement a cumulative probability distribution for three alternatives. The distributions used were typically weighted to add features to the test case, in light of the performance goal given in item 6 above. For example, the weighting to add the air conditioning option that drives the inclusion of HVAC diagnostics was 71 percent, comprised of air conditioning at 49% and air conditioning with automatic climate control at 22%. The weighting for the heater/defroster option was 29%.

Where needed, additional logic was included to assure that compatible systems were being ordered on the vehicle based on the specific model selected for the test chase. Caution was exercised to insure that a random sample was not used for multiple feature selection purposes. In specific cases the model was driven by the selection of the tandem rear axle with is only available on selected MD 107 Models.

The result of the logic placed sales codes in the individual cells of the test cases contained in the spreadsheet. Additional logic was created that modeled the software feature selection process within the CFC. This logic provided an estimate of the expected content within each test case. These estimates were stored in a database to drive the execution of the software feature tests in the CFC. A Visual Basic macro was written to convert the spreadsheet format into the file format used for input into the CFC.

Section 4 analyses the test case design against these criteria and discusses the results from executing the test cases and evaluating the performance of the individual configuration files produced by the CFC.

4. TEST CASE DESIGN ANALYSIS

COMPARISON OF TEST CASE DESIGN CRITERIA - Table 1 compares the count of the number of features in the set of test orders. These statistics were generated from the Configuration File Compiler log file for the test cases, using SQL queries written for a relational database. The queries provide data to compare the minimum, average, maximum, and standard deviations of the number of features provided for each individual vehicle or job. The columns compare the feature content count of the test cases to the feature content count of the MD 107 orders as of February 1, 2001.

	Test Orders	MD 107 Order Board
Number of Jobs	352	2756
Minimum Job Feature Count	41	12
Average Job Feature Count	$\overline{X}_t$, 50.409	$\overline{X}_{4p}$, 41.850
Maximum Job Feature Count	58	54
Standard Deviation Feature Count	s_t, 3.3914	s_{4p} 2.5801

Table 1 – Job Feature Count Statistics

The statistics shown, match the desired criteria discussed in section 3, as $\overline{X}_t > \overline{X}_{4p}$ and $s_t > s_{4p}$. The greater variation in the number of features in the test cases, is believed to reflect the construction of option feature content using uniform random distribution. The order board provides the best population estimate available, at this time; however it statistics are believed to be skewed due to the low number of features that were build into some vehicles, during the manufacturing process validation, as demonstrated by the minimum feature count of 12. This corresponds well to the data reported in Table 2 below, where more unique combinations were generated in the tests cases than existed on the order board for the MD 107 model.

COMPARISON OF TEST CASE DESIGN TO VEHICLE POPULATION SAMPLE - After it was made available, white box test level information was used to develop a set of binary interactions among features. Table 2 compares the potential complexity faced by the Black Box Test Case design and its coverage of this complexity.

	Test Case Content	NGV Order Board	Software Feature Design Data Extract MD107
Observed Number of Feature Codes, a	85	83	66
Theoretical Number of Combinations = (a*(a-1)/2)	3,570	3,403	2,145
Observed Number of Unique Combinations	3,455, (set T)	3,107, (set O)	167, (set H)

Note 1: $T \subseteq O \subseteq H$.

Table 2 – ESC Software Feature Code Coverage Summary

The number of observed feature codes in the first row of Table 2 can be considered to be the vertices of a complete graph, k_n. The maximum possible number of combinations of two feature codes is then the number of edges in a complete graph, as shown in row 2. Row 3 shows the observed number of edges contained in each of the subject data sets. The subset relationships in the observed edges are: $T \subseteq O \subseteq H$. In other words, all the observed feature combinations in ESC Software Design Data are expressed in the order board and in the set of test cases for MD 107 features. Note that the number of output – input relationships among software are features, 167, is sparse compared to the observed number of binary combinations

EVALUATION OUTCOMES ENABLED BY THE TEST CASE DESIGN – The following outcomes can be determined from the test case design and construction described in Section 3:

1. Did the Configuration File Compiler provide a configuration file for each test case?
2. Was the intended software content consistent with the expected results?
3. Do all the software features in each ESC configuration operate as determined by the automated test and evaluation sequences programmed for the Electronics Integration Test Station?

The execution of the CFC satisfied these evaluation questions, as discussed in the next section.

CONFIGURATION FILE COMPLIER EVALUATION RESULTS - The test case content that was designed in April 2000 was finalized in March 2001. The test cases were prepared and submitted as order coding tests into the CFC's test facility. Nineteen (19) of the 352 cases were correctly flagged to be in error by the CFC, due to the feature code content that was generated for them. The error was determined to be the content of the spreadsheet, and the operation of the CFC was determined to be correct. No configuration files were produced for these cases. The root cause was determined to be an error in the spreadsheet content.

It was determined that the 333 test cases remaining provided sufficient coverage though additional analysis using SQL queries of the log file database. The remaining 333 test cases were successfully configured and the configurations were subsequently tested using the automated test sequences developed for the ESC using the Electronics Integration Test Station. The CFC satisfied the test objectives outlined in section 3.

5. CONCLUSION

The function provided by CFC, a 'to order' configuration system for a configurable electronic control system was described. The issues associated with its evaluation were discussed. A test case design and construction, featuring the use of Cumulative Probability Distributions was described. The test case design criteria and the performance of the constructed test cases against these design criteria were reviewed. The performance of the CFC was reviewed.

Sections 3 and 4 demonstrate a successful approach for developing test cases for the CFC, a 'to order' configuration system for a configurable electronic control system. The evaluation target, the CFC successfully met this evaluation criteria. This approach leveraged Monte-Carlo simulation techniques to synthesize uncontrolled feature content, while adhering to the test case design.

DEFINITIONS, ACRONYMS, ABBREVIATIONS

ABS: Anti-lock Brake System

A/D: Analog to Digital Conversion

CFC: Configuration File Compiler

D/A: Digital to Analog Conversion

ECM: Engine Control Module

EITS: Electronics Integration Test Station

ESC: Electronic System Controller

HIL: Hardware In the Loop

IC: Instrument Cluster

SQL: Sequential Query Language

VIEWPOINTS ON FUTURE TRENDS

2005-01-1345

Operating System Requirements for Use in Next-generation In-car Infotainment Systems

David N. Kleidermacher
Green Hills Software, Inc.

ABSTRACT

Not long ago, an in-dash CD player was considered state-of-the-art for in-car entertainment. Contemporary infotainment and telematics systems include audio, video, voice recognition and synthesis, wireless networking, navigation, internet access, and safety/security features, sometimes running on a single microprocessor. This complexity puts an increasing burden on the microprocessor's operating system. Next-generation infotainment systems require next-generation operating system architectures that can handle this challenge.

INTRODUCTION

This paper will present major operating system challenges in, and propose architectural solutions for, the next-generation infotainment systems environment. We consider real-time response, security, portability, production cost, and a high-quality tools integration to enable efficient development and analysis of these complicated systems.

REAL-TIME RESPONSE

One of the key requirements of multimedia infotainment systems is a high-quality audio/visual experience. On powerful in-home PCs, the sheer power of the microprocessor (now besting 3 GHz) often makes up for the lack of real-time responsiveness of Windows. Next-generation in-car systems will depend more on the ability of the operating system to guarantee real-time response because the microprocessor may be loaded with more simultaneously-executing, resource-demanding applications. Researchers at Carnegie Mellon University have studied latency problems associated specifically with audio: "Latency is the stumbling block for music software, and the OS is often the cause of it."[1]

SCHEDULING

Consider a scenario where an occupant is requesting directions from the navigation system while simultaneously playing a DVD for children in the rear of the car. The navigation data-processing requires much of the available CPU time but the occupant does not require an immediate response (a few seconds may be typical and expected). On the other hand, the processing of the audio and video samples must occur on time in order to avoid choppiness.

In this example, a real-time operating system can strictly prioritize the audio and video processing above the navigation processing to ensure a high quality DVD experience. A slight delay in the navigation result, due to this prioritization, is not likely to be noticed. In contrast, operating systems with a desktop heritage often use heuristic scheduling algorithms that attempt to dole out CPU time "fairly". While fairness-based scheduling is often good for human-user interactive desktop applications, this style of scheduling does not actually provide a guaranteed response; even one bad guess by the operating system can adversely affect the multimedia experience.

INTERRUPT LATENCY

When a process executes a system call, and the kernel is modifying internal data structures to accomplish the request, preemption is disabled by the kernel in order to ensure that another process cannot be context switched in and execute a system call which could access these internal data structures while they are in an inconsistent state. The problem with this approach is that high-priority interrupt processing can be delayed while the kernel disables preemption on behalf of potentially lower-priority requests (Figure 1). Even simple, self-described "real-time operating systems" can disable interrupt processing in hundreds of locations. In one study of a real-time operating system's interrupt latency, researchers found that some of the disabling sequences contained triply-nested loops with estimated execution times of more than 25,000 cycles [2].

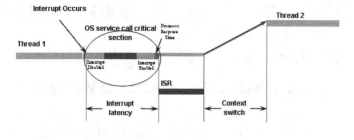

Figure 1

This problem is exacerbated by more complicated, monolithic operating systems. The worst case response time for the high-priority interrupt source (e.g. the audio device) cannot be determined, and thus the quality of the user experience cannot be guaranteed. It is well known that certain operating systems which exhibit good overall throughput have large preemption latencies on the order of milliseconds [3], enough to derail a low-latency device should its interrupt fire at an inopportune time.

A better architecture for guaranteed real-time response is to never disable interrupts in system calls. By always taking the highest-priority interrupt, the kernel ensures smooth performance of the time-critical audio/visual devices. Handling of lower-priority requests is postponed. The kernel must be configurable enough to prioritize interrupt handling appropriately. When a high-priority process is awakened to perform some I/O handling, the kernel's scheduler should automatically inhibit lower-priority interrupts from delaying this processing.

SECURITY

As infotainment systems incorporate internet access and WiFi, security will become a central concern for consumers who know all too well that risk of viruses and other attacks is the price we pay for the convenience of this network connectivity. Although a networked infotainment system can be infected with malicious software, a properly architected operating system can limit the damage of such infiltration and ensure that critical services remain unaffected.

MEMORY PROTECTION

A fundamental requirement for robust infotainment systems is memory-protection. The operating system utilizes memory protection hardware of the microprocessor to isolate unrelated infotainment functions. Malicious code is unable to crash an application or the operating system by corrupting memory. Similarly, an errant pointer caused by a programming error in one application cannot affect another application executing in a separate protected process.

GUARANTEED RESOURCES

Despite memory protection, malicious code can still take down a critical application by starving it of resources. Most operating systems employ a central store for memory resources. As protected processes ask for new resources (e.g. heap memory, threads, etc.), the operating system allocates these resources from the central store. An errant or malicious application can request too many resources, causing the central store to be depleted and critical applications to fail when they attempt to obtain their required resources (Figure 2). A similar denial-of-service attack can occur with CPU time. A malicious process can spawn multiple "confederate" processes that soak up the CPU time, keeping critical applications from accomplishing their tasks.

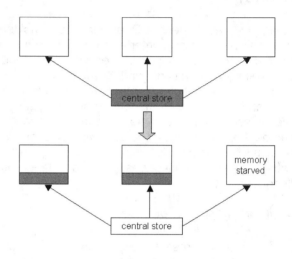

Figure 2

It is possible to architect an operating system such that these critical resources are partitioned according to individual application requirements. An infected process can exhaust its own quota of resources, but it cannot possibly affect the quota of resources held by critical applications (Figure 3).

Figure 3

Despite the compelling security advantage of partitioning, one of its potential drawbacks from a usability perspective is the static nature of quotas. Some complex systems may have varying resource requirements during their lifetimes and may not have enough memory for the union of all the desired quotas. To address this, a partitioning operating system must provide the capability for system designers to create a trusted resource manager that can export memory to, and reclaim memory from, applications on demand. The system designer must be able to specify which critical applications can connect to the resource manager. Malicious software imported into new processes will not have access to the resource manager and are therefore governed by their quotas. Partitioning coupled with a resource manager enables system designers to obtain flexibility where it is needed and security where it is essential.

SECURITY THROUGH SIMPLICITY

As infotainment systems become increasingly complex, integrating more applications from more vendors, the risk of security and robustness problems increases. The operating system, however, is the most critical component since it has direct control of the microprocessor and all of its resources. Whereas a security flaw in an application can make that application unusable, a security flaw in the operating system can be fatal for the entire infotainment system. It follows that use of a secure operating system will increase the overall security and robustness of the infotainment system.

Monolithic Architectures

Monolithic operating systems contain system software, such as networking stacks, file systems, and complex device drivers, that all run in supervisor mode (Figure 4) and provide a plethora of opportunities for hackers to find and exploit vulnerabilities. Examples of monolithic operating systems are Windows, UNIX, and Linux. Most consumers are aware of occasional highly publicized security flaws discovered in these operating systems. What most people may not realize, however, is that many less publicized, yet serious security flaws are being discovered regularly. For example, a quick internet search shows dozens of high-severity Linux security flaws just in the past year; one of the entries is particularly troubling because it reminds us of the fact that no one actually knows how many more flaws are lurking and have yet to be discovered and exploited: "Multiple unknown vulnerabilities in Linux kernel 2.4 and 2.6 allow local users to gain privileges or access kernel memory, as found by the Sparse source code checking tool (CAN-2004-0495)" [4]

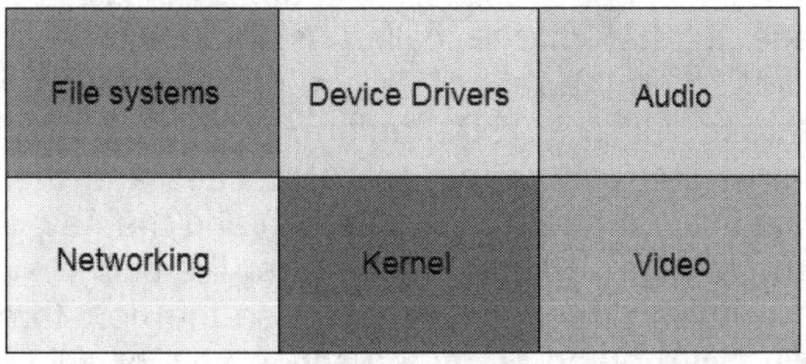

Microkernel Architectures

Microkernel operating systems provide a better architecture for security. A microkernel operating system runs only a minimal set of critical system services, such as thread management, exception handling, and interprocess communication, in supervisor mode (Figure 5) and provides an architecture that enables complex systems software to run in user mode where they are permitted access only to the resources deemed appropriate by the system designer. A virus in one component can not cause damage to the hard drive because the infected component simply does not have access to that resource. Because the microkernel is simple, it can be more easily modeled and verified.

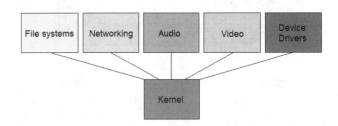

Figure 5 - Microkernel OS

Virtual Device Drivers

Arguably the most important components of the operating system to virtualize are device drivers. Device drivers are frequently compromised by hackers. Monolithic operating systems commonly allow users and processes to dynamically install device drivers into the kernel. A malicious device driver can execute a buffer overflow attack where it intentionally overwrites the runtime stack in order to install malicious code into the kernel. Virtual device drivers prevent these types of attacks because an infiltrated device driver can only harm the process containing the driver, not the kernel itself. In order to facilitate the development of virtual device drivers, the operating system needs to provide a flexible mechanism for I/O control to the virtual driver process. The virtual driver, however, must be provided only access to the specific device resources that the driver is authorized to control (Figure 6).

Figure 4 - Monolithic OS

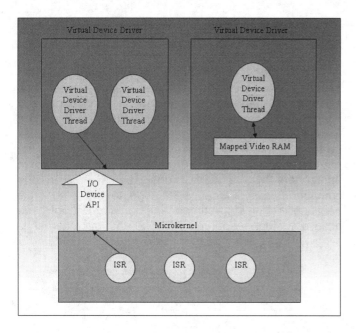

Figure 6 - Virtual Device Drivers

COMMON CRITERIA

Common Criteria is the internationally recognized standard for security evaluation. Common Criteria specifies evaluation assurance levels (EAL), 1 through 7. The higher the EAL for a product, the less likely the product contains security flaws. EAL 7 requires formal modeling of the software and a formal mathematical proof that the software upholds its security functional requirements. As the assurance level increases, the development and life-cycle process requirements (e.g. design, configuration management, testing, maintenance, flaw remediation) become increasingly rigorous.

Linux has achieved an EAL-2 evaluation and Windows has achieved an EAL-4 evaluation. The higher EALs, 5 through 7, are required in order to have any reasonable assurance about security. An NSA analyst said it best when he described EAL-4 as "secure as long as nothing is connected to it."[5] It is simply not possible to achieve high assurance with legacy monolithic operating systems. Not only are they too complex, but they lack a fundamental security architectural design. Systems not designed for high assurance cannot have high assurance retrofitted. According to Common Criteria: "EAL4 is the highest level at which it is likely to be economically feasible to retrofit to an existing product line."[6] Not even Microsoft, with its practically unlimited financial resources, can find and remove all of the security flaws in Windows.

By selecting an operating system with better security, infotainment systems developers also gain a compelling marketing advantage: bragging rights to a more reliable and secure infotainment system that automobile manufacturers and consumers will value.

PORTABILITY

As infotainment systems grow in complexity, integrators will rely increasingly on third parties to provide many of the constituent software packages. Third-party suppliers will, in turn, provide functionality that will be integrated by a variety of systems integrators for all flavors of infotainment systems. Third-party suppliers can get products shipped faster and reach a larger target market if they write their software using open-standard APIs and run on operating systems that support these standards. Conversely, systems integrators can get systems integrated and delivered faster by using an operating system that supports open standards since more third-party software will run unmodified. The universal open standard for operating systems APIs is POSIX.

Recently, the main POSIX standard for operating system APIs, POSIX.1, has been revamped and released as POSIX.1-2001. In addition to many improvements to the API based on years of use and feedback, the new standard comes with a certification program administered by the Open Group that enables operating system vendors to obtain independent acknowledgement of conformance, which in turn gives developers confidence that their POSIX-conformant application code will run unmodified on these operating systems. The list of certified products can be found on the internet [7], and many vendors have announced future plans to obtain certification.

POSIX-like

Infotainment systems developers need to be careful in their use of operating systems that claim to be POSIX-compliant but are not certified conformant. Running POSIX-conformant application code on top of a non-conformant operating system can be problematic. For example, on Linux (which has a POSIX-like API that is not conformant), the **setuid** function only affects the initial thread [8], not the entire process as mandated by POSIX. Consider an application which performs some non-privileged operation (e.g. playing an audio file) whose execution must be logged to a critical log file that only has superuser write permission (e.g. a log of infotainment system usage read by technicians in case of a failure). The application is created setuid root, and when the application starts, it writes an entry to the critical log file, spawns a new thread, lowers the process privilege to that of the caller using setuid, and then proceeds to do the rest of its non-privileged work in the new thread. On a POSIX-conformant operating system, the setuid call affects the entire process and therefore both the initial thread (calling setuid) and the new thread both run in non-privileged mode after the call. On Linux, only the calling thread is affected, so the new thread is actually still running as superuser. A flaw or subversion in the non-privileged code path can now have disastrous repercussions.

436

PRODUCTION COST

The automotive market is characterized by extreme sensitivity to production cost due to large volumes and intense price competition. One dollar of cost can translate into millions in profit (or loss) over a car's production lifetime. Choice of operating system can have a significant impact on the production cost of the infotainment system and, transitively, the automobile itself.

ROYALTIES

Many commercial operating system vendors charge a royalty for each infotainment unit that embeds the operating system. Although royalty charges may be acceptable in certain industries, high volume, low margin products, such as automotive infotainment systems, are intolerant of this drag on profits.

A better business model for infotainment systems is to charge fixed fees up front for licenses, tools and services that help infotainment systems developers bring their products to market.

FOOTPRINT

Infotainment systems developers are also sensitive to per-unit hardware costs and therefore choose the minimum speed microprocessor and minimum size memory components that can still meet feature requirements. Operating system characteristics can affect the choice of microprocessor memory. Monolithic operating systems with a desktop heritage (e.g. Windows, UNIX, Linux) typically have a significantly larger footprint than operating systems designed and optimized specifically for resource-constrained embedded systems. For example, the INTEGRITY operating system from Green Hills Software occupies as little as 30 KB of ROM and 3 KB of RAM in a minimal configuration. In contrast, a basic Linux profile occupies 4.6 MB of memory [9]. When all of the infotainment applications are added on top of the operating system, the resulting image may require a larger flash device or larger RAM chip. This of course increases per-unit production cost in much the same way as a royalty. Even if everything fits into the desired memory chips, developers will often need to spend significant time configuring, tuning, and making functionality-space tradeoffs in order to stay within size limits.

OS-AWARE TOOLS

Next-generation infotainment systems will consist of many complex software applications controlling the multitude of advanced system capabilities. This software will be created by large teams of programmers. These programmers need next-generation development tools to help develop, comprehend, optimize, and debug these complex applications.

Contemporary programmers use integrated development environments (IDEs) to manage software projects. The IDE contains tools for build management, editing code, compiling, debugging, performance analyzing, and more. Ideally, these tools are integrated not only with each other (e.g. the performance analyzer can display performance analysis information in the context of the debugger's source view) but also with the underlying operating system. Examples of operating system awareness include object state awareness, multi-thread debug, and operating system event analysis.

OBJECT STATE AWARENESS

Infotainment applications are often multi-threaded and synchronize using operating system mutual-exclusion (mutex) primitives, message queues, and other kernel objects. Improper use of these objects can lead to deadlock, priority inversions (depicted in Figure 7), and other failures whose causes are difficult to diagnose within a complex application. The IDE should provide a capability for viewing the state of these kernel objects. For example, if a thread is unexpectedly blocked waiting for a semaphore, then the object state of the semaphore can be displayed. The object state display shows the current thread owner of the semaphore. The developer can then interrogate the state of the thread, perhaps discovering that the owning thread is suspended because of an exception (and thus was unable to release the semaphore). Now the developer can focus on why the owning thread crashed, and so on.

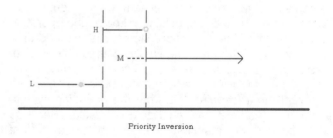

Priority Inversion

Figure 7

MULTI-THREAD DEBUG

There are a variety of important thread-aware features that the debugger must include in order to help find programming errors in complex multi-threaded applications.

- Ability to debug multiple threads simultaneously, each in its own debug window
- Ability to set breakpoints that affect a single thread, an entire process, or a user-defined group of threads (potentially from multiple processes)
- Ability to set breakpoints that gather data on the target system asynchronously (also known as probes or tracepoints).

- Ability to set a breakpoint that when hit causes another thread, process, or thread group to halt
- Ability to trigger a complete system halt from a breakpoint

What all of these features have in common is providing a fine-grained and precise control over thread execution so that bugs can be diagnosed more efficiently.

Consider an audio system that consists of a low-level audio control application (plays and records audio) and a higher-level audio application (e.g. music player) that interacts with the audio control application. The music player exhibits an intermittent failure in which playback spontaneously and unexpectedly suspends. The developer is able to place a breakpoint in the music player at the execution point at which playback is stopped. The developer suspects that the suspension is caused by a bug in the audio control application and thus adds a trigger to the breakpoint which causes the audio application to stop when the breakpoint is hit. Since the audio control application buffers multiple playback operations, this synchronous trigger is required in order for the developer to examine state at the precise time of the failure, before the audio control application has a chance to handle more buffers whose processing could elide the anomalous state. Subsequent analysis of the state involving the halted audio control threads may show a buffer manipulation error that caused the audio stream to be interrupted.

Tracepoints enable the programmer to set breakpoints at strategic locations in the code without halting threads that hit the breakpoints. Tracepoint locations are conveyed to the operating system by the debugger. When a thread hits this special breakpoint, the operating system handles the exception, stores user-defined data into a buffer, and then immediately resumes the thread. This allows interesting data to be collected in real-time with minimal intrusion. The developer uses the debugger to collect the tracepoint data from the operating system at some later time and analyze the results. Because some failures only manifest when running the system at full speed, this kind of operating system integration with the debugger can prove invaluable.

OS EVENT ANALYSIS

Similar to tracepoints, event logging is a mechanism by which interesting events on the target system can be logged and saved away for future analysis. Operating system level events, such as interrupts, context switches, system calls, and exceptions, are logged to target memory while the system runs at full speed. Unlike tracepoints that are user-defined on the fly, operating system events are pre-instrumented by the operating system developers who understand all of the interesting points of kernel execution. Like tracepoints, event logging does not require user interaction, making it a low-intrusion yet effective analysis technique. When a problem occurs in the infotainment system, the event log can be uploaded to the host computer and analyzed with a powerful graphical user interface (Figure 8). Since the

operating system generates the event log, a tight integration between the IDE's event viewer and the operating system is required.

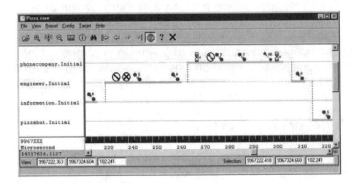

Figure 8 – OS Event Analysis

One of the most famous success stories of using event logging to debug a complex embedded system occurred with the Mars Pathfinder mission in 1997. Soon after landing, the Pathfinder began experiencing system resets that caused loss of scientific data. A failure was captured in an event log and downloaded and analyzed by JPL engineers. The priority inversion bug [10] that was causing the system resets was easy to see in the event viewer. JPL uploaded a software patch to the Pathfinder spacecraft that fixed the problem. Developers of next-generation infotainment systems will find this type of event analysis tool to be critical in understanding system behavior and locating flaws that only manifest at full speed and in the midst of complex system-wide interactions.

TRACE

One of the potential problems with instrumented event analysis and tracepoints is that despite running at full speed, the act of executing the logging code itself can cause subtle timing differences that prevent the manifestation of a bug. In addition, the extra memory overhead of the instrumentation may be prohibitive for resource-constrained systems. Not too long ago, hardware and firmware developers commonly used in-circuit emulators (ICE) to help find problems intolerant of traditional software debugging techniques. One of the most cherished features of the ICE was *trace*. The ICE would record into a trace buffer all instruction and data accessed made by the CPU during its execution. This unprecedented view of execution made some of the most difficult bugs easy to find. Tics Realtime presents several realistic case studies of the use of in-circuit emulator trace to find some insidious bugs [11].

Next-generation hardware trace technology

Unfortunately, ICE technology has not been able to keep up with changes in microprocessor technology, in particular the speed of microprocessors [12]. The good news is that on-chip microprocessor trace is on the uptrend. Many flavors of ARM, PowerPC, and other microprocessors are being outfitted with an on-chip capability to output the same type of instruction and data information to an external hardware debug port. Connecting to on-chip trace ports is much easier, external debug hardware (to collect the trace and transfer it to the IDE) is simpler and less expensive than traditional ICEs, and semiconductor vendors are coming up with clever compression mechanisms that enable the debugging hardware to keep up with high-frequency processors. For example, the latest ARM trace implementation outputs on average only a single byte of instruction trace data for every seven core cycles [13].

Next-generation trace analysis tools

Trace data is historically presented in the IDE as a sequential list of instruction and data accesses. As infotainment systems gain in complexity, large trace buffers are required to store enough history for developers to gain insight into complex sequences of events that led to a failure. But developers are unable to make sense out of a trace listing containing millions of instructions. So the trace tool must mine the trace data to present a higher-level system view of execution. For example, the trace tool can map traced instructions to code functions and show the historical execution path by function name (Figure 9).

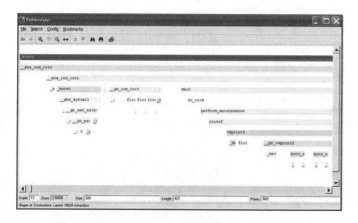

Figure 9 - Execution Path Analysis

If the trace tool is operating system-aware, it can search the trace data for context switches, system calls, and exceptions in order to present the same event analysis display as discussed earlier, except this display is accomplished without any software instrumentation. Some vendors are even using the trace data to drive a virtual multi-threaded debugging session that can run both forwards and backwards in time. A memory corruption can be uncovered simply by setting a watchpoint on the corrupted memory location and running the debugger backwards until the watchpoint is hit. In addition, the debugger can display the values of variables and registers as if the developer was debugging a live system.

CONCLUSION

Next-generation operating system architectures are required in order to ensure that future complex, feature-rich infotainment systems are delivered with the reliability, security, real-time performance, and controlled production cost that the automotive industry and consumers alike will demand. Along with the right operating system, infotainment systems developers need operating system-aware development tools, particularly for debugging and analysis of these complex systems, in order to reduce development cost and time to market.

REFERENCES

1. *Low-latency Music Software Using Off-the-shelf Operating Systems*; Eli Brandt and Roger B. Dannenberg; http://www-2.cs.cmu.edu/~rbd/papers/latency98/latency98.htm
2. *Worst Case Execution Time Analysis, Case Study on Interrupt Latency, For the OSE Real-Time Operating System*; Martin Carlsson; http://www.astec.uu.se/publications/2002/Carlsson-WCET_Exjobb.pdf
3. *Linux extends reach with real-time capabilities*; Charles Murray; http://www.eetimes.com/showArticle.jhtml?articleID=18309476
4. ICAT Metabase - CVE Vulnerability Search Engine; http://icat.nist.gov/icat.cfm
5. Technology Conference; at Green Hills Software, Santa Barbara, CA; 2003
6. *Common Criteria for Information Technology Security Evaluation Version 2.1*; Part 3: Security assurance requirements ; http://csrc.nist.gov/cc/Documents/CC%20v2.1/p3-v21.pdf
7. IEEE POSIX Certification Register; http://get.posixcertified.ieee.org/cert_prodlist.tpl?CALLER=index.tpl
8. *The Native POSIX Thread Library for Linux*; section 8 – Remaining Challenges, p. 16; Ulrich Drepper and Ingo Molnar, Red Hat, Inc.; http://people.redhat.com/drepper/nptl-design.pdf
9. Montavista frequently asked questions; http://www.mvista.com/products/faq.html#q2
10. *What really happened on Mars*; Michael B. Jones; http://research.microsoft.com/~mbj/Mars_Pathfinder
11. *Debugging with an In-Circuit Emulator*; http://concentric.net/~tics/tics0297a.htm
12. *Introduction to In-Circuit Emulators*; section "Practical Realities"; Jack Ganssle; http://www.embedded.com/story/OEG20011126S0065
13. *How CoreSight Technology Gets Higher Performance, More Reliable Product to Market Quicker*; William Orme; http://www.arm.com/pdfs/CoresightWhitepaper.pdf

CONTACT

David Kleidermacher is vice president of engineering at Green Hills Software where he has been developing real-time operating systems and tools for embedded systems for the past fourteen years. David has a bachelor of science in computer science from Cornell University. Contact David at davek@ghs.com.

2004-01-0206

The Requirements of Future In-Vehicle Networks and an Example Implementation

Stephen Channon and Peter Miller
Ricardo Consulting Engineers

ABSTRACT

Electronics is driving 90% of the functional innovation in vehicles which is generating a demand for more (single function) control units to realise the new feature content. Adding such extra ECU's cannot be supported without limit due to packaging space on the vehicle and the significant increase in electrical system complexity. Also, the need for functional integration (between ECU's) is necessary to satisfy key market trends for improved vehicle safety and drive-by-wire capability. This need would not be met by lots of single function ECU's which is leading to a demand for new in-vehicle network architectures. At the same time, the OEM's are looking to define the vehicle "brand image" through advanced software applications which need to be integrated with supplier software within multiple ECU's.

This paper describes the impact on vehicle electrical networks of the multiplying ECU problem (driven by the multiple sensor and vehicle control system technologies) and describes possible system architectures to support the increasing demand for functional integration. An approach to support the evaluation of network architectures shall be illustrated using the development of a prototype network vehicle controller that provides a high performance microprocessor (>500MIPs) and high speed network interfaces.

INTRODUCTION

Technology trends are showing that the demand for more processing power and more electronic control units (ECU's) in the vehicle is growing at a significant rate [Figure 1]. This trend is a consequence of the speed at which new applications are being developed to support the driver where each associated sensor or function is typically demanding a dedicated ECU. The application areas driving this trend are increased safety and infotainment.

SAFETY - New active safety applications are making use of the capabilities of radio frequency, laser and digital video sensors to monitor the local external environment around the vehicle. In order that these sensors work effectively they need to be located around the vehicle

perimeter. Many of these locations already have some electronics, however the additional sensors are increasing the complexity of the installation, for example requiring dedicated communications busses in the interior headlining, door mirrors, front and rear bumpers.

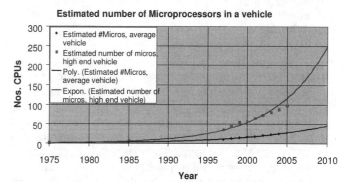

Figure 1 – Projected growth in number of processors/vehicle

INFOTAINMENT - Infotainment is rapidly developing to become a standard installation requirement at vehicle build to include provision for mobile phone and navigation features alongside the CD/DVD. Most of the associated ECU's interconnect using a dedicated communications bus and link into the overall vehicle electrical architecture via a gateway. More advanced feature content is being developed by considering what extra information could be made available by using the positioning and communication systems as sensors to enhance vehicle safety and control.

DRIVE-BY-WIRE – Brake-by-Wire and Steer-by-wire in particular offer significant opportunities for vehicle "branding" as well as environmental (no hydraulic fluid) improvements. They are also enablers for many "integrated safety" applications however the sub-systems are themselves extremely safety critical. Electrical system integration for all these applications is a complex task. New approaches, technologies, tools and processes are needed and the electrical architectures will need to change to meet the challenge.

FUNCTIONAL INTEGRATION - Enhanced safety and additional features can be provided by combining the functionality of the individual sensors. This functional integration needs significant processing power and there are several options as to how and where this could be provided. Further functional integration with the vehicle dynamic control systems offer future applications such as autonomous collision avoidance.

IMPACT FOR ELECTRICAL ARCHITECTURES

Vehicle electrical architectures will naturally evolve from the existing systems mainly due to the number of legacy control systems and the need to manage risk. New Drive-by-wire systems can replace existing vehicle dynamics control sub-systems on a functional basis, provided that the sub-system safety requirements are met, similar to the way that today's ABS/ESP controller meets the vehicle dynamics safety case.

DEMAND FOR NEW BUS PROTOCOLS

Functional integration requiring interaction between the sub-systems is driving the need for new network requirements:

- Interaction between Drive-by-wire sub-systems will require a deterministic communications protocol to replace the existing high speed CAN bus and there are a number being developed [1] that include Flexray, TTP/C and TTCAN. To ensure the most cost effective approach for the whole network the challenge shall be to determine which network nodes need to have a deterministic communications bus connection, as the new protocols will initially have an associated cost premium.
- Integrated safety applications requiring detailed information from the wide range of new sensor systems will need much higher communications bandwidths than provided by existing bus protocols. The functional partitioning of the sensors and safety applications is thus crucial to be able to fit in with the existing bus structure. For example, it would be a radical departure to suggest that the external monitoring sensors link into the (higher capacity) MOST bus protocol that is typically used for the Infotainment features, given the respective locations of the control units and the need to translate information through a gateway to access either the Body control bus for driver warning or the vehicle dynamics control bus for autonomous vehicle control. Safety considerations may also rule out the use of MOST, even though it's speed is attractive.

FUTURE CONTROL UNIT NEEDS

A key requirement for new architectures is to ensure that they are cost effective while at the same time meeting both current and future needs within the vehicle lifetime. As well as providing the right network topology and communications busses, the control units technology will need to provide significantly enhanced processing power

to support the range of new control algorithms. Design of a more capable electronic control unit (ECU) will also need to consider

- Hardware partitioning between application and safety checking microcontrollers and circuits and the protection of interfaces
- Real Time Operating Systems for scheduling and management of complex tasks and data memory
- Software partitioning between the application level code and the low level driver software and programming structures to enable integration of both supplier and OEM control applications
- Safety strategy for hardware fault tolerance and how redundancy can be built in
- Which network links are needed and thus which communications busses, so as to access the appropriate control functions.

Such considerations will already have been made for existing control systems. As the new requirements of the advanced drive-by-wire and integrated safety are captured, so a value analysis will be required to assess whether to introduce a new ECU for the required function or to integrate the function into an existing controller with expanded capability.

NETWORK CONFIGURATIONS

In particular, value analysis is required when looking at the higher level vehicle control systems that require a combination of and interaction between existing sub-system controllers. Should a combined vehicle control function (e.g a command to brake and steer at the same time) be integrated between existing ECU's (a distributed approach) or should a separate controller (a centralized approach, which may be a modified existing sub-system controller) be provided purely to implement the higher level functionality? Combination of integrated safety and drive by-wire for collision avoidance is a good example of a representative feature set.

Consider the case of integrating a new function onto an existing controller, an approach towards centralized network control. While the concept may be quite straightforward, the implementation will be significantly affected by the impact on overall system safety and the need to avoid single points of failure. Equally the system and ECU need to be designed for scalability such that there can be a graceful degradation in performance. It will be possible to design an ECU to meet such criteria, however the controller will be large, dedicated and may lack flexibility to cope with future system expansion.

In comparison, an approach towards distributed network control shifts the emphasis for vehicle system safety from a single sub-system/centralized controller to a network concept. The network will need to enable the synchronization, fault tolerance, redundancy and checking processes that are necessary and for this case are managed and operated between a number of nodes to provide the system scalability.

The core requirements for a controller to satisfy either a centralized or distributed network approach are very similar:

- More processing power
- More memory capability
- Multiple communications protocols (some of these might be internal for the centralized approach)
- Operating systems and software partitioning

Yet the approach to meet the vehicle safety case can differ according to which network topology is chosen. The safety requirements will demand that there is a robust level of checking between microcontrollers, whether a single centralized ECU comprises multiple microprocessors or whether the processing and checking is distributed between processors in a number of network nodes. The diagrams below indicate some techniques to increase the level of safety in ECU systems:

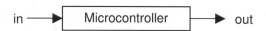

Figure 2: No fault tolerance

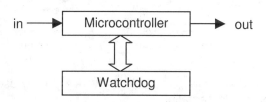

Figure 3: Watchdog, attempts to fail silent

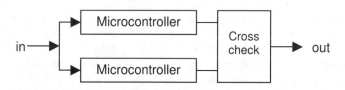

Figure 4: Duplex – Fail Silent

A simple implementation of the triplex system as shown in Figure 5 does not solve the problem of faults in the input (in), the voter or the output (out), where the system configuration is not protected against a single fault in either the input or the output. A more comprehensive system uses multiple sensors/actuators and a distributed voting system to avoid such single point failures, as shown in Figure 6. In this case the individual control units use a separate and redundant high speed communications network over which to share data for the distributed voting protocol. Basic system functionality is still provided by the individual control units while the system protection against faults will need redundant network links from multiple control units to the actuators and the addition of redundant power sources etc. [2]. It should be noted that distributed voting implies a synchronous system, with the high speed communication network also having to be used to provide time synchronisation.

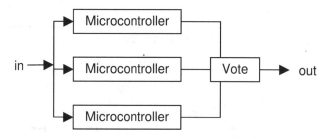

Figure 5: Triplex – fail operational on a single fault

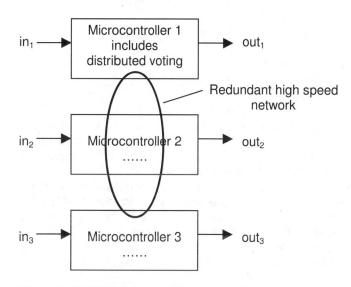

Figure 6: Distributed voting via a redundant network

In redundant systems, common mode failures occur when a single failure affects more than one module at the same time [3]. Design mistakes (in the specification, hardware or software) are potentially the worst type of common mode failures.

In a distributed network, the voting and time synchronisation need to be implemented over high speed communication busses between the network nodes. In a centralized network, the data sharing between the processors can be implemented using different hardware solutions, such as shared memory or dual port RAM instead of the (serial) bus communications. Therefore the additional ECU requirements for a centralized controller design could be said to be a specific case of an ECU that meets the distributed network requirements i.e. a design for a distributed system would work in a centralized system, but the converse is not true.

EXACT AND INEXACT VOTING

In a distributed system, each microcontroller in a single node has to reach the same decision as all the other nodes, using only data available to it via the network (Figures 6). This approach avoids a separate "voter" which would become a single point of failure (Figure 5).

Exact voting can be used for example on digital inputs (just 0/1), or data that is the result of calculations done on previously voted data. Inexact voting is used for analog inputs, etc, which may not be exactly identical. Table 1 below illustrates the results that a distributed voting system would need to achieve.

Case	In 1	In 2	In 3	Out 1,2,3	
				Exact Vote	Inexact Vote
1	1	1	1	1	1
2	0	1	1	1	1
3	0	1	2	Error	1
4	0	1	200	Error	0.5
5	0	100	200	Error	Error

Table 1: Examples of Exact and Inexact Voting results

Case 1 shows the simple voting case when each input is identical. For case 2 when one of the inputs is different, the vote can still be determined on a straightforward majority basis. However in case 3 when all three inputs are different, an exact vote will return an error while an inexact vote could still return a result agreed on a consensus basis. In case 4, one input value is significantly different to the other two (different) inputs, so the exact vote will return an error while an agreed result could similarly be returned for an inexact vote on a consensus basis. However in case 5, no result can be returned from either exact or inexact voting as each of the three input values are significantly different to the other two such that no output value can be agreed.

Ricardo used this approach to voting in the distributed example implementation, where the inexact voting is used to achieve the time synchronisation.

EXAMPLE IMPLEMENTATION

Ricardo have developed an advanced Network Vehicle Controller (NVC) [4] and have used three of these units in a distributed network configuration to demonstrate the concepts of

- Time synchronisation between network nodes and
- Distributed voting algorithm implemented in software

over high speed communication links. The concepts implemented as part of the development of the network controller are not intended to represent the architecture for every vehicle node, rather to suggest an approach to satisfy the safety requirements of future vehicle integrated dynamic control between relatively fewer high power nodes. The diagram in figure 7 shows the high level topology of the network, where an NVC has been used for each node and the indicated functions represent the potential ECU's which might contain the future applications and algorithms.

NETWORK PERFORMANCE

So as to assess the needs for the synchronisation and voting algorithms in the network, two key performance

criteria are required in a prototype distributed network node:

- Higher processing power
- High speed inter-processor communications.

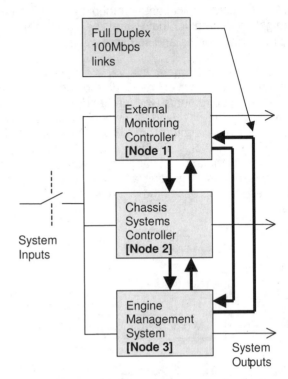

Figure 7: High level distributed network topology used in example implementation

ECU Processor performance

A high performance processor is required to provide the raw computing performance and also to manage the overhead associated with data handling and data memory. Existing state-of-the-art microcontroller performance is typically found in powertrain ECU's, one of today's most widely used devices being based on the Motorola Power PC Oak family. The NVC uses a much more powerful processor from the Power PC family. Table 2 compares the typical processing power performance (based on manufacturers data of performance in MIPs) of the NVC prototype to a typical powertrain ECU today:

Description	NVC	Powertrain ECU
Microcontroller performance	500MIPs	50 MIPs
Average time per instruction	2nsec	20nsec

Table 2: Comparison of microcontroller performance

It should be noted that three NVC units provide the equivalent processing power to the 20+ ECU's found in a typical vehicle today. Figure 8 illustrates the trend toward the demand for increased processing power in vehicles.

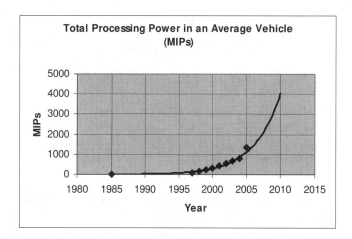

Figure 8: Increase in processing power/average vehicle

Communications bus speed

So as to assess the overhead of data checking between network nodes, a bus speed greater than the widely used high speed CAN bus in powertrain applications was used on the prototype example implementation. Although the NVC includes CAN bus capability, the data checking between network nodes is conducted via dedicated links using 100Mbps Ethernet. Here the Ethernet is configured as point to point between nodes for full duplex operation. Table 3 compares these data transmission rates:

Description	Ethernet	High Speed CAN
Max. Communication speed	100Mbps	1Mbps
Data byte transmission speed	10Mbyte/s	0.1Mbyte/s
Data transmission time/byte	100ns	100µs

Table 3: Comparison of data transmission speeds

The transmission time will be longer than suggested by the network rate due to the protocol overheads and cable propagation delays of ~6ns/m etc. Ethernet was chosen because of it's wide availability (even integrated into microcontrollers), high performance and also as it has been subject to rigorous analysis, including formal proofs [5].

Why is this important?

A comparison of the ratios of processor speeds (instructions per second) against data transmission speed (bytes/sec) is of interest in that the lower the ratio, the less time the processor is waiting for data transfer to take place before further (relevant) tasks can be performed. For the NVC the ratio is $500*10^6:10*10^6$ or 50:1 and for a powertrain controller it is 500:1. Thus the combination of NVC high performance processor and 100Mbps Ethernet gives an order of magnitude improvement in "throughput".

This becomes important when considering data transfer over a network that is managing a regular task at 1msec intervals. A 1ms task represents 500,000 instructions of a 500MIPs processor and 50,000 instructions at 50MIPs, both allowing a significant amount of data processing to be undertaken. The available time for "real work" is, though reduced for distributed network nodes as time synchronisation and voting tasks to validate system safety and fault tolerant checks first need to performed, as illustrated in figure 9

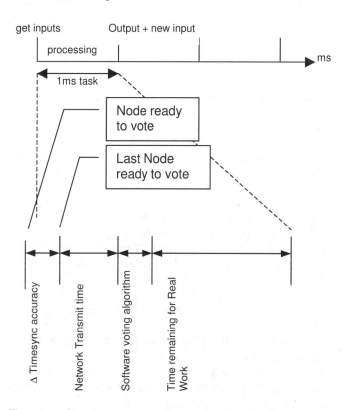

Figure 9: Synchronisation and voting timing during 1ms task

The Network Transmit Time period shown in Figure 9 is the time taken for each node to receive the data for voting from the other two nodes. To achieve this the network transmit time period is split into two separate stages. Stage 1 is when each node sends out its own data to the other 2 nodes (Figure 10). The notation D1 is the voting data for node 1 etc. and each node transmits and receives data in parallel.

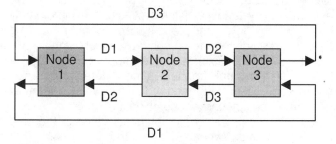

Figure 10: Stage 1 transmission of voting data

445

Stage 2 is when each node retransmits the data received from each other node and passes it forward *in the same direction*, Figure 11.

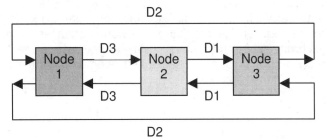

Figure 11: Stage 2 re-transmission of voting data

Thus at the end of each transmission stage, each node has received the voting data from the other two nodes twice, once from each direction around the network ring. This means that the voting data transfer is tolerant of any single network link fault and further, the faulty link can be identified by comparison of data transmission time and direction.

The data to be voted on has been organized into 2 bytes from each node. Thus the total number of bytes to be transmitted during the Network Transmit Time is four which takes 400ns across the 100Mbps Ethernet links.

The time synchronisation accuracy at the start translates into a number of "lost" instructions. This lost time is typically a fraction of the Network Transmit Time period, as too is the time taken for the processor to calculate the voting algorithm. Therefore it is not unrealistic for the total overhead timing associated with the synchronisation and voting to be less than 1µs, giving a ratio of overhead to real work of 1:1000 for a repetitive 1ms task.

Assuming the same task and data voting procedures but using the processing performance and communications bus speeds based on today's powertrain controllers, the Network Transmit time becomes 400µs over a 1Mbps high speed CAN bus. The further overhead from time synchronisation and software voting on the CAN means that very little processor time would be left for real tasks, a typical ratio of overhead to real work being significantly reduced to around 1:2.

It should be noted that while inaccuracies in time synchronisation cause wasted processor time, the extreme of "cycle locking" the processors (which is only really possible in a centralized system) would mean that the probability of common mode faults would significantly increase with a possible adverse impact on the system safety. With a small (random) time difference between processors, common mode faults (which normally happen when an unlikely combination of events occur together) will only be seen on at most one processor at a time, a situation which is corrected by voting. This concept potentially allows the use of (well designed and tested) common hardware and software in each node which dramatically reduces the costs of creating such systems. As an example of suitable testing, the NVC voting software was tested for every possible condition, as well as ensuring that the test case covered every possible executable statement in the code.

COMPARISON TO INDUSTRY DEVELOPMENTS

The example implementation described used microprocessors and higher speed communications busses available today that are not specifically targeted at the automotive industry, yet provide a significant step up in the next level of performance. There are several parallel developments currently that are specifically targeting higher power processors and higher speed networks for the automotive industry. However, the levels of performance are still an order of magnitude lower than provided in the example implementation. Next generation automotive microcontrollers being introduced in 2004 from Motorola in the Power PC family will provide up to 150MIPs (e.g. MPC5554) with 500MIPs forecast to be available around 2008. Flexray [6], a protocol proposed for drive-by-wire systems is only designed to work up to 10Mb/s.

CONCLUSION

This paper has presented an approach toward meeting future requirements in vehicle networks using redundant high speed communications busses with software voting and time-synchronisation. The example implementation demonstrates true distributed control, a target that many OEM's are seeking, to provide the levels of advanced functionality with a minimum number of high performance ECU's.

This approach will have a significant impact on the architecture of future vehicle networks. These are still evolving and there is currently no one standard approach to meet the varied demand for functional integration and vehicle safety requirements. It is, however, clear that both processing performance and communication bus speed in the future networks need to step up significantly from today to meet the requirements. What is less clear is whether the current industry developments provide a sufficient step up in performance to facilitate such a change in architecture as described in this paper, or whether the roll out of architecture changes and integrated features and functions will take longer.

The levels of processing power and communications bandwidth available in 2004 may not support true distributed control that meets the vehicle safety requirements. The question then is; will the OEM's be forced down a centralized network architecture route?

REFERENCES

1. R. Johansson, P, Johansson, K. Forsberg, H. Sivencrona, J. Torin, "On Communication Requirements for Control-by-wire applications", Proceedings of the 21st International System Safety Conference 2003.

2. R. Essermann, R. Schwarz, S. Stolzl, "Fault Tolerant drive by wire systems", IEEE Control Systems magazine Oct 2002 pp64-81.
3. S. Mitra, N. Saxena, E. McCluskey, "A design diversity matrix and analysis of redundant systems", IEEE Trans on Computers May 2002 Vol. 51 no. 5 pp498-510.
4. S. Channon and P. Miller, "An advanced Network Vehicle Controller (NVC) to support future technology applications", AMAA2003, Berlin, Germany
5. F. Wang, Pao-Ann Hsiung, "Efficient & User friendly verifications", IEEE Trans on Computers, Jan. 2002 Vol. 51 no.1 pp61-83.
6. C. A. Lupini, "Multiple Bus Progression 2003", SAE Technical Paper Series 2003-01-0111.

CONTACT

Stephen Channon
Ricardo UK Ltd,
Bedford Heights, Manton Lane, Bedford MK41 7PA
stephen.channon@ricardo.com

Peter Miller
Ricardo UK Ltd,
Bedford Heights, Manton Lane, Bedford MK41 7PA
peter.miller@ricardo.com

www.ricardo.com

Dynamic Discovery Service Protocols for Next Generation Vehicle Network

Rami Baroody, Nizar Al-Holou and Asif Rashid
University of Detroit Mercy

ABSTRACT

The widespread deployment of inexpensive communications technology, computational resources in the networking infrastructure, and network-enabled end-devices pose an interesting problem: how does one locate a particular network service or a device out of millions of accessible services and devices? Traditionally, these services are accessed through well-known URLs or retrieved through search engines. However, these results tend to be very general and may not satisfy the requirements of the user. Moreover, the traditional model cannot react to the dynamic changes in service attributes and client parameters. As a solution, a secure directory tool that tracks services in the network and allows authenticated users to locate them through expressive queries can be used. This paper discusses the Next Generation Vehicle Network (NGVN), an architecture based on Java/Jini and Dynamic Discovery Service (DDS) technology. This paper also discusses what features should be modified to adapt this architecture to implement the required auto NGVN Internet services. Finally, some discovery protocols will be discussed and the application for discovery services will be presented.

INTRODUCTION

Electronics are used in automobiles to improve vehicle performance, serviceability, pollution control, product differentiation, reliability, safety, and convenience features. As customers continue to demand additional functions and features in their automobiles (e.g., navigation system, intelligent highway system support, automated collision avoidance system, Internet access, remote diagnostics), there arises a need for a number of devices, such as sensors, mobile phones, and various computers to be deployed in future cars. These devices need to be interconnected on appropriate networks with appropriate protocols that will allow automobiles to be able to perform these functions at a reasonable cost and with high reliability and availability. Introducing new

technological features in automobiles, such as Dynamic Discovery Service (DDS), becomes of major importance in the case of mobile automobile networks. Dynamic Discovery Service is a system in which clients search through registries to first discover and then invoke services supporting the capabilities they require. A passenger or any electronic device contained in the vehicle would not know which services are available at a particular moment, unless a discovery protocol is used to find them. This situation is rather extreme and requires some special considerations. In this paper, the Dynamic Discovery Service Protocols for Next Generation Vehicle Network have been proposed. In Section II, the Next Generation Vehicle Network architecture is introduced. Section III presents a brief overview of the most important service discovery protocols currently under development. Section IV compares them in more detail, according to their functionality, available implementations, and dependency on operating system, platform, and network transport. Next, section V presents the implementation of Dynamic Discovery Service for Next Generation Vehicle Network. Finally, Section VI concludes this paper.

NEXT GENERATION VEHICLE NETWORK ARCHITECTURE

In automobiles, many sub-systems need extensive interactions for information exchange. For example, an engine management system and transmission controller may work closely together for proper gear change. These kinds of interactions are becoming so numerous that the traditional view of vehicle electronics as independent sub-systems may not be adequate [1-3]. The concept of local area network (LAN) has been introduced in automobiles to improve communications and interactions among different electronics systems. In applying a LAN to automotive electronics systems, an optimal protocol has been adopted for each system and each system may run different protocols that satisfy the data rate for their applications. Multiple sub-networks (CAN, MOST and the internet protocol) have been introduced [4-7]; figure 1 shows multiple network schemes for automotives application. However, these approaches have failed to adequately address and exploit the emerging Internet technologies and services. The main goal of the NGVN project is to design and implement a mobile website for each vehicle such that it is universally accessible (with appropriate security mechanisms) as any contemporary website. Such a website would allow one to access dynamic information about the vehicle almost real-time, such as speed, internal temperature, and fuel supply, as well as live video and audio information from within the automobile.

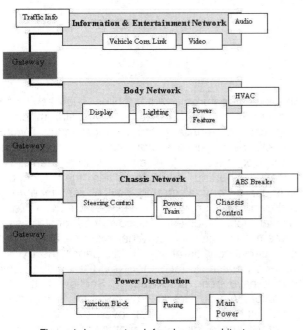

Figure 1. In-car network four Layers architecture

This data is obtained from sensors and other mechanisms placed around the car and connected to the in-car networks. Figure 2 shows the Basic NGVN architecture that is broadly divided into four logical domains:

• The User Interface Domain provides the required communication interfaces to enable all devices (e.g., Cellular, PDA, Laptop) used by a mobile user to interact and exchange information with the NGVN information services

• The Network Interoperability Domain provides all the functions required in order to achieve seamless interactions and exchanges of data among different networks such as different types of cellular networks GPRS [8], CAN [5], and MOST [6].

• The Mobility Support Domain enables the ubiquitous access to the vehicle website, allowing users with Internet access to interact with the passengers in the vehicle by visiting the vehicle website.

• The Auto Information Domain provides the next generation of auto applications and services targeted towards auto users as well as manufactures. This domain will utilize the services offered by an Auto Dynamic Discovery to be developed in this project to support the current and emerging auto information and entertainment services.

The Network Interoperability Domain provides all the functions required in order to achieve seamless interactions and exchanges of data among different networks such as different types of cellular networks GPRS [8], CAN [5], and MOST [6].

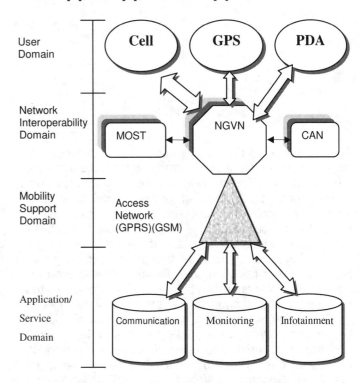

Figure 2: NGVN – Basic Architecture

DYNAMIC DISCOVERY PROTOCOLS

The number of services that will become available in networks (in particular in the Internet) is expected to grow enormously in the next few years. Besides classical services such as those offered by printers, scanners, and fax machines, many novel and revolutionary services will become available. Examples are services for information access, NGVN Internet, music on demand, and services that use computational infrastructures that are being deployed within the network. Following this trend, it becomes increasingly important to give users the possibility of finding and making use of services that are available in a network. What is needed is a functionality that enables users to effectively search for available services that are appropriate to solve a given task. Ideally, users would like to obtain access to services automatically, without needing to reconfigure their system. It would not be convenient or practical for users to search for the IP address of the desired service and manually upload the device drivers. With the widespread deployment of network enabled mobile devices (such as notebooks, PDAs, and enhanced cellular phones), dynamic discovery of services in a visited foreign network and automatic system configuration will be very useful features. This task is addressed by newly emerging *service discovery protocols*, like SLP (Service Location Protocol [9]), Jini [10], UPnP (Universal Plug and Play [11]), Salutation [12], and Bluetooth [13]. In the car environment, passengers could bring their mobile devices into the car and connect them to the car network. These devices could use equipment and services that are installed inside the car. For example, in a mobile office car, there would be a fax machine, a printer, a hard disk, and a color display. A functionality that enables mobile devices to discover and use these services is needed. Furthermore, drivers would be able to locate a service on the Internet by its service type (e.g., restaurants, hospitals) and then make an intelligent selection in case multiple services of the desired type are available based on their attributes (e.g., location, fees). Service discovery in a car environment would simplify the task of maintaining and updating the car network, especially when new services and devices are introduced. The industry and the researching community have recognized that many efforts to create service discovery architectures and protocols have been made. Some of the most important ones are introduced in this section. These protocols are: Jini, Salutation, UPnP, Bluetooth and SLP.

SERVICE LOCATION PROTOCOL

The Service Location Protocol is the service discovery protocol proposed by the IETF. It is being developed by the Svrloc Working Group and is vendor independent. SLP is designed for TCP/IP networks and is intended to become the standard in the Internet community. The current version of SLP is the SLPv2 [14]. The SLP architecture as shown in figure 3 is composed basically of three members: The user agent (UA) is responsible for discovering services on behalf of a client, sending service requests (SrvRqst) to service agents or directory agents. An application (or a user) asks the UA for a service, and the UA eventually returns a service provider's address. The service agent (SA) advertises the service that a service provider supplies, answering the UA's requests with service replies (SrvRply). In environments, in which a directory agent is present, the SA can register the service with it. The directory agent (DA) is an optional central repository of services. A SA would register the service it announces with the DA, and a UA could ask the DA for this information. In this way, the traffic put on the network is reduced and the architecture is more scalable. If no DA is present, the UA will access the SA directly.

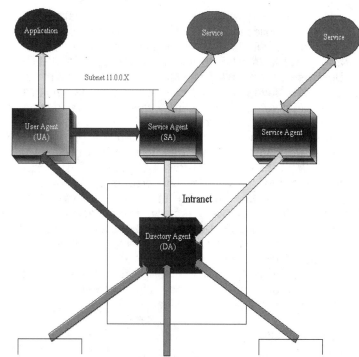

Figure 3: SLP Architecture (Source [22])

In an architecture with DAs, the UAs always try first to discover a DA and, when done, they unicast their requests to it. If no DA is found, they try to find SAs to query. UAs and SAs may know the location of a DA by static configuration (manual or using DHCP's option 78) or by using passive or active discovery. For passive discovery, the DAs announce their presence periodically to the multicast SLP address. For active discovery, the UAs or SAs multicast DAs discovery messages, and the DAs unicast replies. The location of a service provider is made using the so-called service: URL [8]. This is a special URL with the following format: first the prefix "service:" then, the type of service, followed after a colon by the network address and optionally a path. The type of service can contain an abstract type of service and a concrete one, separated by a colon. If an abstract type is requested, any concrete service belonging to that type can be included in a reply. If a concrete type is used, only services that match that type can be included. Services must be registered with a naming authority, which by default is IANA. If they are registered with another naming authority, the name of this authority is appended to the abstract type of service after a period.

JINI

Jini technology is an extension of the programming language Java and has been developed by Sun Microsystems. It addresses the issue of how devices connect with each other in order to form a simple ad hoc network (a Jini "\community"), and how these devices provide services to other devices in this network. Jini consists of an architecture and a programming model. Each Jini device is assumed to have a Java Virtual Machine (JVM) running on it. The Jini architecture principle [10] is similar to that of SLP. Devices and applications register with a Jini network using a process called *Discovery and Join*. To join a Jini network, a device or application places itself into the lookup table on a lookup server, which is a database for all services on the network (similar to the directory agent in SLP). Besides pointers to services, the lookup table in Jini can also store Java-based program code for these services. This means that services may upload device drivers, interfaces, and other programs that help the user access the service. When a client wishes to utilize the service, the object code is downloaded from the lookup table to the JVM of the client. Whereas a service request in SLP returns a service URL, the Jini object code offers direct access to the service using an interface known to the client. This code mobility replaces the necessity of pre-installing drivers on the client. The Jini specifications are open source and may be used freely. However, Sun charges a licensing fee for commercial use. A reference implementation may be downloaded at *http://www.sun.com/jini*. The Jini code can be implemented in 46K of Java binaries.

SALUTATION

Salutation is another approach to service discovery. The Salutation architecture [11] is being developed by an open industry consortium, called the Salutation Consortium (see *http://www.salutation.org*). The Salutation architecture consists of *Salutation Managers (SLMs)* that have the functionality of service brokers. Services register their capabilities with an SLM, and clients query the SLM when they need a service. After discovering a desired service, clients are able to request the utilization of the service through the SLM. Salutation is a rather settled approach, with some commercial implementations, including fax devices and Windows enablers (95/98 and NT), e.g., from IBM and Axis. An example is IBM's NuOffice, a Salutation enhancement for Lotus Notes. Further implementations for Palm OS and Windows CE are planned for the near future.

UNIVERSAL PLUG AND PLAY

Universal Plug and Play (UPnP) is being developed by an industry consortium (see *http://www.upnp.org*), which has been founded and is lead by Microsoft. One can say that it extends Microsoft's Plug and Play technology to the case where devices are reachable through a TCP/IP network. Its usage is proposed for small office or home computer networks, where it enables peer-to-peer mechanisms for auto configuration of devices, service discovery, and control of services. In UPnP's current version (release 0.91), there is no central service register, such as the directory agent in SLP or the lookup table in Jini. The Simple Service Discovery Protocol (SSDP) [12] is used within UPnP to discover services. SSDP uses HTTP over UDP and is thus designed for usage in IP networks. For a description of the UPnP system architecture refer to [13].

BLUETOOTH

Bluetooth is a new short range wireless transmission technology. The Bluetooth protocol stack contains the Service Discovery Protocol (SDP), which is used to locate services provided by or available via a Bluetooth device. SDP is described in the Bluetooth specification part E [14]. It is based on the Piano platform by Motorola and has been modified to suit the dynamic nature of ad hoc network communications. It addresses service discovery specifically for this environment and thus focuses on discovering services; it supports the following inquiries: search for services by service type, search for services by service attributes, and service browsing without *a priori* knowledge of the service characteristics. SDP does not include functionality for accessing services. Once services are discovered with SDP, they can be selected, accessed, and used by mechanisms out of the scope of SDP, for example by other service discovery protocols such as SLP and Salutation (see, e.g., [15], which shows a mapping from Salutation to SDP). SDP can coexist with other service discovery protocols, but it does not require them.

COMPARISON OF THE PROTOCOLS

The presented service discovery protocols are the most likely to succeed in the next years. Now, their main advantages and disadvantages are summarized and their expected future deployment is compared (partly based on the discussion in [16]). Jini is a powerful architecture that gives full functionality, especially considering that it can interoperate with SLP using bridges. The main achievement of Jini is code mobility possibility, but that also produces more traffic in the network and raises the latency of the transactions. It also has the disadvantage that it needs the JLS to be used, while, in ad hoc networks, operating without a central register can be necessary. Jini is platform independent but, as said before, everything must be programmed in Java. That can be considered a drawback itself due to some developers' reluctance to use it. Additionally, every device is required to have a JVM running on it. This can be too expensive for some small devices and may force the use of proxies. The Salutation architecture is rather mature. It is supported by an important consortium, which makes it able to have great acceptance. Its major deficiency is defective scalability, although that can be solved when interworking with SLP. This possibility of operating with SLP, and also with Bluetooth, and its platform and transport protocol independency can make Salutation become the union glue between the different existent protocols and allow interoperability. Universal Plug and Play has great industry support, which suggests that it will succeed. There already exists a certification standard for UPnP enabled Internet gateways. The architecture presents some lacks though. Those are the impossibility of having a central repository of services – which makes scalability worse – and the absence of an advanced searching mechanism. On the other hand, it allows some extra features, such as events notification, device auto-configuration and remote services control. The Bluetooth technology is expected to greatly succeed. Its Service Discovery Protocol is very simple, since it is designed to work in small devices and in ad hoc environments. Unfortunately, this makes the protocol rather limited. It lacks the functionality of the service agents – such as service registration or aggregation – and the lease mechanism, although clients may poll for the availability and estimated life time of services. Furthermore, SDP provides no access to services, only information about them. SDP is restricted to the Bluetooth context, and is not of general purpose. Thus, it is not expected to have full functionality and it will not compete with other service discovery mechanisms, but will rather collaborate with them. The Service Location Protocol is the proposed standard of the IETF. This makes it on the one hand accepted by all the developers but on the other hand not especially supported by any important company. Every technology tries to be able to interwork with SLP, and that and the use of standardized service templates guarantees interoperability. The SLP architecture is of general purpose and is very flexible, as it can work with or without a central register of services and many of its features are optional. Weaknesses of SLP are its dependency on TCP/IP and the lack of some added functionalities. It does not directly provide access to services or any of the UPnP extras such as events notification or remote devices control. These capabilities have to be provided by other protocols or applications themselves, or can be obtained from the interworking of SLP and other service discovery protocols.

Summing it all up, SLP is the more general-purpose protocol, presents flexibility and scalability, and will probably be able to interwork with all the other protocols. Jini is a powerful approach but is limited to certain environments, and its deployment may be difficult. Bluetooth's SDP is a rather simple protocol and its use is limited to a Bluetooth environment, so it will not compete with the rest of the protocols. UPnP and Salutation are industrially supported and will contend for a dominant place, although they present some deficiencies. In the near future all (or most) of the protocols will most probably have to learn to collaborate with each other.

NGVN – BASED SERVICES

NGVN DYNAMIC DISCOVERY SERVICE

At the Computer Networking Laboratory, a proof-of-concept prototype of the Next Generation Vehicle Network has been implemented using Java and Jini Lookup Services [18-19] as well as Dynamic Discovery Service technology. This project adopts the current prototype to implement the required auto NGVN Internet services. Figure 4 illustrates the architecture of a NGVN DDS, which consists of 6 components: Network Service Proxy, Network Service Attributes, Service Manager, Lookup Service, Service Templates and Access Control Manager. These components, their roles in the system,

and their interaction to provide DDS system functionality are described in the following sections.

Network Service Proxy

NGVN information services access auto information/entertainment services through objects called *proxies*, which provide all the code needed to connect to a particular network service. Proxies are analogous to device drivers as they allow an application program to interact with a service while encapsulating it from the implementation details of that service. However, unlike device drivers, which are typically installed by a system administrator before the device can be used, proxies are carried by the services themselves, and are dynamically downloaded by the clients when they invoke a particular network service. Client applications do not have to know the implementation details of these proxies and they do not have to be "compiled in" when the applications are written. These proxy objects typically communicate over the network with the backend portion of the service using whatever protocols supported by the backend system.

Network Service Attributes

Attributes are Java objects that are attached to service proxies. Services attach these attributes when they publish their proxies, and NGVN services (clients) can search for proxies by looking for certain attribute patterns. Attributes are used to associate descriptive information and state with a service. Attributes are important to distinguish between similar services (i.e. services that implement the same interface). Name, address, and location are some of the standard attributes that are associated with services. But sometimes we need to create our own attributes to reveal service specific information. There are some attributes which are static, like name, that do not change often with time, but there are some attributes, like state, that are dynamic and can be updated by the service itself. These changes have to be reflected in all the lookup services that the proxy object is registered with, to ensure that clients have access to the most updated services. Figure 4 details the DDS architecture.

Service Manager

The service manager maintains updated information about the available NGVN services. The service manager detects the changes in the service attributes, compares it against threshold values and if the threshold is exceeded, it updates this information in all the lookup services where the proxy object had previously registered. The services can also have properties, which are not service attributes. These properties can be dynamic and come into effect during the service lookup process. The service manager tracks these dynamic properties and helps the client by providing the most updated values of these dynamic properties.

Lookup Service

The lookup service is a special service that keeps track of all other NGVN services in the community. When a provider of a NGVN service wishes to make itself available, it "publishes" its proxy by storing it in a lookup service. Clients (NGVN users) can then connect to the lookup service and ask what services are available. The lookup service can also inform interested parties when new services appear or when services leave the community through remote events [20-22]. Lookup services announce their existence using the *multicast announcement protocol*. In multicast announcement, interested parties listen on well-known multicast address for announcements about the existence of lookup services. Periodically all lookup services send a multicast message to this address. Once an interested party has received an announcement of the existence of one, it can ask it for its service proxy. It contacts the lookup service using the direct unicast connection, and the lookup service replies with the proxy object that the client can use to communicate with it. The lookup service is the main module in the DDS.

Service Templates

Service templates are Java objects that NGVN clients use to construct their search criteria. The service template matching system is extremely flexible and services can be found based on the set of attributes associated with them using type-based and content-based attribute matching rules. NGVN users/clients are the consumers of services. These clients discover the lookup service using the *multicast request protocol*. Once clients find the lookup service, they search it for services that implement particular services with certain attributes, using templates to construct their search criteria. The client downloads the service proxy and then uses it for communication with the service.

Access Control Manager (ACM)

Every lookup service has an ACM that maintains and updates the list of NGVNs (cars that have subscribed to certain NGVN services) that can access the services. When the clients try to download the proxy object from the lookup service, the ACM checks the list to see if the client is authorized to use the service. If the client is authorized, the ACM allows the proxy to be downloaded; otherwise an "access denied" message is sent back to the NGVN client. The client can request the lookup service, which in turn requests the service for access. If the service accepts the request it updates the access control list through the ACM. Thus, the ACM provides security in the DDS architecture.

Dynamic Lookup

In traditional architectures, services are registered with the lookup service once during a session and continue to use the resources till they are freed explicitly. Also, these architectures do not reflect the dynamic characteristics in the system, which is very important in the NGVN mobile environment. Two methods have been developed to achieve dynamic lookup: Dynamic Update and Dynamic Filtering. In the dynamic update scheme, the service manager detects the changes in the service attributes and compares them with certain threshold values. These threshold values determine that the Jini Lookup Service does not support any inequality matching (like greater than, less than or logical operators) for the attribute values. This is a very rigid constraint in our application. In the dynamic filtering scheme, properties of services that change frequently with time are maintained separately and not defined as service attributes. The service manager has functionalities to retrieve the current values of the dynamic properties when requested by the client.

IN-VEHICLE NGVN SERVICES

The NGVN clients/users will have access to a wide range of auto related information and entertainment services that can be added and removed dynamically. The various suppliers of NGVN services advertise their products through an interface that defines the type and features of their services. There are one or more lookup services running in the network waiting for the NGVN service providers to register with them. A provider locates the lookup service using a combination of multicast announcement and unicast response protocols. This process is known as *discovery*. Then the supplier sends the service proxy object to the lookup service to register itself. This is the *join* process. For a car user, there will be a number of NGVN services and every supplier registers its proxy object with the lookup service it discovers. There can be a number of attributes associated with each service. Some of the important attributes are location, service ID, subscription fee/price, and period. There is a HTTP server running so that the proxy object class can be downloaded on demand when the NGVN client/user makes a request to access one of the provided services. The NGVN service provider that satisfies the request is discovered, and then the lookup service sends the matching provider's proxy to the car NGVN SOAP server. The NGVN server establishes a connection with the service provider and the transaction is started.

CONCLUSION

Service discovery will be an important feature in future network scenarios (e.g., in self-organizing ad hoc networks). With service discovery, devices may automatically discover network services including their properties, and services may advertise their existence in a dynamic way. From the user's point of view, service discovery is about "coming to an unknown network environment with a mobile device and then detecting the available services according to one's need." Presently, there exist a variety of service discovery protocols for this. The most important protocols are SLP, Jini,

Salutation, and UPnP. This paper discussed each of these approaches and compared their respective advantages and disadvantages. Each of these protocols approaches the vision of service discovery from a different perspective. It is likely that in the future, there will continue to be various kinds of service discovery protocols. This makes it important to have bridges between the different protocols to enable service discovery with various devices.

REFERENCES

1. Syed Musbahuddin, Syed Mahmud, Nizar Al-Holou, "Development and Performance Analysis of a Data Reduction Algorithm for Automotive Multiplexing," *IEEE Transaction on Vehicular Technology, Vol. 50, No. 1, January 2001, pp. 162-169.*

2. Nizar Al-Holou, Syed Musbahuddin, "Availability Modeling of Hierarchical Distributed Processing Systems," International Journal of Modeling and Simulation, International Association of Science and Technology for Development (IASTED), Vol. 19, No. 2, August 1999, pp. 137-143.

3. Syed Musbahuddin, Nizar Al-Holou, Syed Mahmud, "Data Reduction Algorithm for Automotive Applications," *SAE 1998 Transactions, Section 6 - Vol. 107, pp. 1667-1670.*

4. Friedrick H.Phail, David Arnet, "In-Vehicle Networking-Serial Communication Requirements and Directions", SAE paper, 8603901998.

5. Bosch, "CAN Specification", ver2.0, Robert Bosch GmbH, Stuttgart 1991.

6. R. Tappe, C. Thiel, R. Konig, "MOST Media Oriented Systems Transport". Elektronik. July 2000.

7. Rami Baroody, Nizar Al-Holou, Salim Hariri, "Development of Car Intranet Infrastructure," Detroit, MI, SAE Technical Paper Series, Paper # 02-0438_112801, March 2002, pp. 61-66.

8. Goddman, D. J., Wireless Personal Communications Systems, Addison-Wesley, Reading, Massachusetts, 1997.

9. Erik Guttman, Charles E. Perkins, Michael Day. Service Location Protocol, Version 2. Internet RFC 2608, June 1999.

10. Sun. Technical White Paper, "Jini Architectural Overview", http://www.sun.com/jini/, 1999.

11. UPnP Forum, "Universal Plug and Play Device Architecture", ver0.91, March 2000.

12. Salutation Consortium, White Paper, "Salutation Architecture: Overview", 1998.

13. Bluetooth SIG. Specification of the Bluetooth System, Volume 2, Profiles, Version 1.1, www.bluetooth.com, February 2001.

14. RFC 2608 E. Guttman, C. Perkins, J. Veizades, M. Day, "Service Location Protocol", ver2, IETF Standards Track, www.ietf.org/rfc/rfc2608.txt, June 1999.

15. Brent Miller, Robert Pascoe, "Mapping Salutation Architecture APIs to Bluetooth Service Discovery Layer", http://www.bluetooth.com, July 1999.

16. C. Bettstetter, C. Renner, "A Comparison of Service Discovery Protocols and Implementation of the Service Location Protocol", Institute of Communication Networks, Munich University of Technology, 2000.

17. Weiser M, "The Computer for the 21st Century", *Scientific American* 265, 3 (September 1991), pp. 94-104.

18. Erik Guttman, James Kempf, "Automatic Discovery of Thin Servers: SLP, Jini and the SLP-Jini Bridge", *IEEE*, Los Alamitos, CA, pp. 722-727.

19. W. Keith Edwards, Core Jini, Second Edition, Prentice Hall, 2001, ISBN 0-13-089408-7.

20. Van Steen, M., Hauck, F., Homburg, P., Tanenbaum, A., "Locating Objects in Wide-Area Systems", *IEEE Communications Magazine*, January 1998, pp104-109.

21. Steven E. Czerwinski, Ben Y. Zhao, Todd D. Hodes, Anthony D. Joseph, Randy H. Katz, *Fifth Annual International Conference on Mobile Computing and Networks, MobiCom '99*, Seattle, WA, August 1999, pp. 24-35.

22. Laliberte, D., Braverman, A. 1995, "A Protocol for Scalable Group and Public Annotations", Computer networks and ISDN systems: *Proceedings of the Third International World-Wide Web Conference*, Vol. 27, No. 6, Elsevier, pp. 911-918.

CONTACT

Rami Baroody

Rbaroody@ecs.syr.edu

Professor: Nizar Alholou

Alholoun@udmercy.edu

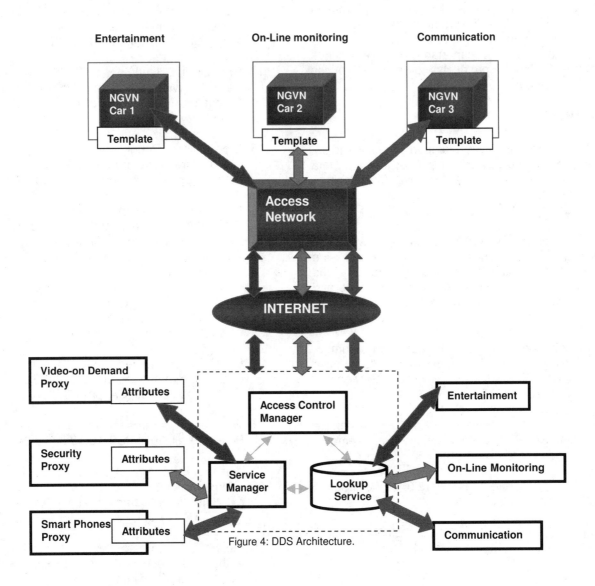

Figure 4: DDS Architecture.

2003-01-3738

Future Trends in Networking

Cyrilla Jane Menon
Dearborn Group, Inc.

Sergio Shimura
Dearborn Group Brazil

ABSTRACT

There are many changes occurring in the automotive industry regarding the usage of serial data communication, or networking. These changes are structural, technical and functional. Structural changes occur in the topology of the networks, technical changes focus on the adaptation of the existing networks, and functional changes incorporate new networks for specific applications. The progression of time, along with the desire to expand the functionality of networks in automobiles, has caused most car companies to realize that multiple independent networks are needed across a vehicle to handle specific functionality. Today there is strong agreement that there is no one single vehicle network that will fill all needs. This paper will address all the different networks and their proposed usages.

STANDARDS DEVELOPMENT

Knowledge of the standard development process is necessary in understanding network usage. Early networks were developed by individual companies trying to solve their specific need. But with the inclusion of legislation by goverments, networks began to be created by o rganizations.

Organizations in international venues are the Society of Automotive Engineers (SAE) and the International Standards Organization (ISO). The decision to standardize an SAE specification lies on the committee's members, comprised of individuals from various companies in the industry, who each have an equal vote; while the final vote for ISO specifications is done by member countries. There are other country-based organizations, such as DIN (Germany), that also create or work on standards as well; and some standards are formally created by one organization and adopted by another. An example of this is ISO 15765 (Diagnostics on CAN), which is also referred to as SAE J2480.

Working within a committee progress slowly, and can be delayed by outside political issues. The recent trend is to have interested parties create stand-alone organizations outside the formal SAE and ISO realm. The purpose for these groups is to quickly get an idea to market. After which, these groups may take their specification to an organization for "standardization".

INTRODUCTION

Transporting information from one electronic control unit (ECU) to another ECU on a vehicle is an ongoing challenge for vehicle manufacturers. In the early days, wires were connected point-to-point. These wire harnesses were vulnerable to high failure rate and added huge weight to the vehicle.

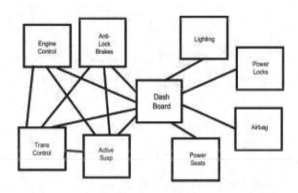

The networking, or multiplexing, of automotive electrical data onto communication buses began in the late 1970s. It was originally a single bus protocol primarily required for diagnostics, but was expanded to incorporate more functionality – a "one-size-fits-all" solution that included Class A, B, and C functions.

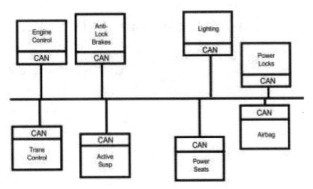

As time has progressed, along with the desire to expand the functionality of networks in automobiles, most car companies have realized that multiple independent networks are needed across an automobile to handle specific functionality. Today there is strong agreement that there is no one single bus that will fill all needs.

There are trade-offs for implementing these multi-tiered solutions - the costs of maintaining expertise on the several buses as well as the additional gateway functionality is a significant one. But the overall benefit to the consumer in achieving so many functions makes this implementation worthwhile.

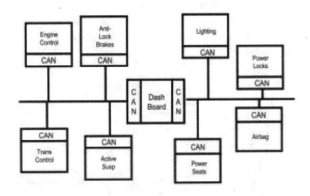

Chris Lupini, of Delphi Delco Electronics Systems, outlined six areas of network functionality in a paper presented at the 2001 SAE Congress. These areas are: Class A, Class B, Class C, Emissions / Diagnostics, Mobile Media, and X-by-Wire / Safety.

Some networks do not encompass just one type, but multiple functionalities. As an example, the most popular protocols being designed into car platforms today for Class A, B, and C functions is Controller Area Network (CAN).

CLASS A

SAE Specification J1213/1 describes Class A networks as "a multiplex system whereby vehicle wiring is reduced by the transmission and reception of multiple signals over the same signal bus between nodes that would have be traditionally accomplished by individual wires in a conventionally wired vehicle. The nodes used to accomplish multiplexed body wiring typically did not exist in the same or similar form in a totally conventionally wired vehicle." Although not officially adopted in any specification, the unofficial rule is that their data rate is less than 10 kps.

Earlier Class A networks were proprietary and used as low-end, non-emission diagnostic, general-purpose communication networks – with the emphasis on diagnostics. Some newer Class A networks are still proprietary, but are used more for information sharing applications. An example is GM's CAN network used in their body electronics application (LSCAN). This application is a single-wire CAN implementation that passes a lot of status information.

It is more popular to create Class A networks with UART-based hardware because of cost effectiveness. Ford's proprietary UART-based Protocol (UBP), is currently being used for body electronics for this reason.

A non-proprietary application under development and adoption is an international standard called Local Interconnect Network, or LIN. The LIN organization is independent and does not officially belong to ISO or SAE. LIN, utilizing an ISO 9141 physical layer, is typically referred to as the CAN subbus. Their website, www.lin-subbus.org, incorporates that terminology.

LIN's intended applications are short distance area control, like, roofing systems, steering wheel controls, a single door interface, etc. LIN is a master-slave network that allows only short packets of data of fixed length to be transmitted. These small local networks would be connected to a larger vehicle network.

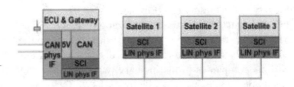

LIN's popularity is increasing. BMW released vehicles in 2003 using LIN. Currently, there is a SAE LIN committee that is attempting to work with the LIN organization to incorporate changes they have envisioned.

CLASS B

SAE J1213/1 describes Class B networks as "a multiplex system whereby data is transferred between nodes to eliminate redundant sensors and other system elements. The nodes in this form of a multiplex system typically pre-existed as stand-alone modules in a conventionally wired vehicle."

Class B networks are used primarily for non-diagnostic, non-critical communication. Their speeds unofficially range between 10 and 125 kbs. They support event-driven and some periodic message transmission in addition to sleep/wakeup.

In the US, SAE J1850 currently dominates this network type eventhough the industry is slowly adapting CAN. There are three flavors of J1850 – one for each of the major US automotive companies. GM refers to it internally as Class 2, Ford calls it Standard Corporate Protocol (SCP), and DaimlerChrysler uses the J1850 name. Ford uses the Type 2 physical layer, while the other two use Type 1. The basic messaging is outlined in SAE J2178. It is fairly easy to reference this specification to decode messaging.

Worldwide, however, CAN is the standard. With CAN defined simply as a Data Link Layer per the ISO Open Systems Interconnect (OSI) model (ISO IEC 7498), implementators can add their own unique physical and application layer (messaging) to create a solution. ISO 11898-2, the most popular physical layer, is used in both the passenger car and heavy truck markets.

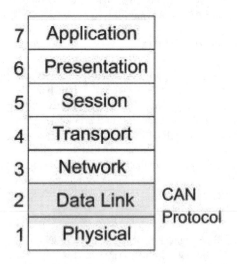

In Europe most car companies utilize CAN on one or more vehicle. Companies like Mercedes produce cars that have up to five separate CAN networks. CAN's expansion into the passenger car segment in North America and Asia is occurring. General Motors (GM), who released cars with CAN on their Saturn platform a few years ago, is gearing up to release many cars with their CAN-based GMLAN applications beginning in 2004.

Other changes for Class B networks are occurring because of new physical layers for CAN. The ISO 11519-2 "fault-tolerant" low speed 2-wire CAN interface is becoming popular in some car applications. This specification is being re-written as ISO 11898-3. Fault tolerant physical layers are slower and cost more than the traditional ISO 11898-2, but provide a more robust implementation.

Another CAN physical layer comes from GM as specification SAE J2411. This utilizes a single-wire interface with a much slower bus speed that normally operates at 33.3 kbps (and a high-speed mode of 83.33 kbps). Called SWCAN, or LSCAN, it captures similar properties as GM's version of J1850.

CLASS C

SAE J1213/1 describes Class C networks as "a multiplex system whereby high data rate signals associated with real time control systems, such as engine controls and anti-lock brakes, are sent over the signal bus to facilitate distributed control and to further reduce wiring."

Class C networks are fast networks used for applications that need high bandwidth such as real-time controlled powertrain implementations. Their speeds are typically 125 kbps to 1 Mbps. Because of their use in powertrain, these networks are typically used for diagnostics as well.

Many car companies have already or will be implementing Class C networks with CAN and ISO 11898-2 physical layers. The US car companies implement this under the specification SAE J2284. In these cases there is no standardization for messaging, because it is proprietary across car manufacturers.

Though there will not be any new Class C networks coming in the near future, there will be many changes or additions to existing implementations. For information regarding updates and additions, SAE J1939, Recommended Practice for Heavy Duty Trucks, and its sister specification, ISO 11783, Implementation for Agriculture, are good references. These CAN-based networks have added specific messaging to cover laser surfacing, navigation, and virtual terminal. Work has begun in Europe, and more recently the US, to commercialize ISOBUS as benefit to the enduser.

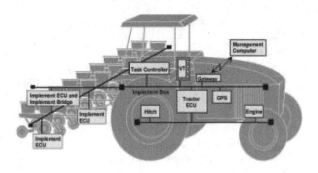

DIAGNOSTICS

Diagnostic applications have greatly influenced the development of networks within vehicles. Diagnostic can be divided into two components: legislated and non-legislated. The non-legislated diagnostics revolves around service repair applications, while legislated diagnostics is the emission-related aspects driven by individual governments for cars sold in their countries. Europe, US and other nations generally follow OBD-II, OBD-III, or E-OBD, but are not in sync with their implementation.

These networks typically use master-slave communication, with the off-board tool acting as the master, or the requestor. The messages that are used in these networks are referred to as services. As an example, a tool will use a specific service to request information from an ECU.

There are some physical networks used for diagnostics and normal communications while others are just exclusively used for diagnostics. These diagnostic-only networks include ISO 9141 and ISO 14230-4 (Keyword Protocol 2000). The latest network edition for the US in 2003 is CAN utilizing ISO 15675, referred to as Diagnostics on CAN. Europe legislated Diagnostics on CAN in the year 2000. Cross references of SAE and ISO standards is common and encapsulated in ISO 15031.

For the future, legislation has been accepted in Europe and the US to only allow passenger vehicles produced from 2008 and onward to use ISO 15765 as the diagnostic network. The debate as to whether or not the trucking industry in the US will be required to use ISO 15765 or remain with J1939-73 is still ongoing.

MOBILE MEDIA

Mobile Media network applications are sometimes referred to as telematics. Telematics is loosely defined as remote connectivity to a vehicle with an emphasis on consumer-driven applications. At least two different networks and protocols may be necessary.

Since this is a relatively new category, there are no networks that are dominant. One early CAN-based entrant that could not find a suitable market is IDB-C. CAN is an excellent choice for powertrain or body networks, because it allows short fast data packets. Unfortunately it does not allow transfer of large amounts of data efficiently – a critical requirement for these applications.

Fiber-optic based solutions seem to be the physical networks of choice for this area. The IDB Forum has added a newer network, IDB-F. This is a based on Firewire, or IEEE-1394. But the leader is MOST. BMW, DaimlerChrysler, and others established the MOST consortium. It is an object model solution that covers most mobile media applications.

For wireless applications, Bluetooth is already being installed in vehicles to interface with cellular phones. However, 802.11 also has it's proponents; and many groups and suppliers are studying a possible co-location so that the two standards can exist near each other. Ultrawideband (UWB) is the wild card that some think will "blow Bluetooth and 802.11 away". Approved by the Federal Communications Commission (FCC) in February 2002, it is essentially "white noise" communication. Using precise clocking, tiny amounts of information are transported across a very wide range of frequencies at very low power (perhaps 1/10000 that of a cell phone). Compared with spread spectrum which uses a small range of frequencies one at a time, UWB uses a wide range of frequencies all at once. UWB can be used for global position sensing, works indoors, and easily penetrates obstructions.

X-BY-WIRE / SAFETY

X-by-wire applications incorporate brake-by-wire, throttle-by-wire, steer-by-wire, etc. The typical bit rate is between 1 Mb/s and 10 Mb/s. and utilizes fiber optics due to high speed. The main concerns for these networks are reliability, performance, and real-time capability. The challenge to create a high-speed fault tolerant network remains. Most implementations utilize a dual bus topology for redundancy, thus cost becomes an issue.

The first drive-by-wire car was released this year using ByteFlight by BMW. Time-triggered Protocol (TTP) has a few applications too, but will be faced with the challenge by Time-triggered CAN (TTCAN), a new version of CAN that requires new silicon that was released this year.

Presently, the leading contender is FlexRay. It was established by DaimlerChrysler and BMW as a way to join ByteFlight and TTFlex. A decision was made that they could create one standard with mutual benefits.

The FlexRay consortium has more major automotive companies with GM signing up this year. BMW released a drive-by-wire car last year utilizing ByteFlight, but intends to produce future applications with Flexray. Flexray's current schedule for completion is within the 2003-04 time frame.

For safety-based networks, Safe-by-wire is an up-and-coming protocol. With companies such as Philips, GM, and TRW, this organization is creating a network that ties into the airbag deployment.

CONCLUSION

The use of multiple networks on a vehicle has been a reality for some time now. Primarily this has been due to the re-use, and carryover of existing networks and protocols. However, in the near future, new functions such as smart connectors, drive-by-wire, and mobile multi media will be forcing the need for additional protocols and networks. The need to provide better functionality to consumers will be the driving force in vehicle network innovation.

REFERENCES

1. ISO 7498 - Data Processing Systems, Open Systems Interconnection Standard Reference Mode.
2. www.lin-subbus.org
3. www.flexray.com
4. www.sae.org
5. www.iso.org
6. www.idbforum.org
7. SAE 2001-01-0060 "Multiplex Bus Progression" Lupini

CONTACT

Cyrilla Jane Menon
Marketing / Training Manager
Dearborn Group, Inc.
cyrilla@dgtech.com
http://www.dgtech.com

Cyrilla Jane Menon holds a master's degree in engineering from Wayne State University and a MBA from the University of Michigan. She has worked at Dearborn Group for over 9 years. Serving on many SAE and ISO multiplex committees, she has also co-chaired several CAN conferences. Mrs. Menon spends much of her time training engineers and technical personnel on vehicle networks, including in SAE's technical seminar.

Sergio Shimura
Dearborn Group Brazil
sshimura@dgtech.com

Graduated in 1987 from Escola Politecnica da Universidade de Sao Paulo in electronic engineering, Sergio Shimura holds a master's and doctor's degree from the same institution. With over 15 years of professional life, his experience includes embedded systems design, consulting and training. Recently Mr. Shimura joined Dearborn Group as its representative for Brazil.

2003-01-0111

Multiplex Bus Progression 2003

Christopher A. Lupini
Delphi Delco Electronics Systems

ABSTRACT

A previous SAE 2001 Congress paper, "Multiplex Bus Progression" [1] introduced the idea of categorizing vehicle serial data protocols into additional areas beyond the traditional SAE Class A, B, and C. This paper will expand on that idea, and provide a 2003 update to the Diagnostics, SafetyBus, Mobile Media, and X-by-Wire categories. All existing mainstream vehicular multiplex protocols (approximately 40) are categorized using the SAE convention plus the new groupings. Top contenders will be pointed out along with a discussion of the protocol in the best position to become the industry standard in each category at this time.

INTRODUCTION

CURRENT MULTIPLEX CATEGORIES –The multiplexing of automotive electrical data onto communication buses dates back to the late 1970s. It was originally hoped that a single bus protocol could handle the needs of any vehicle. Gradually that expanded to the SAE categorization of Class A, B, and C and the realization that up to three protocols and/or networks may be necessary.

By 1995 the need for multiple buses per vehicle was becoming apparent [2]. The cost tradeoff, especially, was studied – do you put everything on one bus or split it up into several buses? Which is more economical? Which is more efficient?

This paper proposes the idea that at least eight in-vehicle networks may be necessary – mainly on high-end vehicles in the next ten years. These categories include (besides the existing SAE classes) diagnostics, airbag, mobile media, X-by-Wire, and wireless. Each area needs its own protocol and one or more networks running that protocol. Sometimes this is for safety reasons, such as with airbags or X-by-Wire. But regardless of vehicle function partitioning, we now have distinct classes of signals that will communicate over their own network, or networks (i.e. multiple sub-buses for smart connector) [3].

This paper may have a similar look to the previous one, but many of the protocol contenders have been updated. The comparison tables remain so as to retain a single source of reference.

Although not discussed at length in this paper, there is a distinction between protocol and network. Conceivably one might have the same protocol running on several networks – say CAN for both a body bus and a powertrain bus. So even though there could be eight or more networks, there may actually be fewer protocols used. Also, not all protocols are complete – meaning they specify attributes of all seven layers of the OSI model [4]. Some are only physical layers (i.e. GM UART, J1708). Some are only higher layers (i.e. TTP).

CURRENT STATUS OF THE OLD CATEGORIES

CLASS A – Usage is for low-end, non-emission diagnostic, general-purpose communication. Bit rate is generally less than 10 Kb/s and must support event-driven message transmission. Cost is generally about "x" adder per node. This cost includes any silicon involved (i.e. microprocessor module or transceiver, etc.), software, connector pin(s), service, etc. The "cost" data discussed in this paper is very crude and is only to be used to compare with other categories.

Not much has changed in the last couple of years in this area. LIN continues to make strides as a common subbus protocol. Proprietary protocols continue to disappear.

Some examples of Class A protocols are listed in Table 1a.

NAME:	USER:	USAGE:	MODEL YEARS:	COMMENTS:
UART	GM	Many	1985 - 2005+	Being phased out
Sinebus	GM	Audio	2000+	Radio steering wheel controls
E&C	GM	Audio/HVAC	1987 - 2002+	Being phased out
I^2C	Renault	HVAC	2000+	Used little
J1708/J1587/J1922	T&B	General	1985 - 2002+	Being phased out
CCD	Chrysler	HVAC, audio, etc.	1985 - 2002+	Being phased out
ACP	Ford	Audio	1985 - 2002+	
BEAN	Toyota	Body	1995+	
UBP	Ford	Rear backup	2000+	
LIN	many OEMs	Smart Connector	2003+	LIN Consortium developing

Table 1a: Some Class A Protocols

Most of these Class A protocols are UARTs. UART is very simple and economical to implement. Most microcontrollers have the necessary SCI module built-in, or it can be implemented without a microprocessor. The transceiver is smaller and cheaper than those of other protocols. The transceiver IC may be a custom chip combining multi-protocol capability with regulators, drivers, etc. Right now the leading candidate for a Class A world standard is LIN.

Table 1b compares some of the major attributes of some of the Class A protocols from Table 1a.

FEATURE	UART(ALDL)	SINEBUS	E & C	I^2C	SAE J1708	ACP	BEAN	LIN
				BUS NAME				
AFFILIATION	GM	DELCO	GM	PHILIPS	TMC - ATA	FORD	TOYOTA	Motorola
APPLICATION	GENERAL & DIAGNOSTICS	AUDIO	GENERAL		CONTROL & DIAGNOSTICS	AUDIO CONTROL	BODY CONTROL & DIAGNOSTICS	SMART SENSORS
MEDIA	SINGLE WIRE	SINGLE WIRE	SINGLE WIRE	TWISTED PAIR	TWISTED PAIR	TWISTED PAIR	SINGLE WIRE	SINGLE WIRE
BIT ENCODING	NRZ	SAM	PWM	AM	NRZ	NRZ	NRZ	NRZ
MEDIA ACCESS	MASTER/ SLAVE	MASTER/ SLAVE	CONTENTION		MASTER/ SLAVE	MASTER/ SLAVE	CONTENTION	MASTER/ SLAVE
ERROR DETECTION	8-bit CS	NONE	PARITY	ACK bit	8-bit CS	8-bit CS	8-bit CRC	8-bit CS
HEADER LENGTH	16 BITS	2 BITS	11 - 12 BITS		16 BITS	12 - 24 BITS	25 BITS	2 BITS/BYTE
DATA LENGTH	0 - 85 BYTES	10 - 18 bits	1 - 8 BITS			6 - 12 BYTES	1 - 11 BYTES	8 BYTES
OVERHEAD	Variable	75 %	Variable	45 %	Variable	25 %	28 %	2 BYTES
IN-MESSAGE RESPONSE	NO	NO	NO		NO	NO	NO	NO
BIT RATE	8192 b/s	66.6 KHz – 200 KHz	1000 b/s	1 - 100 Kb/s	9600 b/s	9600 b/s	10 Kb/s	1 - 20 Kb/s
MAXIMUM BUS LENGTH	Not Specified	10 METERS	20 METERS	Not Specified	Not Specified	40 METERS	Not Specified	40 METERS
MAXIMUM NODES	10		10			20	20	16
µ NEEDED?	YES	NO	YES		YES	YES	YES	NO
SLEEP/WAKEUP	NO	NO	NO		NO	NO	NO	YES
H/W AVAIL?	YES	NO	YES		YES	YES	YES (?)	NO
COST	LOW	LOW	LOW		MEDIUM	LOW	LOW	LOW

Table 1b: Comparison of Class A Protocols

CLASS B – Usage is for the vast majority of non-diagnostic, non-critical communication. Speed is between 10 Kb/s and approximately 125 Kb/s. Must support event-driven and some periodic message transmission plus sleep/wakeup. Cost is around 2x per node. Protocols used for Class B networks are listed in Table 2a.

NAME:	USER:	USAGE:	MODEL YEARS:	COMMENTS:
GMLAN (low)	GM	Many	2002+	GM only user; J2411 single wire CAN
GMLAN (mid)	GM	Infotainment	2002+	95.2 Kb/s– might be IDB-C
Ford MSCAN	Ford	Various	2004+	125 Kb/s; J2284
DCX LSCAN	Chrysler	Various	2004+	125 Kb/s; ISO 11519
ISO 11898	Europe	Many	1992+	Various speeds – 47.6 Kb/s to 500 Kb/s in use
J2284	GM,Ford, DC	Many	2001+	500 Kb/s; based on ISO 11898
Fault-tol CAN	Europe	Many	2000+	ISO 11519 CAN
Class 2	GM	Many	Until 2002+	J1850; being phased out
PCI	Chrysler	Many	Until 2002+	J1850
SCP	Ford	Many	Until 2002+	J1850
J1939	T&B	Many	1994+	Replacing J1708/1587/1922
Intellibus	tbd	tbd	tbd	In use for aircraft industry

Table 2a: Some Class B Protocols

The world standard in this area is still CAN. In particular, ISO 11898 at around 100 Kb/s for car applications and J1939 at 250 Kb/s for Truck & Bus applications. Both of these use the same digital circuitry and transceiver in many cases. J1850 continues its usage, and in fact may not have peaked yet in annual volume.

The ISO 11519-2 "fault-tolerant" low speed 2-wire CAN interface is becoming popular in some car applications, but is still a small percentage of implementations. This CAN physical layer is slower and costs more than an ISO 11898 interface, but the bus fault detection capability is enticing.

Intellibus is shown for the first time in this survey. It has not been picked up by any automotive applications yet, but holds promise due its low cost and flexibility.

Table 2b compares some of the major attributes of the Class B protocols from Table 2a.

FEATURE	CAN 2.0 ISO 11898 ISO 11519-2 ISO 11992 J2284	J1850 ISO 11519-4			SAE J 1939	Intellibus
AFFILIATION	BOSCH/SAE/ISO	GM	FORD	CHRYSLER	TMC - ATA	Boeing/SAE
APPLICATION	CONTROL & DIAGNOSTICS	GENERAL & DIAGNOSTICS	GENERAL & DIAGNOSTICS	GENERAL & DIAGNOSTICS	CONTROL & DIAGNOSTICS	CONTROL & DIAGNOSTICS
TRANSMISSION MEDIA	TWISTED PAIR	SINGLE WIRE	TWISTED PAIR	SINGLE WIRE	TWISTED PAIR	TWISTED PAIR
BIT ENCODING	NRZ-5 MSb first	VPW MSb first	PWM MSb first	VPW MSb first	NRZ-5 MSb first	Manchester Bi-phase
MEDIA ACCESS	CONTENTION	CONTENTION	CONTENTION	CONTENTION	CONTENTION	Master/Slave
ERROR DETECTION	CRC	CRC	CRC	CRC	CRC	CRC, Parity
HEADER LENGTH	11 or 29 BITS	32 BITS	32 BITS	8 BITS	29 BITS	16 - 48 Bits
DATA FIELD LENGTH	0-8 BYTES	0-8 BYTES	0-8 BYTES	0-10 BYTE	8 BYTES	0 - 32 Bytes
MESSAGE OVERHEAD	9.9 % - 22 %	33.3 %	33.3 %	8.3 %	9.9 % - 22 %	28% - 75%
IN-MESSAGE RESPONSE	NO	Optional Normally NO	Optional Normally YES	Optional Normally YES	NO	Optional
BIT RATE	10 Kb/s to 1 Mb/s	10.4 K b/s	41.6 K b/s	10.4 K b/s	250 Kb/s	12.5 Mb/s
MAXIMUM BUS LENGTH	Not Specified 40 (Typical)	35 METERS (5 Meters for scan tool)	35 METERS (5 Meters for scan tool)	35 METERS (5 Meters for scan tool)	40 METERS	30 METERS
MAXIMUM NODES	Not Specified 32 (Typical)	32	32	32	30 FOR STP 10 FOR UTP	64
µ NEEDED?	YES	YES	YES	YES	YES	NO
SLEEP/WAKEUP	NO	YES	NO	NO	NO	YES
H/W AVAIL?	YES	YES	YES	YES	YES	FPGA
COST	MEDIUM	LOW	LOW	LOW	MEDIUM	MEDIUM

Table 2b: Comparison of Class B Protocols

CLASS C –Usage is for somewhat fast, higher bandwidth systems such as engine timing, fuel delivery, etc. Bit rate is between 125 Kb/s and 1 Mb/s. Must support real-time periodic parameter transmission (perhaps in the few milliseconds range). Unshielded twisted pair is the medium of choice instead of shielded twisted pair or fiber optics. Cost is about 3x to 4x per node, unless STP or fiber optics is involved – which is typically necessary above 500 Kb/s. Typical protocols used are listed in Table 3a.

NAME:	USER:	USAGE:	MODEL YEARS:	COMMENTS:
GMLAN (high)	GM	All	2002+	500 Kb/s; J2284
HSCAN	Ford	Various	2004+	500 Kb/s; J2284
HSCAN	Chrysler	Various	2004+	500 Kb/s; J2284
ISO 11898	Europe	Most	1992+	Various speeds of CAN
J1939	T&B	Most	1994+	250 Kb/s CAN

Table 3a: Some Class C Protocols

J1939 is commonly used for Class B and Class C applications for truck & bus, construction, agriculture, marine, and other industries. Most passenger car applications run ISO 11898 at 500 Kb/s for their Class C network. The big difference from CAN in Class B applications is the type of nodes that are connected.

Total CAN usage, according to CAN in Automation (CiA) is in the hundreds of millions of nodes worldwide.

Table 3b compares some of the major attributes of the Class C protocols from Table 3a. GMLAN is not shown due to its confidentiality to GM.

	BUS NAME		
FEATURE	CAN 2.0 ISO 11898 ISO 11519-2 ISO 11992 J2284	SAE J1939	Intellibus
AFFILIATION	BOSCH/SAE/ISO	TMC - ATA	Boeing/SAE
APPLICATION	CONTROL & DIAGNOSTICS	CONTROL & DIAGNOSTICS	CONTROL & DIAGNOSTICS
TRANSMISSION MEDIA	TWISTED PAIR	TWISTED PAIR	TWISTED PAIR
BIT ENCODING	NRZ-5 MSb first	NRZ-5 MSb first	Manchester Bi-phase
MEDIA ACCESS	CONTENTION	CONTENTION	Master/Slave
ERROR DETECTION	CRC	CRC	CRC, Parity
HEADER LENGTH	11 or 29 BITS	29 BITS	16 - 48 Bits
DATA FIELD LENGTH	0-8 BYTES 11 or 29-bit ID	MOST ARE 8 BYTES 29-bit ID	0 - 32 Bytes
MESSAGE OVERHEAD	9.9 % - 22 %	9.9 % - 22 %	28% - 75%
IN-MESSAGE RESPONSE	NO	NO	Optional
BIT RATE	10 Kb/s to 1 Mb/s	250 Kb/s	12.5 Mb/s
MAXIMUM BUS LENGTH	Not Specified 40 (Typical)	40 METERS	30 METERS
MAXIMUM NODES	Not Specified 32 (Typical)	30 W/ SHIELDED TWISTED PAIR 10 W/ UNSHIELDED TP	64
μ NEEDED?	YES	YES	NO
SLEEP/WAKEUP	NO	NO	YES
H/W AVAIL?	YES	YES	FPGA
COST	MEDIUM	MEDIUM	MEDIUM

Table 3b: Comparison of Class C Protocols

NEWER CATEGORIES

EMISSIONS DIAGNOSTICS – Usage is to satisfy OBD-II, OBD-III, or E-OBD. Must be a legally acceptable protocol. Protocols used today (or soon) are listed in Table 4a. There is overlap with some of the other categories.

NAME:	USER:	USAGE	MODEL YEARS:	COMMENTS:
ISO 15765-4	Europe	E-OBD	2000+	E-OBD CAN
ISO 15765-4	All	OBD-III	2007+	OBD harmonized
J 1850	GM, Ford, DC	OBD-II	1994+	Not accepted in Europe
ISO 9141-2	Europe, Asia, some U.S.	OBD-II, general	1994+	Old OBD-II UART
ISO 14230-4	Many	OBD-II, OBD-III	2000+	Keyword 2000

Table 4a: Some Emission Diagnostics Protocols

Since this data link is only needed between the engine controller and the off-board connector, a simple approach is sufficient. Most automakers and truck makers are using KW2000 (ISO 14230) already so this is rapidly becoming the emissions diagnostic standard. In the U.S., high-speed CAN will be phased-in as the "OBD-III" emissions test interface beginning in MY04.

This will be the only legally acceptable protocol by MY07. SAE J2480 was an initiative to develop a CAN emissions diagnostic interface, but it was found to overlap with ISO 15765 so it has been abandoned. General information on these protocols is shown in Table 4b.

FEATURE	BUS NAME					
	ISO 15765	J1850 ISO 11519-4			ISO/DIS 9141 ISO/DIS 9141-2	KEYWORD xx (71, 72, etc.)
AFFILIATION	ISO	GM	FORD	CHRYSLER	WORLD	Various
APPLICATION	EMISSIONS DIAGNOSTICS	GENERAL & DIAGNOSTICS	GENERAL & DIAGNOSTICS	GENERAL & DIAGNOSTICS	DIAGNOSTICS ONLY	DIAGNOSTICS
TRANSMISSION MEDIA	TWISTED PAIR	SINGLE WIRE	TWISTED PAIR	SINGLE WIRE	SINGLE WIRE	1-WIRE
BIT ENCODING	NRZ	VPW MSb first	PWM MSb first	VPW MSb first	NRZ (strt, 7D, P, stop) LSb first	NRZ
MEDIA ACCESS	TESTER/ SLAVE	CONTENTION	CONTENTION	CONTENTION	TESTER/SLAVE	MASTER/ SLAVE
ERROR DETECTION	CRC	CRC	CRC	CRC	PARITY (odd)	x-bit CS
HEADER LENGTH		32 BITS	32 BITS	8 BITS	Not Specified	16 BITS
DATA FIELD LENGTH		0-8 BYTES	0-8 BYTES	0-10 BYTE	Not Specified	0 - 85 BYTES
MESSAGE OVERHEAD		33.3 %	33.3 %	8.3 %	Variable	Variable
IN-MESSAGE RESPONSE		Optional Normally NO	Optional Normally YES	Optional Normally YES	NO	NO
BIT RATE		10.4 K b/s	41.6 K b/s	10.4 K b/s	<10.4 Kb/s	5 b/s - 10.4 Kb/s
MAXIMUM BUS LENGTH		35 METERS (5 Meters for scan tool)	35 METERS (5 Meters for scan tool)	35 METERS (5 Meters for scan tool)	Limited by total impedance to ground	Not Specified
MAXIMUM NODES		32	32	32	Limited by total impedance to ground	10
μ NEEDED?		YES	YES	YES	YES	YES
SLEEP/ WAKEUP		YES	NO	NO	NO	NO
H/W AVAIL?		YES	YES	YES	YES	YES
COST		LOW	LOW	LOW	LOW	LOW

Table 4b: Comparison of Emission Diagnostics Protocols

MOBILE MEDIA – Usage is for "PC-on-wheels" applications. At least two different networks and protocols may be necessary. These sub-categories will be referred to as low speed and high speed. Beginning with this paper wireless is now in its own category. The necessary bit rate for mobile media applications is between 250 Kb/s and 100 Mb/s+.

Low Speed - Usage is for telematics, diagnostics, and general information passing. Cost is around 3x per node. IDB-C, a token-passing form of CAN at 250 Kb/s, has fallen out of favor. Most OEMs already have a mid-speed bus based on CAN to handle low-end telematics communication functions. There has been recent interest in a lower cost "high-end" network that can handle digital audio streams, but not necessarily video. Toward that end, the D2B and MOST developers are working on copper-based solutions. Depending on the EMC performance, 10 Mb/s, 25 Mb/s, or even 50 Mb/s bit rates are being proposed. Table 5a lists some possible low-speed mobile media protocols.

NAME:	USER:	MODEL YEARS:	COMMENTS:
IDB-C	none long term	2002+	250 Kb/s CAN; **www.idbforum.org**
D2B SmartwireX	tbd	2005+	**www.candc.co.uk/candc_company**
MOST over copper	tbd	2005+	**www.mostcooperation.com**

Table 5a: Some Low-Speed Mobile Media Bus Protocols

High Speed - Usage is for real-time audio and video streaming. Cost is around 15x to 25x, mainly due to fiber optics. Fiber optics will be necessary due to the high speed required to pass real-time video streams from multiple sources to multiple outputs. Will probably have to be compatible with industry-standard systems such as Connected Car PC, or AutoPC. D2B has seen the first usage (Mercedes 1999 S-class) but MOST appears to be the top contender at this time. The IDB Forum is leaning toward Firewire via an effort called "IDB-1394". The Automotive Multimedia Interface Collaboration (AMI-C) has set their sights on MOST and Firewire. Table 5b lists some possible high-speed mobile media protocols. Table 5c is a summary of details on some of the low-speed and high-speed protocols.

NAME:	USER:	YEAR:	COMMENTS:
D2B	Mercedes, Jaguar	1999+	**www.candc.co.uk/candc_company**
MOST	BMW, GM, DC, Ford, VW, Toyota	2000+	**www.mostcooperation.com**
Firewire	DC	2000+	**www.1394ta.org**
USB	Clarion	1998+	**www.autopc.com**
IntelliBus	Boeing	2004+	**www.intellibusnetworks.com**

Table 5b: Current High-Speed Mobile Media Bus Protocols

	BUS NAME						
FEATURE	IDB-C	Intellibus	MOST	SmartWireX	MML	USB	IEEE 1394
AFFILIATION	SAE	Boeing/SAE	Oasis	C&C	DELCO	Commercial	IEEE
APPLICATION	Aftermarket Entertainment	CONTROL & DIAGNOSTICS	Stream Data & Control	STREAM DATA & CONTROL	STREAM DATA & CONTROL	PC DEVICES	PC DEVICES
TRANSMISSION MEDIA	2-Wire	TWISTED PAIR	Optical	TWISTED PAIR	OPTICAL FIBER	SHIELDED TWISTED PAIR	SHIELDED TWISTED PAIR
BIT ENCODING	NRZ	Manchester Bi-phase	BiPhase	PWM	NRZ	NRZ	NRZ
MEDIA ACCESS	Token-slot	Master/Slave	Master/Slave	Master/Slave	Master/Slave	Contention	Contention
ERROR DETECTION	15-bit CRC	CRC, Parity	CRC	Parity	CORRECTING (optional)	CRC	CRC
HEADER LENGTH	11 BITS	16 - 48 Bits			1 BYTE		
DATA LENGTH	8 BYTES	0 - 32 Bytes			1 - 200+ BYTES		
MESSAGE OVERHEAD	~ 32 BITS	28% - 75%			5 - 10 %	25 %	25 - 30 %
IN-MESSAGE ACK.	1 ACK BIT	Optional	No	No	No		
BIT RATE	250 Kb/s	30 Mb/s	25 Mb/s	tbd kb/s	110 Mb/s	12 Mb/s	98 - 393 Mb/s
MAXIMUM BUS LENGTH	TBD	30 METERS	TBD	150 METERS	10 METERS		72 METERS
MAXIMUM NODES	16	64	24	50	16	127	16
μ NEEDED?	YES	NO	YES	YES	YES	YES	YES
SLEEP/WAKEUP	YES	YES	YES	YES	YES	NO	NO
H/W AVAIL?	NO	FPGA	YES	YES	NO	YES	YES
COST	LOW	LOW	HIGH	HIGH	HIGH	MEDIUM	MEDIUM

Table 5c: Comparison of Mobile Media Protocol

Wireless – Usage somewhat undetermined at this time. Will be necessary (initially) for cell phones and palm PCs (PDAs). Eventual use may include cameras, pagers, etc. Cost target is around 5x per node.

Much of the advertisement attention has been with Bluetooth. However, 802.11 also has its proponents, so many groups and suppliers are studying co-location so that products containing either standard can exist near each other. Ultrawideband (UWB) is the wild card that some think will "blow Bluetooth and 802.11 away". Approved by the U. S. Federal Communications Commission (FCC) in February 2002, it is essentially "white noise" communication. Using precise clocking, tiny amounts of information are transported across a very wide range of frequencies at very low power (perhaps 1/10000 that of a cell phone). Compared with spread spectrum which uses a small range of frequencies one at a time, UWB uses a wide range of frequencies all at once. UWB can be used for global position sensing, works indoors, and easily penetrates obstructions.

Leading wireless protocols are listed in Table 6.

NAME:	USER:	MODEL YEARS:	COMMENTS:
Bluetooth	tbd	2005+	**www.bluetooth.com**
IEEE 802.1	tbd	tbd	**www.ieee802.org/11**
UWB	tbd	tbd	**www.uwb.org**

Table 6: Current Wireless Mobile Media Bus Protocols

SAFETYBUS – Usage is for airbag systems. There may be two, or more, buses such as one for firing and one for sensing. Must support at least 64 nodes consisting of squibs, accelerometers, occupant sensors, seatbelt pretensioners, etc. Cost is (hoped to be) 1x to 2x per node. The USCAR "SafetyBus" committee was attempting to standardize on a suitable protocol, but degraded into separate, independent, camps. The two main ones are Safe-by-Wire and BST. Byteflight is in production in at least one BMW vehicle. Table 7a is the list of current airbag network protocols.

NAME:	USER:	YEARS:	COMMENTS:
Safe-by-Wire	tbd	2002+	Delphi-TRW-Philips-Autoliv-SDI
BST	tbd	2002+	Bosch-Siemens-Temic
DSI	tbd	2002+	Motorola/AMP
Byteflight	BMW	2002+	"ISIS", SI

Table 7a: Current SafetyBus Protocols

Many issues here involving packaging constraints, existing mechanical envelop, legalities, etc. The winning protocol may well be a hybrid of several existing proposals. In fact, BST and Safe-by-Wire are actually hybrids of earlier protocols. For now there is no clear industry direction. Table 7b compares these protocols.

FEATURE	BUS NAME			
	BST	SafeByWire	DSI	Byteflight
AFFILIATION	Bosch-Siemens-Temic	Delphi-Philips-TRW-Autoliv-SDI	Motorola	BMW
APPLICATION	Airbag	Airbag	Airbag	Airbag
TRANSMISSION MEDIA	2-WIRE	2-WIRE	2-WIRE	2-WIRE or 3-WIRE or optical
BIT ENCODING	Manchester Biphase	3-level voltage	3-level voltage	
MEDIA ACCESS	MASTER/ SLAVE	MASTER/ SLAVE	MASTER/ SLAVE	MASTER/ SLAVE
ERROR DETECTION	Odd Parity and/or CRC	8-bit CRC	4-bit CRC	16-bit CRC
HEADER LENGTH	Various	Various	Various	Various
DATA FIELD LENGTH	1 byte	1 byte	1 - 2 bytes	1 byte
MESSAGE OVERHEAD				
IN-MESSAGE ACK.	NO	NO	NO	NO
BIT RATE	31.25 Kb/s, 125 Kb/s, 250 Kb/s	150 Kb/s	5 Kb/s - 150 Kb/s	10 Mb/s
MAXIMUM BUS LENGTH	TBD	25 - 40 m	TBD	TBD
MAXIMUM NODES	12 squibs, 62 slaves	64	16	
μ NEEDED?	NO	NO	NO	NO
SLEEP/WAKEUP	NO	NO	NO	NO
H/W AVAIL?	YES	YES	YES	YES
COST	LOW	LOW	LOW	LOW

Table 7b: Comparison of SafetyBus Protocols

DRIVE-BY-WIRE – Usage is for brake-by-wire, throttle-by-wire, steer-by-wire, etc. applications. Bit rate is between 1 Mb/s and 10 Mb/s. Fiber optics will be necessary due to the increased speed. The utmost in reliability, performance, and real-time capability is required. Cost is around 15x+ per node. Some possible candidate protocols are given in Table 8a.

NAME:	USER:	YEARS:	COMMENTS:
TTP/C	BMW, Audi	2004+	**www.tttech.com**
TTCAN	tbd	tbd	**www.can-cia.de**
FlexRay	BMW, DC, GM, Audi	2004+	**www.flexray-group.com**

Table 8a: Current Drive-by-Wire Protocols

TTP has the momentum, but work is underway to see if CAN is capable of doing the job (i.e. TTCAN). Meanwhile, FlexRay continues to win support. At this time, none of these protocols are in an automotive application. A major issue is how much fault tolerance is really required. Any scheme will require dual bus interfaces, dual microprocessors, bus watchdogs, timers, etc. Cost is a big problem. The level of fault-tolerance needed requires a lot of silicon and software which, of course, is expensive. The consortium TTAgroup (www.ttagroup.org) is trying to standardize on a protocol, but (like other consortiums) will apparently allow a choice of protocols - either TTP/C or FlexRay. Who knows, in a couple more years TTCAN may be chosen too! Table 8b is a comparison of these protocols' details.

FEATURE	BUS NAME			
	TTP	Intellibus	FlexRay	TTCAN
AFFILIATION	U-VIENNA	Boeing/SAE	Motorola	CiA
APPLICATION	Safety Control	Safety Control	Safety Control	Safety Control
TRANSMISSION MEDIA	Not Specified	Twisted pair	Twisted pair or fiber	Twisted pair
BIT ENCODING	Not Specified	Manchester Bi-phase	NRZ	NRZ-5
MEDIA ACCESS	Isochronous	Master/Slave	Time or Priority	Time or Contention
ERROR DETECTION	16-bit CRC	CRC, Parity	24-bit CRC	15-bit CRC
HEADER LENGTH	1 Byte	16 - 48 Bits	40 Bits	11 - 29 Bits
DATA FIELD LENGTH	16 Bytes	0 - 32 Bytes	0 – 246 Bytes	0 - 8 Bytes
MESSAGE OVERHEAD	18.75 %	28% - 75%	3% - 100%	9.9 - 22%
IN-MESSAGE ACK.	YES	Optional	NO	YES
BIT RATE	Not Specified	12.5 Mb/s	10 Mb/s	1 - 2 Mb/s
MAXIMUM BUS LENGTH	Not Specified	30 meters	Not Specified	40 meters
MAXIMUM NODES	Not Specified	64	Not Specified	32
µ NEEDED?	YES	NO	YES	YES
SLEEP/WAKEUP	NO	YES	YES	YES
H/W AVAIL?	NO	FPGA	NO	YES
COST	HIGH	LOW	MEDIUM	LOW

Table 8b: Comparison of Drive-by-Wire Protocols

CONCLUSION

Multiple buses per vehicle have been a reality for some time. Primarily this has been because of re-use, and carryover of existing networks and protocols. However, in the near future new functions such as smart connectors, drive-by-wire, and mobile multi media has been forcing the need for additional protocols and networks. Despite the best intentions of OEMs, this will continue. There is no "one bus fits-all". Instead, there are different buses for different things.

REFERENCES

1. SAE 2001-01-0060 "Multiplex Bus Progression" Lupini
2. SAE 950293 – "Aspects and Issues of Multiple Vehicle Networks" Emaus
3. SAE 2001-01-0072 "LIN Bus and its Potential for Use in Distributed Multiplex Systems" Ewbank, Lupini, Perisho, DeNuto, Kleja
4. ISO 7498 - Data Processing Systems, Open Systems Interconnection Standard Reference Mode.
5. www.lin-subbus.org

CONTACT

Christopher (Chris) A. Lupini
Staff Research Engineer
Delphi Delco Electronics Systems
Christopher.A.Lupini@delphiauto.com
http://www.delphiauto.com

Mr. Lupini has a BSCompE degree from the University of Michigan and an MSEE degree from Purdue University. He is a registered Professional Engineer and an ASE certified Master Automotive Technician. He has worked in the data communications industry for 15 years, and has been teaching seminars for 10 years. He has spent most of his career assisting in the design and development of various protocols including Class 2, as well as the specification and test of numerous J1850, CAN, and other ICs. He is the lead engineer for the Delphi Delco Serial Data Center of Expertise and has authored many technical papers and articles, along with a textbook, and holds one patent.

2000-01-0146

Current Vehicle Network Architecture Trends – 2000

Bruce D. Emaus
Vector CANtech

ABSTRACT

Vehicle network architectures are migrating toward the use of multiple vehicle networks. Cost, application differences, lessons learned, and industry experiences have pushed aside the original philosophy of using the single vehicle-wide network that was attached to J1850. This paper discusses the current state of vehicle network architectures, the rationale behind the use of multiple vehicle networks, and discusses the current business and technical issues.

PULLING THE PLUG ON J1850

In the beginning, there was only one concept and it was "only one network protocol is needed". And amazingly because this basic philosophy originated at each automotive company independently and at slightly different times, a very large assortment of vehicle network protocols has been created over the past ten years. Several companies even made more than one protocol. Some companies continue such internal protocol development activities to this day.

Each of these reasonably powerful, well-designed, small area network (SAN) solutions represented major corporate investment. The resources used to engineer a near-perfect protocol, to design the protocol chips, to develop a physical layer, to write the protocol software, to create the messaging database, to educate the application developers, to create the protocol tools, and to continue managing the protocol became excessive. So excessive that for some companies "the plug was pulled".

While several American car and truck programs continue to be based on J1850 because of corporate inertia, the funeral is basically over, only the official announcement that "J1850 is dead" is missing.

Based on few technical reasons and primarily on key business reasons, the leading small area network solution has emerged to be the CAN protocol. While the technical debating continues to this day, few are listening and even fewer understand the atomic details of any of today's protocols. Since there is no higher technical authority when it comes to small area networks, most decisions are essentially made at the business level.

THE INDUSTRY SHIFT

The question is not which protocol - but which architecture. Is it really the case, that the CAN protocol has dominated all of the other small area network solutions? As one carefully examines this industry-wide experience, one discovers that it was not a protocol vs. protocol question, but an architecture vs. architecture question that has motivated the major shift to CAN.

The industry's experience-based system-level answer is the multiple vehicle network architecture.

REQUIREMENTS CHANGED – As the vehicle level requirements began to change, the architectural choice of the single vehicle wide network began to lose its following.

DEMAND FOR MORE FEATURES – The demand for new features increased as vehicle networking helped to fuel the proliferation of more and more body electrical functions. Even when early voice recognition systems were experimentally interconnected to network-based body electrical subsystems several years ago, the concept people began to see an ever expanding universe of features with amazing possibilities.

DEMAND FOR MOTION CONTROL FUNCTIONS –

Designers became more interested in distributing motion control functions. Why not make a separation between the engine and transmission? Why not consider changing the throttle control from mechanical to being electronic?

Questions like these coupled with the positive experiences with J1850, redirected a portion of the electrical and electronic systems effort. Networking applied to vehicle motion control functions seemed quite possible, but many new issues were raised. A higher speed network solution also seemed necessary.

CONCERN WITH CRITICAL FUNCTIONS – A concern grew when system designers realized that grouping critical functions with non-critical functions might be inappropriate. Is mixing human-controlled audio volume information with ABS or traction control data on the same network a good idea?

DEMAND FOR LOWER COST NETWORKING – The desire to lower the cost per node also played a major role.

Early J1850 solutions were expensive. Each of the big three attempted to lower costs by funding their own protocol chip development. But these three different chip architectures severely limited the opportunity to obtain an industry-wide high volume low-cost solution. Instead, three medium-cost solutions were the best that could be achieved.

In some cases, the multiple vehicle network architecture provided an opportunity to employ a second low-cost network solution instead of the usual J1850 solution. Ford's Audio engineering (now Visteon) developed ACP, a UART-based protocol, as a suitable alternative. In this experience, the business goal of lowering overall system cost was met by utilizing more than one vehicle network.

The demand for more features, the demand for distributed motion control functions, the concern about mixing critical and non-critical functions, and the demand for lower-cost networking solutions all influenced the migration towards multiple vehicle network architectures.

TOTAL SYSTEM-LEVEL REQUIREMENTS

From the simple to the complex, vehicle system-level requirements continue to impact the vehicle network architecture. At the system level, the vehicle network architecture and its supporting protocol(s) must satisfy the following list of requirements –

- all vehicle subsystem functions including motion control and human control
- all critical vehicle functions including brakes, cruise control, air bag deployment
- the worst case vehicle electrical and physical environment
- the entire volume and traffic rate of all vehicle network data
- the entire range of network speeds up to the maximum bit rate requirement
- all required control methods including real-time, peer-to-peer, master/slave, and TDM
- all required forms of conversation including event, periodic, and conditionally periodic
- all operational modes including normal operation, diagnostic, and manufacturing

Satisfying the entire list of requirements with a single vehicle-wide network creates the need for a network solution pushed to the extreme in both maximum technical robustness and maximum cost per node. However, for a multiple network architecture, the ability to partition the system provides an opportunity to lower the overall system cost while maintaining the necessary level of robustness on a per subsystem basis.

J1850 CANNOT COVER ALL REQUIREMENTS – The original 'one network fits all' approach and its burdensome responsibility of satisfying all requirements became the Achilles Heel of the single network architecture. Detroit's three proprietary versions of J1850 are not capable of handling today's complete list of vehicle system-level requirements.

J1850 VS. CAN – For today's multiple vehicle network architectures the leading protocol choice is CAN.

System Coverage – CAN provides one of the best solutions for today's multiple vehicle network architectures. While lacking some of the protocol advances made during the development of J1850, the older CAN protocol is still more than adequate. CAN's error handling strategy which virtually guarantees data consistency across the network one of its most robust features.

Flexibility – CAN may be used for high speed (up to 1 MBPS), medium speed, or low speed functions. Its relative independence from network speed and physical layer make CAN a flexible protocol.

More Available Physical Layers – Several CAN-based physical layers, each with substantial industry experience, are available. High-speed transceivers, medium speed transceivers with fault-tolerant capability, and new single wire transceivers provide a wide range of interconnection choices.

Large Accessible Knowledge Base – CAN knowledge is available through many sources. Corporations can choose to build their own internal knowledge base or use the existing industry infrastructure to get information. CAN tool companies, industry standards groups, consultants, and a variety of other sources are available.

More Available Silicon – The number of semiconductor companies that are supporting the CAN protocol continues to grow. Even though the old "11 bit only" parts are being retired, the replacement and new silicon, all CAN 2.0 B compliant, give the industry a wide range of choices.

From low-end 8 bit micros up to 32-bit high-performance processors with multiple integrated CAN controllers, the semiconductor industry continues its commitment to this leading small area network solution.

More Tools – A wide variety of off-the-shelf CAN tools are available. Developers have choices that range from low-cost tools with limited features up to programmable high-performance tools with oscilloscope-like graphical interfaces.

Even the distributed system designers have powerful tools capable of node emulation. Vector's CANoe tool which was co-developed with DaimlerChrysler, allows complete system-level evaluation with any number of nodes being real or virtual. Tools like this provide developers the opportunity to do simultaneous engineering and have dramatically decreased the time to market for CAN-based product architectures.

Another important development in the tool arena has been the emergence of CAN-based calibration tools. With an existing CAN network connection, this type of tool provides an opportunity to do module development, test, and evaluation while the module is resident in the vehicle. Changing parameters and evaluating the results while driving down the road is now possible. Module software development no longer needs to be confined to the engineer's desk or lab. What has provided this opportunity has been the CAN Calibration Protocol or CCP. Developed by the European standards group known as ASAM, CCP (also known as ASAP1a) looks like a high-powered monitor program that includes the capability of flash programming.

CURRENT MULTIPLE VEHICLE NETWORK ARCHITECTURES

Many OEMs are currently using multiple vehicle network architectures, especially for some of their high-end vehicle platforms. Figure 1 shows an example of a typical vehicle network architecture that uses three separate networks.

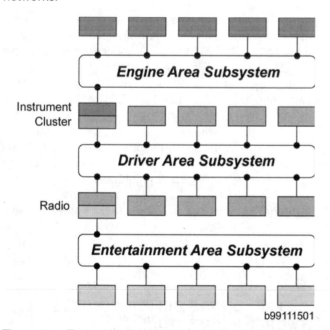

Figure 1. Typical Multiple Vehicle Network Architecture

Using this organization, high speed motion control requirements have been distinctly separated from medium speed human control requirements. A third bus is used to separate the driver (body electrical) functions from the audio/entertainment subsystem.

The module that supports the instrument cluster functions is typically a gateway between the motion control network and the driver area network. Information sharing requirements for the instrument cluster might include moving selected engine information onto the driver area network. For example, the feature of "automatic door locking" will typically involve receiving the vehicle speed from the motion control network and retransmitting this data onto the driver area network so that modules related to door locks can handle the function.

The radio is typically used as a gateway between the driver area network and an audio/entertainment subsystem network.

While many German multiple vehicle network architectures use the CAN protocol for both the motion control and driver control networks, some are considering using other high-speed data plus audio network solutions to support the audio/entertainment subsystem.

KEY ARCHITECTURAL INFLUENCES

Some of the major influences that have redirected the system toward multiple networks include –

- Current Types of Vehicle Subsystems
- Spectrum of Low End to High End Vehicles
- Reusable Network Strategies

CURRENT VEHICLE SUBSYSTEMS – From the system's point of view, the vehicle is a integrated set of mechanical, electrical, chemical, and energy conversion functions or subsystems.

Today's electrical-related vehicle subsystems can be divided into three major functional areas –

- Engine Area Subsystems
- Driver Area Subsystems
- Entertainment Area Subsystems

While some functions such as electrical power distribution and climate control may cross subsystem boundaries, such partitioning does not create too difficult of a problem for the distributed systems engineer.

Engine Area Subsystems – The major engine area subsystems of engine, transmission, brakes, and vehicle dynamics essentially handle vehicle motion functions. This area's harsh underhood environment includes wide temperature extremes, shock, vibration, repetitive high energy spark events, and corrosive elements. The engine area includes module and network design requirements substantially more stringent when compared to most other vehicle subsystems.

For the engine area subsystems, two key networking requirements emerge –

- The processing data rate must be relatively fast to support motion control. If such data is placed onto a network, the relative communication speed will be much faster than that required of the driver area.

- The harsh engine area environment requires additional consideration and usually requires higher performance components.

Driver Area Subsystems – Driver Area Subsystems support driver and passenger functions. Typically these functions directly interface with the customer, using human activated commands. In general, the vehicle's option content heavily influences what collection of functions are implemented.

Typical driver area functions include –

- Window and mirror controls
- Door access controls
- Climate control
- Interior and exterior lighting
- Seat controls
- Window wiper controls

The operation of many of the driver area subsystem functions is rather slow and the expected response time from human input to function activation need only match the customer's expectation.

For the driver area subsystems, three key in-vehicle networking requirements are –

- The data rate is relatively slow to support human actuated control functions when compared to motion control. The approximate 100 millisecond "human expected response time" is appropriate for most of the driver area functions.
- The total number of humans interacting with all driver area subsystems is limited. The driver and a possible front seat passenger are the two prime input sources.
- The primary initiator of network-related conversation for most driver area functions is the human.

Entertainment Area Subsystems – While the audio subsystem can be easily considered a driver area subsystem, many vehicle designers are recognizing that this functionality can be separated. The use of dedicated networks to support distributed audio/entertainment subsystems is becoming more common. Some automotive manufacturers use independent entertainment area networks. Such system partitioning is relatively easy to accomplish, especially when an entertainment module can also be used as a gateway to provide information sharing. The response of the Audio subsystem to input command must also match the "human expected response time" and acceptable response times range from approximately 60 to 100 milliseconds.

SPECTRUM OF LOW END TO HIGH END VEHICLES – Although both the engine area and driver area subsystems are common to all vehicles, the feature content of these subsystems varies across the manufacturer's car platform spectrum from "low-end" vehicles to "high-end" vehicles. Across the entire spectrum from the "low-end" vehicle to the "high-end" vehicle the range of variability in subsystem functionality is wide. Thus, one additional requirement of a vehicle network architecture is to handle this diversity.

REUSABLE NETWORK STRATEGIES – Many companies now use reusable software building blocks to contain their system-specific network strategy. Starting with simple common protocol drivers, the industry is seeing many network-related tasks moving away from the application and into the network software. For example, early implementations had periodic message timing activities handled by the application software, but now many new network software implementations include this repetitive message functionality.

OSEK – OSEK is another of the new reusable network strategies. Although its roots began in Europe, OSEK has now entered the U.S. automotive scene as a small portion of DaimlerChrysler and General Motors network strategies. While OSEK's Communication (COM) and Network Management (NM) blocks establish essential network related requirements, it is the OSEK Operating System (OS) that provides a task management architecture to handle both the application and network territory.

GMLAN – Using its experience with J1850, General Motors has created a comprehensive multi-network strategy called GMLAN to handle its new designs based on the CAN protocol. The GMLAN strategy encompasses both a J2284-compliant high-speed motion control network and its new J2411 Single Wire CAN network used for body control functions. The GMLAN systems strategy requirements have been implemented in software for a variety of microcontrollers and these implementations are given to module suppliers for integration with their application software.

Other Varieties – In Europe, DaimlerChrysler has also been using a vehicle network strategy that has been implemented in software. These software implementations are also given to its module suppliers for integration into the module software.

THE CASE FOR THE MULTIPLE VEHICLE NETWORK ARCHITECTURE

But the CAN protocol with its advantages and flexibility, the knowledge base, the available silicon, and supporting tools do not make the case for the multiple vehicle network architecture. While the choice of a particular architecture is motivated by both business and technical reasons, the decision primarily focuses on a comparison of the advantages to the disadvantages.

ADVANTAGES OF THE MULTIPLE VEHICLE NETWORK ARCHITECTURE – Moving toward a multiple vehicle network architecture provides several business and technical advantages, including lowering costs and reducing complexity.

Lower average node cost per vehicle possible – The multiple vehicle network architecture provides an opportunity to lower the average node cost per vehicle. Rather than an attachment to an expensive network solution for each and every node, alternate low cost solutions may be suitable for a portion of the vehicle's functionality. The addition of another network may reduce costs. Of course, it's also possible to add another network that increases costs. Using sound system engineering principles with heavy emphasis on the business cost issues has helped several car companies to reach their economic targets.

Easier to manage speed requirements – The multiple vehicle network architecture makes it easier to manage speed requirements. The opportunity to partition by speed requirements is possible; high speed motion control-type functions may be separated from slow human events. Even the choice of utilizing a high-speed data network with integrated audio capability is a new opportunity when embracing the multiple network architecture.

Opportunity to partition by functional area – The multiple vehicle network architecture provides the opportunity to partition by functional or geographical area. Most German car companies already have two CAN-based vehicle networks, one to support engine and another for the driver area (referred to in Europe as the body bus).

Opportunity to partition by data criticality – The multiple vehicle network architecture provides the opportunity to partition by data criticality. Perhaps a braking control subsystem should not share the same communication channel as an audio entertainment subsystem. This data criticality issue is the main concern motivating a new industry-wide effort to create an air bag bus separate from the existing vehicle networks.

Opportunity to partition by control architecture – The multiple vehicle network architecture provides the opportunity to partition by control architecture or by communication method. While CAN's use of peer-to-peer communications is adequate for most vehicle functions, perhaps the same communication method found in TTP (Time Triggered Protocol), that is time-division multiplexing will be found to be more appropriate for the control of high speed distributed functions.

Opportunity to partition by subsystem – The multiple vehicle network architecture provides the opportunity to partition by subsystem. Perhaps an audio subsystem should always be separated from the remaining vehicle networks in order to allow the OEM the opportunity to obtain complete audio subsystems from their suppliers.

Rather than buying the individual audio-based network modules and managing the subsystem integration, partitioning by subsystem may allow the OEM a business opportunity to purchase a complete audio subsystem without attending to the complexity of its supporting network.

DISADVANTAGES OF THE MULTIPLE VEHICLE NETWORK ARCHITECTURE – While the advantages are quite significant, it is important to list and consider the business and technical disadvantages.

The architecture of multiple vehicle networks has disadvantages which include –

- Additional resources are required
- May need to support different protocols
- Increased complexity to share information across different networks
- Usually requires gateway functionality
- Servicing may be more difficult

Additional resources are required – Additional resources are required to support each different network. The systems engineering design, development, integration and system-level validation effort plus the support for production and diagnostics may require adding engineering staff for each vehicle network.

May need to support different protocols – Each network may use a different protocol. This can also add another layer of complexity which requires more resources to manage. Multiple protocols require multiple knowledge bases, multiple message and database tools, and multiple protocol tools.

Increased complexity to share information across different networks – The sharing of information between multiple networks adds complexity. One of the biggest advantages of the single vehicle network architecture was the ability to easily access virtually any piece of information across the network. In some instances, the shift to a multiple vehicle network architecture may remove this advantage and create a potential need for moving information from one network to another.

Usually requires gateway functionality – The sharing of information across different networks may easily increase system requirements such that a gateway will be required. Gateway (or bridge) functionality typically becomes an additional burden for one of the vehicle electronic modules above and beyond the module's normal set of application requirements. Supporting two network connections while handling network-to-network message timing requirements makes gateway implementation complex. Late changes to the vehicle system functionality might even be impossible if sufficient micro-controller resources are not available.

Servicing may be more difficult – With the single vehicle network architecture, a single diagnostic method and strategy could be used. But the multiple vehicle network architecture will likely cause diagnostics and the servicing of the different buses to become more complicated.

CURRENT TECHNICAL ISSUES

Many of the current technical issues facing multiple vehicle network architectures are carryovers from the earlier single network solutions. Continuing issues include –

- EMC
- physical layer
- network speed and bandwidth
- data criticality
- the mixing of information with control functions
- continuous demand for new vehicle features

New issues that require attention include how to flash program modules, how to partition functions across multiple networks, how to handle information sharing across multiple networks, and how to address the diagnostics issues that grow with the increasing number of networks.

THE INFLUENCE OF CHANGING REQUIREMENTS –

Requirements for vehicle network architectures are not stable. Once requirements change to a significant enough extent such that a change in the architecture is needed, the choice between a new architecture or re-partitioning the existing system must be revisited. And how often are requirements changing? Continually. Vehicle system complexity continues to increase with the demand for more features and functions. This creates a never-ending discussion of architecture versus architecture and forces the technical/business analysis to be examined on a recurring basis.

Do changing requirements further the case for multiple vehicle network architectures? To a large extent it does, using a multiple network solution provides more opportunity to manage change.

CURRENT BUSINESS ISSUES

The current business issues that surround MVNA can be concentrated into 4 key areas –

- human resources
- time to market
- cost
- cross company variation

HUMAN RESOURCES – To manage a multiple vehicle network architecture requires additional resources. While supporting multiple networks does not necessarily carry twice the resource burden, resource expenses are much higher when standard " off-the-shelf" protocols are not used. If each vehicle network uses the same or similar network implementation, the resource drain can be highly minimized.

Electrical engineering, systems engineering, test engineering, and management are all necessary to handle a company's multiple vehicle networks. A maintained knowledge base is also necessary.

TIME TO MARKET – Time to market may be of little interest to the engineering side of vehicle networking, but to management it is an important issue. When examining the product development time associated with implementing a multiple vehicle network architecture, one recognizes that the time to market –

- increases if each OEM perpetuates its own unique network solution
- increases when cross industry variation increases
- increases when standardization is avoided
- increases when companies need to make their own tools

When each electronic module supplier must develop special circuitry or software for each OEM's unique network solution, the overall industry time to market increases. One major tier 1 supplier reports the need to support over 30 different protocols to satisfy its entire customer base. Even if this supplier is managing several of its own proprietary protocols, this number is staggering.

COST – For each OEM and module supplier, the business cost of developing and managing multiple vehicle network architectures is a key issue. However, with no industry cost figures available and few accountants assigned to examine this area, little is known about this issue. While the engineering staff adequately handles the various networking technical tasks, their expertise does not cover the cost issue. Perhaps when upper management understands the business issues in managing vehicle network engineering, the cost issues will be carefully scrutinized.

Several factors affect the issue of cost. When developing and managing multiple vehicle network architectures, cost –

- increases when cross company variation increases
- decreases when using off-the-shelf solutions
- decreases through standardization

Because there is no general overseer of industry costs for this vehicle networking area, this business issue will continue to remain in the dark.

Some of the factors that influence this lack of attention to cost include –

- management not knowing of the problem
- the engineering staff's inexperience in handling business issues
- the inability to recognize the overall system and business environment that affects cost

CROSS COMPANY VARIATION – When one examines the spectrum of multiple vehicle network architectures across the industry, one sees considerable cross company variation. Each company's vehicle network implementation differs. The CAN protocol may be the only thing that is common. Whether based on technical reasons or an interest in creating local fiefdoms, this variation in vehicle network implementations increases complexity for all module suppliers.

While J2284 establishes requirements for a CAN-based high-speed 500K BPS bus, each car company still requires its own different implementation. For each module supplier a circuit change is necessary and this does not reduce the time to market.

The medium or low speed CAN bus used to accomplish driver area functions will be another area of major cross company variation. The chance that DaimlerChrysler, GM, and Ford will each deploy a different CAN network solution seems a certainty.

Even though J1850 has been abandoned and many OEM's have converged to the CAN protocol, each OEM still continues to do whatever it wants. Each OEM has essentially created a company-specific CAN-based network solution that is different from the other OEMs. At one time, we had three versions of J1850, and now, we have at least three different versions of CAN.

Yes, this has also happened in Europe.

But the issue closer to home is - what does this say about our automotive electronics industry?

Could it be the case, that engineers who get control of new technologies that management does not understand will continue to fight like Saturday afternoon college football teams regardless of the business consequences. With no referees to call fouls and no impartial judges to help guide the way, our standardization efforts continue at a snail's pace.

Cross company variation can only be stopped through the efforts of industry standardization. Someday in the future, vehicle networking, just like tires, will become a complete off-the-shelf item. This will happen at an unbelievably slow rate regardless of the number of engineers standing in the way.

CONCLUSION

From a system's point of view, the migration toward multiple vehicle networks has provided the means to lower the overall vehicle networking cost and has helped to manage the increasing complexity that faces the automotive electronics industry.

The development and maintenance of proprietary protocols is a waste of company resources when suitable " off-the-shelf" solutions exist.

It will be the new requirements that will influence the future of vehicle networking architectures.

It is the vehicle networking architecture that establishes the requirements for small area network protocols.

While many of the past vehicle networking decisions were based on reasonably sound technical reasoning, it will be primarily business decisions that determine the future direction of vehicle networking architectures.

REFERENCES

1. OSEK/VDX NM Specification V2.5, OSEK/VDX NM Working Group, May 1998
2. OSEK/VDX COM Specification V2.1r1, OSEK/VDX COM Working Group, June 1998
3. OSEK/VDX OS Specification V2.0r1, OSEK/VDX OS Working Group, October 1997
4. CAN 2.0B Specification – 1991, Robert Bosch GmbH
5. CCP CAN Calibration Protocol, ASAP Standard, Version 2.1, February 1999
6. SAE 950293 "Multiple Vehicle Networks" by Bruce D. Emaus
7. SAE 941650 "Rethinking Multiplex" by Nigel Allison
8. SAE J1213/1 "Glossary of Vehicle Networks for Multiplexing and Data Communications".

ADDITIONAL SOURCES

The Bosch CAN 2.0B Specification is available at the following website http://www.vector-cantech.com look in "CAN Information".

To contact ASAM, visit the ASAM website at: http://www.asam.de.
The CCP CAN Calibration Protocol V2.1 can be found using the link:
http://www.asam.de/Asap_Gen.htm

CONTACT

The author has over 20 years of distributed embedded system engineering experience with both proprietary and "off-the-shelf" protocols and may be contacted at emaus@vector-cantech.com.

DEFINITIONS, ACRONYMS, ABBREVIATIONS

ASAP: Arbeitskreis zurStandardisierung von Applikationsystemen (Standardization of Application/ Calibration Systems)

ASAM: Association for Standardization of Automation and Measuring Systems

CAN: Controller Area Network

CCP: CAN Calibration Protocol

ECU: Electronic Control Unit

UART: Universal Asynchronous Receiver Transmitter